Lecture Notes in Electrical Engineering

Volume 688

The book series *Lecture Notes in Electrical Engineering* (LNEE) publishes the latest developments in Electrical Engineering - quickly, informally and in high quality. While original research reported in proceedings and monographs has traditionally formed the core of LNEE, we also encourage authors to submit books devoted to supporting student education and professional training in the various fields and applications areas of electrical engineering. The series cover classical and emerging topics concerning:

- Communication Engineering, Information Theory and Networks
- Electronics Engineering and Microelectronics
- Signal, Image and Speech Processing
- Wireless and Mobile Communication
- Circuits and Systems
- Energy Systems, Power Electronics and Electrical Machines
- Electro-optical Engineering
- Instrumentation Engineering
- Avionics Engineering
- Control Systems
- Internet-of-Things and Cybersecurity
- Biomedical Devices, MEMS and NEMS

For general information about this book series, comments or suggestions, please contact leontina.dicecco@springer.com.

To submit a proposal or request further information, please contact the Publishing Editor in your country:

China

Jasmine Dou, Associate Editor (jasmine.dou@springer.com)

India, Japan, Rest of Asia

Swati Meherishi, Executive Editor (Swati.Meherishi@springer.com)

Southeast Asia, Australia, New Zealand

Ramesh Nath Premnath, Editor (ramesh.premnath@springernature.com)

USA, Canada:

Michael Luby, Senior Editor (michael.luby@springer.com)

All other Countries:

Leontina Di Cecco, Senior Editor (leontina.dicecco@springer.com)

**** Indexing: The books of this series are submitted to ISI Proceedings, EI-Compendex, SCOPUS, MetaPress, Web of Science and Springerlink ****

More information about this series at http://www.springer.com/series/7818

Ning Zhou · S. Hemamalini
Editors

Advances in Smart Grid Technology

Select Proceedings of PECCON 2019—Volume II

Springer

Editors
Ning Zhou
Electrical and Computer
Engineering Department
Binghamton University
Binghamton, NY, USA

S. Hemamalini
School of Electrical Engineering
Vellore Institute of Technology
Chennai, India

ISSN 1876-1100 ISSN 1876-1119 (electronic)
Lecture Notes in Electrical Engineering
ISBN 978-981-15-7240-1 ISBN 978-981-15-7241-8 (eBook)
https://doi.org/10.1007/978-981-15-7241-8

This Springer imprint is published by the registered company Springer Nature Singapore Pte Ltd.
The registered company address is: 152 Beach Road, #21-01/04 Gateway East, Singapore 189721, Singapore

Preface

Power system is undergoing tremendous developments such as integration of more and more renewable energy resources, advancements in power electronic interfaces, energy efficiency measures, control and introduction of more intelligence in the grid. This book constitutes selected high-quality papers presented in the Second International Conference on Power Engineering, Computing and CONtrol, PECCON 2019, during 12 to 14 December 2019. The main theme of PECCON 2019 is "smart technology for smart grid". The conference provided a forum for practising professionals, academicians, research scholars and students to exchange their innovative ideas, inferences and knowledge gathered through rigorous experiments. This book volume consists of 42 papers and is organized into four parts—Power Engineering, ICT Technologies, Data Science and Analytics, and Smart Eco Structures Systems. This book will be useful to the researchers, academicians, students and professionals working in the field as well as R&D organizations in the domain of electrical power and energy infrastructure.

We, the editors of this book, are thankful to all the authors who have contributed to the paper submission. We also thank all the reviewers and technical committee members for providing their valuable comments on time and help towards the improvement of the quality of papers presented in the conference. We also thank members of advisory board and session chairs for their valuable comments. Our special thanks to Series Editors, Lecture Notes in Electrical Engineering, Springer, for giving us the opportunity to publish this edited volume in the series.

Chennai, India — S. Hemamalini
Binghamton, USA — Ning Zhou

Contents

About the Editors

Ning Zhou is currently with the Electrical and Computer Engineering Department at Binghamton University as an assistant professor. In 2005, he received his Ph.D. in Electrical Engineering with a minor in statistics from the University of Wyoming. From 2005 to 2013, Dr. Zhou worked as a power system engineer at the Pacific Northwest National Laboratory. His research interests include power system dynamics and statistical signal processing. Dr. Zhou is a senior member of the IEEE Power and Energy Society (PES). He has been an associate editor for IET Generation Transmission and Distribution since 2016. He is the lead author of the 2009 Technical Committee Prize Paper from the IEEE/PES Power System Dynamic Performance Committee. He has been the co-Chair of IEEE PES Working Group on Data Access and secretary of IEEE PES Task Force on Oscillation Source Location since 2016. He is the recipient of 2009 Outstanding Engineer of Year Award from IEEE Power and Energy Society (PES) Richland Chapter. He is the recipient of IEEE PES Outstanding Branch Counselor Award in 2017.

S. Hemamalini is currently working as a Professor in the School of Electrical Engineering, Vellore Institute of Technology, Chennai, India. She received the Ph.D and M.Tech degree from National Institute of Technology, Tiruchirappalli, India and B.E degree from Thiagarajar College of Engineering, Madurai Kamaraj University, Madurai, India. She is the member of Institute of Electrical & Electronics Engineers (IEEE), IEEE Smart Grid society and Indian Society for Technical Education. She worked as a Treasurer and Execom Member in IEEE Madras Section in the year 2012 and 2013 respectively and also visited many foreign technical universities in Malaysia, Singapore and USA. She has published her research works in Scopus indexed international journals and conferences. She is also reviewer for peer-reviewed Journals. Her research interest are focused on Reliability and Protection in Microgrids, Grid Integration of Renewable Energy Sources, High gain DC-DC converters for Renewable Energy Applications, Power system Optimization, energy management, and Smart grid.

Power Engineering

Online Insulation Monitoring of Low-Voltage Unearthed Distribution Systems

K. Barkavi and Prithak Kumar Bhattacharyya

Abstract This paper deals with the implementation of online insulation monitoring of low-voltage unearthed distribution systems. The purpose of choosing and applying unearthed distribution systems is the various advantages it possesses in the industrial safety area such as continuity of service and prevention from risk of fire and shock. It is an important task to detect insulation fault in the energized (online) system without affecting the distribution system. The main purpose is to deal with the insulation fault in the energized system and maintain the continuity of the system supply. The idea is to continuously monitor the low-voltage distribution system and detect the insulation/ohmic faults in the ungrounded energized low-voltage distribution system using an insulation monitoring device (IMD). Basically, 'online' suggests real-time measurement and 'insulation monitoring' is the continuous insulation monitoring and leakage fault detection of ungrounded distribution systems. The IMD detects insulation faults between the energized system and earth. An external injection, generated by the IMD, is made in the distribution system to monitor the system. During the healthy condition of the distribution system, the IMD reads and displays high insulation values. When an insulation fault occurs, the IMD reading drops down to the corresponding value of insulation fault, which is proportional to the voltage. Evaluation of various studies has been made, and various cases have been studied and implemented using Multisim software. Based on these case studies, testing of the hardware and hardware implementation of one of the case studies has been showcased.

Keywords Online insulation monitoring · Low voltage · Unearthed distribution system · Insulation faults · IMD · Multisim

---

K. Barkavi · P. K. Bhattacharyya (✉)
Department of Electrical and Electronics Engineering, SRM Institute of Science and Technology, Kattankulathur, India
e-mail: prithak.51197@gmail.com

N. Zhou and S. Hemamalini (eds.), *Advances in Smart Grid Technology*, Lecture Notes in Electrical Engineering 688, https://doi.org/10.1007/978-981-15-7241-8_1

1 Introduction

By the growing demand for electrical devices and equipment, the inherent dangers of electrical power are proportionally posing a threat to people and equipment alike. The industry for instance demonstrates the ultimate dependency on a reliable power supply. The complexities of electrical installations and their high investment costs require adequate measures of protection. Otherwise, the economic loss incurred alone would be immeasurable. The International Electro-technical Commission (IEC) is one such organization which is worldwide known for standardization comprising of many technical committees (TCs). One important series of IEC standard deals with low-voltage electrical installations, which is under Technical Committee TC 64. IEC 61140 applies to protection of persons and animals against electric shock. Low voltage is any rated voltage including and up to 1000 V AC or 1500 V DC. The standard states that hazardous live parts shall not be accessible and accessible conducive parts shall not be hazardous live either under single fault conditions, or under normal conditions.

'Online' suggests real-time measurement and 'insulation monitoring' is the continuous insulation monitoring and leakage fault detection of ungrounded distribution systems when the system is energized. This paper showcases the use of an IMD in which an external low voltage is generated and injected into the distribution system which does not affect the system and a low-frequency injection which does not interfere with the common range of frequencies of the distribution system, which makes the system monitoring and output more accurate and precise.

2 Types of System Earthing

The main distribution systems are:

I. IT system
II. TN system
III. TT system (Table 1).

IT system has a single point connected to earth through impedance and direct electrical connection of the exposed conductive components/parts to earth or all live components/parts isolated from earth, which is not dependent of the earthing of any point of the power system.

Recognized protective devices for IT systems include insulation monitoring devices (IMD), overcurrent protective devices (OPD), and residual current devices (RCD).

TN system has direct electrical connection of the exposed conductive components/parts to the earthed point and direct connection of a single point to earth of the power system.

Table 1 Codes for classifying distribution systems (Ref. [3])

First letter	Relationship of earth with distribution system
I	One point connected to earth via an impedance or no live components connected with earth
T	Direct connection of one point to earth
Second letter	**Relationship of network or earth and electrical device being supplied**
T	Local direct connection to earth, independent of earthing of any point of the electrical system
N	Local direct connection of the exposed conductive parts to the earthed point of the electrical system

TT system has direct electrical connection of the exposed conductive components/parts to earth and direct connection of one point to earth, which is not dependent of the earthing of any point of the power system.

Recognized protective devices for TN/TT systems include OPD and RCD.

3 Unearthed IT Systems

The IT system is usually equipped with equipotential bonding and insulation monitoring devices. IT systems are supplied by a transformer, generator, and battery or by an independent voltage source. The distinctive feature of these AC and DC systems is the fact that no live conductor of the system is directly earthed. The exposed conductive parts of the equipment, in the IT system, are being connected to the protective conductor in such a way that the upstream fuse is triggered in the case of the second fault at another conductor.

However, in order to already receive a warning at the first insulation fault, monitoring of the insulation level of the installation is required. For continuous monitoring of the insulation status of the IT system, insulation monitoring devices (IMDs) are installed between phase conductors and protective conductors which continuously measure the insulation resistance and give visual signals and/or acoustic signals, if the insulation resistance drops under a minimum threshold point/value. For the safety of the operation, it is highly recommended to detect and correct an insulation fault as soon as possible.

4 Features of IT Systems

IT systems are completely insulated from earth and are applied in all areas where a high degree of operating safety, fire safety, and accident prevention is required. In many countries, these IT systems are therefore mandatory, recommended, or used in various applications. Compared to solidly earthed systems, IT systems have the following advantages:

- Higher operational safety
- Higher fire safety
- Better accident prevention due to limited touch currents
- Higher permissible earthing resistance
- Information advance

Another feature of the IT system is that the insulation fault is not automatically indicated by a blown fuse. A certain amount of training and a sound knowledge of the operational conditions are needed for fault location, particularly in large installations.

5 Insulation Monitoring Device (IMD)

IMD monitors continuously the insulation resistance between ground (PE) and the live conductors of the unearthed system (IT system). It indicates audible and/or visually when the ohmic value drops under the minimum threshold points/values. It must be in compliance with the standard for insulation monitoring devices IEC 61557-8. Figure 1 depicts the basic diagram of an IMD and its working during an ohmic insulation fault. The most common principle is the superimposition of a DC measuring voltage between the protective conductor and the system. The measuring voltage is generated in the insulation monitoring device and applied to the system via high ohmic coupling resistances. In the event of an ohmic fault, the measurement circuit associated with system and earth closes over an insulation fault Rf, so that a DC measuring current, proportional to the ohmic insulation fault, is in compliance with Eq. (1). The system leakage capacitances are charged to the DC measuring voltage and have no influence to the measurement after the brief period of transient response. The AC components of the system are eliminated by filters.

where

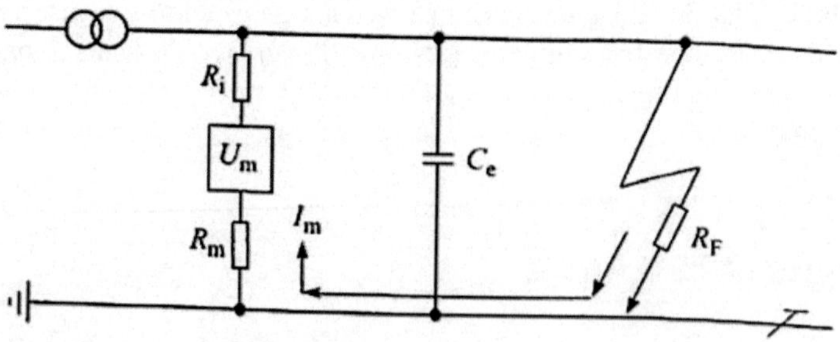

Fig. 1 Basic diagram of an insulation monitoring device

$$Im = \frac{Um}{Ri + Rm + Rf} \tag{1}$$

where

Im	DC measuring current
Um	DC measuring voltage
Ri	internal resistance of IMD
Rm	measuring resistance of the IMD
Rf	sum total of all ohmic insulation faults
Ce	system leakage capacitances

6 Simulations and Observations of Unearthed Systems

The following observations (Figs. 2, 3, 4, 5, 6, 7, 8, 9 and 10) have been deduced using simulation software based on the typical practical low-voltage unearthed industrial distribution systems possible. A transient analysis of each type of distribution system is carried out. Each design contains a source V1, internal resistance R9, load distribution R2 to R6, load leakage capacitances C2 to C11 (based on IEC standards). The online insulation monitor comprises of injecting/measuring voltage V2, measuring resistance R10, and coupling resistances R7 and R8. The output waveforms depict the voltage across R10 and current through R10. Various combinations of sources (DC and AC) and low-voltage, low-frequency injections (DC and AC), with and without ground fault Rf, are studied, and observations are noted. Based on this theoretical study, a suitable hardware is implemented as a prototype. They are showcased as follows:

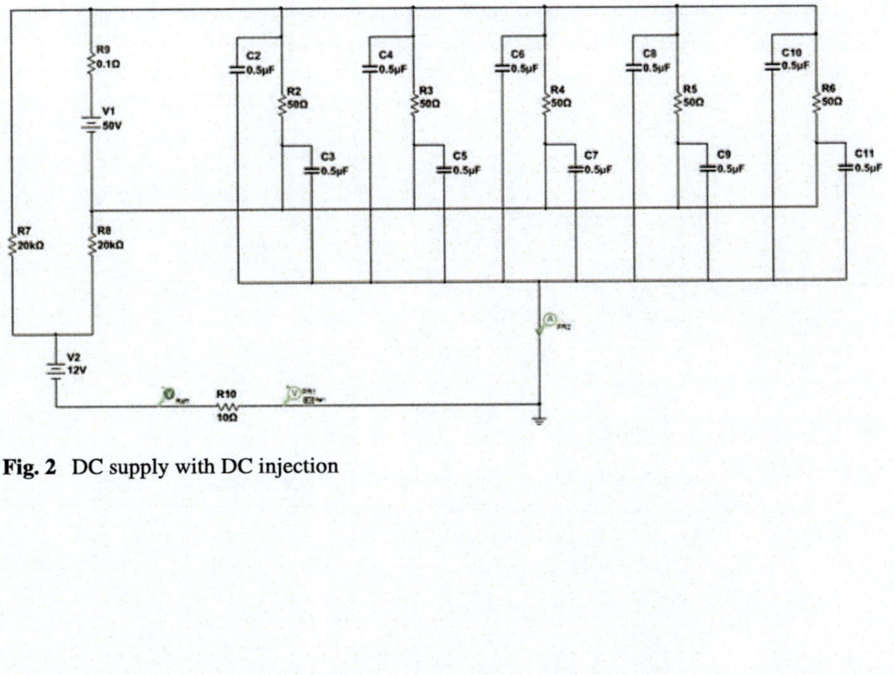

Fig. 2 DC supply with DC injection

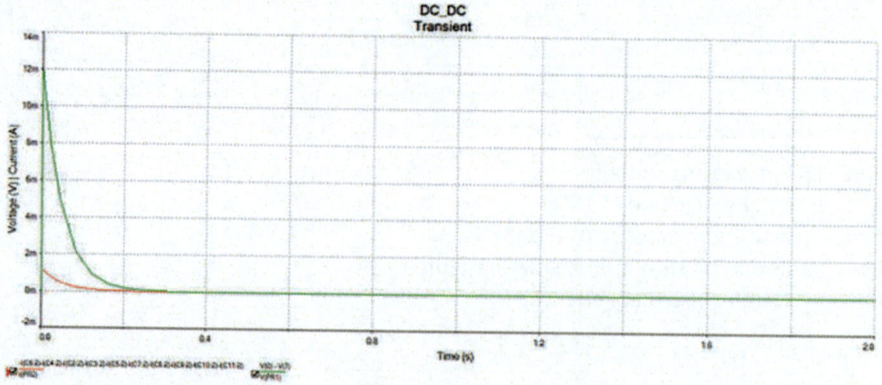

Fig. 3 Voltage (V) versus time (s), green graph and current (A) versus time (s), red graph

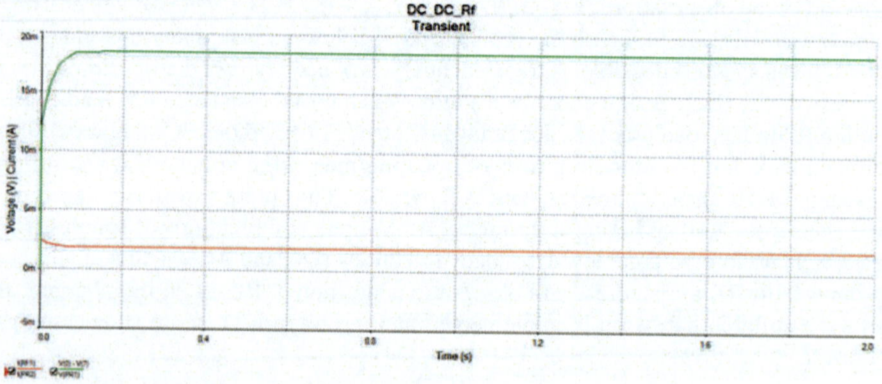

Fig. 4 Voltage (V) versus time (s), green graph and current (A) versus time (s), red graph

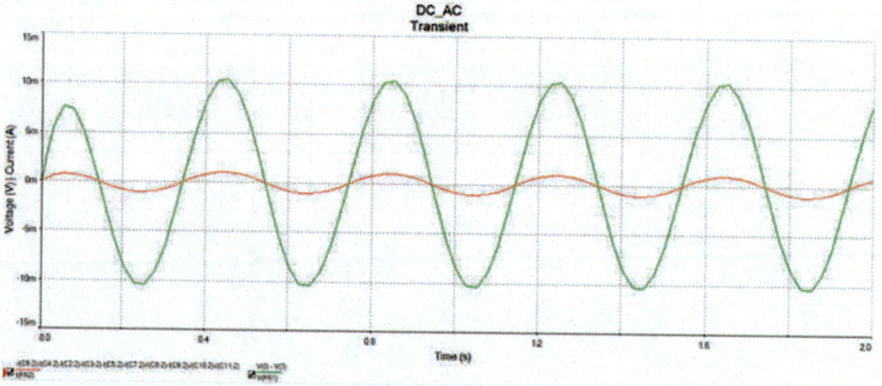

Fig. 5 Voltage (V) versus time (s), green graph and current (A) versus time (s), red graph

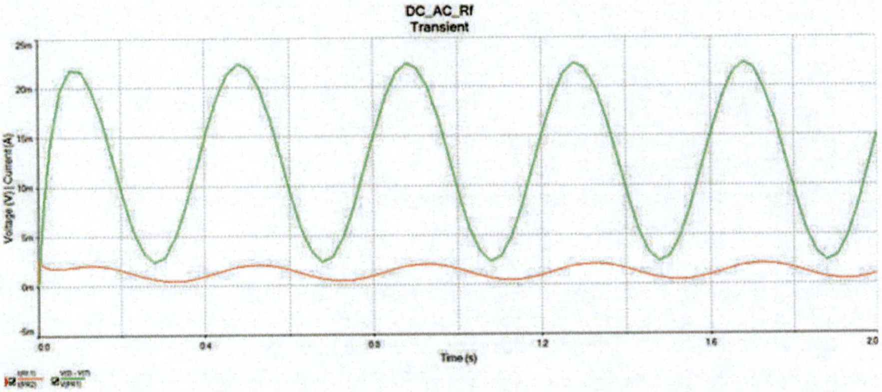

Fig. 6 Voltage (V) versus time (s), green graph and current (A) versus time (s), red graph

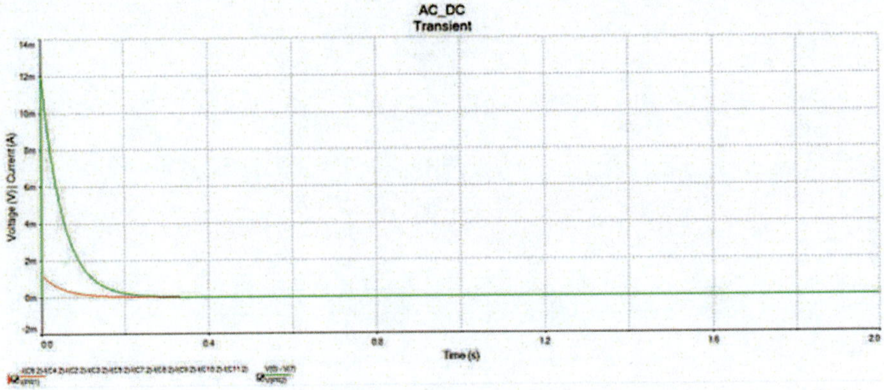

Fig. 7 Voltage (V) versus time (s), green graph and current (A) versus time (s), red graph

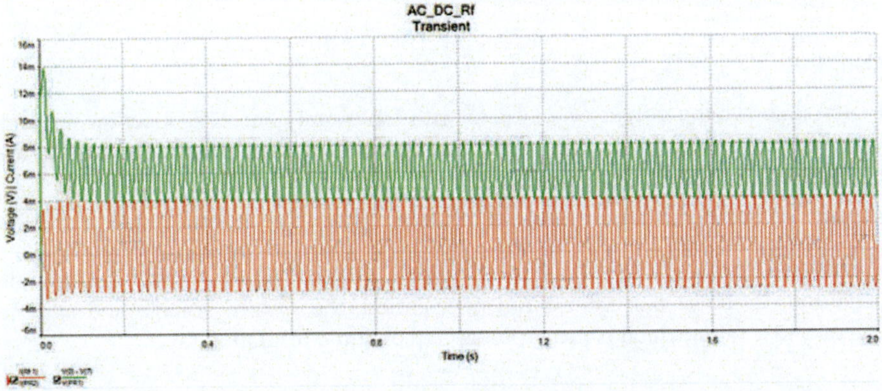

Fig. 8 Voltage (V) versus time (s), green graph and current (A) versus time(s), red graph

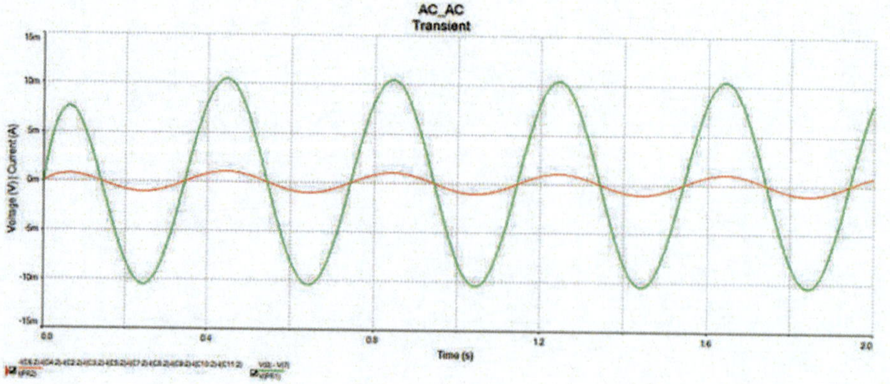

Fig. 9 Voltage (V) versus time (s), green graph and current (A) versus time (s), red graph

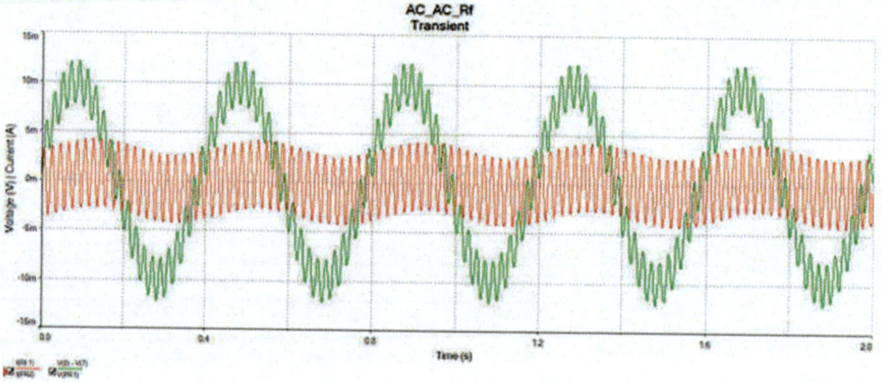

Fig. 10 Voltage (V) versus time (s), green graph and current (A) versus time (s), red graph

Case 1

Figure 3 shows the output of a system with DC supply and DC injection without fault condition. It shows that after a transient period, the system stabilizes and shows no fault values. Hence, system is healthy.

Case 2

Figure 4 shows the output of a system with DC supply and DC injection with fault condition. It shows that after a transient period, the system voltage and current increase to a value which is proportional to the ohmic insulation fault.

Case 3

Figure 5 shows the output of a system with DC supply and AC injection without fault condition. It shows that after a transient period, the AC injection voltage is retrieved and hence system is healthy.

Case 4

Figure 6 shows the output of a system with DC supply and AC injection with fault condition. It shows that after a transient period, the injection AC voltage is shifted proportionally to the ohmic insulation fault present in the system.

Case 5

Figure 7 shows the output of a system with AC supply and DC injection without fault condition. It shows that after a transient period, the system stabilizes and shows no fault values. Hence, system is healthy.

Case 6

Figure 8 shows the output of a system with AC supply and DC injection with fault condition. It shows that after a transient period, the system stabilizes and shows values proportional to the ohmic insulation fault along with AC superimposition due to leakage capacitances.

Case 7

Figure 9 shows the output of a system with AC supply and AC injection without fault condition. It shows that after a transient period, the system stabilizes and the AC injection voltage is retrieved. Hence, system is healthy.

Case 8

Figure 10 shows the output of a system with AC supply and AC injection with fault condition. It shows that after a transient period, the system stabilizes, and the output shows values proportional to ohmic insulation faults along with AC components of supply due to leakage capacitances (Table 2).

Table 2 Observations and conclusions

Case	No.	Insulation fault	Remarks
DC source, DC injection	1	Without fault	DC superimposition
	2	With fault	
DC source, AC injection	3	Without fault	Voltage shift due to DC component
	4	With fault	
AC source, DC injection	5	Without fault	AC components in output due to leakage capacitances and voltage shift
	6	With fault	
AC source, AC injection	7	Without fault	AC components in output due to leakage capacitances
	8	With fault	

7 Proposed Hardware Model

The proposed hardware model consists of a distribution system with DC supply and AC injection. Figure 11 shows the block diagram of the proposed hardware model. The model comprises of a 40–50 V low-voltage distribution system and an external injection of low-voltage and low-frequency of 12 V and 2.3 Hz, respectively, based on the IEC safety standards. The low-voltage injection is done for measuring the insulation fault in the system online without affecting the components of the system and a low frequency, such as a prime number, is chosen to avoid superimposition of system frequencies in the higher ranges. Appropriate range adjustment electronic circuits are used to display the output with maximum accuracy possible in the given range.

7.1 Proposed Circuit Diagram

The proposed model uses a step-down transformer to convert 230 V, 50 Hz AC supply down to 24 V DC supply using a full-wave bridge rectifier. This DC supply is given to the system load which forms the distribution system. A signal generator is formed using op-amp which generates a 12 V, 2.3 Hz sine wave. This signal enters the distribution system through the measuring resistance and coupling resistances. Figure 12 depicts the simulation diagram of proposed model. As and when an insulation ohmic fault occurs between earth and system, measuring current which is directly proportional to the ohmic insulation fault comes into play. The measuring voltage proportional to measuring current across the measuring resistance is then amplified using an op-amp as an amplifier.

A range adjustment circuit using op-amp converts the amplified voltage into required output which is shown through a display. Another step-down transformer

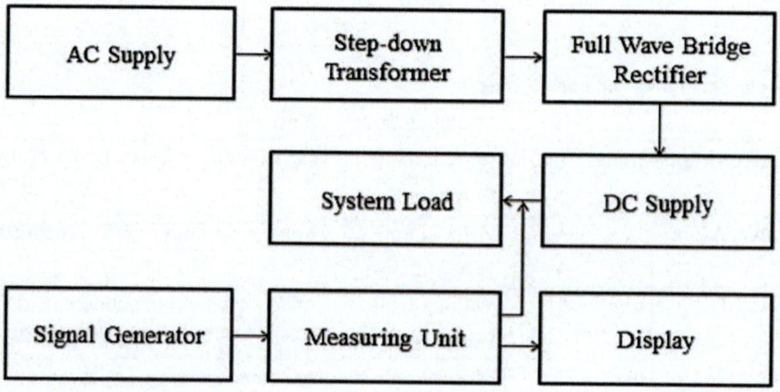

Fig. 11 Block diagram of the proposed hardware model

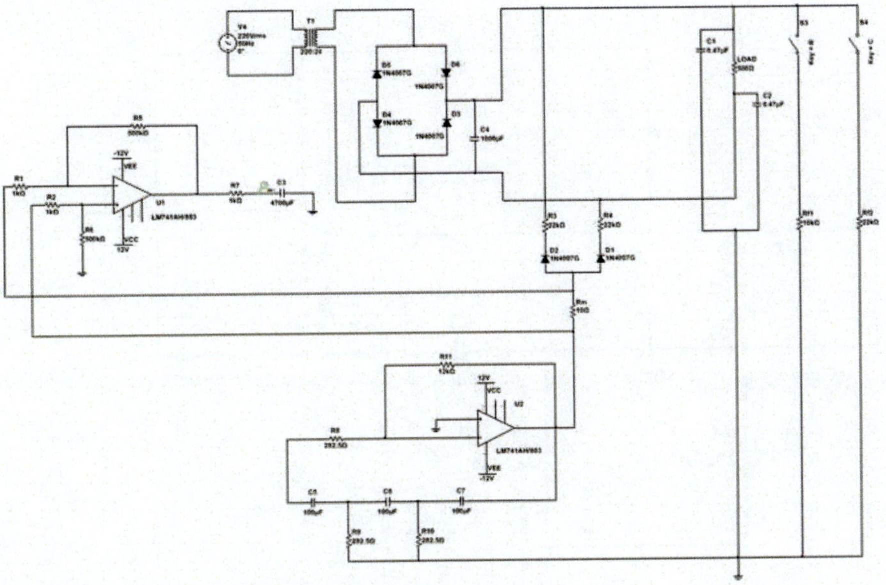

Fig. 12 Simulation diagram of the proposed model

is used to convert 230 V, 50 Hz AC supply down to 24 V DC supply using a full-wave bridge rectifier. This DC supply is given as input to a DC–DC converter/buck converter which gives an output of +12 V DC and −12 V DC for running the op-amps.

7.2 *Signal Generator/Injection*

RC oscillator is built using an operational amplifier and a RC network circuit in feedback. Figure 13 depicts this formation. The oscillation frequency (in Hertz) is given by equation,

$$F = \frac{1}{2\pi RC\sqrt{6}} \tag{2.1}$$

Substituting $R = 282.5\ \Omega$ and $C = 100\ \mu\text{F}$, we get,

$$F = 2.299Hz \approx 2.3Hz \tag{2.2}$$

From Eq. (2.2), we can verify that the signal generated using op-amp is of frequency 2.3 Hz.

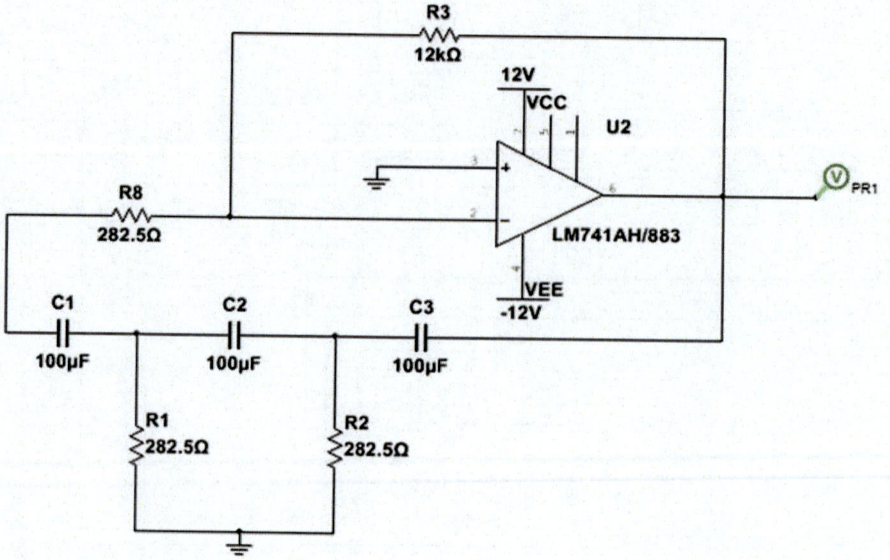

Fig. 13 Simulation diagram of signal generator

This low-frequency and low-voltage sinusoidal wave (Fig. 14) is injected into the distribution system to continuously monitor the distribution system online without affecting the distribution system.

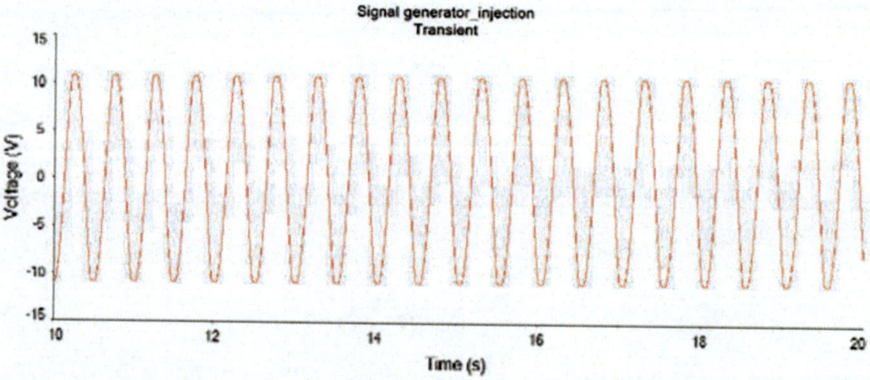

Fig. 14 Output voltage waveform of signal generator

7.3 Testing of Hardware Model

The testing of hardware model is done using breadboards before implementing it on a printed circuit board (PCB), and two cases of ohmic insulation faults have been taken as per suitable for testing for a prototype model. Figure 15 shows an insulation ohmic fault of 5 KΩ, and Fig. 16 shows an insulation ohmic fault of 22 KΩ. These ohmic insulation faults are shown in a display as output, and system is verified.

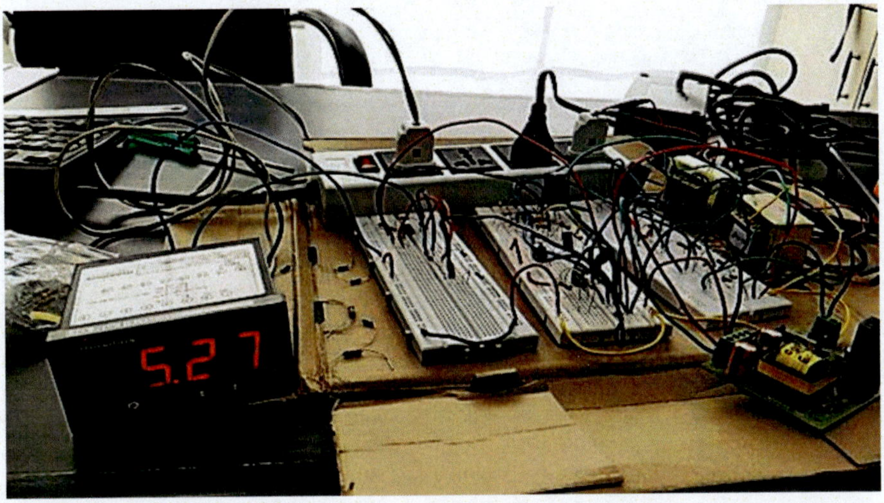

Fig. 15 Hardware model testing of the proposed circuit with 5 KΩ insulation fault

Fig. 16 Hardware model testing of the proposed circuit with 22 KΩ insulation fault

Table 3 Practical observations

S. No.	Theoretical value	Practical outcome	Output error percentage (%)
1	Healthy condition	OVER	–
2	5 KΩ	5.27 KΩ	5.4%
3	10 KΩ	10.64 KΩ	6.4%
4	12 KΩ	12.71 KΩ	5.9%
5	20 KΩ	20.02 KΩ	0.1%
6	22 KΩ	22.87 KΩ	3.9%
7	Combination of 12 KΩ AND 22 KΩ IN same line	7.94 KΩ	2.3%

7.4 *Results and Observations*

From Table 3, we can infer that various values have been studied and output is noted. The output of the hardware has a minimum error percentage of 0.1% and a maximum error percentage of approximately 7% within the given range of ohmic insulation faults, i.e., 0–22 KΩ. Hence, the system is working within the practical limitations, as discussed and mentioned earlier within this paper. When the distribution system condition is healthy, the display shows 'OVER' which means the insulation resistance of the system is high, as per healthy condition. When the system is at first insulation fault, the insulation resistance/ohmic insulation of the system decreases and hence the corresponding values are showcased as per the ohmic fault on the display.

7.5 *Hardware Outcome*

Figure 17 depicts the top view of hardware implementation, and Fig. 18 depicts the front view of the hardware showing 22 KΩ as ohmic insulation fault in the display as one of the observations noted before, in Table 3.

8 Conclusion

Based on various simulation studies and practical observations, an online insulation monitoring device is implemented based on first principles using appropriate electronic circuits to detect and measure insulation fault in the 40–50 V low-voltage energized (online) DC system with a 12 V low-frequency AC injection with maximum accuracy and minimum error within a range of 0–22 KΩ, applicable to clean DC systems without appreciable industrial EMIs. However, further research and development with high accuracy enhanced filters and increased measuring range can lead

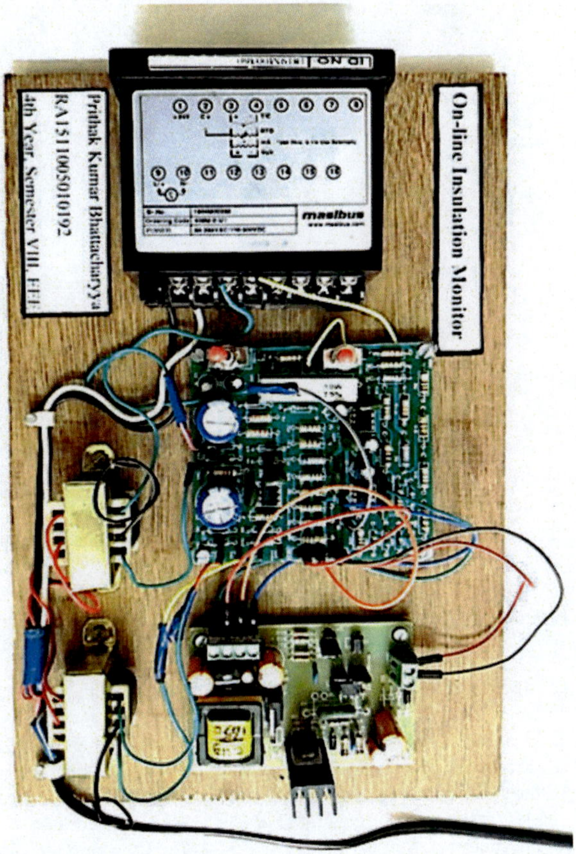

Fig. 17 Hardware outcome (top view)

to a better insulation monitoring device suitable for a broad range of industrial applications and capable of dealing with the prevalent industrial noises, EMI, and EMC tests in the practical industry with minimum error and maximum accuracy at hand.

Fig. 18 Hardware outcome showing 22 KΩ ohmic insulation fault (front view)

References

1. 'Technical aspects affecting the use of insulation monitoring devices' by Bender GmbH & Co. KG, Germany
2. IT Systems-the basis for reliable power supply by Bender GmbH & Co. KG, Germany
3. 'Protective Measures with Insulation Monitoring' by Wolfgang Hofheinz
4. International Standard, IEC 60364-1: Low-voltage electrical installations-fundamental principles, assessment of general characteristics, definitions
5. International Standard, IEC 60364-4-41: Low voltage electrical installations-protection for safety-Protection against electric shock
6. International Standard, IEC 61140: Low voltage electrical installations-protection of persons and animals against electric shock
7. International Standard, IEC 61557-8: Electrical safety in low voltage distribution systems up to 1000 V A.C. and 1500 V D.C.-equipment for testing, measuring or monitoring of protective measures-Part 8: Insulation monitoring devices for IT systems
8. https://www.electronics-tutorials.ws
9. https://learn.sparkfun.com

Soft Switching and Voltage Control for Three Phase Induction Motor

S. Hemalatha and A. Deepak

Abstract The role of induction motors is varied and of much importance in the field of electrical drives. For such induction motors, proper starting is required to improve versatility of the machine, reduce the starting current and mitigate power quality issues caused due to motor starting. There are different methods of starting, such as direct on line starter, electromechanical starting and autotransformer based starter. However star–delta starters are proved to be better in case of low and medium power motors, on account of cost and efficiency; however this gives transients when the circuit configuration is changed. In the proposed system power electronic switches are being used as soft starters for induction motor drives. Different power electronic switches are used and comparisons of the outputs provide a perspective on the effectiveness of the circuit.

Keywords Star-delta starter · GTO · Soft starter · Induction motor · Voltage control

1 Introduction

Induction motors are used in many places, be it industries or in residential areas. They are important in terms of design considerations, taking into account the wide usage and effects of wrong operation. The three phase induction motor during the initial starting condition draws up much higher current than its capacity and the motor instantly reaches the full speed. This results in a mechanical jerk and high electrical stress on the windings of the motor. Sometimes the windings may get burnt. The induction motor should start smoothly and gradually catch up the speed for a safer operation [1]. For instance, improper starting would result in more energy losses and its cumulative effects are not favourable. Hence, proper starting mechanism for induction motors is necessary. There are basically four types of starting induction motors [2]. They are direct on-line (DOL) starting, electromechanical reduced

S. Hemalatha · A. Deepak (✉)
Department of EEE, St Joseph's Institute of Technology, OMR, Chennai, India
e-mail: adeepakceg@gmail.com

N. Zhou and S. Hemamalini (eds.), *Advances in Smart Grid Technology*, Lecture Notes in Electrical Engineering 688, https://doi.org/10.1007/978-981-15-7241-8_2

voltage starting, solid state reduced voltage starting and variable frequency drive starting. Direct on-line starting are cost-effective provided the motor ratings are small, otherwise it is not advised to be used; it has its own demerits like producing surges in the system while starting the motor and so on. In case of star–delta starting, the sudden surge problem is mitigated to a certain extent by using two different configurations of connections to the motor from the supply. However this is applicable only for delta-connected stator with six leads [3–6].

Usages of mechanical switches and contactors have disadvantages like frequent maintenance, issues caused due to frequent switching and so on [7]. Autotransformer starters are suitable for large rating induction motors, but their large size and higher costs contribute to their disadvantages. In addition to this, soft starters apply reduced voltage and torque to the motor [8], which are reliable than most existing techniques for starting of induction motors. But still soft starters aren't fully functioning as an energy saver because of harmonics. Soft starters for three phase induction motors have been designed as in with SCR as the semiconductor switch, [9–14]. Then the applications of Gate Turn off Thyristors (GTOs) are being tested for such applications as in [15]. This provided scope for incorporating different other power semiconductor devices to be used in place of SCRs for induction motor starting. The ac motor starters employing power semiconductors are being increasingly used to replace electromagnetic line starters and conventional reduced-voltage starters because of their controlled soft-starting capability with limited starting current. Thyristor-based soft starters are cheap, simple, reliable, and occupy less volume, and therefore, their use is a viable solution to the induction motor (IM) starting problem [16]. The electronic starter for the single phase induction motor that incorporates both short circuit and overload protection is given in [17]. The starter is used for switching and protecting the electric motor from the dangerous overloads by tripping. It reduces the starting current to the AC induction motors and also reduces the motor torque. The proposed system is a modified version of soft starter applied with star-delta switching configurations with different power semiconductor devices other than SCR, which supports induction motors effectively with better voltage control and torque control. The study of Soft Starter Study for Induction Motors using fuzzy PID control is given in [18]. The main circuit of the voltage controller is consisted of three opposite parallel Thyristors. IM terminal voltage is controlled by adjusting the firing angle. The firing angle is controlled by synchronized 6-pulse generator. Synchronized 6-pulse generator is controlled by intelligent controller. This paper presents a design of a soft switching circuit for induction motors, controlled by a PIC microcontroller and the operation is done in open loop configuration for simplicity purpose. Section 2 gives a basic idea about the conventional star-delta starter with electromechanical switching, gives an idea of the proposed system and its working. Section 3 tells about the various simulation results obtained with the help of the power semiconductor devices—SCR, IGBT and MOSFET. Section 4 tells about the conclusion and future scope of the project.

2 Implementation of a Star-Delta Starter

The structure of conventional star-delta starter is shown in Fig. 1. In a star-delta starter, the supply terminals R, Y and B are connected to the main contactor (KM3) and delta contactor (KM2). The terminals of the contactor KM3 are then connected to the primary terminals of the motor (terminals U1, V1 and W1). The terminals U2, V2 and W2 are connected in delta fashion to the delta contactor (KM2), with U2 connected to W1, V2 connected to U1 and W2 connected to V1. In addition to this the secondary terminals U2, V2 and W2 are connected to the star contactor (KM1).

During operation, the main contactor (KM3) and star contactor (KM1) are closed, so that the motor starts at a reduced voltage and supply current is limited to the motor. Then after a few cycles (after a few seconds), KM1 is opened and KM2 is closed. So, now the induction motor is in delta connection. The closure of contactors should be such that KM1 should be opened and KM2 should be closed simultaneously. Timers used in the starter are responsible for the control of the contactors.

The proposed star-delta starter is shown in Fig. 2. In this work, six GTOs are used, namely G1, G2, G3, G4, G5 and G6. Anode terminals of the GTOs G1, G2, G3 are connected to the terminals U1, V1, W1 of the motor respectively in star configuration. G4 is connected between terminals V1 and U2, G5 between W1 and V2 and G6 between U1 and W2. Hence the GTOs G4, G5 and G6 are connected in delta configuration. During Delta all the six GTO's are utilized. The GTOs G4, G5 and G6 are connected in star configuration.

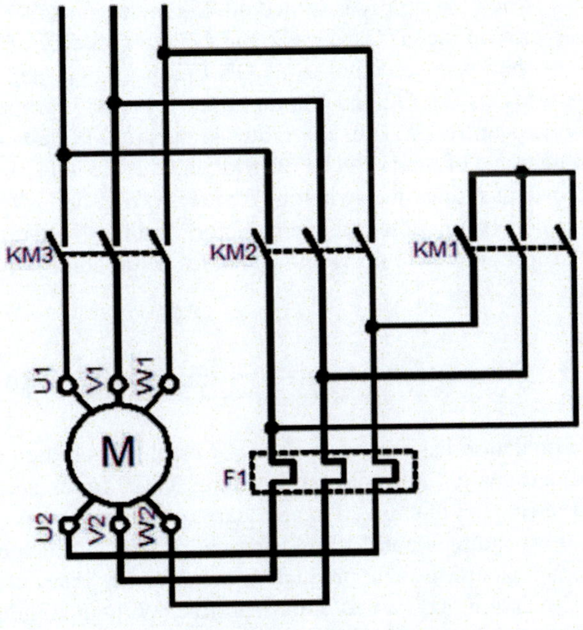

Fig. 1 Conventional star-delta starter

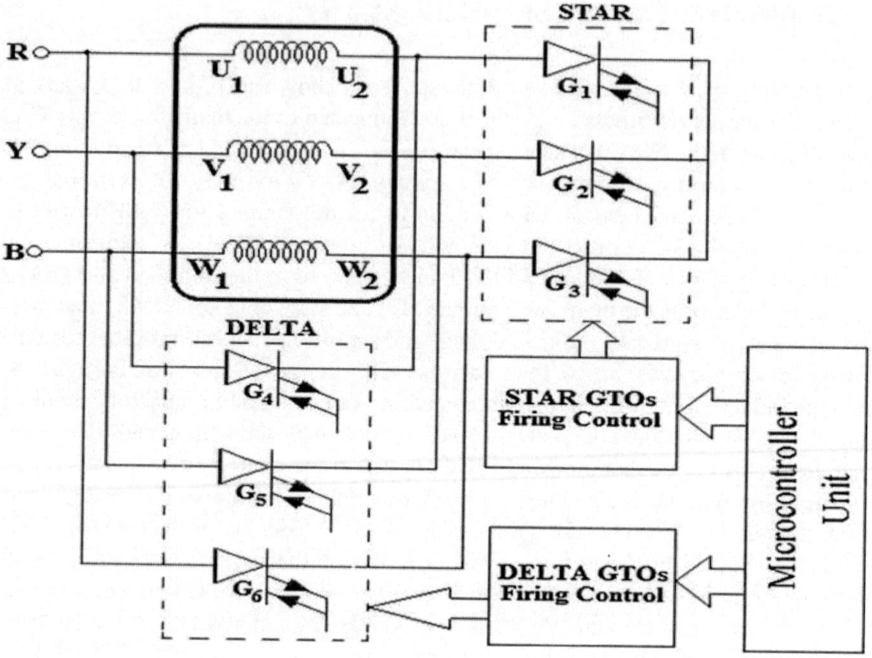

Fig. 2 Proposed soft switching circuit based on star-delta starter configuration

When the supply is turned on, the microcontroller provides gate signal of +5 V to turn on the GTOs G1, G2 and G3 after a delay of 2 s. Then, after an interval of 5 s the pulse generator turns ON GTOs G4, G5 and G6 at the same instant when the GTOs G1, G2 and G3 are turned OFF, with proper phase sequencing using the microcontroller. Thus the motor is operated in delta connection now. In this work the usage of SCRs can be overridden with the help of GTOs, owing to its versatility towards ratings and switching frequency. ZCS and ZVS are available since switching is done using power semiconductors and not by using electromechanical switches. Proper control of firing angle enables better control over the modified soft starter.

3 Simulation of Soft Switching and Voltage Control

Simulation has become a very powerful tool on the industry application as well as in academics, nowadays. It is now essential for an electrical engineer to understand the concept of simulation and learn its use in various applications. Many industries are spending a considerable amount of time and money in doing simulation before manufacturing their product. In most of the research and development work, the simulation plays a very important role. Without simulation, it is quite impossible to proceed further. It should be noted that in power electronics, computer simulation and a proof of concept hardware prototype in the laboratory are complementary to each other. The objective of this chapter is to describe the simulation of impedance source inverter with R, RL and RLE loads using MATLAB tool.

3.1 Existing Star- Delta Starter

In this existing process, three phase programmable voltage source is connected to star delta starter and the output of star delta starter is connected to asynchronous machine and it is connected to the bus and output of the system is used to measure the speed and torque of the machine which is shown in Fig. 3. Initially two breakers are closed and the normal operation of induction machine runs, while starting the machine first it is connected in star and then after attaining speed it changes to delta mode of operation. The internal connection of star-delta starter is shown in Fig. 4.

Initially two breakers are closed and the normal operation of induction machine runs without any connections of starters. The supply is given and breaker 1 is closed and no connection to the starter so that breaker 2 also closed and given to the motor. Figure 5 explains about the speed and torque of the motor without starter conditions. Initially two breakers are closed and the normal operation of induction machine runs, while starting the machine first it is connected in star and then after attaining speed it changes to delta mode of operation. Figure 6 explains about the speed and torque of the motor with starter conditions. From both figures if the starter connected to the circuit it produces more oscillations in motor output. So the performance of the motor with different switches in the starter is given in next section.

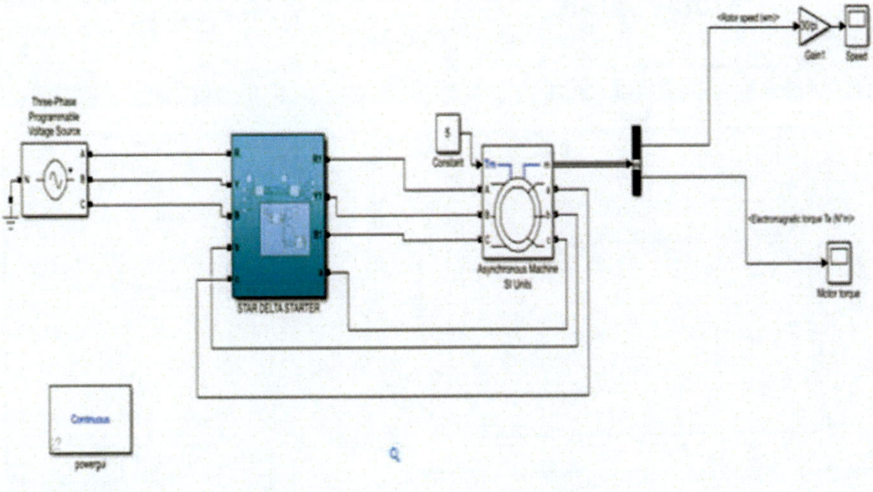

Fig. 3 Conventional star-delta starter

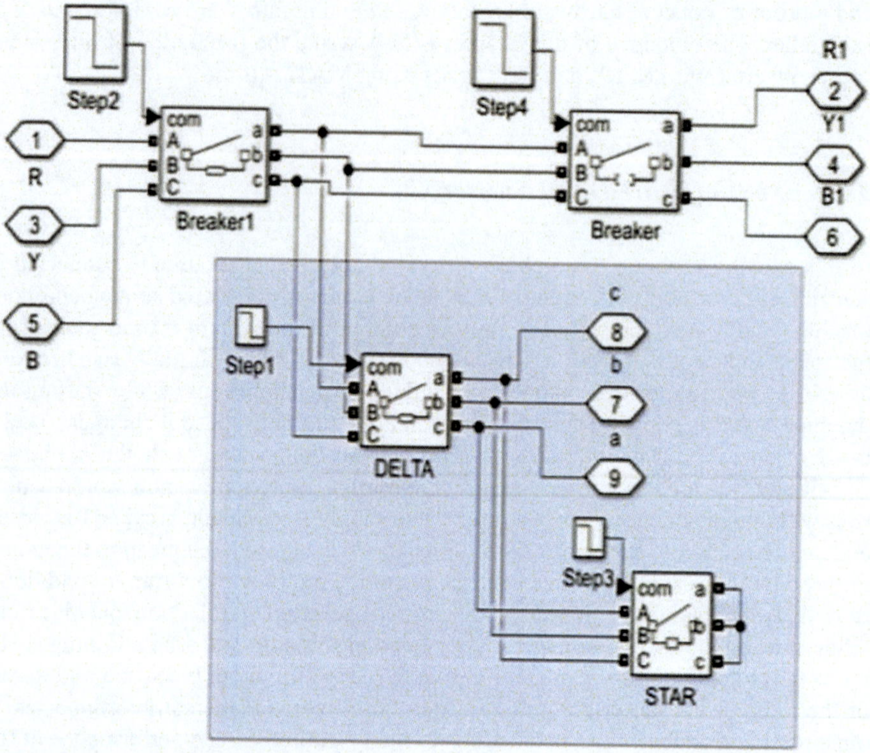

Fig. 4 Internal connection of conventional star-delta starter

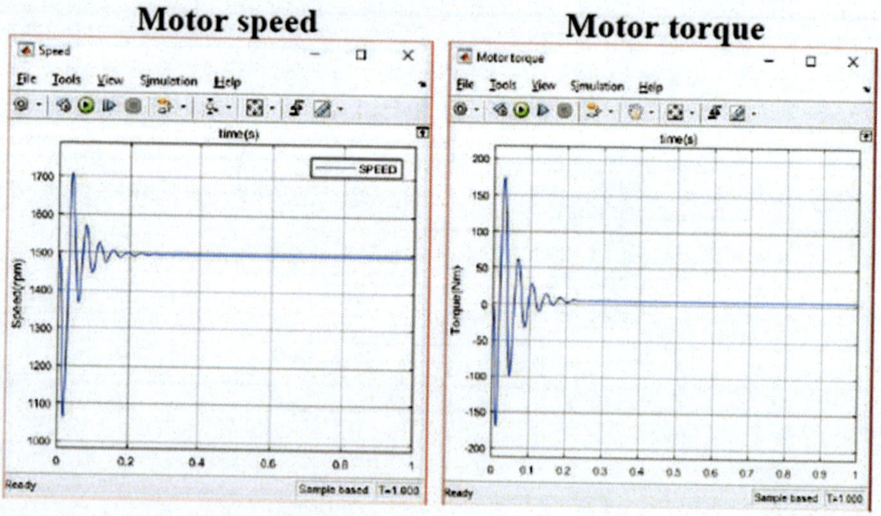

Fig. 5 Motor speed and torque without starter

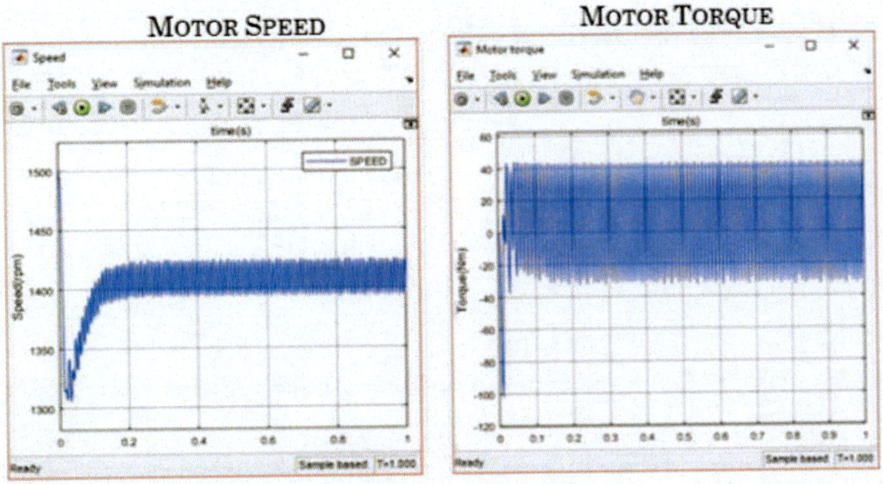

Fig. 6 Motor speed and torque with starter

3.2 *Star–Delta Starter with GTO*

In this simulation of GTO process, three phase programmable voltage source is connected to soft starter and the output of soft starter is connected to asynchronous machine and three phase voltage current measurement is used to measure the voltage of GTO and it is connected to the bus and output of the system is used to measure the speed and torque of the machine. The advantages of GTO are low cost, weight and volume, less noise, low Turn off time, higher blocking voltage capability and high efficiency. The disadvantages of GTO are magnitude of latching and holding current is more, losses are more and on state voltage drop and associated loss are more.

The simulation model of the proposed starter with GTO is shown in Fig. 7. The internal structure of the starter is shown in Fig. 8. Initially two breakers are closed and the normal operation of induction machine runs, while starting the machine first it is

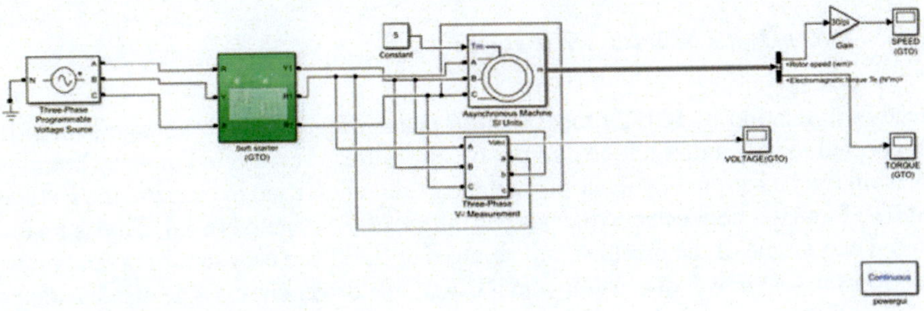

Fig. 7 Simulation model with GTO

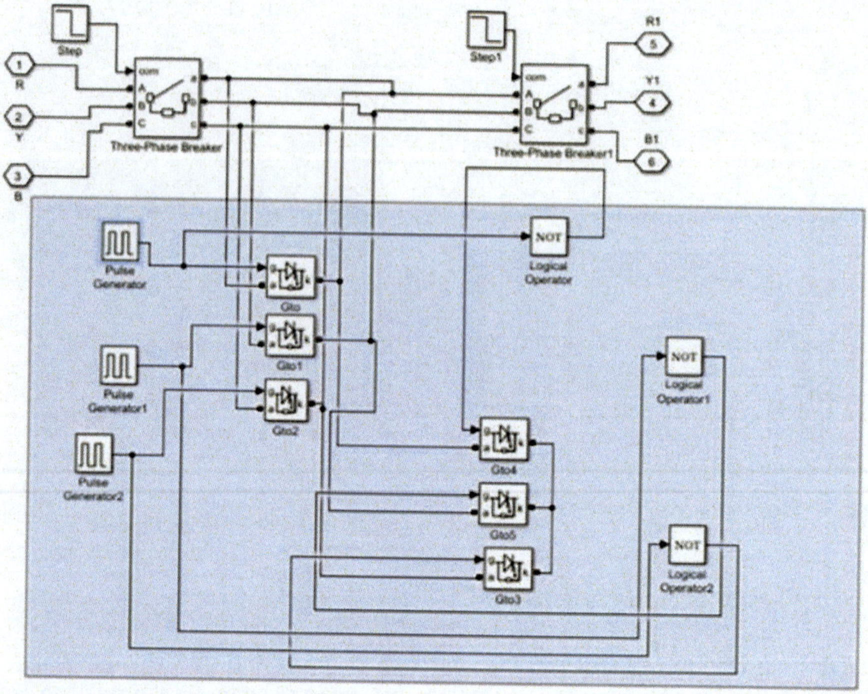

Fig. 8 Internal structure of starter with GTO

connected in star and then after attaining speed it changes to delta mode of operation. Three pairs of GTO are used in the star conditions and after star operations is finished another three pairs of GTO for delta conditions. Figure 9 explains about the speed and torque of the motor in the GTO operation. The voltage control circuit with GTO connected to induction motor is shown in Fig. 10 and the output voltage of the starter is shown in Fig. 11.

3.3 Star–Delta Starter with IGBT

In this simulation of IGBT process, three phase programmable voltage source is connected to soft starter and the output of soft starter is connected to asynchronous machine and three phase voltage current measurement is used to measure the voltage of IGBT and it is connected to the bus and output of the system is used to measure the speed and torque of the machine. Advantages of IGBT are easy to turn on and off, it has low on state power dissipation and it has simple driver circuit. The disadvantages of IGBT is cannot block high reverse voltage, turn off time is high, latching up problem and cost is high.

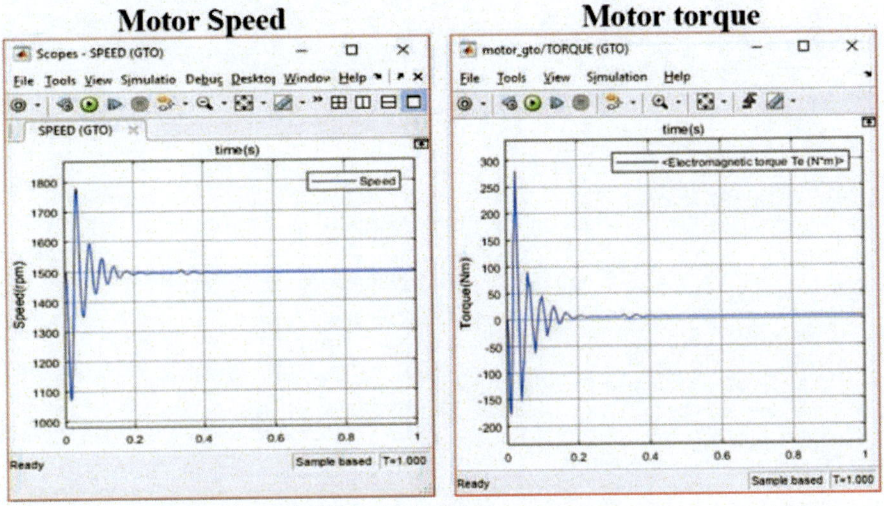

Fig. 9 Motor speed and torque with GTO based starter

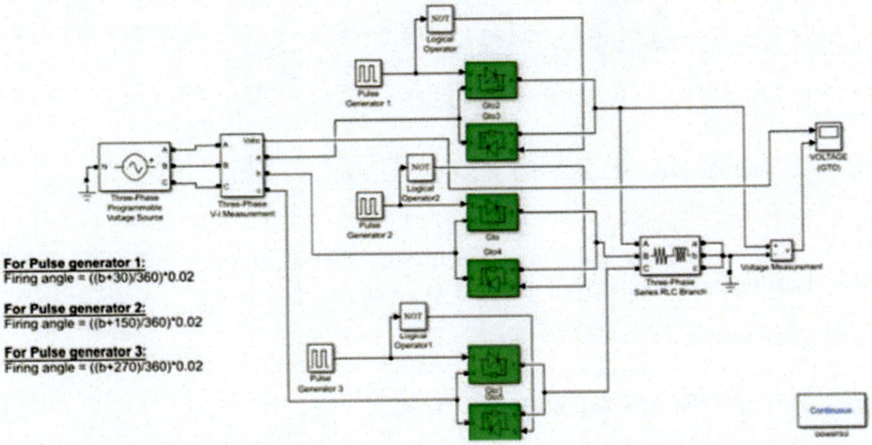

Fig. 10 Voltage control circuit with GTO

The simulation model of the starter with IGBT is shown in Fig. 12 and the internal connection of the starter with IGBT shown in Fig. 13. Initially two breakers are closed and the normal operation of induction machine runs, while starting the machine first it is connected in star and then after attaining speed it changes to delta mode of operation. Three pairs of IGBT are used in the star conditions and after star operations is finished another three pairs of IGBT for delta conditions. The motor speed and torque characteristics are given in Fig. 14. It has better character compare to GTO based starter. Similarly the voltage control circuit of three phase induction motor with IGBT switches is shown in Fig. 15. The connected induction motor has 7.5 kW,

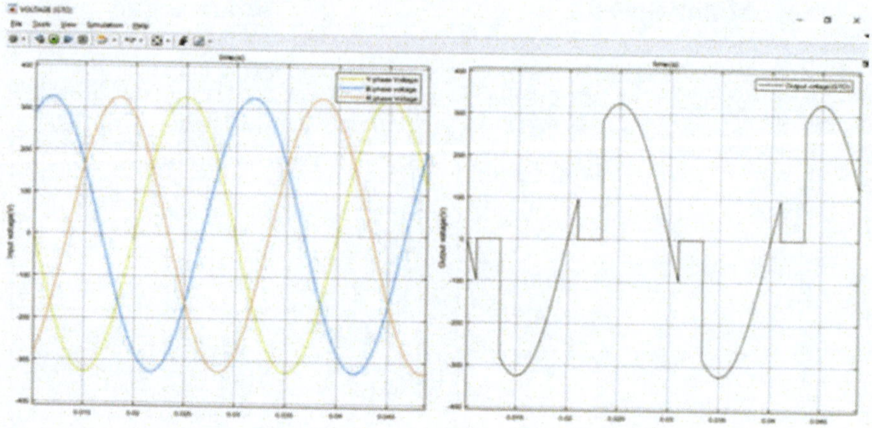

Fig. 11 Output voltage with GTO

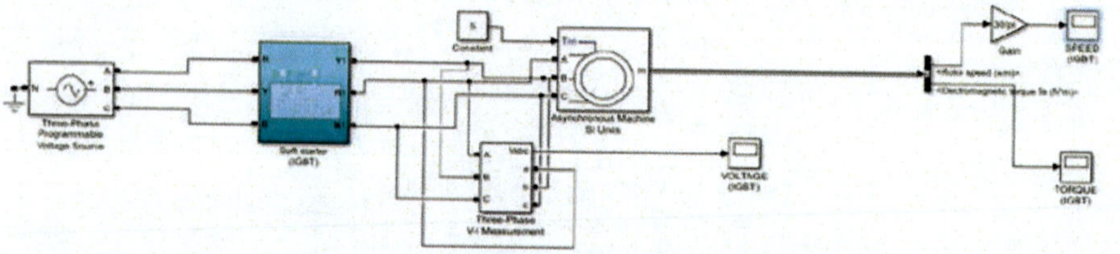

Fig. 12 Simulation model of starter with IGBT

400 V, 1500 rpm, 50 Hz and moment of Inertia of 0.0343 kg m^2. In this the output voltage of starter or input given to motor is given in Fig. 16.

3.4 Star–Delta Starter with MOSFET

In this simulation of MOSFET process, three phase programmable voltage source is connected to soft starter and the output of soft starter is connected to asynchronous machine and three phase voltage current measurement is used to measure the voltage of MOSFET and it is connected to the bus and output of the system is used to measure the speed and torque of the machine. The advantages of MOSFET are high input impedance, they support high speed operation and they have high drain resistance. The disadvantages of MOSFET are very susceptible to overload voltages hence special handling is required and cost is high.

The simulation model of the starter with MOSFET is given in Fig. 17. The internal structure of the starter with MOSFET is shown in Fig. 18. Initially two breakers are closed and the normal operation of induction machine runs, while starting the machine first it is connected in star and then after attaining speed it changes to delta

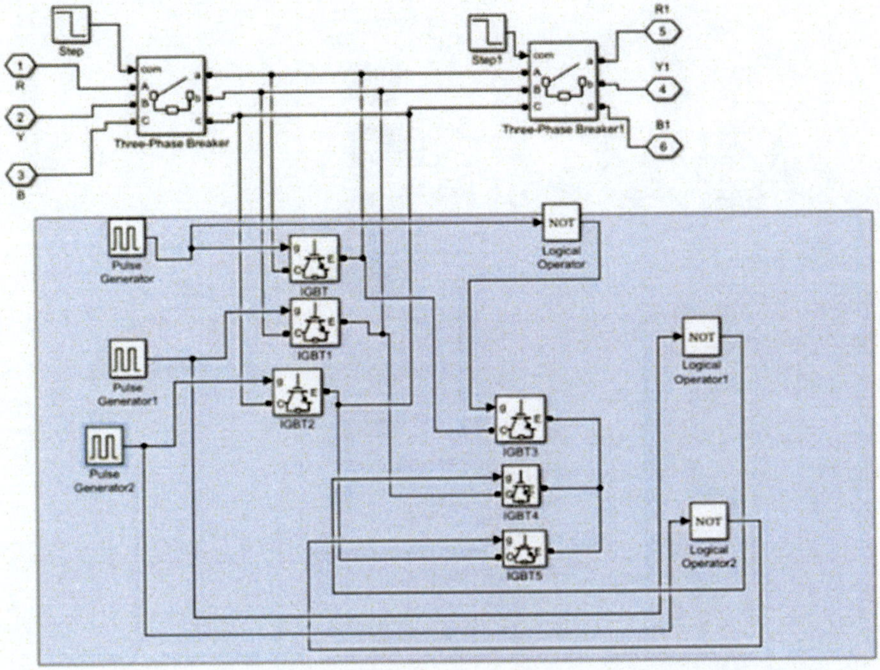

Fig. 13 Internal structure of starter with IGBT

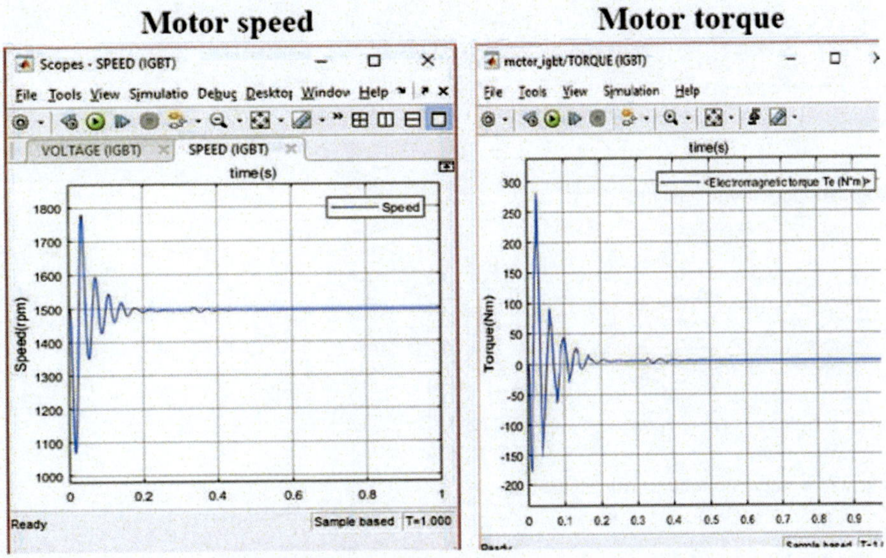

Fig. 14 Motor speed and torque with IGBT based starter

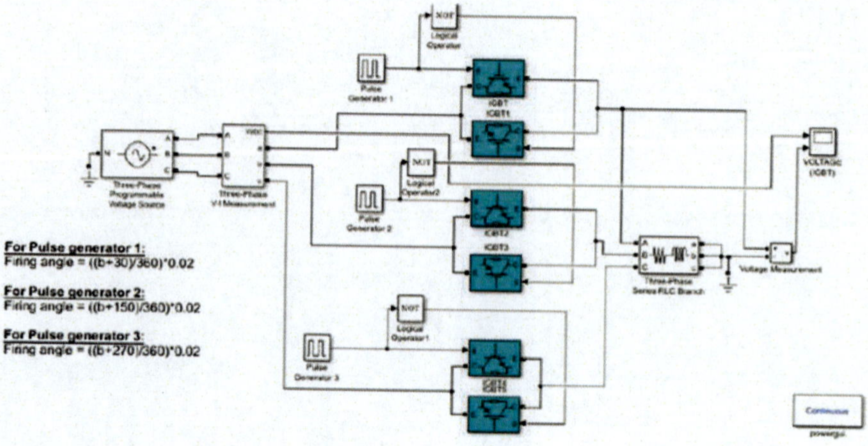

Fig. 15 Voltage control circuit with IGBT

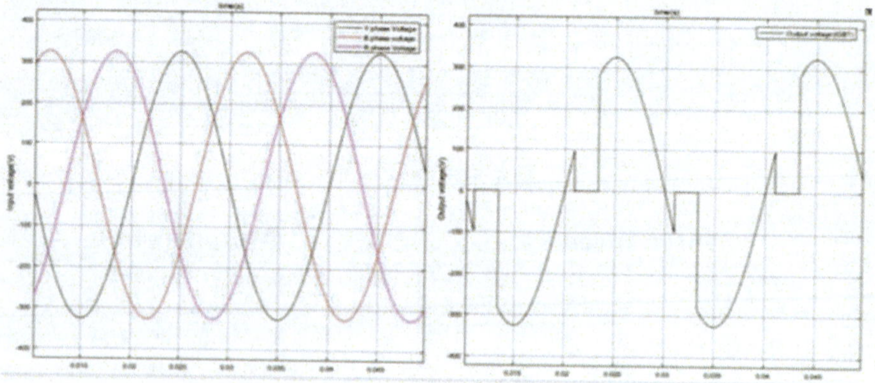

Fig. 16 Output voltage with IGBT

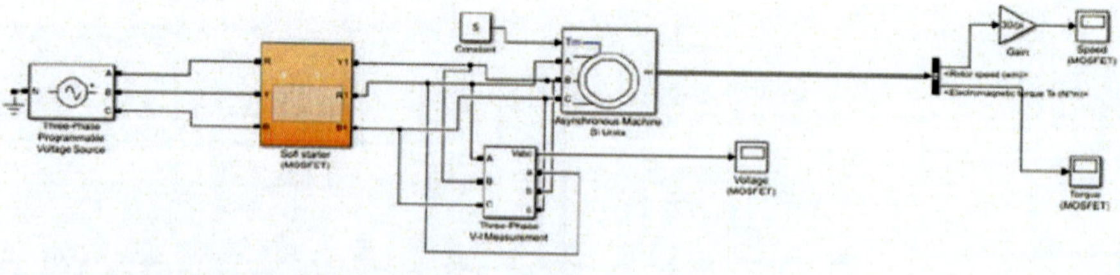

Fig. 17 Simulation model with MOSFET

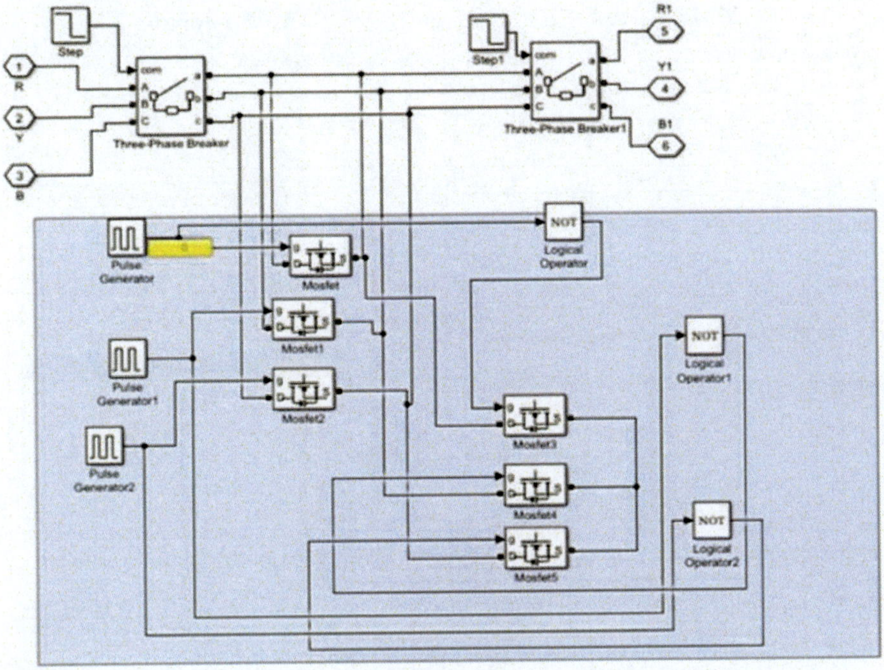

Fig. 18 Internal structure of starter with MOSFET

mode of operation. Three pairs of MOSFET are used in the star conditions and after star operations is finished another three pairs of MOSFET for delta conditions. The speed and torque characteristic of the motor with MOSFET is given in Fig. 19. It has less oscillation and quick settling compare to other two cases. The voltage control circuit of the three phase induction motor with MOSFET is given in Fig. 20. The input voltage applied to the motor from starter is given in Fig. 21.

From the above graphs we can observe that the starting provided by the proposed soft switching mechanism is a better one, owing to smoother torque and speed curves. Also, the compatibility of this type of switching is experimented with other power semiconductor devices, and the obtained values are given in Table 1. In this table, find that MOSFETs have produced faster switching time and motor speed. MOSFET based starter produces more torque than other switches based starters. But still on comparison usage of GTOs are near to the other devices. Also, with the higher ratings of GTO, we can implement this in high rating induction machines too; which is not possible for other two devices.

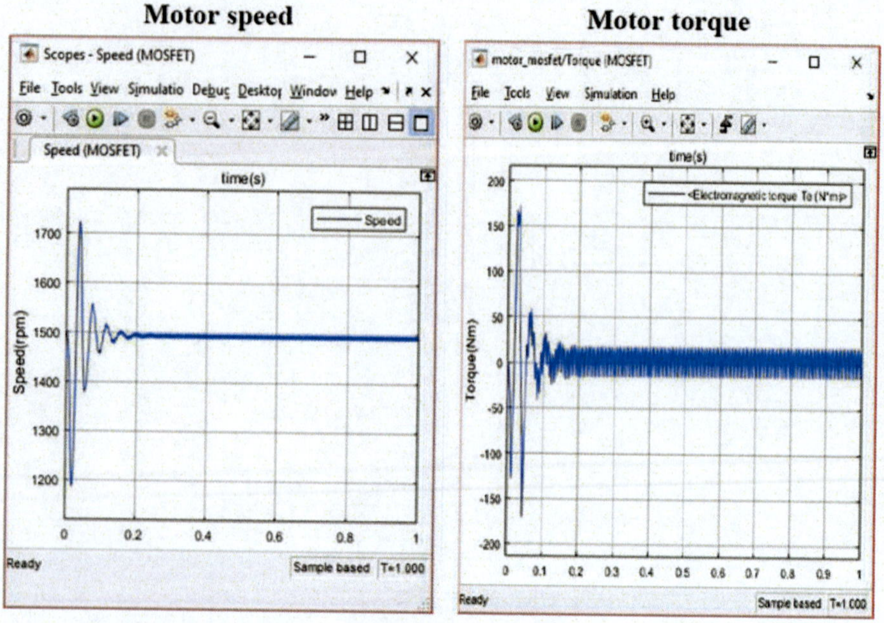

Fig. 19 Motor speed and torque with MOSFET based starter

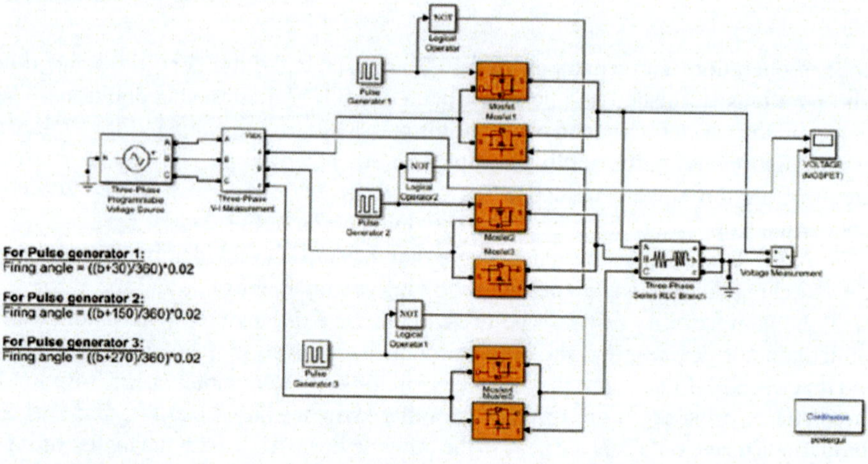

Fig. 20 Voltage control circuit with MOSFET

4 Conclusion

A modified soft starter has been implemented with GTO in the form of star-delta starting. Results show that GTO based starters provide smoother starting, thereby

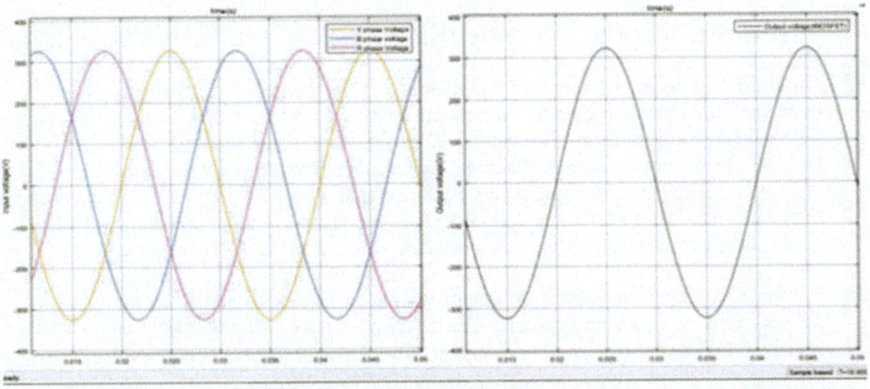

Fig. 21 Output voltage with MOSFET

Table 1 Comparison of various devices used in soft switching scheme

Device Used	Firing angle	Switch output voltage	Motor speed	Motor torque	Switching period (s)
GTO	30	326.52403	1494.20774	5.08443	0.34
IGBT	30	327.28746	1494.20764	5.08445	0.36
MOSFET	30	326.56104	1498.95479	22.87935	0.32

controlling voltage fluctuations to a greater extent. Similar tests have been done using other power semiconductor devices namely IGBT and MOSFET. Out of these, GTO can be a proven semiconductor device for soft starting of three phase induction motors owing to its versatility and energy efficiency. Also low cost microcontroller is used which enables better control. In addition to this parameters can be measured and controlled using the power semiconductor devices.

References

1. Ganar SB, Jodh OV, Gulhane GG (2017) Implementation of soft starter using 3 phase induction motor. Int J Sci Res 693–697
2. Henk de Swardt (2018) Star-delta starting and dual voltage motors. Marthinusen & Coutts
3. Anjitha RS, Ramesan R, George SM, Nazeer S, James G (2017) Soft switch starter for low power three phase induction motor. Int J Adv Res Electr Electron Instrum Eng 6(4)
4. Solveson MG, Mirafzal B, Demerdash NAO (2006) Soft-started induction motor modeling and heating issues for different starting profiles using a flux linkage ABC frame of reference. IEEE Trans Ind Appl 42(4):973–982
5. Sundareswaran K, Jos BM (2005) Development and analysis of novel soft-starter/energy-saver topology for delta-connected induction motors. IEE Proc Electr Power Appl 152(4):922–932
6. Zenginobuz G, Cadirci I, Ermis M, Barlak C (2001) Soft starting of large induction motors at constant current with minimized starting torque pulsations. IEEE Trans Ind Appl 37(5):1334–1347

7. McElveen RF, Toney MK (2001) Starting high-inertia loads. IEEE Trans Ind Appl 37(1):137–144
8. Shaogang H, Huagao H, Guoyong L, Weidong C (2001) Simulation of soft start-up process of induction motors using hybrid variable model. In ICEMS'2001. proceedings of the fifth international conference on electrical machines and systems, pp 459–462
9. Ginart A, Esteller R, Maduro A, Pinero R, Moncada R (1999) High starting torque for AC SCR controller. IEEE Trans Energy Convers 14(3):553–559
10. Deleroi W, Woudstra JB, Fahim AA (1989) Analysis and application of three-phase induction motor voltage controller with improved transient performance. IEEE Trans Ind Appl 25(2):280–286
11. Rowan TM, Lipo TA (1983) A quantitative analysis of induction motor performance improvement by SCR voltage control. IEEE Trans Ind Appl 19(4):545–553
12. Murthy SS, Berg GJ (1982) A new approach to dynamic modeling and transient analysis of SCR-controlled induction motors. IEEE Power Eng Rev 2(9):32
13. Nath G, Berg GJ (1981) Transient analysis of three-phase SCR controlled induction motors. IEEE Trans Ind Appl 17(2):133–142
14. Lipo TA (1971) The analysis of induction motors with voltage control by symmetrically triggered thyristors. IEEE Trans Power Appar Syst 90(2):515–525
15. Despe O, Wang J (2010) Design of a gate turn off (GTO) switch for pulsed power application. Argonne National Laboratory, Argonne, Illinois
16. Akshaykumar S, Sagar T, Vijay S, Mone PP (2017) Soft starting of three phase induction motor. Int J Res Trends Innov 2(5):1–10
17. Gitonga Hilda Wanjiku (2016) Electronic starter for single phase induction motor. The University of Nairobi
18. Ze Z, Ming HH (2018) Soft starter study of induction motors using fuzzy PID control. Mater Sci Eng 1–6

Design and Performance Comparison of Permanent Magnet-Assisted Synchronous Reluctance Motors

D. Pradhap, P. Ramesh, and N. C. Lenin

Abstract Permanent magnet-assisted synchronous reluctance motor (PMa-SynRM) helps to reduce the dependency of rare earth magnets, particularly in electric vehicle applications. In this paper, the significance of magnet position in PMa-SynRM is discussed in detail. By keeping the magnet in appropriate position, it can be utilized effectively, which enhances the power density of the motor. A 1 kW PMa-SynRM is designed for electric two-wheeler, and its performance for various magnet positions is analyzed and compared.

Keywords Electric vehicle · Permanent magnet · Synchronous reluctance motor

1 Introduction

An electric motor is a key component in the electric vehicle (EV) traction system. In general, interior permanent magnet (IPM) motor with rare earth magnets is employed in EVs, due to its high torque and power density, wide speed operating range [1, 2]. But, the usage of rare earth magnets increases the cost of IPM motor to a greater extent [3, 4]. To overcome this issue, synchronous reluctance motor with rare earth free magnet is a better choice. Replacing the high-cost rare earth PM with rare earth free magnets reduces the cost, while the energy product of the non-rare earth free magnet is comparatively low [5]. To overcome this challenge, optimized utilization of permanent magnets is desirable. In order to improve the utilization of permanent magnet, permanent magnet placement in the rotor is vital.

In this paper, the importance of magnet position in the rotor of PMa-SynRM is analyzed. The parameters influencing the reluctance and magnet torque are discussed in Sect. 2. The design specifications are given in Sect. 3. In Sect. 4, different rotor

D. Pradhap · P. Ramesh · N. C. Lenin (✉)
Vellore Institute of Technology, Chennai Campus, Chennai 600127, India
e-mail: lenin.nc@vit.ac.in

N. Zhou and S. Hemamalini (eds.), *Advances in Smart Grid Technology*, Lecture Notes in Electrical Engineering 688, https://doi.org/10.1007/978-981-15-7241-8_3

topologies based on magnet positions are specified. The no-load and load characteristics of the designed PMa-SynRMs are compared in Sects. 5 and 6. Finally, the performance characteristics are discussed in Sect. 7. The electromagnetic performance is done by finite element software MagNet.

2 Permanent Magnet-Assisted SynRM

The torque production in permanent magnet-assisted synchronous reluctance motor (PMa-SynRM) depends on two components, namely reluctance torque and magnet torque [6]. The reluctance torque is produced by rotor saliency due to its magnetic anisotropy. The magnet torque is produced due to the interaction of magnetic fields created by permanent magnet and stator current. In this motor, the flux barrier in the rotor is used to create saliency and also facilitates to keep magnets. The axis with high permeance is referred as *d*-axis, while the axis with high reluctance is termed as *q*-axis, as shown in Fig. 1.

The torque developed in PMa-SynRM [7] can be expressed as,

$$T = \frac{3}{2}\frac{P}{2}\left(\lambda_{\text{pm}} i_d + \left(L_d - L_q\right) i_d i_q\right) \tag{1}$$

where P is the number of pole pairs, i_d and i_q are the d-axis and q-axis components of the stator current. L_d and L_q are d-axis and q-axis inductance, respectively, and λ_{pm} is the flux linkage due to permanent magnets. From Sect. 3, PMa-SynRM with ceramic magnet has been discussed.

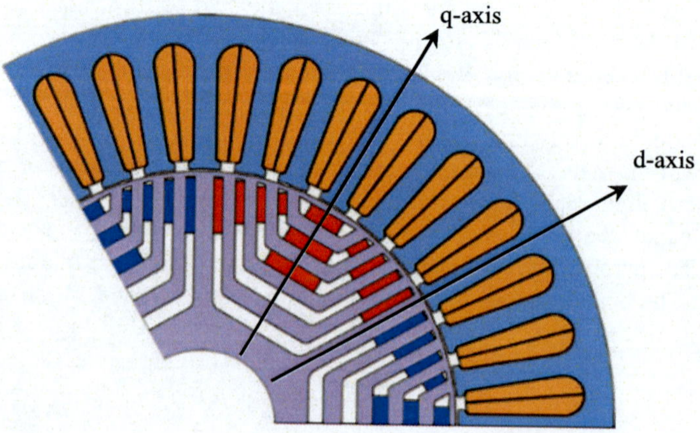

Fig. 1 PMa-SynRM

Table 1 Specification and dimension

Items	Value
Rated power	1 kW
Supply voltage	48 V
Rated torque	3.8 Nm
Rated speed	2500 rpm
Maximum speed	10,000 rpm
Stator outer diameter	150 mm
Stack Length	40 mm
Number of stator slots	36
Number of poles	6
Number of turns per coil	7
Winding type	Distributed

3 Design of CMa-SynRM

A 1 kW, ceramic magnet-assisted synchronous reluctance motor (CMa-SynRM) is designed to satisfy the requirement of electric two-wheeler traction application. The target specifications are listed in Table 1. The design process of CMa-SynRM is shown in Fig. 2.

4 Designed Rotor Topologies of CMa-SynRM

The reluctance torque of the motor is influenced by the structure of flux barrier in the rotor, whereas the magnet torque depends on the permanent magnet flux linkage (λ_{pm}). The position of permanent magnet in the rotor flux barrier decides the amount of flux linkage with the stator windings. So, various magnet placement positions, as shown in Fig. 3, are considered to study the effective utilization of permanent magnet. According to the rotor topology, the motors are labeled as CMa-SynRM I, CMa-SynRM II, and CMa-SynRM III. In order to have a better comparison, some of the parameters like stator design, excitation, and volume of permanent magnets are kept same for all topologies.

5 No-Load Characteristics

The designed CMa-SynRMs has been modeled in finite element platform and initially analyzed without load. The airgap flux density obtained from the motors is depicted in Fig. 4. The airgap flux density in CMa-SynRM I is 22% higher than CMa-SynRM II and nearly two times of CMa-SynRM III. In CMa-SynRM I, all the permanent

Fig. 2 Design procedure

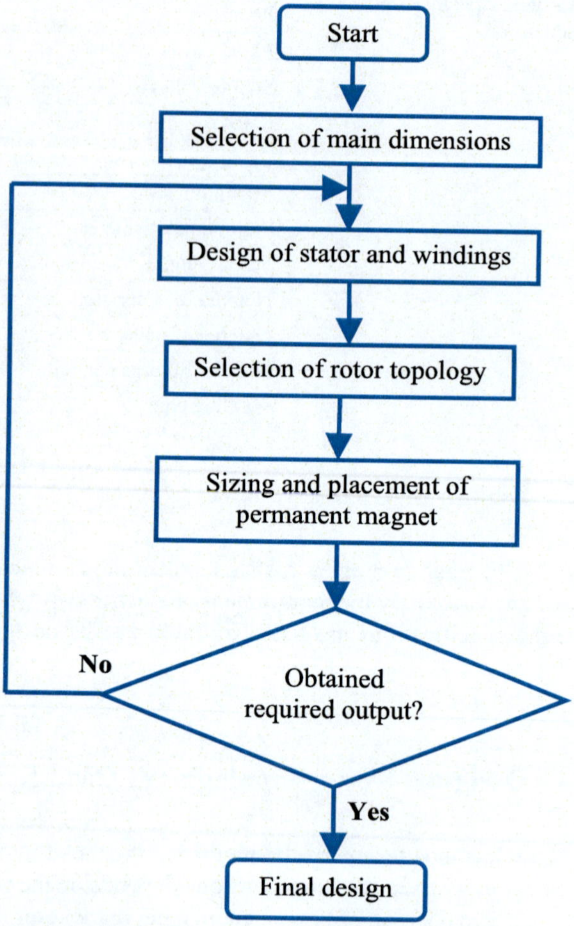

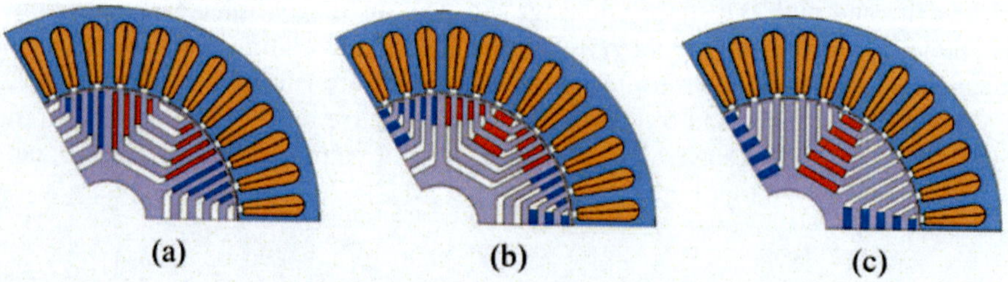

Fig. 3 Rotor topologies, **a** CMa-SynRM I, **b** CMa-SynRM II, **c** FMa-SynRM III

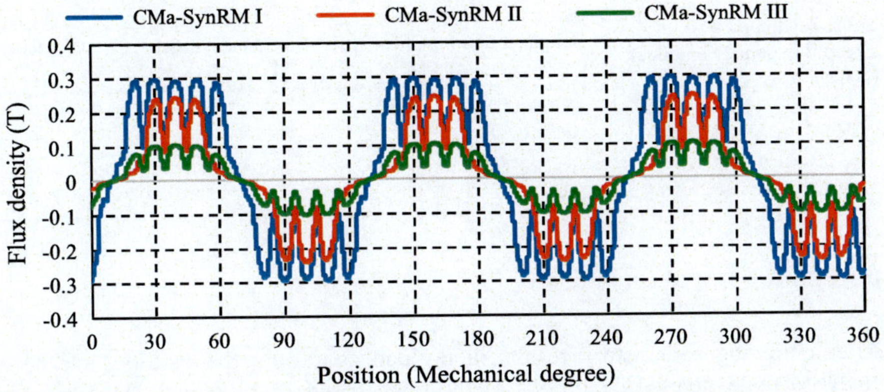

Fig. 4 Airgap flux density at no-load condition

magnets are positioned along the *d*-axis, which has minimum reluctance and also closer to airgap [8]. Hence, the resulting airgap flux density is improved.

The back EMF induced for one electric cycle is depicted in Fig. 5. The peak value of back EMF obtained in CMa-SynRM I is 22.05 V, which is 55% higher than CMa-SynRM II and more than two times of CMa-SynRM III. This shows that the permanent magnet flux linkage is higher in CMa-SynRM I [9, 10]. The comparison of airgap flux density and back EMF under no-load condition for three rotor topologies are listed in Table 2.

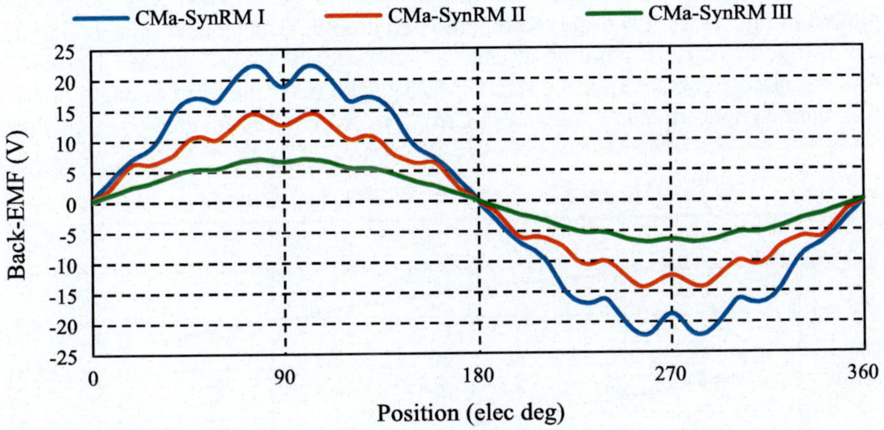

Fig. 5 Back EMF under no load

Table 2 Comparison of airgap flux density and back EMF

Characteristic	CMa-SynRM I	CMa-SynRM II	CMa-SynRM III
Airgap flux density (T)	0.30	0.24	0.10
Back-EMF (V)	22.05	14.38	6.99

6 Load Characteristics

After obtaining satisfactory results in no-load condition, the motors have been analyzed under loaded condition. At the rated speed of 2500 rpm, 28 A current is drawn to deliver the rated torque. The *d*-axis and *q*-axis components of the rated current are plotted for various current angles in Fig. 6. These currents can be calculated by [11],

$$i_d = i_p \cos\theta \tag{2}$$

$$i_q = i_p \sin\theta \tag{3}$$

The reluctance torque depends on the *d*-axis and *q*-axis inductances (L_d and L_q). Figure 7 shows the L_d and L_q variation of the designed motors. Compared to other motors, CMa-SynRM III provides better saliency.

The reluctance torque and magnet torque obtained for various current angles are plotted in Fig. 8a, b. It is observed that the variation in reluctance torque among the motor topologies is very minimal. Due to high saliency, CMa-SynRM III provides 4% increased reluctance torque than CMa-SynRM II, at the current angle of 60°. But, the magnet torque is 7% lesser in CMa-SynRM III. The magnet position along

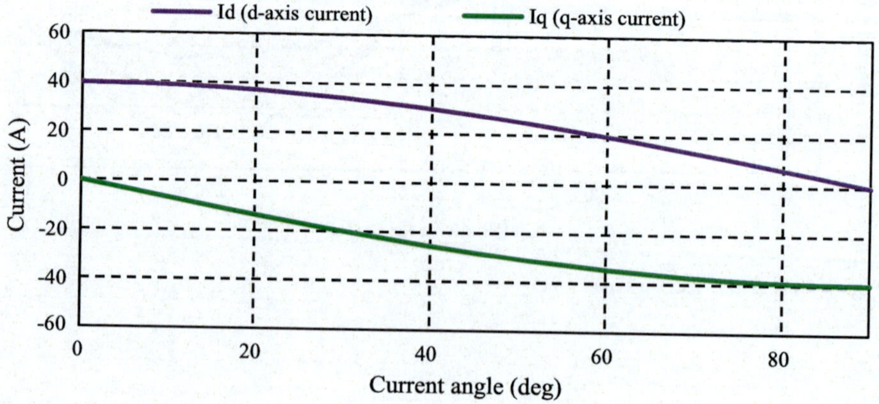

Fig. 6 *d*-axis and *q*-axis current

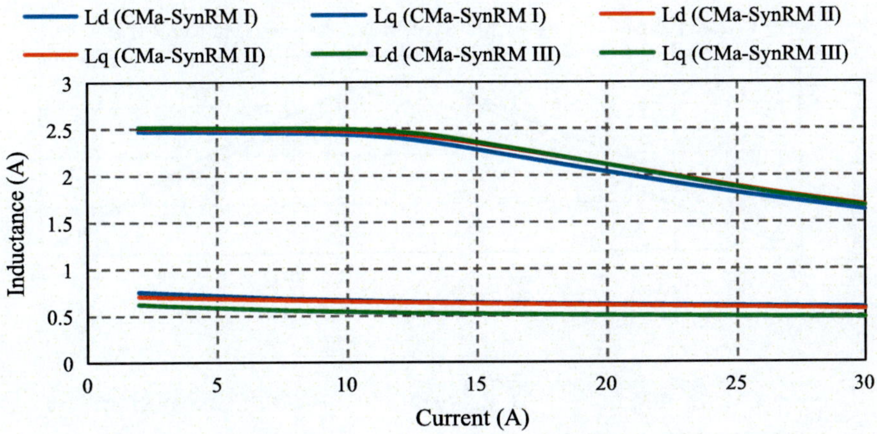

Fig. 7 *d*-axis and *q*-axis inductance

d-axis makes CMa-SynRM I to offer high magnet torque, which is 14% higher than CMa-SynRM II. This helps to enhance the net torque to a greater extent [12], as shown in Fig. 8c.

7 Performance Characteristics

The torque–speed characteristics of the designed motors are depicted in Fig. 9. Because of the magnet placement in *d*-axis, CMa-SynRM I provides better torque–speed characteristics than other motors [13]. At the rated speed of 2500 rpm, CMa-SynRM I delivers 14 and 22% higher torque than CMa-SynRM II and III. This enhances the rated power to 1.2 kW, as shown in Fig. 7. Due to increased torque at higher speeds, CMa-SynRM I also has extended constant power–speed range.

The power–speed characteristics of the three motor topologies are shown in Fig. 10. It is observed from Fig. 10, the power developed by CMa-SynRM I is comparatively better among three motors.

Cogging torque is originated due to the interaction between permanent magnet of the rotor and stator teeth [14, 15]. Figure 11 shows the cogging torque developed in all three motors. CMa-SynRM III has the lowest cogging torque, 0.003 Nm, whereas CMa-SynRM I has 0.042 Nm due to high magnet flux linkage. Cogging torque is the main source for torque ripple [16, 17]. Torque ripples in three rotor topologies are listed in Table 3.

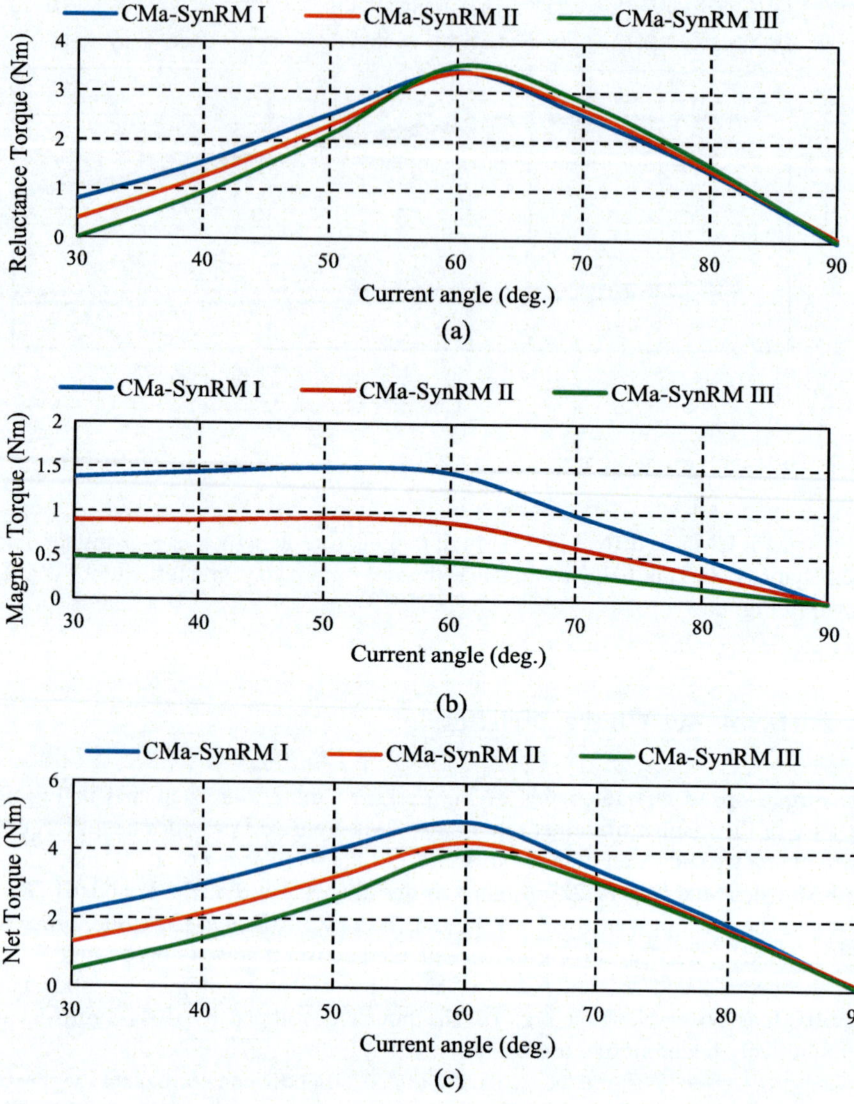

Fig. 8 Torque characteristics of three CMa-SynRM **a** reluctance torque-current angle, **b** magnet torque-current angle, **c** net torque-current angle

8 Conclusion

In general, interior permanent magnet motor (IPM) with rare earth magnets is employed in EVs. In order to avoid the usage of rare earth magnet, the low-cost ceramic magnet-assisted synchronous reluctance with three rotor topologies is

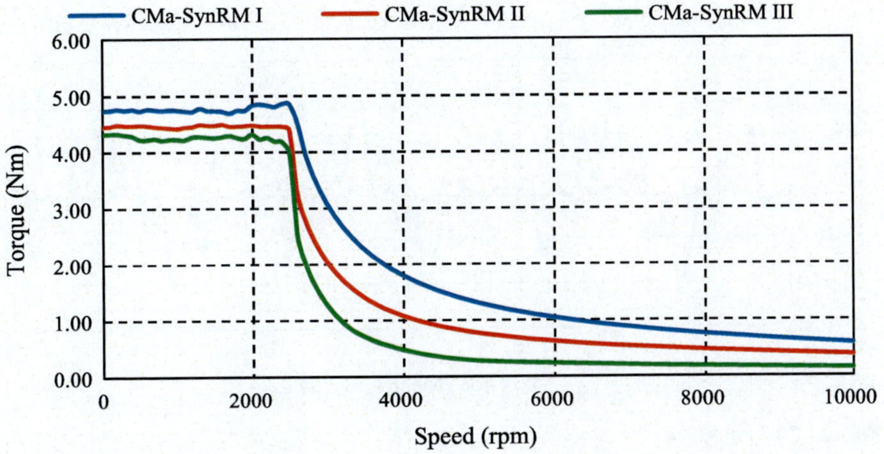

Fig. 9 Torque–speed characteristics

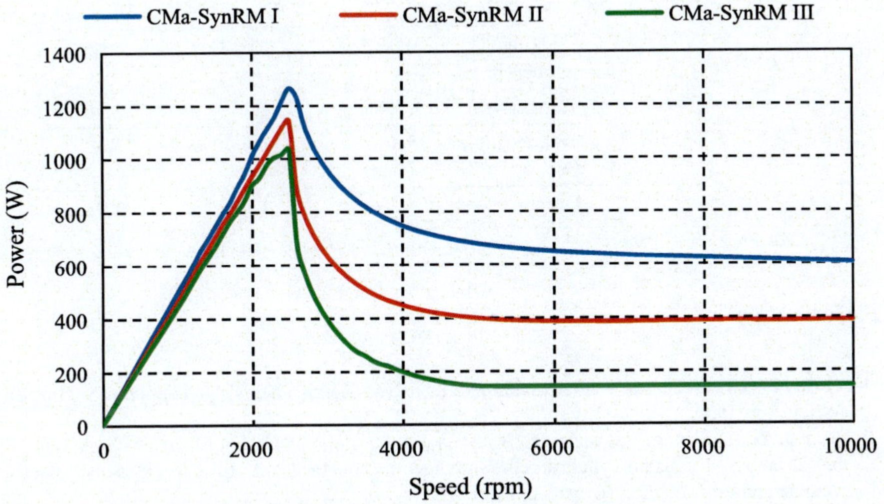

Fig. 10 Output power versus speed

compared. From the analysis, it can be inferred that, by positioning the magnet along the d-axis improve the magnet torque to a greater extent, which also reflects on the resultant output power. Preferring this machine would help to reduce the cost and avoid the usage of rare earth magnet.

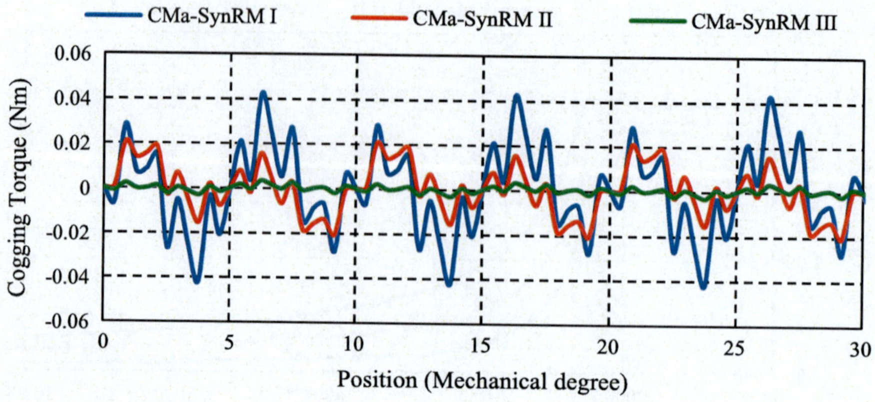

Fig. 11 Cogging torque

Table 3 Comparison of torque and output power

Characteristic	CMa-SynRM I	CMa-SynRM II	CMa-SynRM III
Cogging torque(Nm)	0.042	0.013	0.003
Torque(Nm)	4.84	4.26	3.96
Torque ripple (%)	15.4	18.6	11.3
Output power (kW)	1.26	1.14	1.03

References

1. Chau KT (2015) Electric vehicle machines and drives design, analysis and application, 1st edn. Wiley, Singapore
2. Bianchi N, Fornasiero E, Carraro E, Bolognani S, Castiello M (2014) Electric vehicle traction based on a PM assisted synchronous reluctance motor. In: 2014 IEEE international electric vehicle conference (IEVC), pp 1–6, Florence
3. Mitch O (2011) Final report on assessment of motor technologies for traction drives of hybrid and electric vehicles. A Report submitted to Energy Efficiency and Renewable Energy Vehicle Technologies Program, Washington D.C
4. Bianchi N, Fornasiero E, Ferrari M, Castiello M (2016) Experimental comparison of PM-assisted synchronous reluctance motors. IEEE Trans Ind Appl 52(1):163–171
5. Guan Y, Zhu Z, Afinowi I, Mipo J, Farah P (2016) Design of synchronous reluctance and permanent magnet synchronous reluctance machines for electric vehicle application. COMPEL—Int J Comput Mathem Electr Electron Eng 35(2):586–606
6. Nattuthurai S, Neelamegham R (2017) Design and performance evaluation of SynRM and ferrite assisted-SynRM for EV application using FEA. In 2017 international conference on smart technologies for smart nation (SmartTechCon), pp 492–497, Bangalore
7. Ngo D, Hsieh M, Huynh TA (2019) Torque enhancement for a novel flux intensifying PMa-SynRM using surface-inset permanent magnet. IEEE Trans Magn 55(7):1–8

8. Wang B, Wang J, Griffo A, Sun Z, Chong E (2016) A fault tolerant machine drive based on permanent magnet assisted synchronous reluctance machine. In: 2016 IEEE energy conversion congress and exposition (ECCE), pp 1–8. Milwaukee, WI
9. Ge L, Zhu X, Wu W, Liu F, Xiang Z (2017) Design and comparison of two non-rare-earth permanent magnet synchronous reluctance motors for EV applications. In: 2017 20th international conference on electrical machines and systems (ICEMS), pp 1–5. Sydney
10. Huynh TA, Hsieh M, Shih K, Kuo H (2018) An investigation into the effect of PM arrangements on PMa-SynRM performance. IEEE Trans Ind Appl 54(6):5856–5868
11. Duane Hanselman (2006) Brushless permanent magnet motor design (2nd edn). Magna Physics Publishing, Ohio
12. Li P, Ding W, Liu G (2018) Sensitivity analysis and design of a high performance permanent-magnet-assisted synchronous reluctance motor for EV application. In: 2018 IEEE transportation electrification conference and expo (ITEC), pp 406–411, Long Beach, CA
13. Liu C, Luo T, Shih P, Yen S, Lin Hsu Y, Hwang C (2017) On the design and construction assessments of a permanent-magnet-assisted synchronous reluctance motor. IEEE Trans Magn 53(11):1–4
14. Lian G, Li H, Ban F, Chen B (2018) Optimization design of cogging torque and back EMF waveform for PMSM. In: 2018 IEEE international conference on applied superconductivity and electromagnetic devices (ASEMD), pp 1–2, Tianjin
15. Naveen Kumar E, Naveen Deepak V, Ragavan K (2018) A compact expression for cogging torque considering both radial and tangential fields for surface-mounted PM motors. IEEE Trans Magn 54(9):1–12
16. Kong Y, Lin M, Yin M, Hao L (2018) Rotor structure on reducing demagnetization of magnet and torque ripple in a PMa-synRM with ferrite permanent magnet. IEEE Trans Magn 54(11):1–5
17. Hwang M-H, Lee H-S, Cha H-R (2018) Analysis of torque ripple and cogging torque reduction in electric vehicle traction platform applying rotor notched design. Energies 11(11):3053

Performance Analysis of Fuzzy Logic Control-Based Classical Converter Fed 6/4 SRM Drive for Speed Precision

Indira Damarla and M. Venmathi

Abstract Switched reluctance motor (SRM) has become admirable to conventional electrical machines due to its robust construction, a wide range of speed control, ruggedness, low cost, high torque to weight ratio and highly reliable. SRMs are gaining most attractive with these excellent features in the field of industrial applications such as electric vehicles, electric traction, aerospace, mining drives. By cause of harmonics in air gap flux leads to periodic speed pulsations and large torque ripple which leads to vibration and acoustic noise. Hence to overcome this problem, the intelligent controller like a fuzzy logic controller (FLC) is employed. In this paper, FLC-based classical converter fed SRM drive is presented. The main aim of this work is to control the speed of SRM drive with a fuzzy controller in the MATLAB/Simulink platform. To show the effectiveness of fuzzy controller, its performance is correlated with the conventional proportional integral (PI) controller. The statistical parameters like rise time, settling time for various speeds with constant load torque are reported. The comparative simulation results between the FLC and PI are also shown to validate the effectiveness of FLC to diminish periodic speed ripples and settling time.

Keywords Switched reluctance motor · Classical converter · Proportional integral controller · Fuzzy logic controller · Speed pulsations · Dynamic response

1 Introduction

It is well known from many years ago SRM has been modernized attraction as a research carrier. The enhancement in power electronic converters, modern controllers along with tremendous speed embedded controllers with dynamic ability to perform calculation SRM drives is preferred in many applications such as electric vehicles, robotic control, automobile applications, etc. Both stator and rotor of SRM consist of salient poles; the rotor does not carry any winding and is known as a double salient pole machine. SRM gives many merits due to its winding free rotor like low

I. Damarla (✉) · M. Venmathi
Department of EEE, St. Joseph's College of Engineering, Chennai, India
e-mail: indira.damarla@gmail.com

N. Zhou and S. Hemamalini (eds.), *Advances in Smart Grid Technology*, Lecture Notes in Electrical Engineering 688, https://doi.org/10.1007/978-981-15-7241-8_4

inertia, no maintenance, fault-tolerant capability, ability to run on a wide range of speed, good performance and simple construction, etc. [1–3]. The essential characteristics of adjustable speed drive consist of reduced torque pulsations, acceptable transient and steady-state response, and minimum speed ripple. Nevertheless due to harmonics in air gap flux, nonlinearity in magnetic structure and nonlinear correlation among the position of the rotor, torque and electromagnetic flux linkages, it is not so simple to regulate the speed of SRM [4, 5]. Practically, it is very difficult to bring out pure sinusoidal flux around the air gap. Torque ripples originate by the interaction of nonsinusoidal flux with sinusoidal phase currents [6]. Therefore to control SRM drive, nonlinear controllers are necessary. In [7], some of the controllers are highlighted. Nowadays, the modernization and advancements in SRM such as modified construction design [8, 9], high-performance magnetic core [10] have been conferred. To improve the performance, a controller is introduced depend on feedback in [11]. The robust control technologies are recommended in [12, 13]. A new sliding mode controller has been established in literature to produce perfect control under abrupt disturbances in load. Currently, in many industrial applications, the proportional, integral and derivative controllers are good candidates for improving the performance [14]. A set of space vectors that are used to improve the torque in SRM is proposed in [15]. By properly tuning the parameters of P, PI, PD, PID controllers deliver high efficiency if we know the precise mathematic model [16, 17]. An artificial intelligence-based controls like FLC, artificial neural networks (ANN) and genetic algorithms are good enough for improved operation if the mathematical modeling is painful [18, 19]. Out of these, FLC plays a dominant role in various applications due to its flexibility. At present, fuzzy controllers are emerging trends in variable speed applications. A C-dump converter fed SRM with FLC was proposed in [20]. Under dynamic behavior, FLC gives better response compared to traditional controllers. As compared to traditional controllers, FLC reduces the cost of hardware [21]. The self-tuning FLC has the merits of large capability of control, simple and greater adaptability and it is presented in [22]. The PI controller uses pulse width modulation (PWM) technique to control the phase current of the SRM and turn on, turn off angles are controlled by the current controller. In [23], a stable, fast digital PWM current controller is developed for SRM drive. The drawbacks of the PI controller are overcome by using FLC and the comparative study was carried out in [24, 25]. In this paper, FLC has been designed for a wide range of speeds including low and high speeds with classical converter fed SRM to regulate the speed ripples and to achieve smooth control at variable speed operation. The FLC is compared with a conventional controller in aspects like rise time and settling time.

2 Dynamic Modeling of SRM

The electromagnetic torque developed in SRM is directly proportional to the square of phase current and change of reluctance to rotor position. Inductance profile plays a major role in the production of torque. Mathematical modeling of SRM is difficult due to its nonlinear relation between flux linkage, inductance and electromagnetic torque with rotor position and phase current. The cross-sectional view and equivalent circuit for one phase of 6/4 SRM are depicted in Fig. 1.

According to mathematical modeling [26], the state space dynamic equations are listed below:

$$v(t) = Ri(t) + \frac{d\psi(\theta, i)}{dt} \tag{1}$$

$$\psi(\theta, i) = L(\theta, i) * i \tag{2}$$

From Eqs. (1) and (2),

$$v(t) = Ri(t) + L(\theta, i)\frac{di}{dt} + \frac{dL(\theta, i)}{d\theta}i\omega_m \tag{3}$$

The EMF induced in motor is

$$e = \frac{dL(\theta, i)}{d\theta}i\omega_m = K_b i\omega_m \tag{4}$$

Where,

$$K_b = \frac{dL(\theta, i)}{d\theta} \tag{5}$$

The addition of winding resistance loss, rate of change in field energy and air gap power will give the instantaneous input power and it is given below:

$$P_i = v(t)i(t) = Ri^2 + i^2\frac{dL(\theta, i)}{d\theta} + L(\theta, i)i\frac{di}{dt} \tag{6}$$

Air gap power

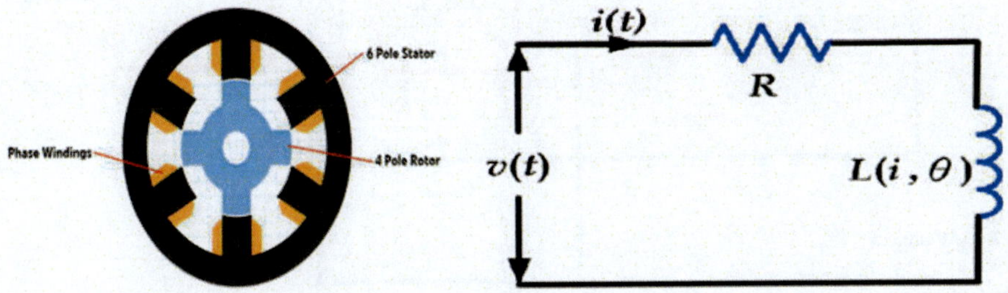

Fig. 1 Cross-sectional view and equivalent circuit of 6/4 SRM

$$P_{ag} = \frac{1}{2} i^2 \frac{dL(\theta, i)}{dt} \tag{7}$$

$$P_{ag} = \frac{1}{2} i^2 \frac{dL(\theta, i)}{d\theta} \frac{d\theta}{dt} \tag{8}$$

$$P_{ag} = T_e(\theta, i)\omega_m \tag{9}$$

From Eqs. (8) and (9)

$$\omega_m = \frac{d\theta}{dt} \tag{10}$$

Electromagnetic torque developed in SRM is

$$T_e(\theta, i) = \frac{1}{2} i^2 \frac{dL(\theta, i)}{d\theta} \tag{11}$$

The mechanical expressions can be expressed as

$$T_e - T_l = J_m \frac{d\omega_m}{dt} + B_m \omega_m \tag{12}$$

By rearranging Eq. (12), the speed expression can be written as

$$\frac{d\omega_m}{dt} = \frac{1}{J_m}[T_e - T_l - B_m \omega_m] \tag{13}$$

The electrical model of 6/4 SRM for one phase is shown in Fig. 2.

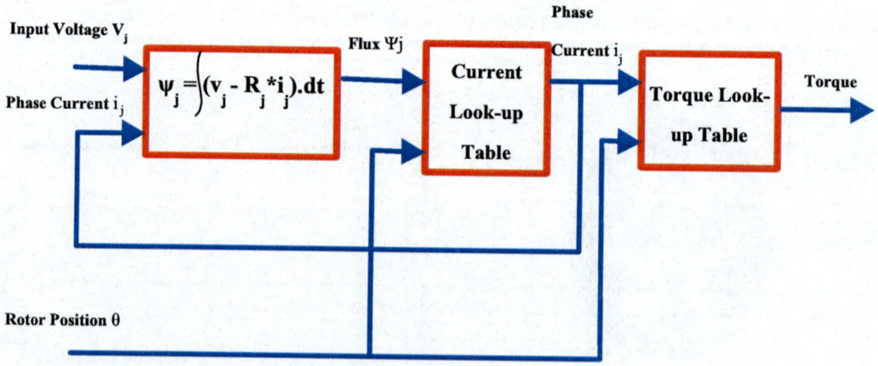

Fig. 2 Electric model of one phase of 6/4 SRM

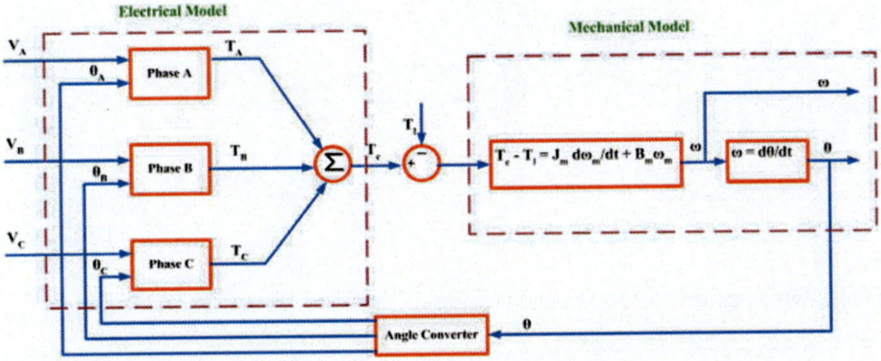

Fig. 3 Dynamic model of 6/4 SRM

In the dynamic model, current lookup and torque lookup tables are used to find the phase current and torque by knowing the parameters of input voltage and rotor position. The electric model shown in Fig. 2 is obtained by combining current and torque lookup tables. The dynamic model of SRM is obtained by combining a mechanical and electrical model of SRM and is pictured in Fig. 3.

3 Control Schemes for Speed Regulation

The elementary block diagram of the 6/4 SRM drive is shown in Fig. 4. To regulate the speed of SRM, the reference speed ω^* is compared with the actual speed ω produces an error which is the input to the speed controller. Here, speed controller is either PI or FLC. Reference current signals I^* are generated from the speed controller and compared with actual phase current I which gives an error. This error is used as the input to the hysteresis current controller which limits the phase currents within the hysteresis band. The output of hysteresis current controller gives the control pulses to the asymmetric bridge converter which fed to the SRM.

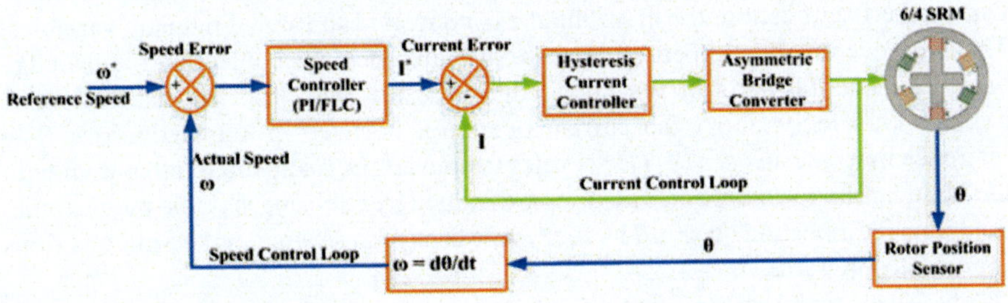

Fig. 4 Speed control of 6/4 SRM

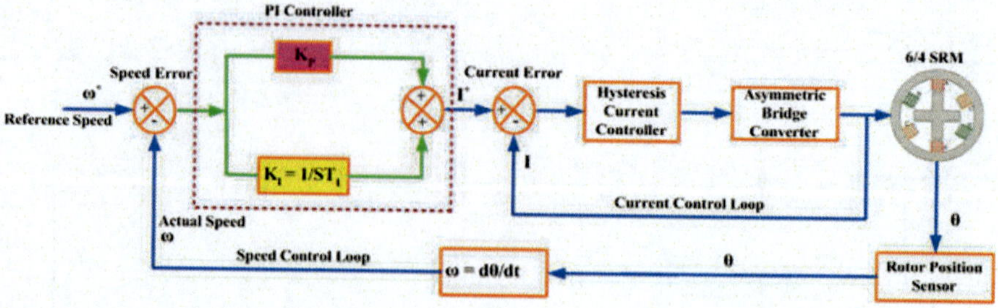

Fig. 5 Speed control of 6/4 SRM with PI controller

3.1 PI Controller

The speed control of SRM with a traditional PI controller is shown in Fig. 5. To minimize the periodic speed pulsations and steady-state error, it is necessary to combine both proportional and integral controllers. To upgrade the performance of the converter and to overcome the disturbances, the feedback signal is mandatory. But by the use of an integral controller causes a negative impact on plant stability and speed output. With the PI controller, the speed response will not improve. Therefore, PI controllers are sitting in the areas where speed is not a benchmark [27]. The K_p and K_i values in traditional PI controller are tuned using Ziegler and Nichols tuning method.

3.2 Fuzzy Controller

Nowadays fuzzy controllers are most attractive in many industrial applications due to its excellent quality that does not require any mathematical modeling of the plant. It is an intelligent control that mimics the human brain based on the expert's knowledge. Any nonlinear system can be handled efficiently by shifting the operating point, and hence, it can absorb any sudden disturbances of the system. The basic structure of FLC consists of mainly three stages such as fuzzification, rule base and defuzzification. During the fuzzification, the crisp input r is converted to fuzzy linguistic variables. The components of an expert system are the inference mechanism and knowledge base. The internal parts of the knowledge base are rule base and database. The expert's knowledge about the current operation is saved in knowledgebase. The inference mechanism uses IF-THEN rules to convert the fuzzy input into output with the Mamdani-type controller. The rules are framed by choosing suitable membership functions for input and output. The fuzzy rule base with seven membership functions is presented in Table 1.

The seven membership functions which are used for forming the rule base are negative large (NL), negative medium (NM), negative small (NS), zero (ZE), positive small (PS), positive medium (PM), positive large (PL).

Table 1 Rule base for FLC

Error	Change in error						
	NL	NM	NS	ZE	PS	PM	PL
NL	NL	NL	NL	NL	NM	NS	ZE
NM	NL	NL	NL	NM	NS	ZE	PS
NS	NL	NL	NM	NS	ZE	PS	PM
ZE	NL	NM	NS	ZE	PS	PM	PL
PS	NM	NS	ZE	PS	PM	PL	PL
PM	NS	ZE	PS	PM	PL	PL	PL
PL	ZE	PS	PM	PL	PL	PL	PL

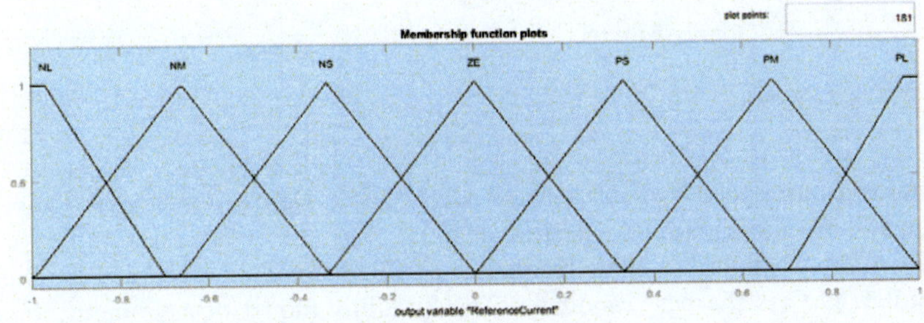

Fig. 6 Membership function for reference current

The membership function for reference current is shown in Fig. 6. The dependency of the output variable on input variables is estimated by the use of surface viewer. The reference current as a function of error and change in error is depicted in Fig. 7. The speed control of SRM with an intelligent fuzzy controller is shown in Fig. 8.

4 Simulation Results and Discussion

The performance of asymmetric bridge converter fed 6/4 SRM embedded with PI and fuzzy controllers are evaluated by using MATLAB/Simulink environment. The Simulink model of asymmetric bridge converter fed SRM encapsulated with PI controller is shown in Fig. 9. The specifications of three phase 6/4 SRM used for simulation are highlighted in Table 2.

The output waveforms for instantaneous flux, phase current, torque developed and speed are displayed in Figs. 10, 12, 14 and 16, respectively. The zoomed view of flux, phase current and torque is represented in Figs. 11, 13 and 15, respectively.

The Simulink model embedded with PI controller is tested based on the parameters of 0.5 Nm load torque and the reference speed of 1800 rpm. The settling time required

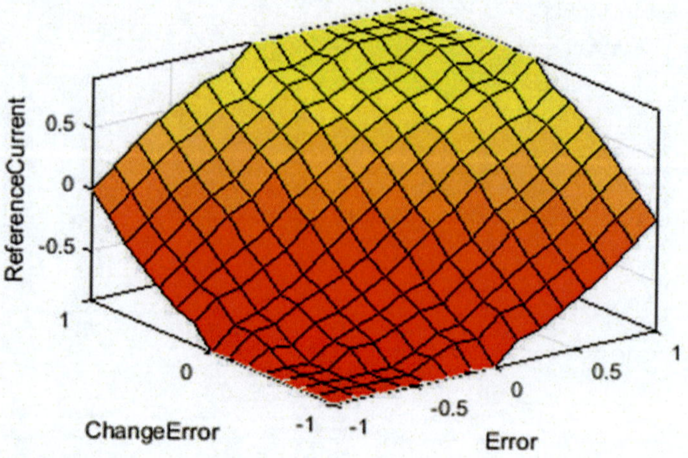

Fig. 7 Surface view

to attain the base speed is too large about nearly 0.35 s. Hence, it reduces the stability of the system and degrades the performance.

Therefore, to improve the performance of SRM drive, an advanced intelligent controller FLC is the best candidate. The Simulink model of asymmetric bridge converter fed SRM incorporated with FLC is shown in Fig. 17. The output waveforms for instantaneous flux, phase current, the torque developed and speed with FLC are shown in Figs. 18, 20, 22 and 24 respectively. The zoomed view of flux, phase current, torque is shown in Figs. 19, 21 and 23 respectively.

From the speed response, it is observed that time required settling and reaches to its base speed is reduced large amount as compared to PI controller and it is equal to 0.1 s. Therefore, the FLC gives the best results and the performance of drive is excellent for an intelligent controller. The enhancement of speed response for PI and FLC is compared and demonstrated in Fig. 25. The comparative analyses of both controllers are evaluated with parameter like rise time, settling time for various speeds with constant load torque is tabulated in Table 3. It is realized that the SRM speed with the PI controller has excessive settling time as compared to FLC subjected to fixed load torque with adjustable speed.

Simulation results are also verified with a step change in reference speed from 1000 to 2000 rpm and the load torque is applied as 1.2 N-m for both the controllers and the speed response of SRM using the PI controller and FLC are shown in Fig. 26a, b.

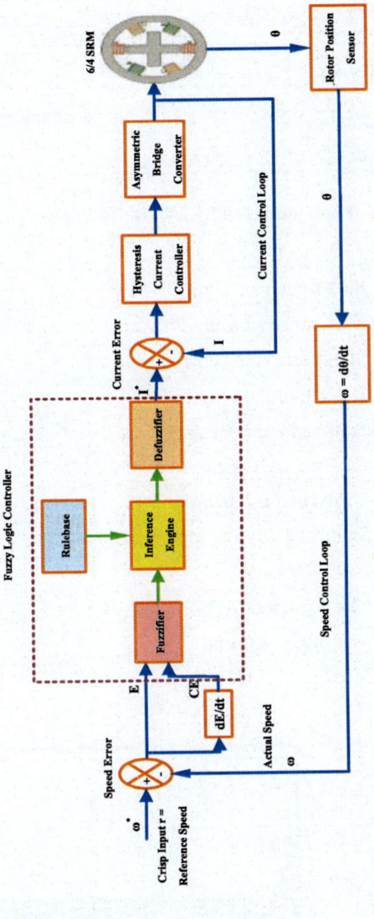

Fig. 8 Speed control of 6/4 SRM with FLC

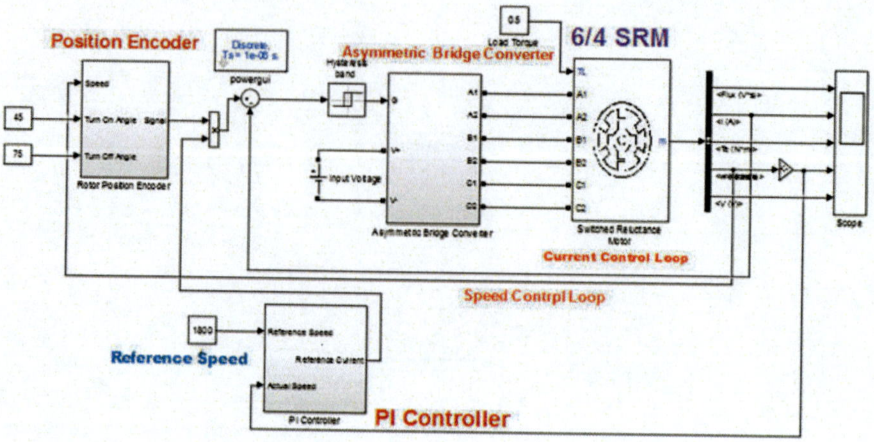

Fig. 9 Simulink model of classical converter fed SRM with PI controller

Table 2 Specifications of 6/4 SRM

Parameter	Value
Number of phases	3
Stator and rotor poles	6 and 4
Rotor poles	4
Stator phase resistance	0.5 Ω
Aligned inductance	23.6 mH
Unaligned inductance	0.67 mH
Saturated aligned inductance	0.15 mH
Inertia constant	0.005 Kgm^2
Viscous friction coefficient	0.0001 Nm s
Maximum current	30 A
Maximum flux linkages	0.486 Vs

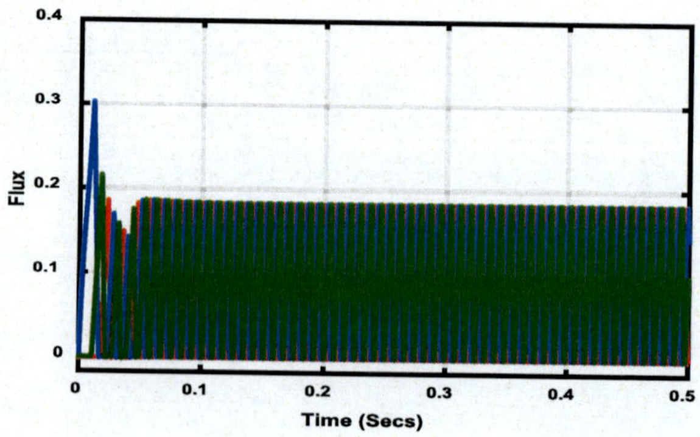

Fig. 10 Flux of SRM using PI controller

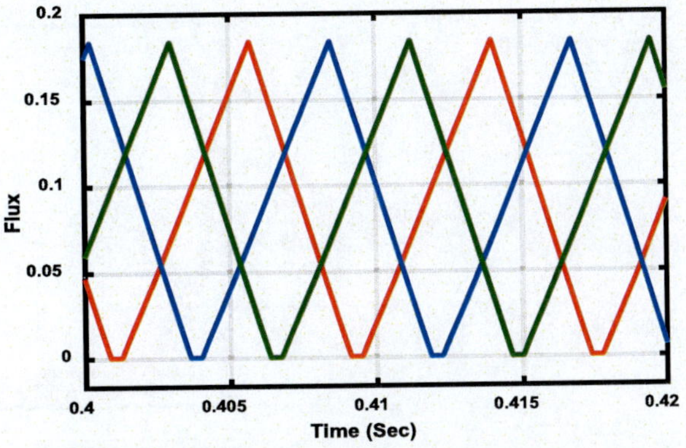

Fig. 11 Zoomed view of flux using PI controller

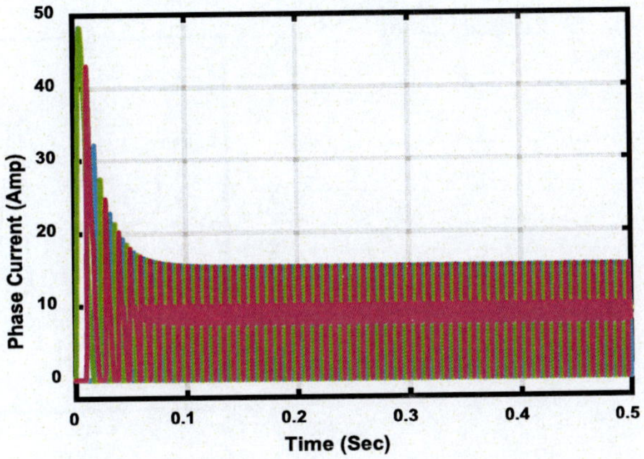

Fig. 12 Phase current of SRM using PI controller

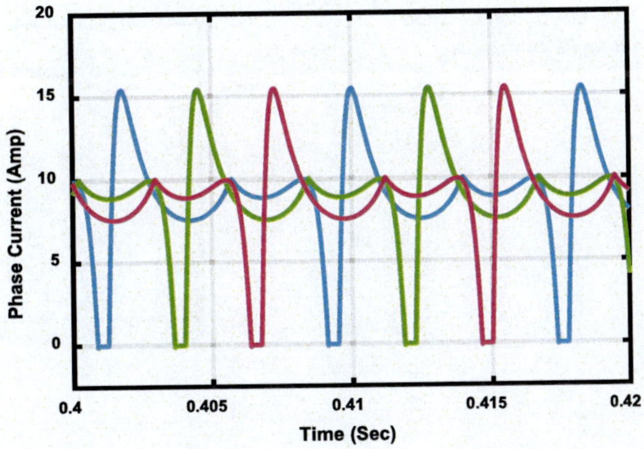

Fig. 13 Zoomed view of phase current

Fig. 14 Torque of SRM using PI controller

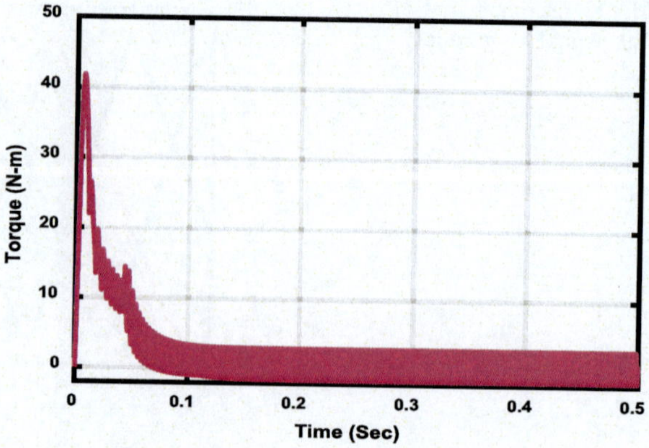

Fig. 15 Zoomed view of torque

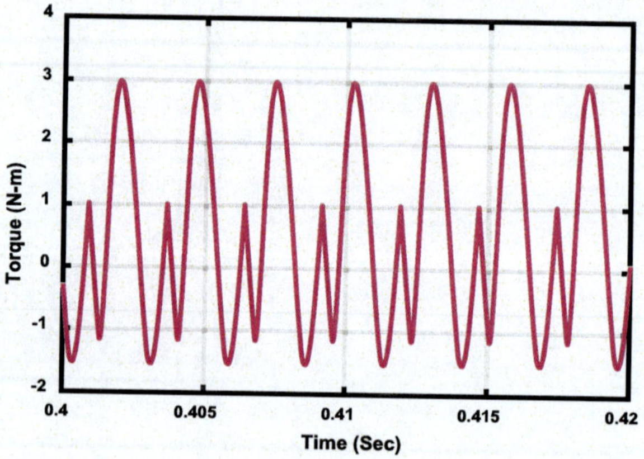

Fig. 16 Speed of SRM using PI controller

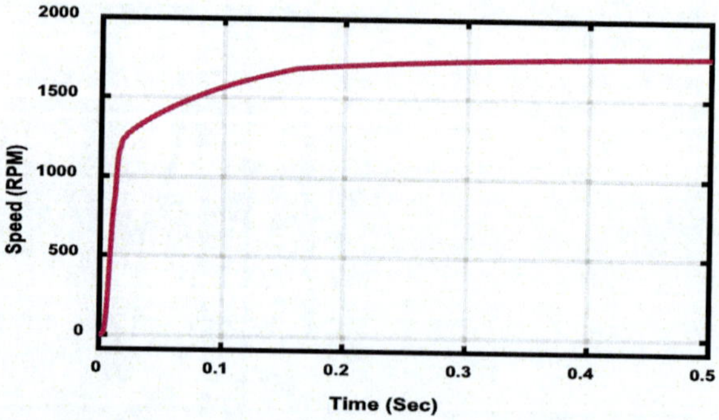

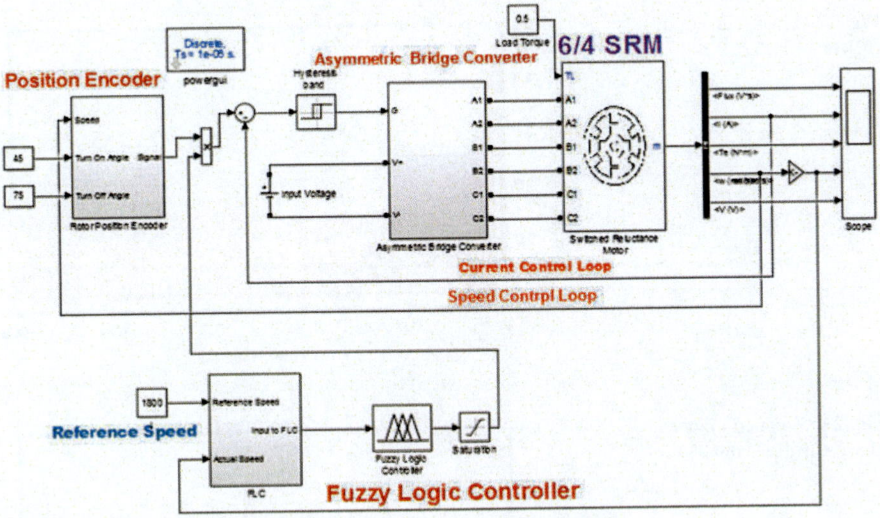

Fig. 17 Simulink model of classical converter fed SRM with FLC

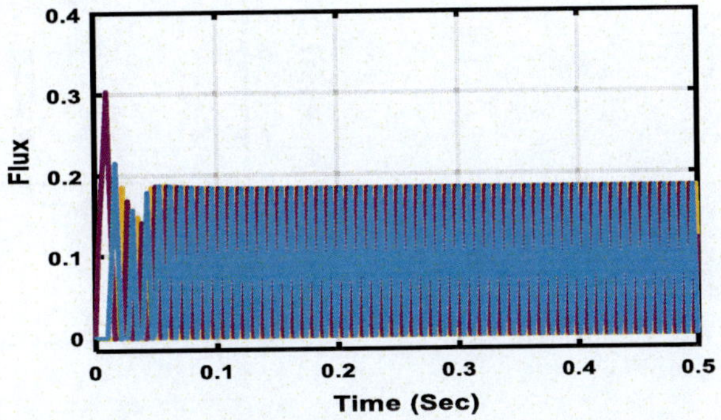

Fig. 18 Flux of SRM using FLC

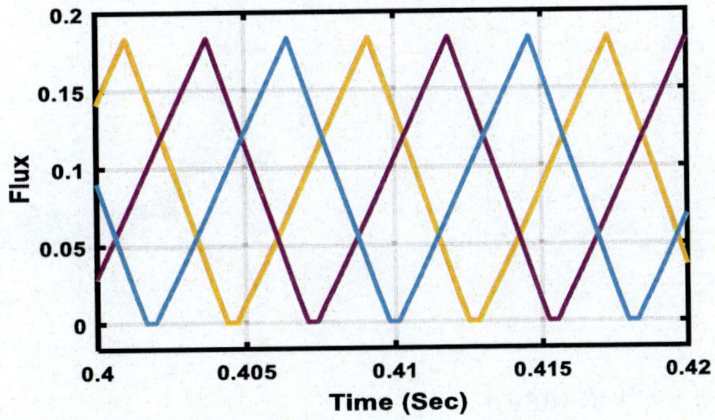

Fig. 19 Zoomed view of flux using FLC

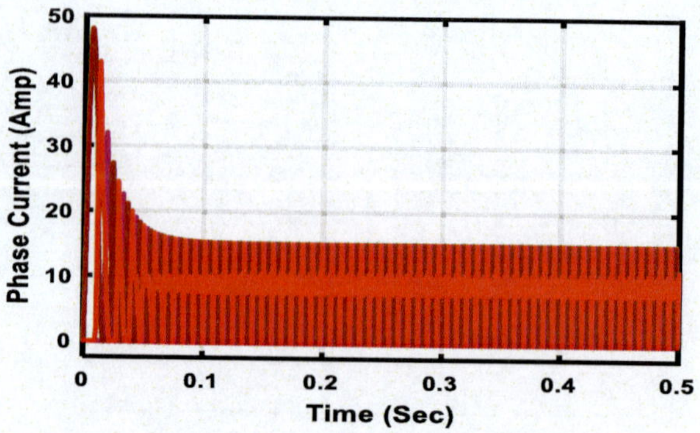

Fig. 20 Phase current of SRM using FLC

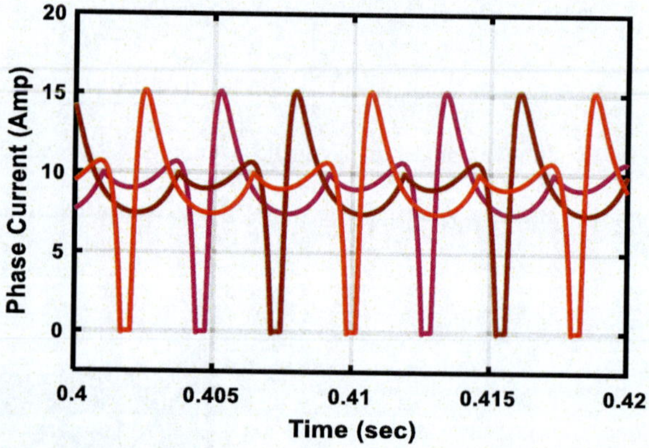

Fig. 21 Zoomed view of phase current using FLC

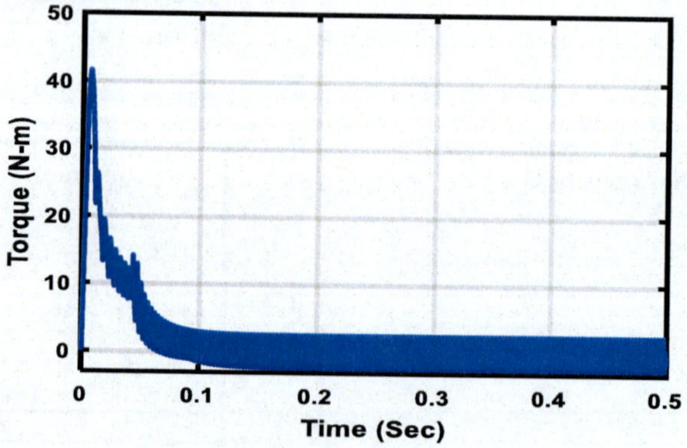

Fig. 22 Torque of SRM using FLC

5 Conclusion

To characterize the performance of SRM drive, this paper is focused on the controller. In this paper, two controllers are demonstrated and tested in the MATLAB/Simulink

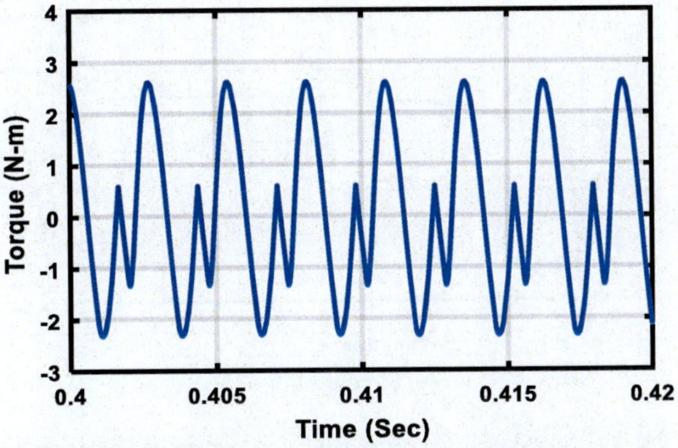

Fig. 23 Zoomed view of torque using FLC

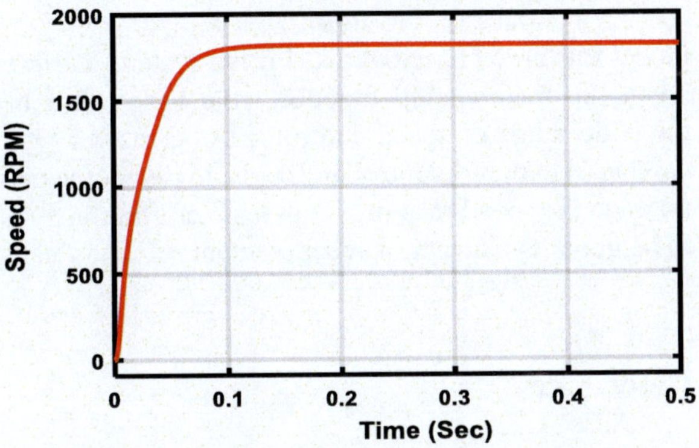

Fig. 24 Speed of SRM using FLC

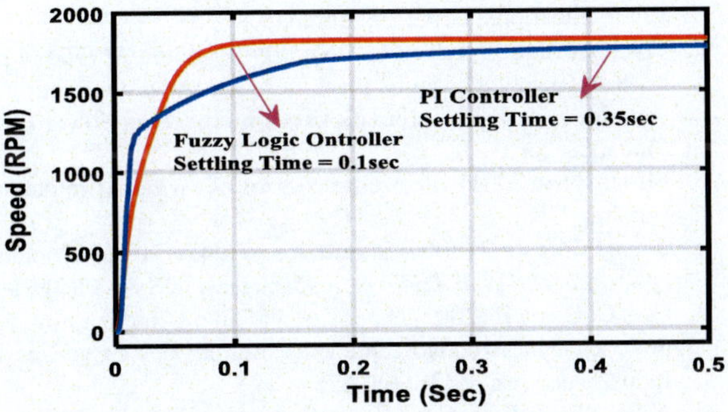

Fig. 25 Enhancement of speed response for PI and FLC

Table 3 Estimation of PI and FLC under different speeds

Specification	Load torque	At 1800 RPM		At 1000 RPM	
		PI	FLC	PI	FLC
Rise time	0.5	0.3	0.09	0.1	0.05
Settling time	0.5	0.35	0.1	0.16	0.08

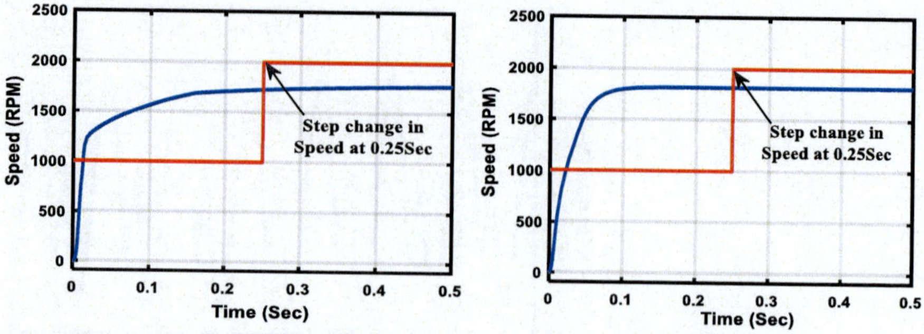

(a) PI controller (b) FLC

Fig. 26 Speed of SRM using with step change in reference speed using PI and FLC

model for three phase 6/4 SRM drive system. Different aspects of the design study about the membership function, rule base, have been explained in detail. Over the wide range of speed control, FLC gives very good stability and robustness as compared to the PI controller. Lastly, to defeat the periodic speed oscillations and to upgrade the speed response with less settling time, an intelligent controller is found to be good its conventional supplement.

References

1. Krishnan R (2001) Switched reluctance motor drives: modeling, simulation, analysis, design and applications. CRC, London, UK
2. Yuan Y, Auger F, Loron L, Debrailly F, Hubert M (2011) Design of a lying sensor for permanent magnet synchronous machine torque ripple reduction using the iterative learning control technique. In: IEEE International Conference on Power Electronics and Drive Systems PEDS, Singapore, pp 5–8
3. Miller TJE (2001) Electronic control of switched reluctance machines. Oxford Univ. Press, London, UK
4. Soares F, Costa B (2001) Simulation of 6/4 switched reluctance motor based on matlab/simulink environment. IEEE Trans Aerospac electron Syst 37(3):989–1009
5. Luis O, Costa B (2002) Proposition of an off-line learning current modulation for torque-ripple reluctance in switched reluctance motors: design and experimental evaluation. IEEE Trans Industr Electron 49(3):665–676
6. Anvari B, Toliyat HA (2017) Simultaneous optimization of geometry and firing angles of in-wheel switched reluctance motor. In: Proceedings of 9th annual IEEE energy conversion congress exposition (ECCE), Oct. 2017
7. Xue XD, Cheng KWE, Bao YJ (2014) Control and integrated half bridge to winding circuit development for switched reluctance motors. IEEE Trans Industr Inform 10(1):109–116
8. Amoros JG, Andrada P (2010) Sensitivity analysis of geometrical parameters on a double-sided linear switched reluctance motor. IEEE Trans Ind Electron 57(1):311–319
9. Desai PC, Krishnamurthy M, Schofield N, Emadi A (2010) Novel switched reluctance machine configuration with higher number of rotor poles than stator poles: concept to implementation. IEEE Trans. Ind. Electron. 57(2):649–659

10. Corda J, Jamil SM (2010) Experimental determination of equivalent circuit parameters of a tubular switched reluctance machine with solid-steel magnetic core. IEEE Trans Ind Electron 57(1):304–310
11. Miller TJE (2002) Optimal design of switched reluctance motors. IEEE Trans Ind Electron 49(1):15–27
12. Huang HP, Yan JL, Cheng TH (2010) Development and fuzzy control of a pipe inspection robot. IEEE Trans Ind Electron 57(3):1088–1095
13. Orlowska-Kowalska T, Kaminski M, Szabat K (2010) Implementation of a sliding-mode controller with an integral function and fuzzy gain value for the electrical drive with an elastic joint. IEEE Trans Ind Electron 57(4):1309–1317
14. Zhang H, Trott G, Paul RP (1990) Minimum delay PID control of interpolated joint trajectories of robot manipulators. IEEE Trans. Ind. Electron. 37(5):358–364
15. Srinivas P, Prasad PVN (2013) PWM control of asymmetrical converter fed switched reluctance motor drive. In: Proceedings of the world congress on engineering and computer science, 23 Oct 2013, San Francisco, USA
16. Skoczowski S, Domek S, Pietrusewicz K, Broel-Plater B (2005) A method for improving the robustness of PID control. IEEE Trans Ind Electron 52(6):1669–1676
17. Zhang H (2000) Speed control of two-inertia system by PI/PID control. IEEE Trans Ind Electron 47(3):603–609
18. Mudi RK, Pal NR (1999) A robust self-tuning scheme for PI and PD type fuzzy controllers. IEEE Trans Fuzzy Syst 7(1):2–16
19. Vas P (1999) Artificial-intelligence-based electrical machines and drives: application of fuzzy, neural, fuzzy-neural, and genetic-algorithm-based techniques. Oxford University Press, London, UK
20. Abut N, Cakin B, Inanc N, Yildiz AB, and Bilgin MZ (1997) Switched reluctance motor drive by using fuzzy controller. In: Proceedings IEEE IECON, pp 348–353
21. Luo A, Tang C, Shuai Z, Tang J, Xu XY, Chen D (2009) Fuzzy-PI-based direct-output-voltage control strategy for the STACOM used in utility distribution systems. IEEE Trans Ind Electron 56(7):2401–2411
22. Chung HY, Chen BC, Lin JJ (1998) A PI-type fuzzy controller with self-tuning scaling factors. Fuzzy Sets Syst 93(1):23–28
23. Peng F, Ye J, Emadi A (2016) A digital PWM current controller for switched reluctance motor drives. IEEE T Power Electr 31:7087–7098
24. Dursun M, Boz AF, Karabacak M (2018) Sensorless control application of PMSM with a novel adaptation mechanism. Neural Comput Appl 29:87–103
25. Premkumar K, Manikandan BV, Kumar CA (2017) Antlion algorithm optimized fuzzy PID supervised on-line recurrent fuzzy neural network based controller for brushless DC motor. Electr Power Comp Syst 45:2304–2317
26. Zhihui Z, Yuren L (2012) Numerical and analytical modeling of switched reluctance machines. J Comput 7(12)
27. Guettaf A, Chabane F (2013) Dynamic modeling in a switched reluctance motor SRM using finite elements. J Power Technol 93(2):100–104

Dynamic Economic Dispatch Incorporating Commercial Electric Vehicles

K. Abinaya, Velamuri Suresh, Suresh Kumar Sudabattula, and S. Kaveripriya

Abstract In future, the electric vehicle will be expected to dominate the other transportation networks. Many commercial electric vehicles are being charged at common times due to their usage pattern. This results in sudden rise of power demand on the electricity networks. In this paper, dynamic economic dispatch considering the impact of commercial electric vehicles is proposed. The objectives of this work are minimization of fuel cost and network power loss. A novel methodology for charging the commercial electric vehicles is developed and is utilized in conjunction with dynamic economic dispatch. The proposed work is implemented on IEEE 30 bus system and the results show that this method is very effective.

Keywords Economic dispatch (ED) · Salp swarm algorithm (SSA) · Electric vehicles (EVs)

1 Introduction

Economic dispatch (ED) is a significant problem in power system. The aim of this problem is to schedule the generators in an optimum way to reduce the running cost (fuel cost) of the system [1–3]. Further, environmental concern and depletion of coal reserves, it is essential to consider the renewable energy resources along with thermal generators in the present era. Also, minimization of operating cost over a fixed time interval is not suitable with consideration of renewable sources. So, it is essential to solve dynamic ED because output of these sources changes from time to time [4]. Also, in pollution aspect, both power industry and transportation sector represent 64% of global emission [5]. So, significant step is required to reduce the pollution level to considerable extent. Further, the power generation is majorly depending on coal-based thermal plants and also transport sector depends on conventional internal

K. Abinaya · V. Suresh (✉) · S. Kaveripriya
School of Electrical and Electronics Engineering, SASTRA University, Thanjavur, India
e-mail: velamuri.suresh@gmail.com

S. K. Sudabattula
School of Electronics and Electrical Engineering, Lovely Professional University, Phagwara, India

N. Zhou and S. Hemamalini (eds.), *Advances in Smart Grid Technology*, Lecture Notes in Electrical Engineering 688, https://doi.org/10.1007/978-981-15-7241-8_5

combustion engines. Thus, usage of electric vehicle is an alternative. Although, usage of more EVs increases peak load demand. So, it is crucial to examine the scheduling of EVs while developing the optimum dispatch strategy for power systems.

From the last two decades, different authors solved SED problem using various optimization techniques [6–10]. Considering the load variations throughout the day, various authors have attempted DED problem [11–14]. Recently, few authors attempted different methods for incorporating electric vehicles in DED problem. In [15], DED is solved by considering plug in electric vehicle for peak shaving and valley filling. In [16], the new scheduling method is proposed for integrating multiple electric vehicle charging profiles comparatively into a 24-h load demand in an economic environment. In [17] gives a dual optimization method for the DC optimal power flow with cost reduction. Here, PEVs are used for both energy storage and energy supply device. [18] addresses issues for a conventional unit commitment-mixed integer optimization problem for power system operators. For solving this problem, the meta-heuristic algorithm is employed and the transfer functions are utilized in binary optimization and investigated in a power system with PEVs. In [19], a framework to identify fleet size capability of a power network in both uncontrolled and vehicle-to-grid modes of operation is performed. In this paper, a peak to average ratio (PAR)-based charging scheme for EVs is implemented and DED is performed using SSA optimization.

2 Problem Formulation

The main objective of the DED problem is to minimize the cost while satisfying all constraints.

2.1 *Objective Functions*

The main objective function is to minimize the cost given by,

$$\min \sum_{i=1}^{24} F_c \tag{1}$$

The fuel cost function of the ith generating unit

$$F_c(i) = a_i P_{gi}^2 + b_i P_{gi} + c_i \tag{2}$$

where P_{gi} is the power generated by the ith generator in MW and a_i, b_i and c_i are cost coefficient of ith generators.

The power loss minimization objective is

$$\min P_L = \sum_{i=1}^{24} P_L(i) \tag{3}$$

2.2 System Constraints

The various constraints involved in solving DED are

2.2.1 Power Balance Constraint

Considering the charging and discharging time intervals of commercial EVs, the power balance equation is given by,

$$\sum_{i=1}^{N} P_{g,t} + P_{\text{disch},t} = P_{d,t} + P_{L,t} + P_{\text{ch},t} \tag{4}$$

where $P_{\text{ch},t}$ and $P_{\text{disch},t}$ are the charging and discharging load of EVs at time interval of t.

$P_{g,t}$, $P_{d,t}$, and $P_{L,t}$ are the power generation, demand, and power loss at tth hour, respectively.

2.2.2 Commercial Electric Vehicle Constraints

A. *Battery storage capacity constraint*

The State of Charge (SOC) of the EV at ith hour is

$$\text{SOC}^{\min} \le \text{SOC}_i \le \text{SOC}^{\max} \tag{5}$$

B. *Charging/discharging power constraints*:

The charging and discharging limits of EVs should be within the range of power limits.

$$\begin{aligned} p_{\text{ch},i,t} &\le p_{\text{ch},i}^{\max} \\ p_{\text{disch},i,t} &\le p_{\text{disch},i}^{\max} \end{aligned} \tag{6}$$

3 Proposed Methodology

In this work, an attempt is made to schedule the commercial electric vehicles using PAR method integrated with DED. SSA optimization is utilized for solving the DED problem. This method is very efficient in minimizing the fuel cost and power loss.

Initially, PAR method is used for determining the charging interval for CEVs. This information is passed to the DED problem and the demand for the charging intervals is updated. Further, DED is solved using SSA method and fuel cost is obtained.

3.1 Salp Swarm Algorithm

Salp swarm algorithm is a novel optimization technique which is used for solving optimization problem [20]. SSA is a recent meta-heuristic method and it is inspired by the swarming behavior of salp in oceans. Salps belong to the family of Salpidae that is having transparent barrel-shaped body. Their tissues are similar to the jelly fish, in which the water is pumped through body as propulsion to move forward. In deep oceans, the salp that is in the form of chain is called salp chain. The population is divided into two groups; they are leaders and followers. Front of the chain is considered as leader and the rest of the salp is considered as followers. The flowchart of SSA is shown in Fig. 1.

4 Result and Discussion

The study is performed on a standard IEEE 30 bus system [9]. The system consists of six generators, 41 transmission lines, and 30 buses. The analysis is performed in MATLAB v16. To prove the efficiency of the method, initially, static economic dispatch (SED) is performed and the results are compared with recent methods available in literature. Later, DED is performed considering the CEVs.

4.1 Static Economic Dispatch

SED is performed on IEEE 30 bus test system for a demand of 1263 MW using SSA optimization. The SSA method considers a population of 40 and maximum iterations are set to 300. The results are presented in Table 1. From the results, it can be observed that SSA method is efficient in producing promising results for the considered objective function. The convergence curve for fuel cost minimization is shown in Fig. 2. The fuel cost and power loss obtained using SSA method are better than other methods. So this method is utilized throughout the paper.

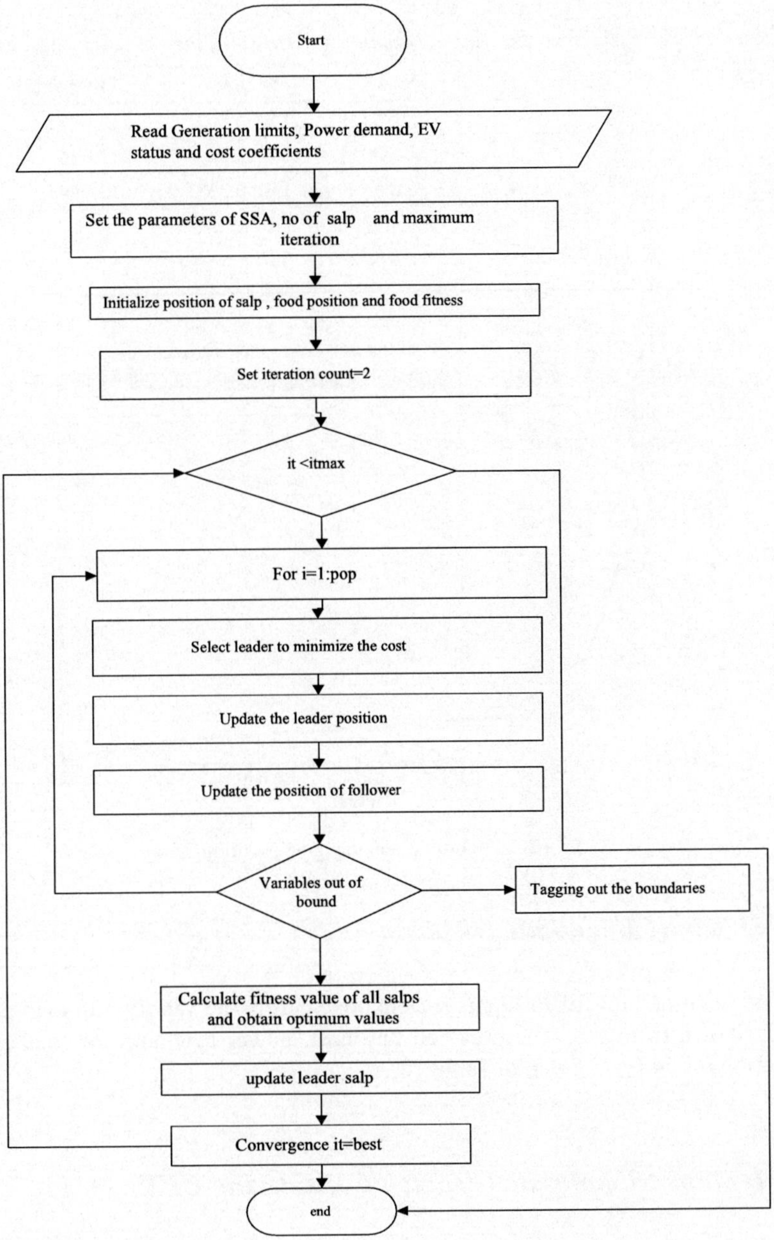

Fig. 1 Flowchart DED with CEVs using SSA

Table 1 Comparison of fuel cost and power loss obtained using SSA

S. No	Algorithm	Fuel cost ($/hr)	Power loss (MW)
1	SSA	15428.97	11.03774
2	DFA [7]	15436.869	11.8952
3	BBO [8]	15443.09	12.446
4	CBA [9]	15450.2381	12.9848

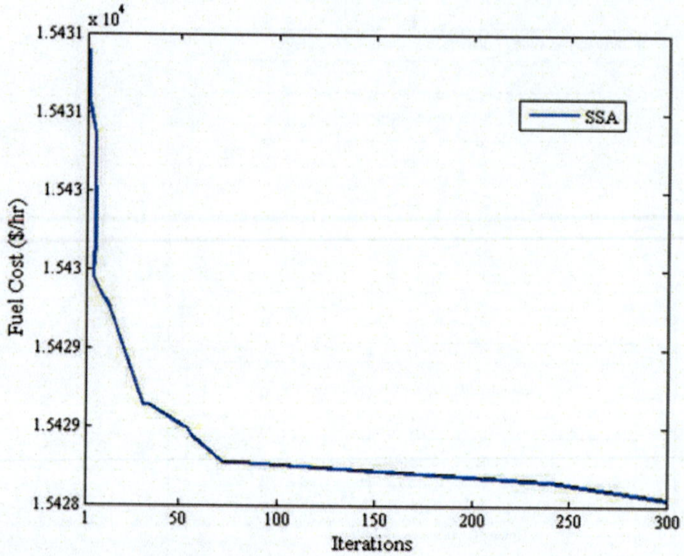

Fig. 2 Convergence curve for IEEE 30 bus system using SSA optimization

4.2 Dynamic Economic Dispatch

DED is performed on IEEE 30 bus system considering the variation in load profile [22] as shown in Fig. 3. The obtained fuel cost, power loss, and the total power generation for 24 h are given in Table 2.

4.3 Dynamic Economic Dispatch Considering CEVs

The proposed method illustrates the commercial electric vehicle charging sequence along with generation scheduling using SSA algorithm. The outcome of this work is to reduce the generation cost as well as power loss in the system.

CEVs are generally public transport vehicles used in intercity. In this paper, BYDK9 model CEVs are considered and the analysis is performed. It is assumed that

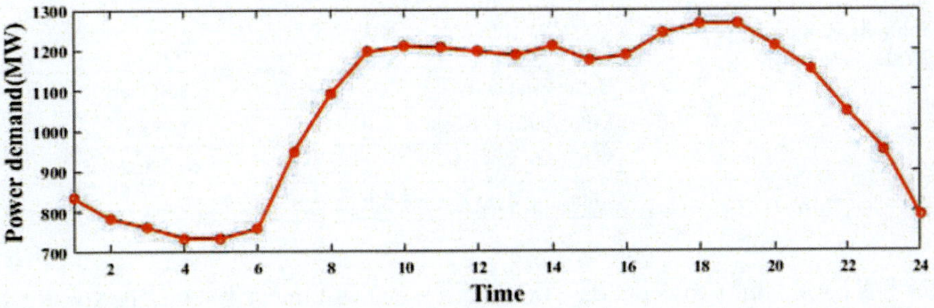

Fig. 3 Variable load profile of IEEE 30 bus system

Table 2 Variation of fuel cost and power loss for 24 h

Hour	P_G (MW)	P_L (MW)	Fuel cost ($/h)
1	840.2401	5.3971	9948.48
2	789.0879	4.7649	9349.097
3	767.3751	4.5231	9091.629
4	739.1754	4.1094	8769.545
5	739.2171	4.1511	8781.112
6	762.2542	4.4542	9038.321
7	953.9709	6.7209	11327.88
8	1099.872	8.6397	13157.1
9	1204.623	9.8246	14515.58
10	1218.743	10.0517	14699.52
11	1214.707	9.8048	14650.16
12	1204.705	9.9067	14516.17
13	1195.521	9.5642	14398.31
14	1218.709	10.0176	14700.56
15	1182.438	9.1105	14230.92
16	1195.471	9.5142	14399.75
17	1250.341	10.0753	15124.36
18	1274.83	11.8295	15428.97
19	1274.83	11.8295	15426.35
20	1218.595	9.9042	14700.77
21	1159.728	9.1345	13934.43
22	1054.967	7.9403	12587.91
23	922.1717	6.4967	10934.02
24	794.3063	4.9313	9398.746

Table 3 Parameters of BYDK9 eV Bus [21]

Specifications	Rating
Battery capacity	300 kwh
Mileage/full charge	135 km
SOC min and max	0.2 and 0.9
Charging time for one full charge	6 h

the bus covers three trips per day. In each trip, bus will travel 40 km. The specifications of BYDK9 eV bus are given in Table 3. The CEVs will operate from 6.00 AM to 6.00 PM daily and they are available for charging during the left hours in a day. In this study, the charging pattern of these CEVs is optimized based on PAR-based strategy and the generation scheduling is performed. The obtained results are compared with traditional method and merits of the same are shown in the following section.

Initially, the CEVs are assumed to be charged at the end of all their trips and get ready for their next day trip. This method is called as dumb charging method, since it does not consider the system power demand conditions. The charging and discharging intervals for dumb charging method are shown in Fig. 4. It can be observed that the demand rises suddenly at 6.00 PM which is usually a peak hour. The generation scheduling with increased load demand is simulated and the results are given in Table 4.

Later, we introduced a smart charging strategy to handle this surge in demand. This method considers the system demand, scheduling cost, and deadline for charging before next trip. The hourly peak-to-average ratio (PAR) is chosen as a parameter for charging the CEVs. This method compares the PAR for each hour and decides

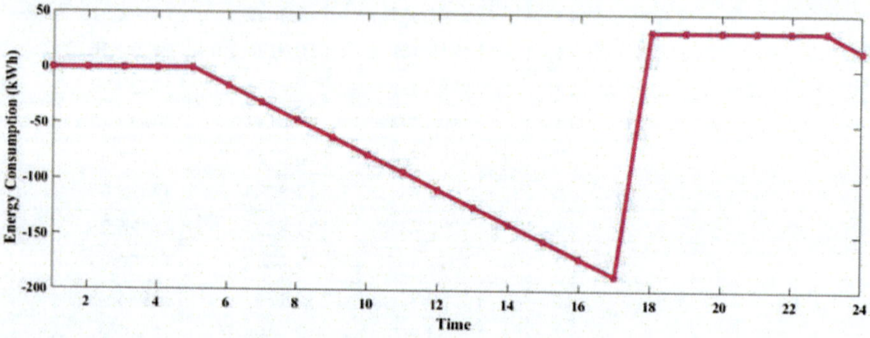

Fig. 4 Charging and discharging of CEVs using dumb charging method

Table 4 Conditions for charging

Condition	Status
If PAR < 1	charging allowed
If PAR > 1	Not allowed

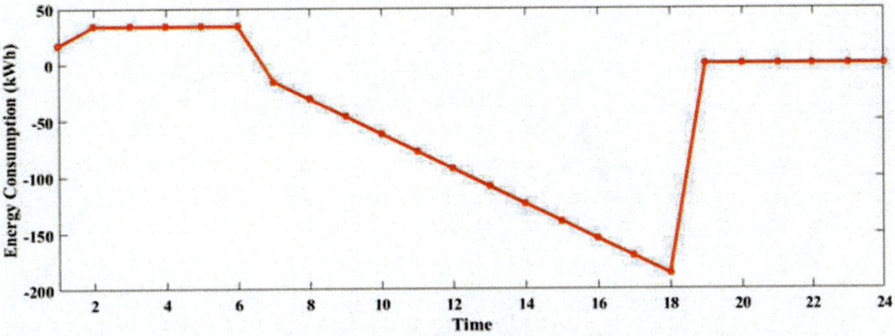

Fig. 5 Charging and discharging of CEVs using smart charging method

Table 5 DED considering CEVs in IEEE 30 bus system

Method	Power loss (MW/day)	Cost ($/day)
Dumb charging + SSA method	196.097	303556.00
SMART charging + SSA method	194.584	29325.91

the charging interval. The rule-based approach is given in Table 4. If the condition is satisfied cumulatively for desired period of charging, then the information will be passed to the SSA algorithm and optimal generation scheduling is produced. The charging and discharging patterns obtained from the strategy are shown in Fig. 5. It is observed that the charging of CEVs has started at 1.00 AM instead of 6.00 PM.

DED is performed on IEEE 30 bus system considering the CEVs charging using dumb and smart charging methods and the obtained power loss and operating cost of generators per day is given in Table 5. It can be inferred that our rule based approach for determining the charging time interval of CEVs aids in minimizing the fuel cost and power loss of the system significantly. The hourly power loss variation using dumb and smart methods is shown in Fig. 6 which indicates that the smart charging strategy is essential for minimizing the operating cost and power loss of the system.

5 Conclusion

In this paper, the impact of CEVs on system power loss and operating cost is studied. The analysis is performed on an IEEE 30 bus system using SSA optimization. SED is initially performed and the obtained results are compared with recent methods. Later, DED is performed without and with CEVs. Dumb and smart charging strategies were implemented for determining the appropriate time intervals for charging the CEVs. The results of DED with CEVs using both methods are simulated and it is observed

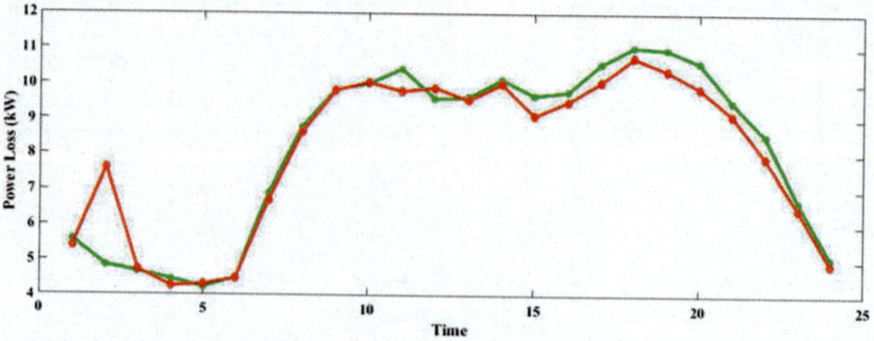

Fig. 6 Variation in power loss using dumb and smart charging methods

that proper charging time helps in minimizing the power loss and operating cost significantly.

References

1. Abdelaziz AY, Ali ES, Elazim SA (2016) Implementation of flower pollination algorithm for solving economic load dispatch and combined economic emission dispatch problems in power systems. Energy 101:506–518
2. Yang T, Lu J, Wu D, Wu J, Shi G, Meng Z, Johansson KH (2016) A distributed algorithm for economic dispatch over time-varying directed networks with delays. IEEE Trans Industr Electron 64(6):5095–5106
3. Liang H, Liu Y, Shen Y, Li F, Man Y (2018) A hybrid bat algorithm for economic dispatch with random wind power. IEEE Trans Power Syst 33(5):5052–5061
4. Alham MH, Elshahed M, Ibrahim DK, El Zahab EEDA (2016) A dynamic economic emission dispatch considering wind power uncertainty incorporating energy storage system and demand side management. Renew Energy 96:800–811
5. Labatt S, White RR (2011) Carbon finance: the financial implications of climate change, vol 362. Wiley, Hoboken, NJ, USA
6. Vennila H, Rajesh R (2019) Combined static economic and emission dispatch by improved moth optimisation with valve point loading. Int J Enterpr Netw Manag 10(2):152–161
7. Suresh V, Sreejith S (2017) Generation dispatch of combined solar thermal systems using dragonfly algorithm. Computing 99(1):59–80
8. Bhattacharya A, Chattopadhyay PK (2009) Biogeography-based optimization for different economic load dispatch problems. IEEE Trans Power Syst 25(2):1064–1077
9. Adarsh BR, Raghunathan T, Jayabarathi T, Yang XS (2016) Economic dispatch using chaotic bat algorithm. Energy 96:666–675
10. Pradhan M, Roy PK, Pal T (2016) Grey wolf optimization applied to economic load dispatch problems. Int J Electr Power Energy Syst 83:325–334
11. Zaman MF, Elsayed SM, Ray T, Sarker RA (2015) Evolutionary algorithms for dynamic economic dispatch problems. IEEE Trans Power Syst 31(2):1486–1495
12. Lokeshgupta B, Sivasubramani S (2018) Multi-objective dynamic economic and emission dispatch with demand side management. Int J Electr Power Energy Syst 97:334–343
13. Elattar EE (2015) A hybrid genetic algorithm and bacterial foraging approach for dynamic economic dispatch problem. Int J Electr Power Energy Syst 69:18–26

14. Suresh V, Sreejith S, Sudabattula SK, Kamboj VK (2019) Demand response-integrated economic dispatch incorporating renewable energy sources using ameliorated dragonfly algorithm. Electr Eng 101(2):421–442
15. Liang Huijun, Liu Yungang, Li Fengzhong, Shen Yanjun (2018) Dynamic economic/emission dispatch including PEVs for peak shaving and valley filling. IEEE Trans Industr Electron 66(4):2880–2890
16. Yang Z, Li K, Niu Q, Xue Y, Foley A (2014) A self-learning TLBO based dynamic economic/environmental dispatch considering multiple plug-in electric vehicle loads. J Modern Power Syst Clean Energy 2(4):298–307
17. Yao Y, Gao DW, Momoh J (2019) Dual-optimisation of power sources including plug-in electric vehicles and renewable energy resources at transmission-level system. J Eng 5:3448–3454
18. Yang Z, Li K, Guo Y, Feng S, Niu Q, Xue Y, Foley A (2019) A binary symmetric based hybrid meta-heuristic method for solving mixed integer unit commitment problem integrating with significant plug-in electric vehicles. Energy 170:889–905
19. Ahmad MS, Sivasubramani S (2018) Optimal number of electric vehicles for existing networks considering economic and emission dispatch. IEEE Trans Ind Inform 15(4):1926–1935
20. Mirjalili S, Gandomi AH, Mirjalili SZ, Saremi S, Faris H, Mirjalili SM (2017) Salp swarm algorithm: a bio-inspired optimizer for engineering design problems. Adv Eng Softw 114:163–191
21. BYD K9 [online] Available at http://www.byd.com/auto/ElectricBus.html. Accessed 1 Oct 2019
22. Cherukuri SHC, Saravanan B, Swarup KS (2018) A new choice based home energy management system using electric springs. In: 2018 20th National power systems conference (NPSC), pp 1–6. IEEE

Optimal Allocation of DERs in Distribution System in Presence of EVs

S. Kaveripriya, Velamuri Suresh, Sudabattula Suresh Kumar, and K. Abinaya

Abstract Usage of electric vehicles by mankind is increasing rapidly. It is observed that most of the electric vehicles in a distribution system are being charged at common time which indirectly reflects as a considerable load. In this paper, the impact of these electric vehicles on distribution system performance is studied and a new methodology for reducing the power loss is implemented. A standard IEEE 33 bus radial distribution system with variable load profile is considered for analysis. This method uses a combined loss sensitivity factor and grasshopper optimization algorithm to determine the optimal location and size of distributed generation throughout the day. Various charging patterns for electric vehicles are analyzed and best possible approach for minimizing the power loss is presented.

Keywords Power loss · Loss sensitivity factor · Grasshopper optimization algorithm (GOA) · Electric vehicles

1 Introduction

The sales of electric vehicles (EVs) throughout the globe are increased significantly from last two decades [1]. This shows a glimpse of the future transportation network. Further, it is necessity to upgrade power sector in order to meet load requirement of EVs. Finally, from pollution point of view, the power industry represents a major share in global emission (i.e., 40% of the global CO_2 production), followed by the transportation sector (24%) [2]. Therefore, a major step has to be taken for curtailment of this pollution level. Also, the situation becomes more complicated when the coal-based thermal power plants constitute approximately 40% [3] of the total electrical energy production. Hence, the future transport network and the power sector will have a strong correlation in order to reduce global pollution. So, for all these, usages of EVs

S. Kaveripriya · V. Suresh (✉) · K. Abinaya
School of Electrical and Electronics Engineering, SASTRA University, Thanjavur, India
e-mail: velamuri.suresh@gmail.com

S. Suresh Kumar
School of Electronics and Electrical Engineering, Lovely Professional University, Phagwara, India

N. Zhou and S. Hemamalini (eds.), *Advances in Smart Grid Technology*, Lecture Notes in Electrical Engineering 688, https://doi.org/10.1007/978-981-15-7241-8_6

and power generation based on renewable energy sources (RES) are the alternatives. Due to these advantages, government and private organizations promote the usage of EVs and power generated through RES. From distribution system perspectives, EVs are connected to the existing system and it draws power from the system and influences the load generation pattern. So, it is significant to study the performance of the system [4]. Next, the losses in the DS is very high comparatively the transmission system [4]. So, it is necessary to reduce these losses as much as possible. For these, some of the authors solved DG allocation problem and considered power loss was the major objective [5–10]. Further, connecting EVs to the existing system further influences the system performance. So, it is significant to consider these sources in appropriate manner. In the literature, some of the authors solved the EV placement problem in the DS and studied the performance. Electric vehicles charging is used effectively in a distribution system based on voltage stability, reliability, and power loss [11]. In this paper, they proposed an innovative smart grid-based Volt-VAR optimization engine. It has able to minimize system power loss cost [12]. As the number of electric vehicles is increased, it causes overloading, power loss, reduced efficiency in the distribution level. So, they proposed coordinated charging strategy to minimize losses in the system [13]. Here, they are using two stage optimization techniques: PSO and TS algorithms for wind DGs and EVs placement [14]. Charging stations are integrated with solar PV modules. Combined EV charging stations and DGs are studied here for increasing the reliability in the system [15]. In this paper, effective methodology is proposed to solve DG allocation problem in DS considering the impact of EVs. The charging patterns of EVs are suitably changed in accordance with the load variance and allocation of DGs is determined for reducing the DS power loss to the considerable extent.

2 Problem Formulation

Increased penetration of EVs in DS causes significant power loss in the system. The main objective of this work is to reduce this power loss while satisfying all the constraints. At first, using peak-to-average ratio (PAR) analysis is considered and suitable charging time for EVs is determined. Later optimal location of DGs is determined using LSF method. Further appropriate size of DGs considering the impact of EVs is determined throughout the day.

2.1 Objective Functions

The power loss minimization objective is

$$\min \sum_{i=1}^{24} P_L(i) \tag{1}$$

where

$$P_L(i) = \sum_{i=1}^{24} I_i^2 R_i \tag{2}$$

2.2 Power Balance Constraint

Power balance equation is represented as

$$\sum_{i=1}^{24} P_G(i) = \sum_{i=1}^{24} [P_L(i) + P_{\mathrm{EV}}(i) + P_D(i)] \tag{3}$$

where P_L, P_{ev}, P_D are the power loss, power consumed by electric vehicles, and load demand for the ith hour, respectively.

2.3 Battery Storage Capacity Constraint

The State of Charge (SOC) of the EV at ith hour is

$$\mathrm{SOC}^{\max} \le \mathrm{SOC}_i \le \mathrm{SOC}^{\max} \tag{4}$$

2.4 Charging/Discharging Power Constraints

The charging and discharging limits of EVs should be within the range of power limits.

$$\begin{aligned} p_{\mathrm{ch},i,t} &\le p_{\mathrm{ch},i}^{\max} \\ p_{\mathrm{disch},i,t} &\le p_{\mathrm{disch},i}^{\max} \end{aligned} \tag{5}$$

3 Proposed Methodology

A combined method of LSF and GOA techniques is used to reduce the power loss in the system. LSF technique is used to find the optimal location of the system for placing DGs and EVs. GOA technique is used to find the sizes of DGs.

3.1 Loss Sensitivity Factor (LSF) Method

Using LSF technique, the vulnerable buses for the placement of DGs are identified [16]. Further, active and reactive power loss is calculated from Eqs. (6) and (7).

$$P_{\text{lineloss}}[b] = \frac{\left(P_{\text{eff}}^2[b] + Q_{\text{eff}}^2[b]\right)R[k]\big)}{(V[b])^2} \tag{6}$$

$$Q_{\text{lineloss}}[b] = \frac{\left(P_{\text{eff}}^2[b] + Q_{\text{eff}}^2[b]\right)X[k]\big)}{(V[b])^2} \tag{7}$$

where, $P_{\text{eff}}[b], Q_{\text{eff}}[b]$ are the total effective active power and reactive power supplied. Now, the LSF for both cases can be obtained as,

$$\frac{\partial Q_{\text{lineloss}}[b]}{\partial Q_{\text{eff}}} = \frac{(2 * Q_{\text{eff}}[b]) * X[k])}{(V[b])^2} \tag{8}$$

$$\frac{\partial P_{\text{lineloss}}[b]}{\partial P_{\text{eff}}} = \frac{(2 * Q_{\text{eff}}[b]) * R[k])}{(V[b])^2} \tag{9}$$

The LSF can be calculated for each and every bus and higher LSF denotes that at particular bus needs to be compensated first. Similarly, we can find the optimal location for various DGs as well as EVs.

3.2 Grasshopper Optimization Algorithm (GOA)

The GOA algorithm was developed by Shahrzad Saremi et al. in 2017. In this paper, it is used to identify the optimal sizes of DGs [17]. Here, 24 h load profile is considered and determines sizes for each and every hour. Also, in GOA, each particle updates the position by the current position, the global best, and by the positions of all buses. The flowchart of optimal DG sizing using GOA is shown in Fig 1.

The equation shown below is used for updating the position of each particle

$$C(t) = C_{\max} - t\frac{C_{\max} - C_{\min}}{t_{\max}} \tag{10}$$

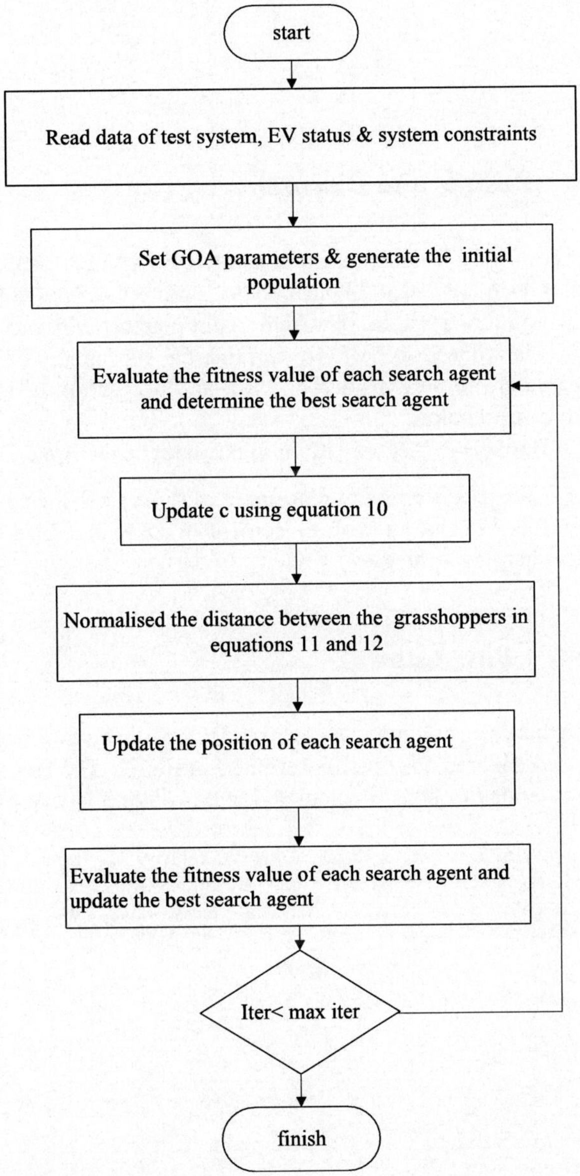

Fig. 1 Flowchart for optimal sizing of DGs using GOA

For calculating the distance between each grasshopper

$$\frac{\sum^{Nb-1} K_i r_i \left(P_i^2 + Q_i^2\right)}{V_i^2} \tag{11}$$

$$s_{ij} \le s_{ij}^{\max}. \qquad (12)$$

4 Results and Discussions

The proposed method is implemented by MATLAB software. To check the efficiency, it is tested on IEEE 33 bus system. Before placing the DGs in the system, the power loss obtained will be very high. After placing DGs and EVs in the system, the power loss is reduced to half. To examine the supremacy of the developed method, three cases are to be considered. Also, an observation of loss reduction in each case is discussed below.

Base case: system without placement of DGs and EVs

Case 1: system with placement of three DGs located optimally

Case 2: system with placement of DGs and EVs optimally with two types of charging strategies.

4.1 Base Case

In this case, without placement of DGs and EVs is considered. Also, 24 h load profile is considered and it is represented in Fig. 2. The total power loss for 24 h without placement of DGs is calculated that is 3520.335 kW.

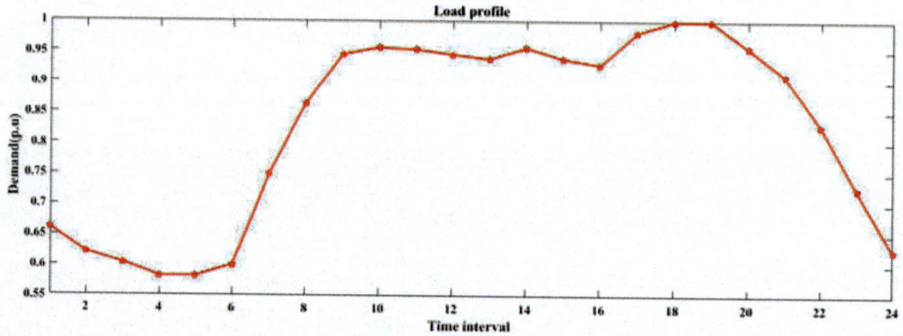

Fig. 2 Load profile of distribution system [18]

Table 1 Comparison of power loss and DG size obtained using GOA

Method	FPA [19]	PSO [20]	GOA
Base case P_L (kW)	210.99	210.9	210.9983
DG size and location (kW)	936.5(12) 1050(30) 1414(33)	863(16) 925(11) 1200(32)	787.6104(13) 506.514(24) 978.314(31)
P_L with DG (kW)	87.4	103.3	73.8009
% P_L reduction	58.55	50.99	65.022

4.2 Case 1

Optimal location and sizing of DGs without considering EVs is solved in this section. At first, the analysis is presented for one interval and the obtained results are compared with recent methods and same is represented in Table 1. This GOA method is efficient in minimizing the power loss to 65% compared to base case. Hence, the method is utilized throughout the paper.

Now, the analysis is extended for a 24 h interval and the results are presented in Table 2. Three DGs are placed at 13th, 24th, and 31st buses, respectively. The overall power loss obtained in this case is 1244.128 kW. It is reduced drastically as compared to base case. Also, comparison of both cases is illustrated in Fig. 3.

4.3 Case 2

In this case, allocation of DGs in presence of EVs with two different strategies is presented. The specifications of EV considered for the study are given in Table 3. The considered vehicle is an office going vehicle which leaves the home at 8.00 AM and returns back by 17.00 PM. It is assumed that the vehicle cannot be charged during the trip.

Charging Strategies Used

In this case, two types of charging strategies have been considered and determined the power loss.

Dumb Charging

In this method, the EVs are connected for charging immediately after they reach home. They are not concerned of the system demand considerations. The pattern followed by EVs is shown in Fig. 4.

Smart Charging

In this method, the EV users are concerned about the system demand. They do not allow the charging of EVs during peak demand. Smart charging follows the load demand curve and EVs are charged during low demand hours. A rule-based strategy considering the PAR is implemented for charging the EVs is given in Table 4. The revised charging pattern of EV considering smart charging method is shown in Fig. 5.

Table 2 DG sizes for every hour

Hour	DG_1 (kW)	DG_2 (kW)	DG_3 (kW)
1	547.3	796.94	604.265
2	508.753	692.531	574.9123
3	495.7525	642.3422	561.0475
4	488.7422	642.3422	538.2093
5	479.722	634.9339	541.9059
6	496.8495	645.358	560.3378
7	622.5009	825.678	703.5536
8	716.1227	956.0751	791.8206
9	798.525	1074.523	871.4466
10	799.131	1080.922	890.523
11	799.5139	1074.196	886.2512
12	790.4775	1040.303	874.3798
13	787.6104	1064.601	885.2137
14	792.2505	1093.352	871.02
15	769.064	1054.47	885.2137
16	774.886	1044.291	871.02
17	821.77	1102.751	862.47
18	827.7741	1142.874	911.017
19	834.4326	1135.035	936.875
20	793.938	1072.288	939.172
21	744.2512	1019.808	886.428
22	684.014	941.5154	857.92
23	596.3692	803.8091	774.0542
24	506.5104	698.48	680.446

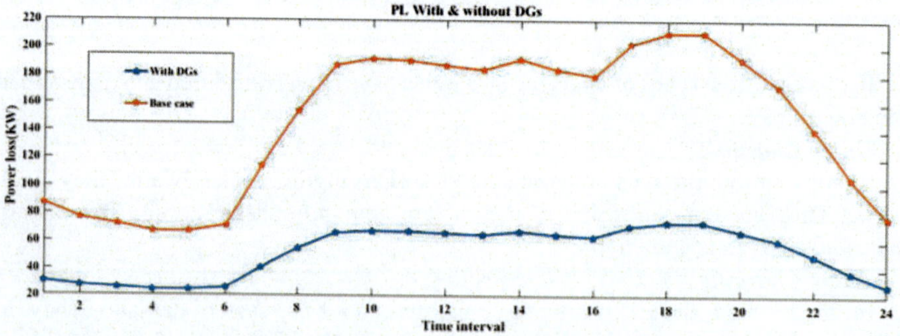

Fig. 3 Power loss with and without DGs

Table 3 Electric vehicle ratings of Chevolt [21]

Specifications	Ratings
Energy consumed	18.5 kWh
Miles (km)	53 miles(85)
SOC min and max	0.2 and 0.9
time for one full charge	4 h
Distance travelled per day	60 km
Charge consumed per km	0.1523 kwh
Charge consumed per day	9.138 kwh
Usage time	8:00 am to 9:00 am 4:00 pm to 5:00 pm
Charging time (after vehicle returns home)	3 h

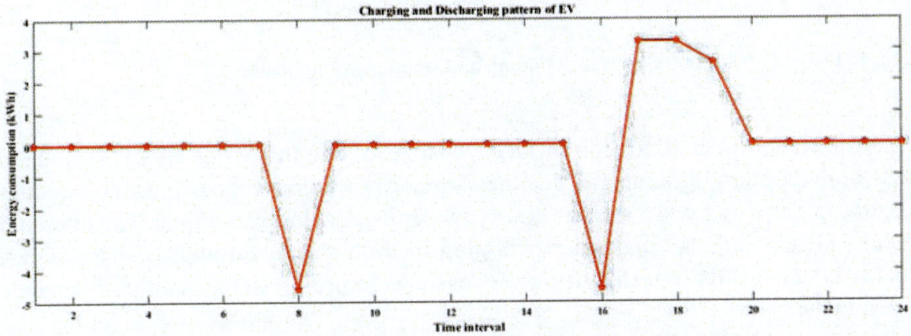

Fig. 4 Dumb charging pattern of EVs

Table 4 Charging strategy adopted by EVs for smart charging

Condition	Status
If PAR < 1	Charging allowed
If PAR > 1	Charging not allowed

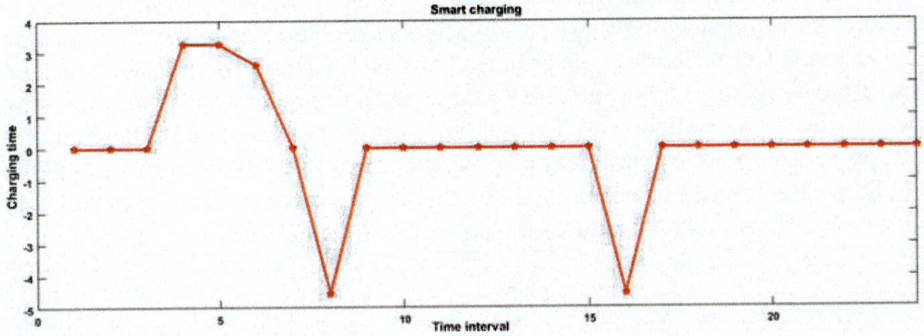

Fig. 5 Smart charging pattern of EVs

Table 5 Power loss during dumb and smart charging

Method	Dumb type	Smart type
P_L (kW)	1251.7723	1248.80

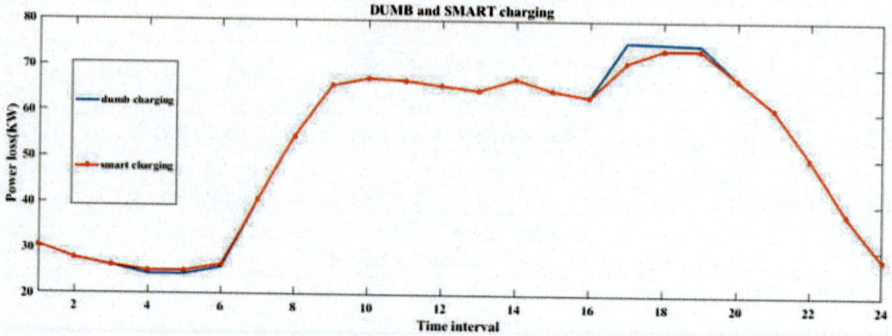

Fig. 6 Power loss for every hour using dumb and smart charging methods

Now, the optimal allocation of DGs considering the impact of EVs is simulated with both the charging strategies and the obtained power loss for each case throughout the day is given in Table 5 and the hourly power loss is plotted in Fig. 6. The obtained results clearly indicate that smart charging method combined with LSF and GOA yields better results in terms of power loss and improves the system performance effectively.

5 Conclusion

In this paper, a combined LSF—GOA algorithm is presented to solve the DG allocation problem in presence of EVs is presented. The objective of the work is to minimize the power loss of DS effectively. First, the sensitivity factors are used to find out the vulnerable nodes for DGs placement problem. After the identification of nodes, the GOA algorithm determines the sizes of DGs. The proposed method was tested by IEEE 33 bus system. EVs are charged using dumb and smart charging methods before DG placement. The observed result for each case is shown. From the comparison point of view, it is obvious that the power loss is very less in EVs and DGs case in smart charging type. Finally, it can be concluded that the optimal allocation of DGs and EVs reduces the power loss proficiently.

References

1. Amirioun MH, Kazemi A (2014) A new model based on optimal scheduling of combined energy exchange modes for aggregation of electric vehicles in a residential complex. Energy 69:186–198
2. Olivier JGJ, Schure KM, Peters JAHW (2017) Trends in global CO_2 and total greenhouse gas emissions. PBL Netherlands Environmental Assessment Agency 5
3. Mishra UC (2004) Environmental impact of coal industry and thermal power plants in India. J Environ Radioact 72(1-2):35–40
4. Putrus GA, Suwanapingkarl P, Johnston D, Bentley EC, Narayana M (2009) Impact of electric vehicles on power distribution networks. In: IEEE vehicle power and propulsion conference, pp 827–831
5. Bayat A, Bagheri A (2019) Optimal active and reactive power allocation in distribution networks using a novel heuristic approach. Appl Energy 233:71–85
6. Sudabattula SK, Muniswamy K, Suresh V (2019) Simultaneous allocation of distributed generators and shunt capacitors in distribution system. ECTI Trans Electr Eng Electron Commun 17(1):35–50
7. Kumar MV, Bilgundi SK, Pradeepa H (2019) Optimal allocation of DG units in distribution system considering variation in active power load. Arch Electr Eng 68(2):265
8. Quadri IA, Bhowmick S, Joshi D (2018) A comprehensive technique for optimal allocation of distributed energy resources in radial distribution systems. Appl Energy 211:1245–1260
9. Yuvaraj T, Devabalaji KR, Ravi K (2018) Optimal allocation of DG in the radial distribution network using bat optimization algorithm. In: Advances in power systems and energy management, pp 563–569
10. Poornazaryan B, Karimyan P, Gharehpetian GB, Abedi M (2018) Optimal allocation and sizing of DG units considering voltage stability, losses and load variations. Int J Electr Power Energy Syst 79:42–52
11. Deb S, Tammi K, Kalita K, Mahanta P (2018) Impact of electric vehicle charging station load on distribution network. Energies 11(1):178
12. Manbachi M, Farhangi H, Palizban A, Arzanpour S (2016) A novel volt-VAR optimization engine for smart distribution networks utilizing vehicle to grid dispatch. Int J Electr Power Energy Syst 74:238–251
13. Sortomme E, Hindi MM, MacPherson SJ, Venkata SS (2010) Coordinated charging of plug-in hybrid electric vehicles to minimize distribution system losses. IEEE Trans Smart Grid 2(1):198–205
14. Ahmadian Ali, Sedghi Mahdi, Aliakbar-Golkar Masoud, Fowler Michael, Elkamel Ali (2016) Two-layer optimization methodology for wind distributed generation planning considering plug-in electric vehicles uncertainty: a flexible active-reactive power approach. Energy Convers Manag 124:231–246
15. Hemakumar Reddy G, Badepalli A, Shivaprasad N, Behera C, Goswami AK, Dev Choudhury NB (2019) Impact of electric vehicles on distribution system performance in the presence of solar PV INTEGRATION. Int J Comput Intell IoT 2.1
16. Prakash K, Sydulu M (2007) Particle swarm optimization based capacitor placement on radial distribution systems. In: IEEE power engineering society general meeting, pp 1–5
17. Ahanch M, Asasi MS, Amiri MS (2017) A Grasshopper Optimization Algorithm to solve optimal distribution system reconfiguration and distributed generation placement problem. In: IEEE 4th international conference on knowledge-based engineering and innovation (KBEI), pp 0659–0666, IEEE
18. Cherukuri SHC, Saravanan B, Swarup KS (2018) A New choice based home energy management system using electric springs. In: 20th national power systems conference (NPSC), pp 1–6. IEEE
19. Sudabattula S, Kowsalya M (2016) Distributed energy resources allocation using flower pollination algorithm in radial distribution systems. Energy Procedia 103:76–81

20. Moradi MH, Abedini M (2012) A combination of genetic algorithm and particle swarm optimization for optimal DG location and sizing in distribution systems. Int J Electr Power Energy Syst 34(1):66–74
21. Chevy volt (online) available at https://www.chevrolet.com/electric/volt-plug-in-hybrid. Accessed 9 Oct 2019

Transmission System Security Enhancement with Optimal Placement of UPFC in Modern Power System

C. H. Nagaraja Kumari

Abstract This paper is aimed to apply the improved particle swarm optimization (IPSO) algorithm using line stability index (LSI) for the optimal positioning of unified power flow controller (UPFC) in standard IEEE test system as it can control real and reactive power simultaneously. It also provides series line compensation and independent controllable shunt compensation, which can maintain a stable and secure operation of modern deregulated power system. In this paper, MATLAB environment was employed to seek the optimum placement of UPFC considering practical constraints. This work expounds the capability of UPFC in terms of loss minimization, voltage stability enhancement, and power flow control. The reduced LSI values can indicate the stress relief over the transmission lines and conjointly indicating the congestion relief. To show the effectiveness, the proposed method was demonstrated on IEEE30 bus test system.

Keywords Line stability index · UPFC · Improved particle swarm optimization

1 Introduction

The electric utility industry is one among the largest and complex industries pushing the electrical engineer to encounter challenging problems in the modern power system in order to efficiently deliver quality power to the consumer. Hence, the electrical engineer has great concern at every step of power generation, transmission, distribution, and utilization. The power system networks are being operated under exceedingly vital conditions due to the continuous energy demand. This has imposed the threat of maintaining the required bus voltage, transmission capability, i.e., system loadability, control of power flow flexibility, transmission congestion management, and thus the power system has been facing problem of inefficiency in the maintenance of modern power system. Incorporation of flexible AC transmission system (FACTS) in the transmission system will assist to enhance the horizons of

---

C. H. Nagaraja Kumari (✉)
Nalla Malla Reddy Engineering College, Hyderabad, India
e-mail: nagarajakumari.eee@nmrec.edu.in

N. Zhou and S. Hemamalini (eds.), *Advances in Smart Grid Technology*, Lecture Notes in Electrical Engineering 688, https://doi.org/10.1007/978-981-15-7241-8_7

usage of the underneath-applied transmission system, i.e., it could lessen the flows in closely loaded lines, ensuing in an increased loadability, decreased system loss, advanced voltage balance, and minimization of congestion management issues within the network [1].

The unified power flow controller (UPFC) is considered as one of the most technically promising devices inside the flexible AC transmission system family [2] and acts as an incredible asset for the financially savvy usage of individual transmission lines by encouraging the independent control of both active and reactive power flows. For controlling the receiving-end voltage in order to serve the consumption, reactive power must control and can be effectively handled by UPFC as it can control both active and reactive power simultaneously. In [3], a steady-state voltage stability evaluation scheme to evaluate the proximity to the voltage falls apart at every load bus and in [4] based on sensitivity analysis and rank correlation concept, the optimal placement of series compensator was identified. A detailed review can be observed in [5]. Researchers have shown keen interest on UPFC because of its dynamic nature and its effective contribution to the system study. The optimal location of UPFC is a complex of combinational analysis. To resolve such kind of problem, heuristic strategies may be used which enables to obtain acceptable solution within a limited computation time. Two heuristic strategies are chosen for comparison: Basic particle swarm optimization (BPSO) and improved particle swarm optimization (IPSO).

In this paper, the size and best place of the FACTS device are selected primarily based on the line stability index (LSI). With this optimal placement, it is found that with the replacement of conventional PSO technique with improved PSO technique has shown considerably good results in terms of loss minimization, voltage stability enhancement, and power flow control within the network hence reducing the associated economic loss. Thus, a more green transmission system may possible.

2 Transmission System Security

According to the North American Electric Reliability Corporation, i.e., NERC's definition, reliability encompasses two ideas: adequacy and security. Adequacy is the capability of the power system to deliver the specified energy call for demanded via the client at any given instance of time. Security is described as the ability of a power system to withstand unexpected disturbances [6]. Transmission system security is one of the primary concerns of the individual system operator (ISO) since its support is necessary for successful operation of any competitive electricity market model. Security is often termed as the ability of a power system to maintain the desired state without violating any of the imposed operational limits(e.g., bus voltages and power flows) against predictable changes(demand and generation evolution) and unpredictable events (called contingencies) likely to occur during real-time operation.

The security of this deregulated power system operation is mainly dependent on the decisions of independent system operator (ISO). For instance, a generating unit may need to be kept off-line because of auxiliary device failure. By retaining the right quantities of spinning reserve, the remaining units on the system can make up the deficit without a too low-frequency drop or want to shed any load. Similarly, a transmission line may be broken by a hurricane and brought out with the aid of automated relaying. If in committing and dispatching technology, right regard for transmission flows is maintained, the remaining transmission traces can take the extended loading and still remain within the limit [7]. This takes place in particular because of the over loading of the transmission lines, voltage deviation, and absence of reactive power support at the load buses. So it is far critical to limit severity and analyze the system situation to improve security. As such, this work considers LSI to come to be aware of the critical line(s) and buses.

3 UPFC Mathematical Model

Due to the high efforts for the voltage source converters and the protection, a UPFC is getting pretty costly, which limits the realistic applications where the voltage and strength flow manage is needed simultaneously [6] (Fig. 1).

The operating mode of UPFC can be altered according to the requirement. This alteration is done without any change in hardware structure; this helps the UPFC to get adapted with the different real-time changes under system operating condition. For realizing and conducting the power flows in the power system containing the UPFC, the mathematical model of the UPFC is very important. UPFC can be modeled based on an ideal voltage sources connected in series or two voltage sources in which, one is connected in series while the other in parallel. The later model is also called as power injection model. In power injection model (PIM), two ideal voltage sources are connected in series with the source impedance, where one set is connected in series with the transmission line and the other in parallel with the transmission line. This model is the most suitable one for representing the UPFC in power flow studies. The representation of PIM for UPFC is as follows.

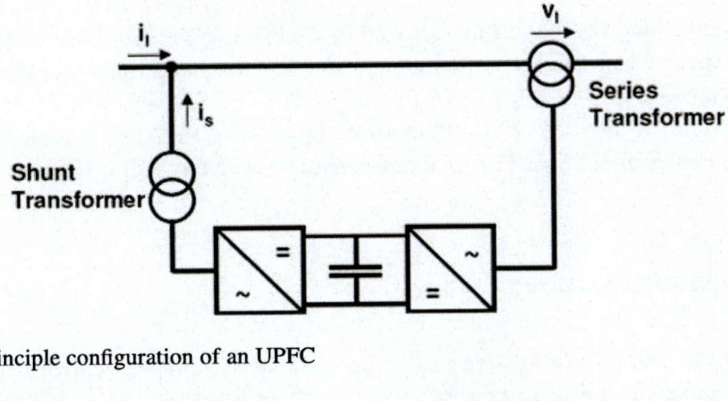

Fig. 1 Principle configuration of an UPFC

The apparent power supplied by the series connected converter, S_{se} can be expressed as:

$$\text{Sse} \begin{aligned} &= \left\{b_{se} r V_i^2 \sin(\gamma) - b_{se} r V_i V_j \sin(\theta_i + \gamma - \theta_j)\right\} \\ &\quad + j\left\{b_{se} r V_i^2 \cos(\gamma) - b_{se} r V_i V_j \sin(\theta_i + \gamma - \theta_j)\right\} \end{aligned} \tag{1}$$

By spiting the above equation into real and imaginary part, the above equation can be written as following:

$$\therefore P_{se} = b_{se} r V_i^2 \sin(\gamma) - b_{se} r V_i V_j \sin(\theta_i + \gamma - \theta_j) \tag{2}$$

$$\therefore Q_{se} = b_{se} r V_i^2 \cos(\gamma) - b_{se} r V_i V_j \cos(\theta_i + \gamma - \theta_j) \tag{3}$$

The reactive power supplied or taken by the shunt connected converter is not considered in this model. So Q_{sh} can be taken or assumed to be zero. The mathematical model of UPFC can be obtained by combining the series and shunt power injection model at both *i*th and *j*th bus.

$$\begin{aligned} P_{i,\text{inj,upfc}} &= P_{sh} + P_{is} = -1.02 P_{se} + P_{is} \\ &= -1.02\left(b_{se} r V_i^2 \sin(\gamma) - b_{se} r V_i V_j \sin(\theta_i + \gamma - \theta_j)\right) - b_{se} r V_i^2 \sin\gamma \\ &= 0.02 r b_{se} V_i^2 \sin\gamma - 1.02\, r b_{se} V_i V_j \sin(\theta_i - \theta_j + \gamma) \end{aligned} \tag{4}$$

$$P_{j,\text{inj, upfc}} = P_{js} = V_i V_j b_{se} r \sin(\theta_i - \theta_j + \gamma) \tag{5}$$

$$Q_{i,\text{ inj, upfc}} = Q_{sh} + Q_{is} = 0 - b_{se} r V_i^2 \cos\gamma = -b_{se} r V_i^2 \cos\gamma \tag{6}$$

$$\begin{aligned} Q_{j,\text{ inj, upfc}} &= Q_{sh} + Q_{js} = 0 + V_i V_j b_{se} r \cos(\theta_i - \theta_j + \gamma) \\ &= V_i V_j b_{se} r \cos(\theta_i - \theta_j + \gamma) \end{aligned} \tag{7}$$

where V_i and V_j are the magnitudes, θ_i and θ_j are the angles of i, j buses respectively and X_{se} is the series branch admittance; r and γ are the magnitude and phase angle series voltage source.

By considering Eqs. (5)–(7), the incident buses of UPFC are modeled as a PQ buses with power injections. The final representation PIM of UPFC is given in Fig. 2.

4 Line Stability Index

Since FACTS devices are passive in nature, the reactive power control is explored through computing the line stability index (LSI) in every operative situation. The

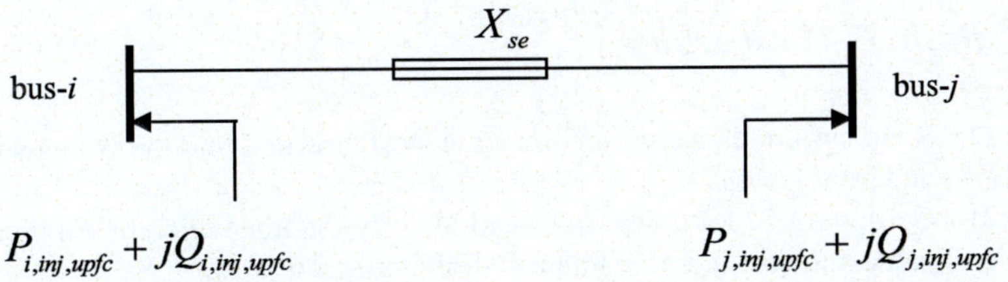

Fig. 2 Representation of PIM for PFC

static security of the network has been assessed with the aid of many techniques inside the literature. One of the brilliant techniques is line stability index [8] and is taken into account during this work.

For a transmission line connected between bus-p and bus-q is given by Eq. (8):

$$\mathrm{LSI}_{pq} = \frac{4x_{pq}Q_{qp}}{\left\{\left|V_p\right|\sin\left(\theta_{pq} - \delta_p + \delta_q\right)\right\}^2} \leq 1.00 \tag{8}$$

where X_{pq} and θ_{pq} are the reactance and impedance angle of transmission line, respectively, Q_{qp} is the reactive power flow of the line at bus-q, $|Vp|$ is the voltage magnitude at bus-p (sending end), δ_p and δ_q are the voltage phase angles at bus-p and bus-q, respectively. For stable operation, the LSI ought to be a great deal less than 1 for all of the lines. The LSI values greater than one suggests the proximity of instability or voltage fall apart. Thus, the power flow management or congestion relief in the network is decided via computing LSI at each case.

If any line running with high values of LSI will challenge to stress operative circumstance. Thus, it is far useful to have FACTS devices in both this sort of line or incident buses. In this exploration, the lines which might be the incident to generator buses, synchronous condensers, and which have tap-changers are excluded from the priority listing. Among the rest of the traces, the line with the highest value of LSI has been selected to locate the series type FACTS device and the buses which might be incident to that line have been chosen for the best placement of shunt FACTS device.

5 Implementation Algorithm

In this section, the implementation procedure of basic particle swarm optimization (PSO) algorithm for achieving the minimum losses while improving the system voltage profile and consequently redistributing the power flows in the transmission lines toward security margin improvement. The steps given in this section are followed for UPFC parameters optimization.

5.1 Basic PSO Algorithm

Step 1) Read system data and run base case load flow and find the voltage profile and total losses.
Step 2) Compute LSI for all the lines and identify the line which has highest LSI value and consider this line for UPFC integration.
Step 3) Modify the load bus as generator bus.
Step 4) Define PSO variables. Number of particles (NP) = 50; Number of variables (size) = 4 (bus voltages v and UPFC incident bus angles δ; initial weight $(w) = 0.9$; specify velocity min and max range.
Step 5) Here, the voltage magnitudes of all buses as [0.9 p.u to 1.1 p.u] and [$-\pi$ to $+\pi$] for UPFC series converter angles.
Step 6) Find initial velocity using $V = V_{\min} + U(V_{\max} - V_{\min})$ for NP times.
Step 7) Find initial solution $v^0 = v_{\min} + U(v_{\max} - v_{\min})$ and $q^0 = q_{\min} + U(q_{\max} - q_{\min})$.
Step 8) Update all equations of system operating constraints with initial solution obtained at step 6.
Step 9) Repeat step (7) for NP times and at each time, find P_{best} and G_{best} values of v and q which gives minimum losses.
Step 10) Set number of iteration (IT) = 50 and $\alpha = 0.5$
Step 11) Compute $V_{\max} = \begin{bmatrix} \frac{v_{\min} - v_{\max}}{\alpha} \\ \frac{q_{\min} - q_{\max}}{\alpha} \end{bmatrix}$
Step 12) Update the velocity and variables. Repeat step (6) to step (10) for IT times and update at each iteration P_{best} and G_{best}. Stop after reaching the convergence criterion. Here, we have taken IT as convergence criterion.
Step 13) Print the load flow solution with optimal ratings and determined new LSI values for all the lines.

5.2 Improved PSO Algorithm

The procedure is same as basic PSO and the major difference is taken place at step (10) while updating dating the velocity with improved inertial weight as follows:

$$\omega^k = \omega\text{max} - (\omega\text{max} - \omega\text{min})\frac{k}{k\text{max}} \quad \text{if} \quad p^k_{gb,i} \neq x^k_i \tag{9}$$

$$\omega^k = \omega\text{max} \ \text{ if } \ p^k_{gb,i} = x^k_i \tag{10}$$

where $\omega_{\max}$ and $\omega_{\min}$ are the initial and final inertia weights respectively; k and $k_{\max}$ are the iteration and maximum iteration numbers respectively. Generally, $\omega_{\max}$ is set to 0.9 and $\omega_{\min}$ is set to 0.4. The steps are as follows:

Step 1) Read system data and run base case load flow and find the voltage profile and total losses.

Step 2) Compute LSI for all the lines and identify the line which has highest LSI value and consider this line for UPFC integration.

Step 3) Modify the load bus as generator bus.

Step 4) Define IPSO variables as number of particles (NP) = 50; Number of variables (Size) = 4 (bus voltages v and UPFC incident bus angles δ; Inertial weight $\omega_{max} = 0.9$ and $\omega_{min} = 0.4$; Specify velocity min and max range

$$V_{min} = \begin{bmatrix} -V_{min} * 0.5 \\ -\delta_{min} * 0.5 \end{bmatrix} \text{ and } V_{max} = \begin{bmatrix} V_{max} * 0.5 \\ \delta_{max} * 0.5 \end{bmatrix}$$

Here, the voltage magnitudes of all buses as [0.9 p.u to 1.1 p.u] and [$-\pi$ to $+\pi$] for UPFC series converter angles.

Step 5) Find initial velocity using $V = V_{min} + U(V_{max} - V_{min})$ for NP times.

Step 6) Find initial solution $v^0 = v_{min} + U(v_{max} - v_{min})$ and $q^0 = q_{min} + U(q_{max} - q_{min})$.

Step 7) Update all equations of system operating constraints with initial solution obtained at step 6.

Step 8) Repeat step (7) for NP times and at each time, finds $P_{best\ and}$ G_{best} values of v and q which gives minimum losses.

Step 9) Set number of iteration (IT) = 50 and $\alpha = 0.5$

Step 10) Compute $V_{max} = \begin{bmatrix} \frac{v_{min} - v_{max}}{\alpha} \\ \frac{q_{min} - q_{max}}{\alpha} \end{bmatrix}$

Step 11) Update the velocity and variables. Repeat step (6) to step (10) for IT times and update at each iteration P_{best} and G_{best}. Stop after reaching the convergence criterion. Here we have taken IT as convergence criterion.

Step 12) Print the load flow solution with optimal ratings and determined new LSI values for all the lines.

6 Results and Analysis

The proposed approach for the optimal placement of UPFC has been tested on the IEEE30 bus device. The IEEE30 bus system represents two generator buses (1 and 2), four synchronous condenser buses (5, 8, 11 and 13). The base case load has 287.452 MW and it has 41 transmission lines [10]. The simulation results for the top five crucial lines are given on the basis of LSI in Table 1 and it may be found that line 4–12 has maximum LSI cost in the base case. Hence, line 4–12 is taken into consideration to have UPFC.

The voltages at the critical buses with the base case and with the implementation of BPSO and IPSO are given in Table 2.

Table 1 IEEE30 bus line flow data for base case

Fb	Tb	LSI		
		Base case	With BPSO	With IPSO
4	12	0.222	0.294	0.219
6	9	0.169	0.131	0.11
6	10	0.148	0.15	0.117
9	11	0.123	0.016	0.018
1	2	0.082	0.079	0.075

Table 2 BPSO-based simulation results of IEEE 30-bus system bus data with UPFC in line 4–12

Bus #	Voltage (p.u)		
	Base case	With BPSO	With IPSO
1	1.06	1.0995	1.0998
2	1.043	1.0842	1.0839
4	1.0129	1.0606	1.0557
6	1.0121	1.0598	1.0553
9	1.0507	1.0914	1.0816
10	1.0438	1.0941	1.0813
11	1.082	1.0958	1.0864
12	1.0576	1.124	1.1024

With the simulation results, we located that the real power loss of 17.528 MW in the base case is reduced considerably to 16.208 MW for BPSO and 16.112 MW for IPSO using UPFC. The reactive energy goes with the flow changes in the transmission system causes to alternate the LSI values as given in Table 3. The widespread decrement in LSI values is indicating the UPFC impact on now not handiest line flow control but also on the progressed voltage stability (Table 2).

The LSI value of line 4–12 for the base case is 0.222 and is accelerated extensively to 0.294 with BPSO and decreased to 0.219 with IPSO. This shows the power

Table 3 Simulation results of IEEE 30-bus system bus data with UPFC in line 4–12

Fb	Tb	P_{ij} (MW)	Q_{ij} (MVAR)	LSI	Total system losses (MW)
Base case					
4	12	44.147	−16.795	0.222	17.528
With BPSO					
4	12	46.096	−26.057	0.294	16.208
With IPSO					
4	12	44.625	−18.621	0.219	16.112

flow increment capability or control in the network successfully. The convergence characteristics of BPSO and IPOS are given in Figs. 3 and 4 respectively.

The improved voltage profiles with BPSO and IPSO are given in Table 2. As compared to BPSO, the IPSO has resulted better result in terms of losses and voltage profile and also fast convergence.

For better insight, the transmission system security in terms of LSI values of all the lines obtained with UPFC in IEEE 30 bus test system using BPOS and IPSO are illustrated in Fig. 5.

Fig. 3 Fitness function of BPSO with UPFC

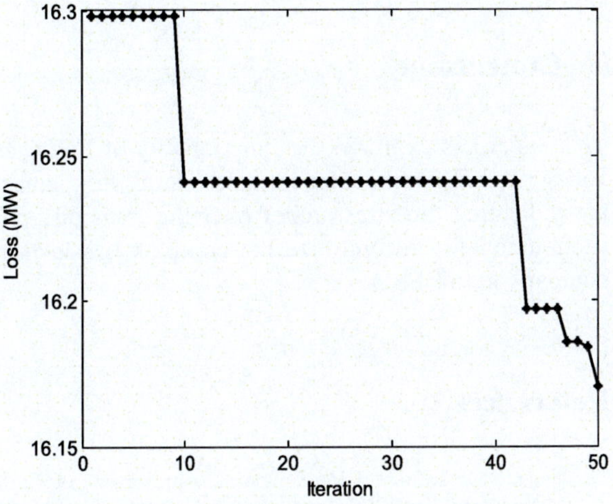

Fig. 4 Fitness function of IPSO with UPFC

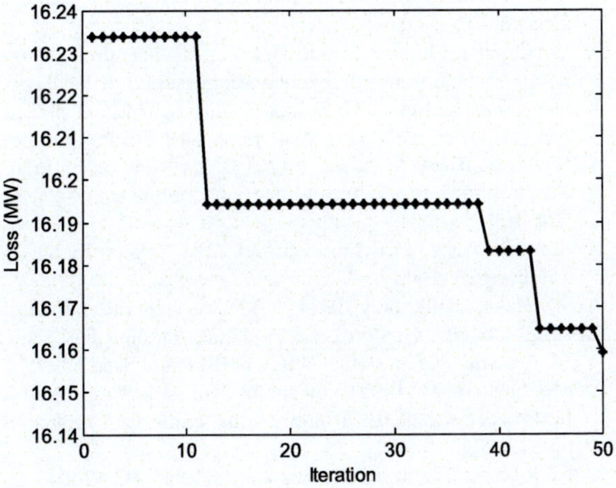

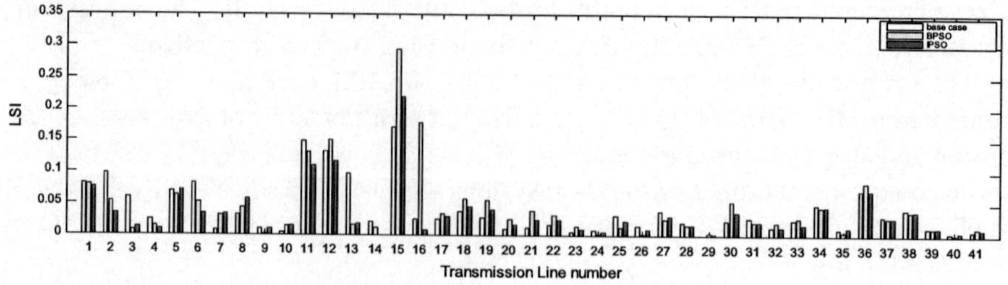

Fig. 5 Transmission system security with UPFC in IEEE 30-bus system

7 Conclusion

This paper has explored the functionality of UPFC in terms of loss minimization, voltage stability enhancement, and energy flow control. The decreased LSI values are indicating the stress relief over the transmission lines. The IPSO has proven its functionality to optimize the complex hassle in power system applications in comparison to BPSO.

References

1. Naga Raja Kumari CH, Sekhar C Application of Sensitivity Approach for Congestion Management by UPFC. Int J Control Theory Appl 10(5):673–678
2. An S, Condren J, Gedra TW (2007) An ideal transformer UPFC model, OPF first-order sensitivities, and application to screening for optimal UPFC locations. IEEE Trans Power Syst 22(1):68–75
3. Liu C, Wang B, Hu F, Sun K, Bak CL (2018) Online voltage stability assessment for load areas based on the holomorphic embedding method. IEEE Trans Power Syst 33(4):3720–3734
4. Naga Raja Kumari CH, Sekhar C (2016) Optimal placement of TCSC based on sensitivity analysis for congestion management. Int J Electr Comput Eng 6(5):2041–2047
5. Rawat N, Bhatt A, Aswal P (2013) A review on optimal location of FACTS devices in AC transmission system. In: International conference on power, energy and control, June 2013, pp 104–109
6. Stott B, Alsac O, Monticelli AJ (1987) Security analysis and optimization. Proc IEEE 75(12):1623–1644
7. Wood AJ, Wollenberg BF (1996) Power generation operation and control. Wiley, New York
8. Hingorani NG, Gyugyi L (1999) Understanding FACTS: concepts and technology of flexible AC transmission systems. Wiley IEEE Press, Dec 1999
9. Moghavvemi M (1997) New method for indicating voltage stability condition in power system. In: Proceedings of IEEE international conference on Power engineering, Singapore, IPEC'97, pp 223–221
10. Enrique A, Claudio RF, Hugo AP, Cesar AC (2004) Modelling and simulation in power networks. Wiley, England

Fuzzy System Approaching on Designing Intelligent Process—A Modelling for Thermal Power Plant

Subodh Panda and Nagesh Deevi

Abstract This research work focuses on power thermal power plant efficiencies, which are associated with many more indirect losses and have been broadly examined for proposeful solution for it. Waste heat can be primary point here, which be suitable to reuse at certain heating process, is a concept of efficiency improvement. Some of most sensitive points like draying of fuel, preheating air for combustion monitoring, and rising of feed water temperature are observed for better approach during this work plan. It is theoretically proved, minimisation of an extra 3% wt. on moisture of fuel and rise up to inlet air of 35 °C. Improving efficiency by 1% and saving of 50 ton of fuel on and average during a year. An intelligent process has been designed that operates, monitors and controls properly the heat recovery. Due analysis of traditional PID with proposed FUZZYPID after the result justifies its utility over nonlinear complex to nonlinear process. So, this proposed intelligent process is accepted for all range of operating condition with most expected developing of efficiencies.

Keywords Intelligent system · Blow down monitoring · Process efficiency · Fuzzy-based controller · Hybrid system · Thermal power plant

1 Introduction

Power generation system and process have a vital role in most production industries. So, in process plant, beside power generation, stem acts as source of heat for many more subunit, which has a demand on waste heat of other terminal. Through this research work, a valuable solution on utilisation of waste heat in optimised way on improvement of operational efficiency. This also minimised indirect losses, not to change operational behaviour with valuable technology adaptive to system needs.

S. Panda (✉) · N. Deevi
Pragati Engineering College, Department of ECE, Kakinada, AP, India
e-mail: Subodh.panda@gmail.com

N. Deevi
e-mail: nagesh.deevi417@gmail.com

N. Zhou and S. Hemamalini (eds.), *Advances in Smart Grid Technology*, Lecture Notes in Electrical Engineering 688, https://doi.org/10.1007/978-981-15-7241-8_8

Basic study on earlier work over this found that many more desired outcome remains unsolved and yet to fulfil for industrial process. Some more character can be added on development of operational goal. On real efficiency of boiler like process unit, some of mostly affecting unit be concentrated to regain from minimise indirect losses. Several works have been tested earlier and some where it has been implemented successfully but it is noticed that the proposed has not being solved.

To fulfil further demands, this research outcome may have great contribution to technological world. The observation over various processes and associate units required redesign by utilising this result directly or indirectly for efficiency enhancing. Present platforms need designed model based on this technology of more adaptive in nature and user-friendly on operation.

. Blow down station is a major area for losses of heat source and it can be reutilised for reheating feed water, drying air pre heating, making fuel free of moisture which has been noticed during our real-time surveying industrial platform. A better structure with possible technology can be associated for combining all available unutilised heat loss for a remarkable outcome that improves efficiency up to certain satisfactory level.

2 Proposed Designing of Heat Recovery Unit

Figure 1 expressing the required subbranch which on integral represents a successful intelligent system on present industrial automation. This also helps on designing an efficient process keeping all this subgroup individually or together. This research work designing an intelligent blow down station where the waste heat is minimised and extra heat out of this can be proposed fully use to develop overall behaviour.

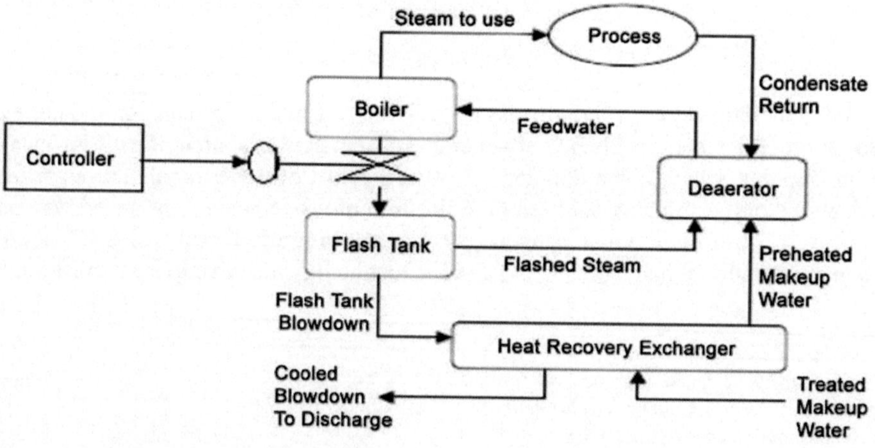

Fig. 1 Conventional method of control

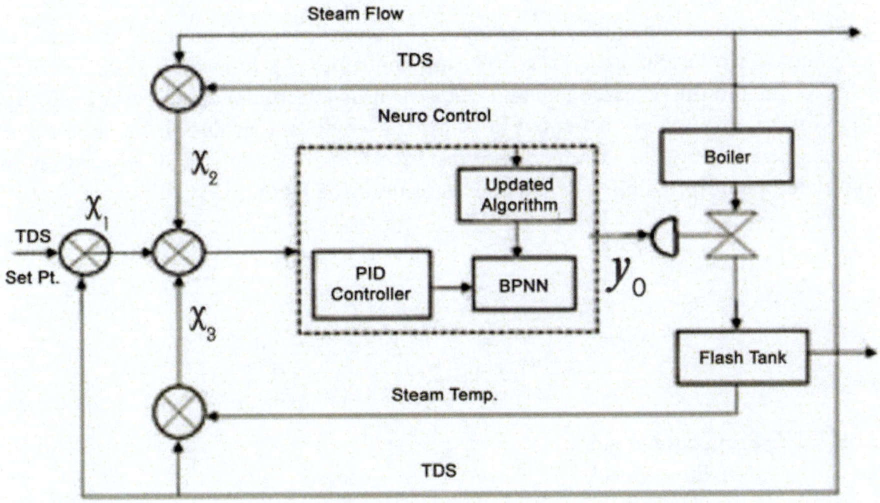

Fig. 2 Blow down losses control with BPNN

And corresponding systematic diagram shown in Fig. 2, This proposed design regulates the discharge of drum output at required interval, corresponding availed heat on discharge part be systematically and stepwise used on heating furnace preair, drying fuel moisture, and in grate helpful on rising the feed water temperature which ultimately responsible [1]. Most important factor about this designing is, since no such crisp data on this is possible at plant flora, designed has been attempted on data on fuzzy nature. It also sequentially optimised the input data for a targeted output. This experimental observation on increasing boiler efficiency, traditional operation of blow down station has been focused on waste heat recovery and its proper utilisation. The possible amount of this unutilised heat can be used through an intelligent heat transfer system. Associative designed part optimised general working of heat utilising unit.

Required amount of heat for fuel drying at conveyer can be extracted from some of the total waste heat at blow down terminal. The proposed heat recovery unit for this require process may have the characteristic as stated in Table 1.

Table 1 Basic designing parameter of heat recovery unit

Parameter under discussion	Value	
Uo, system overall coefficient heat transfer	18.5	W m^2 °C
Air flow rate	3--65	m^{3}s
Heat exchanger (*A*) surface area	158.18	mm^2
Heat exchanger (*Qheat_changer*)	88.7.9	kW
Temperature different algorithm (T_{LMTD})	24–28	

The successful operation of complete system depends on characteristic of designed controller. So, analysis of excising controller is primary job of this work. It helps to find out the research gap and leads towards intelligent controller modelling. PID controller is widely accepted in industrial environment due to its simple and predesigned tuning for assigned work station. The mathematical expression of complete operation with PID controller is as describe further.

$$\begin{aligned} U(K) = U(K-1) + K_P e(K) - e(K-1) \\ + \frac{K_1 T}{2}[e(e(k)) + e(k-1)] \\ + \frac{K_0}{2}[e(k) + 2e(k-1) + e(k-2)] \end{aligned} \quad (1)$$

(kp)	Proportional gain
Kd	derivative gain kd
Ki	interregnal gain
Tis and U	represent sampling period, And discrete time index respectively

For the desire error, the deviation of real plant output
$E = r - y$
So

$$\Delta U(K) = K_P e_P(K) + K_i e_1(K) + K_P e_P(K) \quad (2)$$

$$U(K) = U(K-1) + \Delta U(K) \quad (3)$$

where

$$e_P(K) = e(K) - e(K-1),\ e_1(K) = \frac{T}{2}[e(k) + e(k-1)]$$
$$e_P(K) = \frac{1}{T}[e(k) + 2e(k-1) + e(k-2)] \text{ with } e(K) = 0 \text{ for } K < 0 \quad (4)$$

Designed controller have well-defined algorithm based on automatic control logic. The model proposed for designing is supported with a logarithm composed of linguistic control potentiality. In common, an intelligent system designing has five subingredients associated individual nature [4].

(a) Input and output variables properly define suitably in linguistic nature covering approximate range.
(b) The well-defined membership function useful in converting will convert the process behaviour into expressible form.
(c) A system flow diagram designed with knowledge-based configuration
(d) fuzzy rules that based on experience use in developing architecture of system through inference unit.

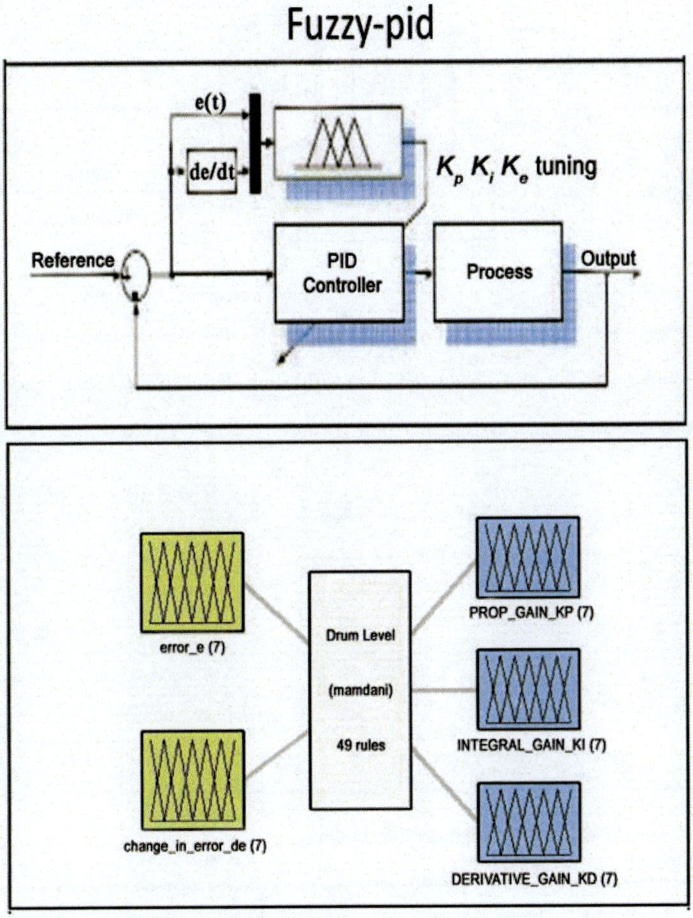

Fig. 3 Systematic model and membership function for input

(e) proportional fuzzy logic control action generated for physical control action needed develop with designed system (Fig. 3).

2.1 *Simulating Model*

See Figs. 4 and 5.

The simulating model of total blow down heat recovery model with tradisational to intelligent, i.e. PID controller, FUZZY and FUZZY-PID controller analysed, individual output has shown in Figs. 6 and 7, been demonstrated and evaluated which gives overall performance of designed controller it FUZZY-PID is consider to be good.

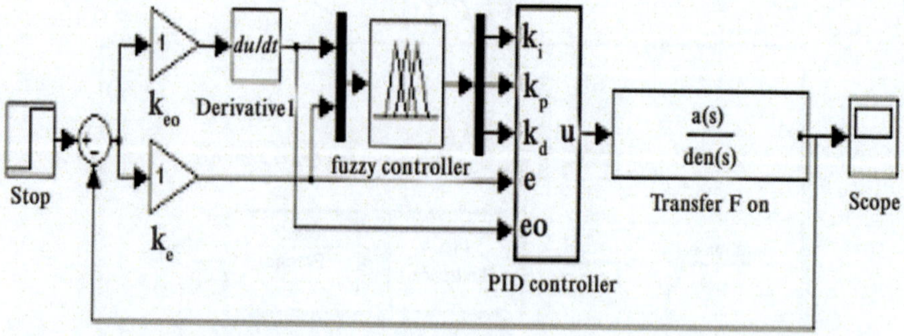

Fig. 4 Simulink model for blow down operation with controller in FUZZY–PID mode

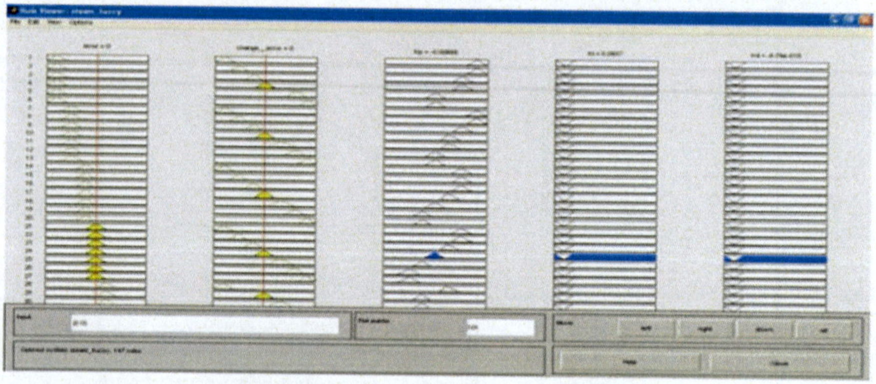

Fig. 5 Fuzzy rule-based for proposed control model

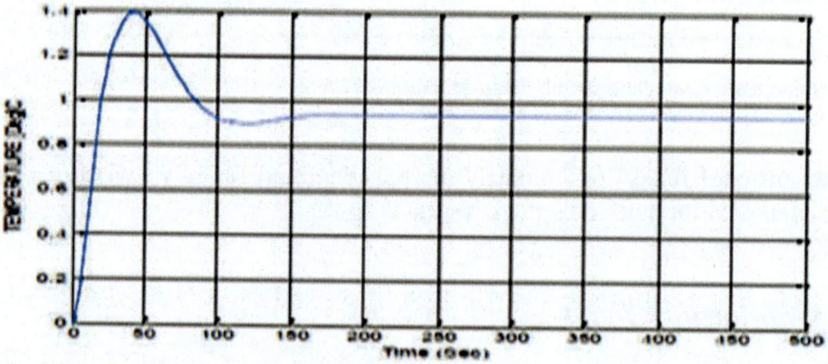

Fig. 6 PID controller responding to process reaction

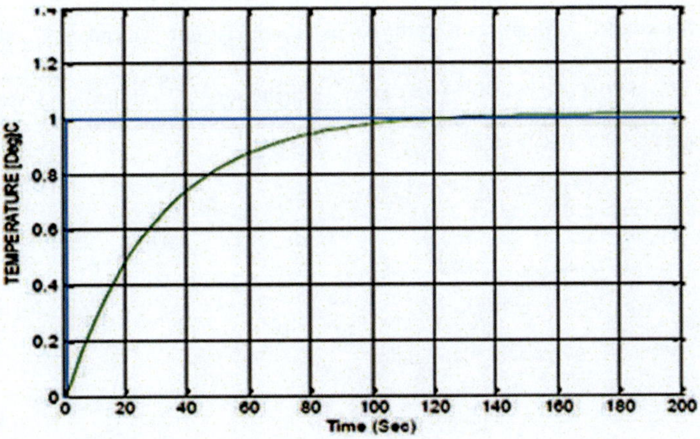

Fig. 7 Fuzzy—PID controller response to process reaction

The surface view of systematic representation of the process expresses the better controllability of FUZZY—PID over tradisational PID operation. Basic two parameters like overshoot and the settling time have been considered for analysis of the overall operational behaviour of the system study over the entire process; after using proposed model, indirect losses have been reduced to some extent that help to improved efficiency of complete process (Fig. 8).

Table 2 describes that there is a remarkable change on operation system that is benefited more on fuel moisture, preheat of inlet air, preparing makeup water with reutilisation of waste heat recovered from this modelling. Theoretical calculation

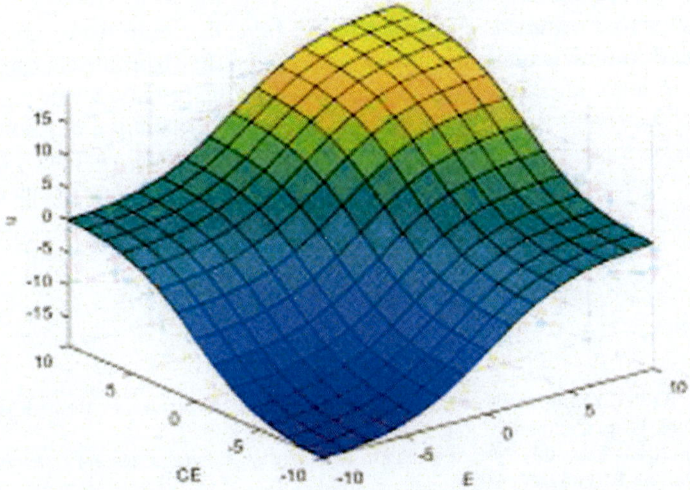

Fig. 8 Surface view of complete response to heat recovery unit

Table 2 characteristic of heat recovery of boiler operational after considering and before the designed system

Parameter under observation	Initial without proposed model	Dryness of fuel	Air preheater	Controlled combustion	Overall outcome with proposed model considering
Efficiency of boiler	76.48	76.89	77.20	80.82	81.63
Overall consumption of energy (kWh)	–	1.73	–	–	1.73
Fuel/stem flow ratio	5.87	5.90	6.12	6.21	6.41
Yearly fuel in Ton saving	–	18.50	32.50	195.88	246.88

indicates that increasing boiler efficiency by approximately 5% with a considerable amount of fuel saving/year when this modelling concept has been adopted over all possible unit.

3 Conclusion

This research leading to modelling of acceptable structure of digital controller with intelligent operational behaviours.

Adaptation for self-tuned behaviour improved over the controller to rapid changes of the error deviation is major support of this. Again, the error rate change by delay effects is no more affecting the proposed model. This proposed model controller designed of such nature improves heat exchanger optimal operation and motivates the researchers on increasing efficiency of overall unit up to certain extent… Extracted heat can be reused for preheating of inlet air and removing moister content of fuel and much more possible way to improved boiler efficiency.

References

1. Chamoso A (2018) Reuse of waste energy from power plants in greenhouses through MAS-based architecture. Wirel Commun Mobile Comput. Hindawi 218
2. Bortolini (2017) Assembly system design in the Industry 4.0 era: a general framework. IFAC Papers OnLine 18(1):5700–5705
3. Bradley DA et al (2018) Mechatronics and the design of intelligent machines and systems. CRC Press

4. Horberry T, Regan MA, Stevens A (2017) Driver acceptance of new technology: theory, measurement and optimisation. CRC Press
5. Liu C, Rao Z (2017) Challenges in various thermal energy storage technologies. Sci Bull 621(4):231–233
6. Liu M et al (2017) Intelligent assembly system for mechanical products and key technology based on internet of things. J Intell Manuf 28:271–299
7. Nihtianov S, Antonio L (2018) Smart sensors and MEMS: intelligent sensing devices and microsystems for industrial applications. Woodhead Publishing
8. Pang CK et al (2017) Intelligent diagnosis and prognosis of industrial networked systems. CRC Press
9. Umbrello S, De Bellis AF (2018) A value-sensitive design approach to intelligent agents, Artificial intelligence safety and security. CRC Press

Performance Enhancement of Permanent Magnet Synchronous Motor Employing Iterative Learning Controller with Space Vector Pulse Width Modulation

N. Subha Lakshmi, S. Allirani, S. Sundar, and H. Vidhya

Abstract The PMSM generates the magnetic flux on its own as the rotor has permanent magnet and thus the motor never depend on any exterior source. The torque and speed ripples are the some of the disadvantages which affect the performance of drive. This paper proposes the performance enhancement of PMSM employing ILC and SVPWM driven by FOC to reduce both the speed and torque ripple. The result is compared with conventional PI controller. The outcome shows the reduction in torque and speed ripple which enhances the drive performance by using the above technique. The hardware result is obtained from DSP controller.

Keywords Field-oriented control (FOC) · Torque pulsation · Speed pulsation · PI · ILC · SVPWM · PMSM

1 Introduction

As the PMSM rotor is constructed using permanent magnets, it can be able to produce magnetic flux of its own and thus does not rely on other source. The efficiency of the system can be improved by minimum power consumption. The various application of PMSM is servo and robot drives because of high efficiency, high torque. The presences of torque and speed ripples affect the usage of motor in low power and speed drive application.

N. Subha Lakshmi (✉)
EEE Department, Sri Krishna College of Engineering & Technology, Coimbatore, India
e-mail: nsubhalakshmi90@gmail.com

S. Allirani (✉) · H. Vidhya
EEE Department, Sri Ramakrishna Engineering College, Coimbatore, India
e-mail: allirani.saminathan@srec.ac.in

H. Vidhya
e-mail: Vidhya.karthik@srec.ac.in

S. Sundar
EEE Department, Bannari Amman Institute of Technology, Coimbatore, India
e-mail: sundars@bitsathy.ac.in

N. Zhou and S. Hemamalini (eds.), *Advances in Smart Grid Technology*, Lecture Notes in Electrical Engineering 688, https://doi.org/10.1007/978-981-15-7241-8_9

The cogging, flux harmonics, and unbalanced phase are the major causes for the production of oscillations in motor. The various methods to solve this problem are by either in construction of motor design which is tedious or the other method is to adapt several schemes to control the oscillations by considering the reference current and time period which promote smooth torque. To maintain the torque to be constant, the current should be controlled using FOC method.

This paper emphasizes on the utilization of iterative learning controller associated with field-oriented control to reduce the speed oscillations in the PMSM. The pulses to the inverter are given by SVPWM technique and the two main controllers employed, namely ILC and PI are used to give high torque performance in transient and steady-state conditions. The ILC has ability to compensate the repetitive disturbances from outside by frequent learning based on preceding iterations.

2 Overview of Field-Oriented Control

The often method used to control the PMSM in variable speed application is vector control otherwise called as FOC. This method will calculate the drive performance in terms of DC machine and thus the control of stator current is represented in terms of vectors. In this method, clarks and parks transformation are used for three-phase to two-phase transformation and it is done for torque and flux component. The FOC control works in both steady-state and transient-state operating conditions. The FOC solves the problem using two components, namely torque and flux component thereby transformed to d and q reference frame. Equation 1 gives the relation between torque and flux component.

$$T \alpha \Psi_R i_{sq} \tag{1}$$

Figure 1 shows the block diagram of FOC of PMSM. This includes PMSM supplied through two-stage inverter, PWM module, Park and Clark transformation blocks, PI controller, and the load.

2.1 Proposed System

Figure 2 shows the proposed FOC with ILC for the PMSM. In this proposed scheme, the conventional PI controller for speed is replaced by the ILC controller. The ILC is an error correction algorithm which stores the previous iterative values in memory and compares the new value and produces the error value, which is easier when compared to other techniques [1–3].

SVPWM The SVPWM is used to generate the desired voltage to be supplied to the PMSM. Thus, the pulses produced for the gate driver circuit are done by eight

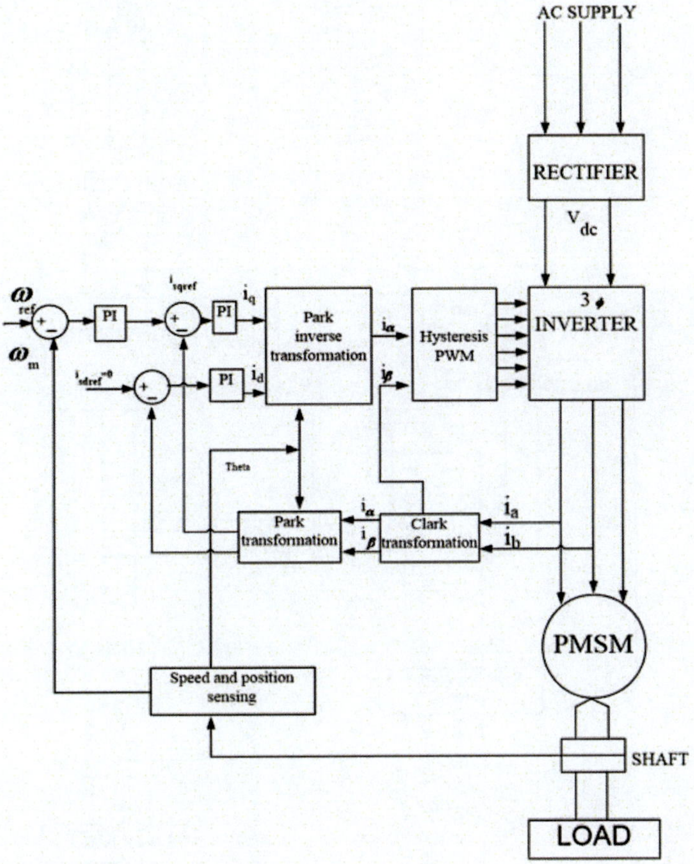

Fig. 1 Block diagram of existing method

possible switching combinations which nearby estimate to the V_{ref}. The vectors are referred by $V1$ to $V6$ which divides the plane to sectors and called as active vectors whereas $V0$ and $V7$ are called as null vectors.

ILC Iterative learning control uses the previous iteration values and error signal which tries to make the better control signal for the concurrent iteration to keep the error value as low as possible. The major advantage is the ability to deal with different system despite its order. Thus, the noise in drive is been reduced as much as possible [1–3]. The flowchart on working of ILC is shown in Fig. 3.

PI Controller Tuning The main output relies on PI controller tuning. The error output is given to the vector control block and the value gives the reference current. The PI controller of current loop compares the values and produces the current component. Thus, the regulation is carried out using testing and analyzing method.

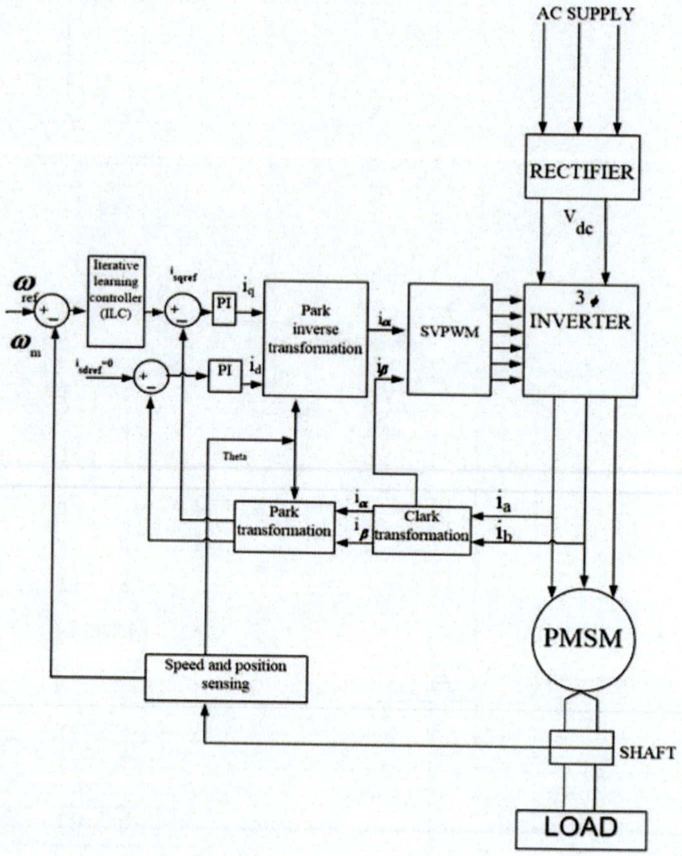

Fig. 2 Block diagram of proposed method

2.2 *Simulation Result*

Figures 4 and 5 show the simulated results obtained using PI controller and ILC controller.

Inference The speed ripple factor is given by the ratio of ω_{p-p} to the ω_{rated}. From the above simulation result, the ripple is reduced from 12.3 to 4.8%.

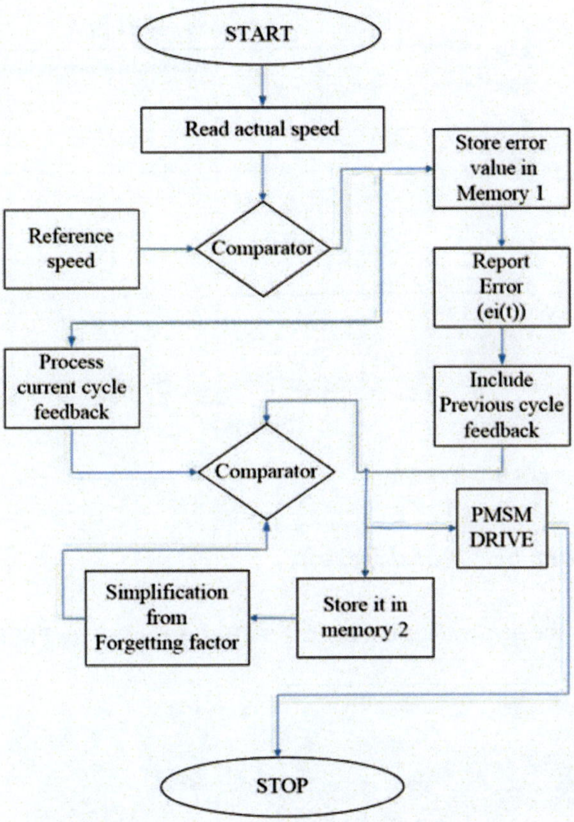

Fig. 3 Flowchart for ILC control

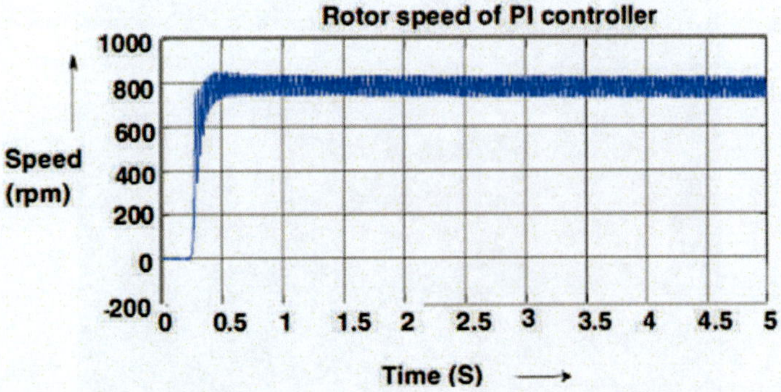

Fig. 4 Output speed by PI-SVPWM controller

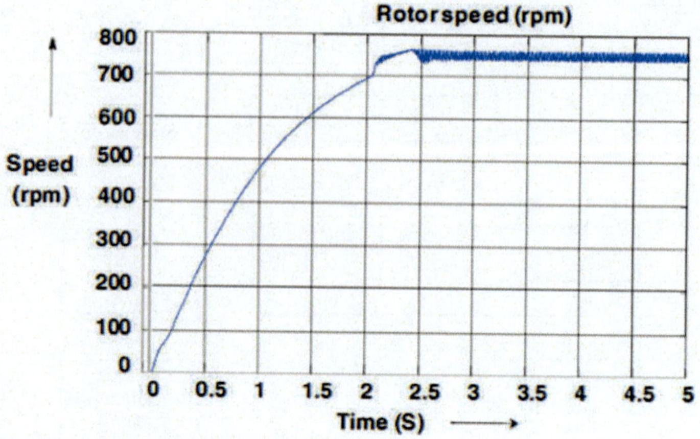

Fig. 5 Speed output by ILC-SVPWM

2.3 *Hardware Setup and Results*

Figure 6 shows the hardware setup of ILC-SVPWM control for PMSM motor.

Fig. 6 Hardware setup of ILC-SVPWM

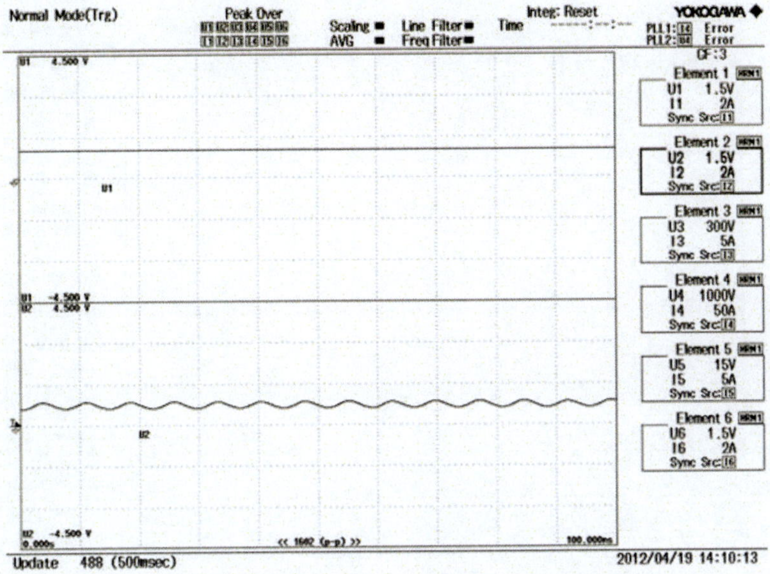

Fig. 7 Waveform of output voltage and currents for 50% loading at speed 2000 rpm

Output Waveform of voltage and current for 50% loading

See Fig. 7.

Waveform of output voltage and current for 100% loading

See Fig. 8.

Waveform of torque using PI-SVPWM

See Figs. 9 and 10.

SVPWM, line-to-line voltage without filtering and with filtering

Figures 11 and 12 shows the SVPWM, phase output voltage of inverter. The voltage value is 220 V at 2000 rpm.

Waveform of torque Using ILC-SVPWM

See Figs. 13 and 14.

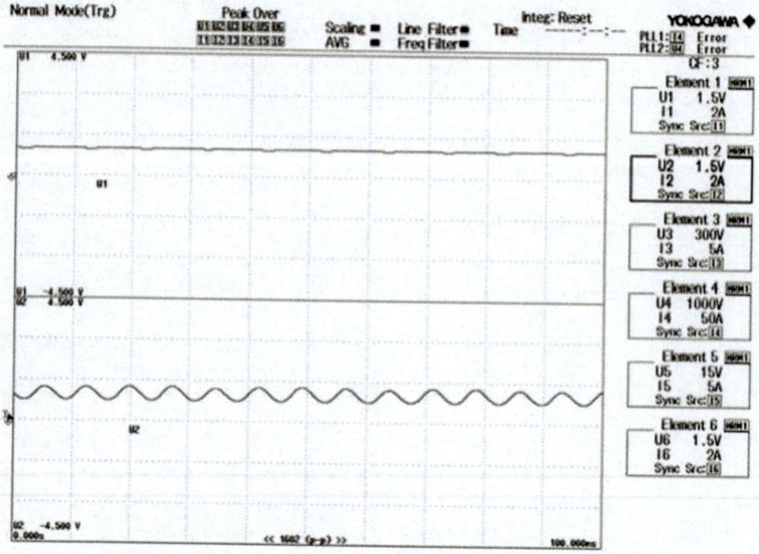

Fig. 8 Waveform of output voltage and currents for 100% loading at speed 2000 rpm

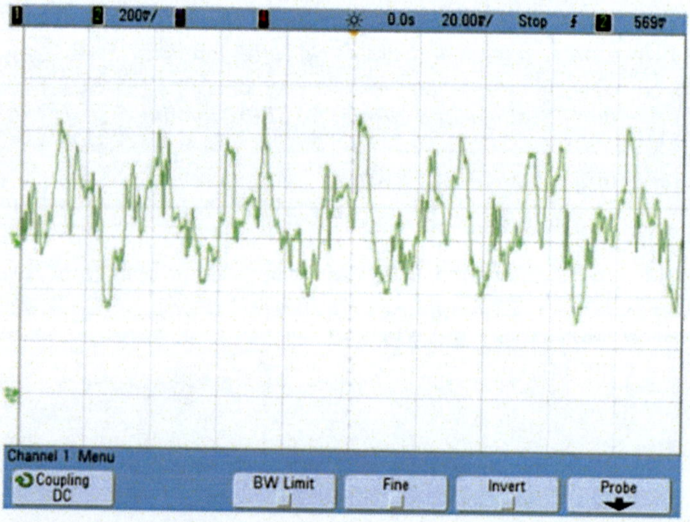

Fig. 9 Torque ripple in PI-SVPWM

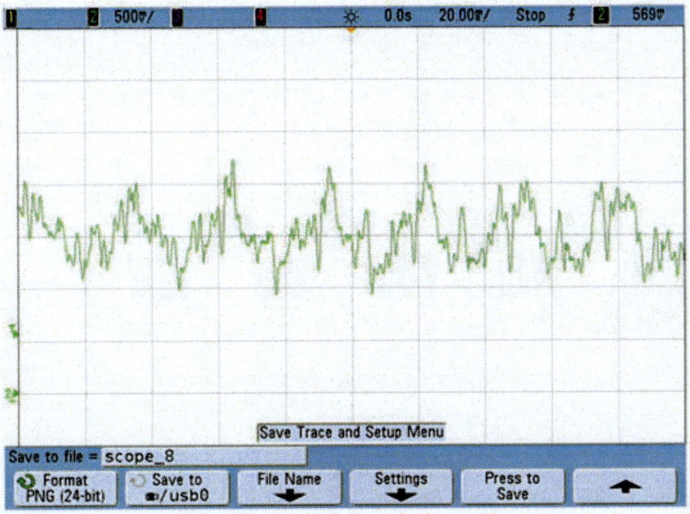

Fig. 10 Torque ripple in PI-SVPWM

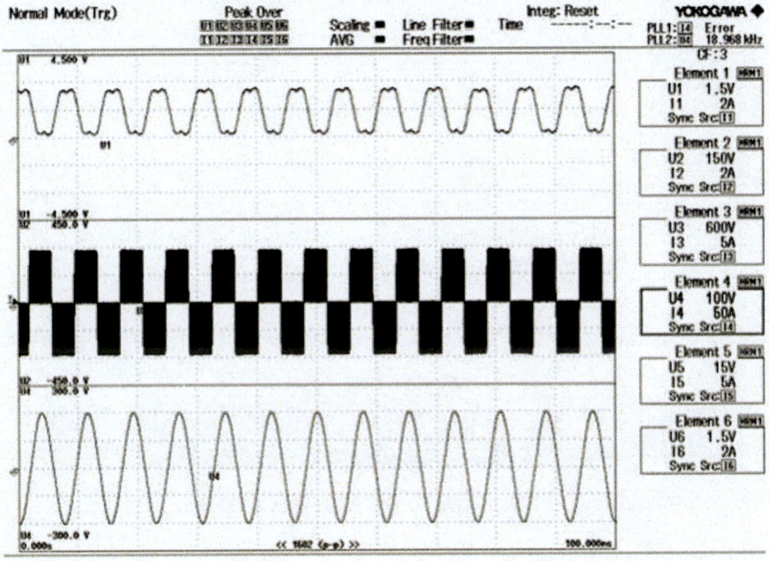

Fig. 11 SVPWM, voltage waveform

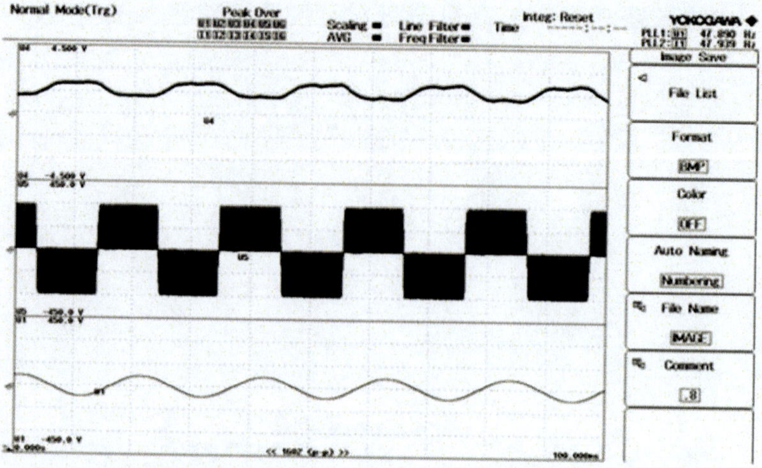

Fig. 12 ILC-SVPWM, voltage waveform

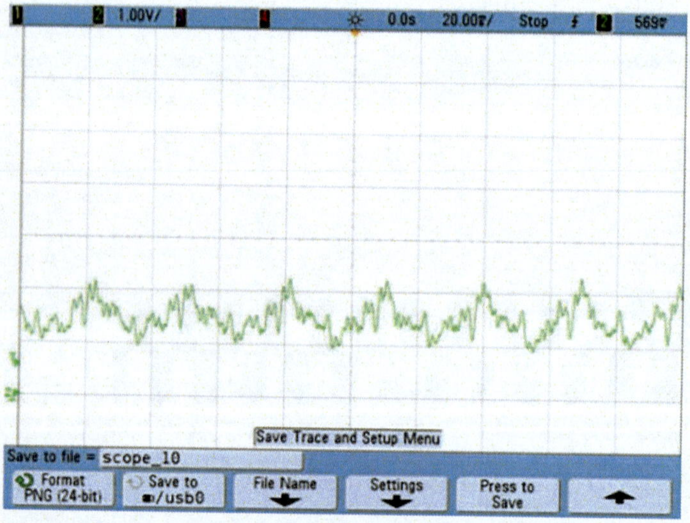

Fig. 13 Waveform of ILC-SVPWM torque at 50% loading at 2000 rpm

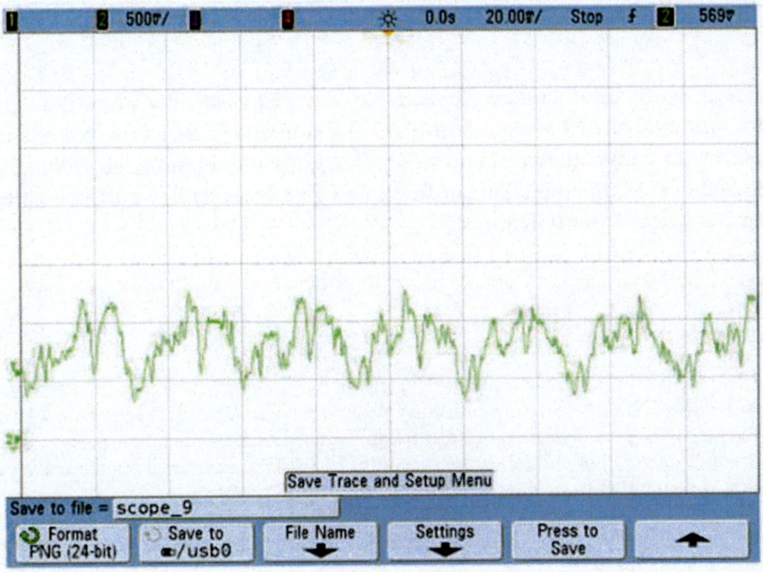

Fig. 14 Torque using ILC-SVPWM at 2000 rpm

System specification

PARAMETERS OF PMSM

Rated power	1 HP
Rated voltage	300 V
Rated current	2.5 A
Rated torque	8 Nm
Rated speed	750 rpm
Pole pairs	4
Stator resistance	0.2 Ω
Inductance l_d	0.0030 H
Inductance l_q	0.0030 H
Magnetic flux	0.175 Weber
Moment of inertia	0.090 Kg/m^2
Friction factor	0.089999(N.m.s)

3 Conclusion

Thus, from the result obtained, the ILC-based speed controller gives the enhanced performance over the PI speed controller. The ripple is reduced up to 7.5% by ILC compared with conventional PI controller. Thus, the model has been simulated. The desired output is implemented in hardware using DSP controller and the results show the improved ripple minimization.

References

1. Subha Lakshmi N, Adhavan B, Jagannathan V, Ravichandran CS (2013) Reduction of Transient and steady state speed pulsation of permanent magnet synchronous motor using space vector pulse width modulation control. In: Proceedings of IEEE international conference on circuit, power and computing technologies, ICCPCT 2013, pp 252–257
2. Zeng Z, Zhu C, Jin X, Shi W, Zhao R (2017) Hybrid space vector modulation strategy for torque ripple minimization in three phase for switch inverter fed PMSM drives. IEEE Trans Industr Electron 64(3):2122–2133
3. Panda SK, Xu JX, Qian W (2018) Review of torque ripple minimization in PM synchronous motor drives. IEEE
4. Subha Lakshmi N (2016) Hardware performance comparison of PI and space vector PWM techniques for speed ripple minimization in permanent magnet synchronous motor with and without iterative learning control. Int Res J Eng Technol (IRJET) 03(10):366–371
5. Subha Lakshmi N (2014) Comparative analysis of speed ripple minimization in permanent magnet synchronous motor using PI-ILC-hysterisis-SVM. Int J Appl Eng Res 9(26):8740–8743 ISSN 0973-4562
6. Subha Lakshmi N, Allirani S (2018) Realization and encompassment of torque and speed pulsations depreciation in permanent magnet synchronous motor employing ILC-Sine PWM. Int J Eng Technol 7(4.19):454–457
7. Lai C, Feng G, Mukherjee K, Loukanov V, Kar NC (2018) Torque ripple modeling and minimization for interior PMSM considering magnetic saturation. IEEE Trans Power Electron 33(3):2417–2428
8. Cho Y, Lee K-B, Song JH, Lee YI (2015) Torque ripple minimization of fast dynamic scheme for torque predictive control of permanent magnet synchronous motor. IEEE Trans Power Electron 30(4):2182–2190
9. Liu Qian, Hameyer Kay (2016) Torque ripple minimization of direct torque control of PMSM with modified FCSMP. IEEE Trans Ind Appl 52(6):4855–4864

Fault Classification in SRM Drive Using Hilbert Transform

Padala Lakshmi Sai Vineetha and M. Balaji

Abstract Switched reluctance motor (SRM) is widely used for variable speed applications due to its enormous advantages like simplicity, ruggedness, and lower cost. In variable speed applications, detection and diagnosis of faults are vital. This paper investigates the performance of SRM drive under faulty condition and proposes a Hilbert transform-based technique to diagnose the faults. The faults considered in this work include phase open-circuit fault, short-circuit fault, interturn fault with uniform and non-uniform turns, and phase-to-phase short-circuit fault. The performance of the motor under normal and fault conditions is analyzed using finite element analysis-based package MagNet 7.5. The features extracted from Hilbert transform will aid in detection and classification of faults in the motor.

Keywords SRM · Open-circuit fault · Short-circuit fault · Interturn fault · Phase-to-phase short-circuit fault · Hilbert transform

1 Introduction

SRM has been widely used in aerospace and automotive applications [1] due to its advantages like robust structure, reliability, and low production cost. In an industrial application, reliability of the drive is very important. One of the important features of SRM is its inherent fault tolerant capability. However, in the event of any malfunction, the performance of the drive needs to be monitored and corrective action should be initiated. The extent of performance deterioration depends on the nature and degree of the fault. The fault can occur in the power converters as well in the motor [2, 3]. In [4], the authors have described a procedure for classifying power converter fault. The remedial action for different types of the fault has been discussed in [5]. The literatures indicate that an effective fault diagnosis system needs to be developed to detect the

P. L. S. Vineetha · M. Balaji (✉)
Department of EEE, SSN College of Engineering, Kalavakkam, India
e-mail: balajim@ssn.edu.in

P. L. S. Vineetha
e-mail: vineedice@gmail.com

N. Zhou and S. Hemamalini (eds.), *Advances in Smart Grid Technology*, Lecture Notes in Electrical Engineering 688, https://doi.org/10.1007/978-981-15-7241-8_10

nature and location of the fault. A fast Fourier transform-based fault classification scheme has been discussed in [6]. In [7], the authors have developed a new algorithm to identify the location of fault. An online-based fault diagnosis method for detecting power converter fault has been discussed in [8]. The methods classify the fault based on the input from current sensors. A lookup table-based method for identifying and locating the faults has been elaborated in [9]. A model-based approach for diagnosing power converter fault has been discussed in [10].

The thorough analysis of literature indicates that reliability, complexity of the system, and response time are the major factors that influence the development of an effective fault diagnosis algorithm. In this context, this paper proposes a Hilbert transform-based procedure for classifying the faults in SRM. The Hilbert features extracted from the current profile form the basis of classification.

2 Electromagnetic Analysis of Switched Reluctance Motor

The structure of the 8/6 SRM is shown in Fig. 1. The main dimensions and specifications of the motor are shown in Table 1. The performance of the SRM under rated operating conditions has been analyzed using FEA-based software package MagNet 7.5. The flux lines at aligned and unaligned position of rotor are shown in Fig. 2. The inductance and torque characteristics are depicted in Figs. 3 and 4, respectively.

The dynamic characteristics have been obtained through circuit coupled simulation in FEA-based software package MagNet 7.5. The dynamic characteristics under normal operating conditions are presented from Fig. 5.

Fig. 1 Structure of 8/6 SRM

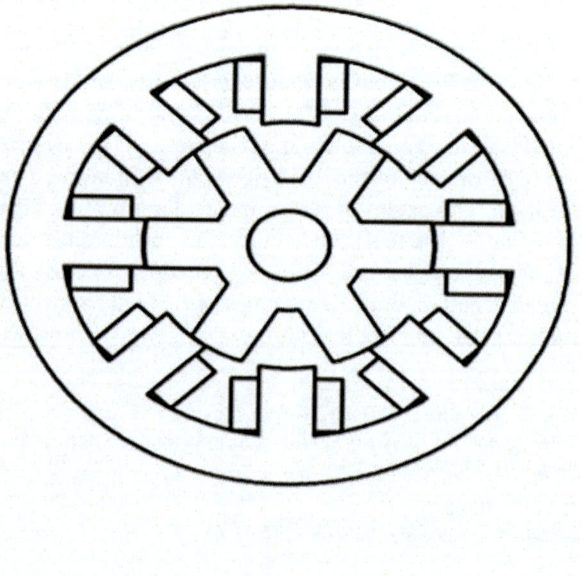

Table 1 Design specifications for 8/6 SRM

Design parameter	Value
Stator arc, β_s	21 degrees
Rotor arc, β_r	24 degrees
Air gap length, g	0.5 mm
Stator diameter, D_o	90 mm
Bore diameter, D	48 mm
Stack length, L	40 mm
Shaft diameter, D_{sh}	8.5 mm
Back iron thickness, C	11.25 mm
Height of stator pole, h_s	9.25 mm
Height of rotor pole, h_r	9 mm
Turns per phase	316
Rated current	4.5 A

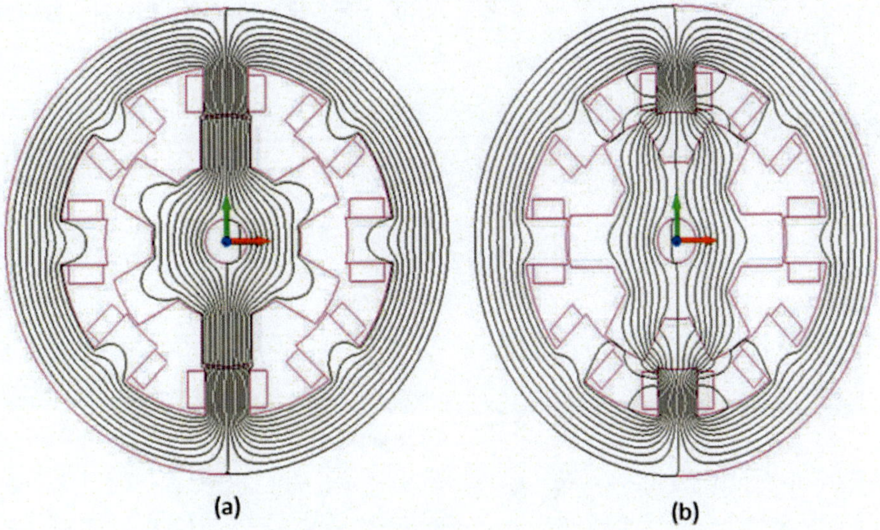

Fig. 2 Flux path at aligned and unaligned position

3 Performance Analysis Under Fault Conditions

The SRM drive is prone to fault in the machine and converter. In this work, the fault in the machine has been considered and the performance is investigated.

Fig. 3 Inductance profile

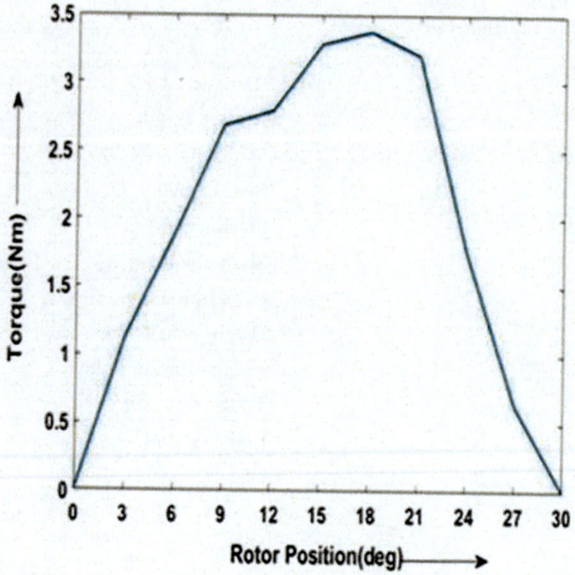

Fig. 4 Torque characteristics

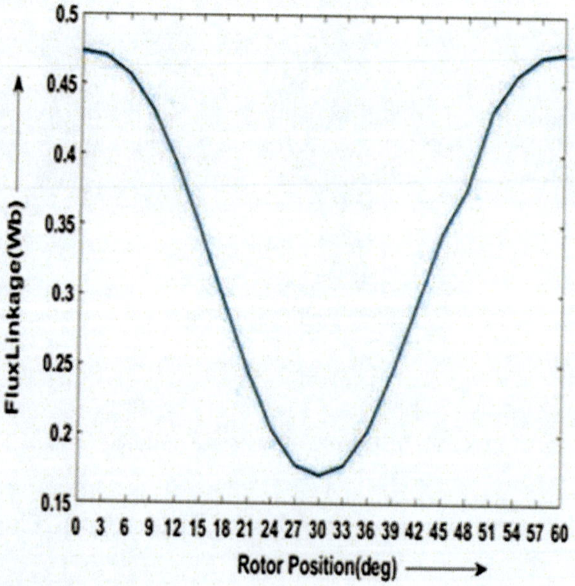

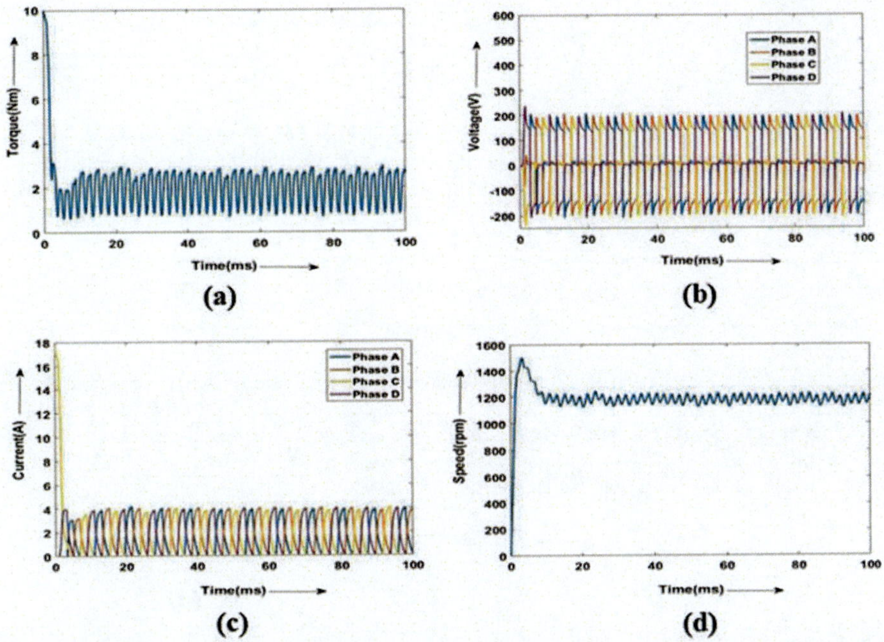

Fig. 5 Torque, voltage, current, and speed characteristics under normal operating condition

3.1 *Open-Circuit Fault*

The performance characteristics of SRM under open-circuit condition of a particular phase have been analyzed. Open-circuit fault has been created in phase C, and characteristics are investigated. The dynamic characteristics are shown in Fig. 6.

By analyzing the characteristics under open-circuit fault condition, it is evident that the speed decreases by 16.66% and torque output decreases by 16.49% with respect to normal operation.

3.2 *Short-Circuit Fault*

The performance characteristics of SRM under short-circuit condition of a particular phase have been analyzed. Short-circuit fault has been created in phase C, and characteristics are investigated. The dynamic characteristics are depicted in Fig. 7.

It is inferred from the characteristics that under short-circuit fault condition the speed decreases by 33.33% and torque output decreases by 27.83% with respect to normal operation.

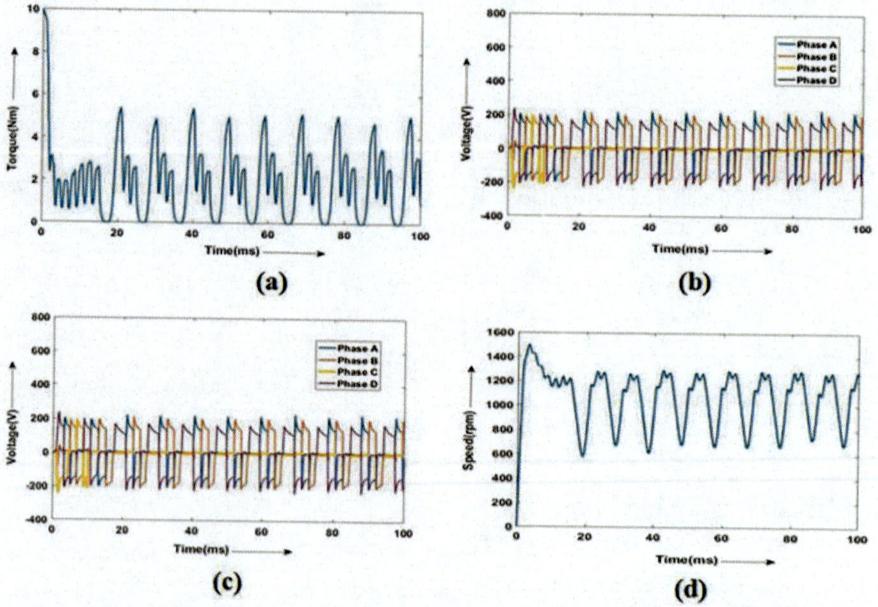

Fig. 6 Torque, voltage, current, and speed characteristics under open circuit fault condition

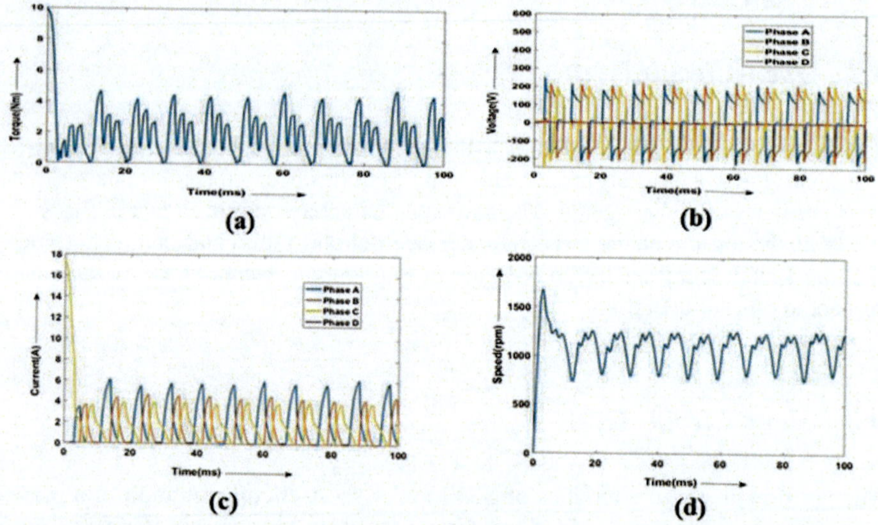

Fig. 7 Torque, voltage, current, and speed characteristics under the short-circuit fault condition

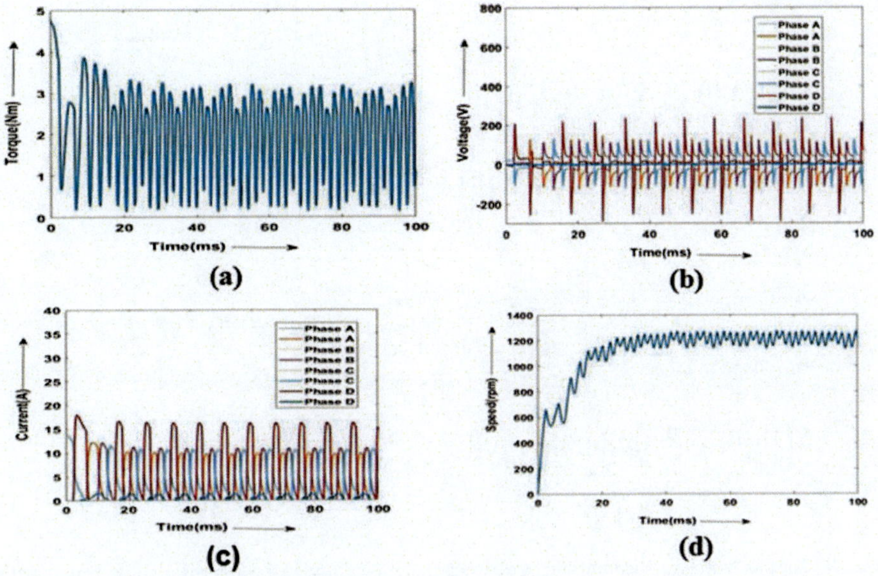

Fig. 8 Torque, voltage, current, and speed characteristics under the interturn short-circuit fault condition with uniform turns

3.3 Interturn Short-Circuit Fault with Uniform Turns

The performance characteristics of SRM under interturn short-circuit fault with uniform turns of a particular phase has been analyzed. Interturn short-circuit fault has been created in phase C, and characteristics are investigated. The dynamic characteristics are depicted in Fig. 8.

By analyzing the characteristics under interturn short-circuit fault condition with uniform turns, it is evident that the speed decreases by 8.33% and torque output decreases by 10.30% with respect to normal operation.

3.4 Interturn Short-Circuit Fault with Non-uniform Turns

The performance characteristics of SRM under interturn short-circuit fault non-uniform turns of a particular phase have been analyzed. Interturn short-circuit fault has been created in phase C, and characteristics are investigated. The dynamic characteristics are presented in Fig. 9.

It is evident from the characteristics under interturn short-circuit fault condition with non-uniform turns, the speed decreases by 25% and torque output decreases by 12.37% with respect to normal operation.

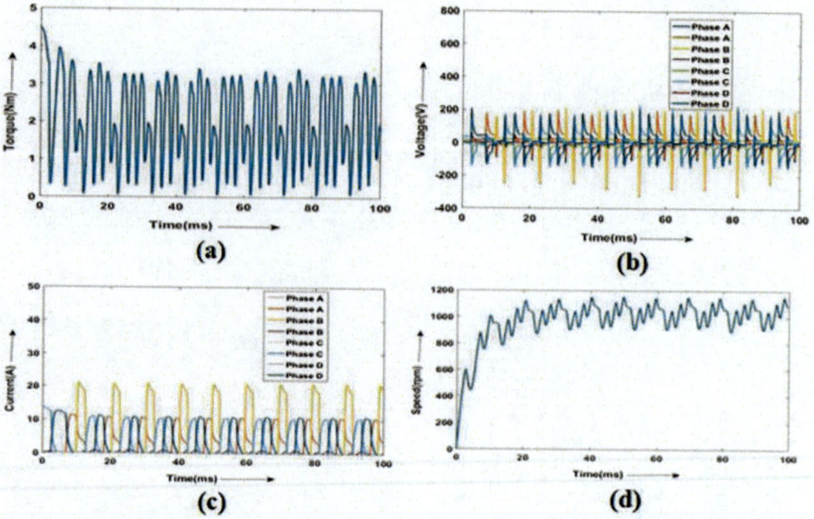

Fig. 9 Torque, voltage, current, and speed characteristics under the interturn short-circuit fault condition with non-uniform turns

3.5 Phase-to-Phase Short-Circuit Fault

The performance characteristics of SRM under phase-to-phase short-circuit fault of a particular phase have been analyzed. Phase-to-phase short-circuit fault has been created in phase C, and characteristics are investigated. The dynamic characteristics are shown in Fig. 10.

It is evident from the characteristics under short-circuit fault condition the speed decreases by 20.83% and torque output decreases by 22.68% with respect to normal operation.

The above analysis reveals that the performance of the motor deteriorates in the presence of fault and fault diagnostic routine needs to be designed to improve the performance of the motor.

4 Proposed Method of Fault Diagnosis Procedure

The fault signals are time-varying, and it is essential to apply transformation techniques to characterize the faults. This work engages the Hilbert transform to extract the features that help to characterize the different type of fault. The Hilbert transform is an adaptive method that produces physically meaningful representation of non-stationary and nonlinear signals. The final result is obtained by the convolution operation of Original signal $x(t)$ and Hilbert transform. The frequency component of

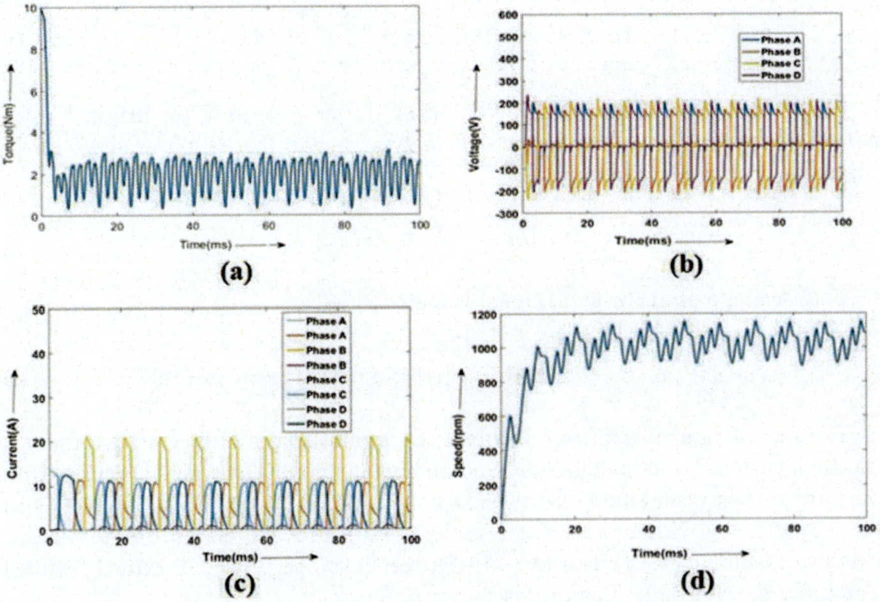

Fig. 10 Torque, voltage, current, and speed characteristics under the phase-to-phase short-circuit fault condition

the resultant signal lags the frequency of the original by 90° with the same amplitude of the original signal.

Hilbert transform can be defined as the convolution of the signal with the function $\left(\frac{1}{\pi t}\right)$ which emphasizes the local properties of the signal can be expressed as shown below:

$$H(x(t)) = \hat{y}(t) = \left(\frac{1}{\pi t}\right) * x(t) \tag{1}$$

$$H\big(\hat{y}(t)\big) = x(t) = \left(\frac{1}{\pi t}\right) * \hat{y}(t) \tag{2}$$

As $\left(\frac{1}{\pi t}\right)$ cannot be integrated, the integrals which define the convolution do not converge. Hilbert transform can be expressed as below:

$$H(x(t)) = \frac{1}{\pi} PV \int_{-\infty}^{\infty} \frac{x(\tau)}{(t-\tau)} d\tau = \frac{1}{\pi} PV \int_{-\infty}^{\infty} \frac{x(t-\tau)}{\tau} d\tau \tag{3}$$

where PV represents the Cauchy's principal value of the singular integral. The mathematical process used to generate complex signals from real signals is the discrete Hilbert transform.

$$x_H(t) = x(t) * \frac{1}{\pi t} = \frac{1}{\pi} \int_{-\infty}^{\infty} \frac{x(\lambda)}{t - \lambda} d\lambda \tag{4}$$

Since the output of the DHT is 90° phase shifter version of the original signal $x(t)$, a complete signal (also known as analytic signal) that is associated with the original signal can be written as

$$x_c[k] = x[k] + j x_H[k] \tag{5}$$

The envelope of the original signal is then defined as

$$|x_A[k]| = \sqrt{x^2[k] + x_H^2[k]} \tag{6}$$

The fault diagnosis approach involves the application of Hilbert transform. The transient current is a non-stationary and nonlinear signal. The signal is sensed from the current sensor, and the Hilbert transform is applied for that sensed signal. From the output of the Hilbert transform, features are extracted. Based on the features extracted, fault has been classified. The Hilbert envelope is used to extract features. The procedure for fault diagnosis is shown in Fig. 11.

The features considered for fault diagnosis are given below

- Feature 1: Mean

$$\text{Mean}, \mu = \frac{1}{\pi \sum_k x_A[k]} \tag{7}$$

- Feature 2: Standard deviation

$$\text{Standard Deviation, SD} = \frac{1}{N \sum_k \left(x_A[k] - 1/N \sum x_A[k]\right)^2} \tag{8}$$

- Feature 3: Maximum value

$$\text{Maximum Value, MX} = \max(|x_A[k]|) \tag{9}$$

- Feature 4: Kurtosis

$$\text{Kurtosis, KRT} = \frac{E(x_A[k] - \mu)^4}{\text{SD}^4} \tag{10}$$

Fig. 11 Flowchart of fault diagnosis approach

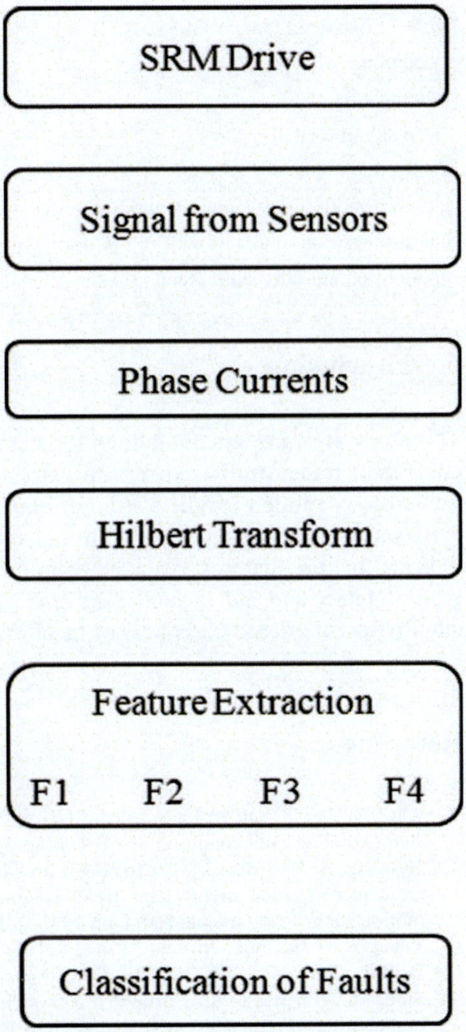

The values of various features extracted for the detection of type of fault are shown in Table 2. The values are tabulated with respect to phase current.

Table 2 presents the features extracted by applying Hilbert transform on the stator current. From the table, it is evident that the extracted features vary with respect to the fault and this aids in classification of the fault in SRM drive. The classification of faults can be done by employing rule-based classifiers or machine learning classifiers.

Table 2 Extracted features of SRM

Condition	$F1$	$F2$	$F3$	$F4$
Without fault	1.8489	1.7182	21.7338	34.9783
Open circuit fault	0.5640	1.4649	21.8876	96.29
Short-circuit fault	2.5688	2.1874	24.1772	27.48
Interturn short-circuit fault with uniform turns	2.0171	1.8116	19.4552	21.6615
Interturn short-circuit fault with non-uniform windings	5.030	3.897	15.389	1.7700
Phase-to-phase short-circuit fault	4.6136	3.7283	18.9712	2.0861

5 Conclusion

This study presents a methodology for stator winding fault detection in SRM based on Hilbert transform based feature extraction. In the proposed approach, Hilbert transform is applied for the stator currents and features are extracted. The features extracted are used to classify the normal and fault condition. The simulation results indicate that the features extracted from the Hilbert envelope differentiate different type of faults and aid in fault diagnosis. The fault classification can be done by employing rule-based classifiers or machine learning techniques.

References

1. Chen H, Han G, Yan W, Lu S, Chen Z (2016) Modeling of a switched reluctance motor under stator winding fault condition. IEEE Trans Appl Superconduct 26:4
2. Hye-Ung S, Kyo-Beum L (2016) Fault diagnosis method for power transistors in switched reluctance machine drive system. In: Proceedings of IEEE international power electronics and motion control conference, Hefei, pp 2481–2486
3. Zheng Z, Yong Zhi L, Jin Long S et al (2016) A method for power converter fault diagnosis. Beijing, Chine, October, Proc. Int. Conf. Aircraft Utility Systems, pp 62–67
4. Suresh G, Omekanda AM, Bruno L (2006) Classification and remediation of electrical faults in the switched reluctance drive. IEEE Trans Ind Appl 42(2):479–486
5. Nutan S, Choudhury DS (2016) Analysis of different types of faults exhibited in switched reluctance motor drives. In: Proceedings international conference intelligent systems and control, Coimbatore, India, pp 1–5
6. Chun G, Jianhua W, Shiyou Y et al (2015) Fault diagnosis scheme for open circuit faults in switched reluctance motor drives using AST Fourier transform algorithm with bus current detection. IET Power Electron. 9(1):20–30
7. Marques JF, Estima JO, Gameiro NS et al (2014) A new diagnostic technique for real-time diagnosis of power converter faults in switched reluctance motor drives. IEEE Trans. Ind. Appl 50(3):1854–1860
8. Hao C, Shengli L (2013) Fault diagnosis digital method for power transistors in power converters of switched reluctance motors. IEEE Trans Ind Appl 60(2):749–763
9. Sheng Li L, Hao C, Hui Z et al (2010) On-line fault diagnosis of power converter in switched reluctance motor. Proc. CSEE 30(3):63–70

10. Yang W, Gou B, Lei Y, Wang J (2018) Short switch fault diagnosis method for power converter using a model-based approach in switched reluctance motor drives. The 14th IET international conference on AC and DC power transmission (AC DC 2018)

Impact of Distributed Generation on Distribution System Under Fault and Islanding Condition

Pujari Harish Kumar, R. Mageshvaran, Guru Mohan Baleboina, and Koppola Vasavi

Abstract The energy drawn from the wind power plant and photovoltaics is considered to be essential sources in the recent distributed generation (DG) of electrical power. Since a day-to-day load demand for electricity is increasing rapidly and therefore the generation of power, needs to be improved for balancing the current demand. Because of this growing demand, non-renewable energy sources are on the brink of extinction. In order to solve this problem, the concept of integration of renewable energy sources has been introduced in the electrical distribution system and thereby improves the power quality and reliability to meet out the existing needs of the customers. Hence, the concept of installation of the DG has an impact on the operation and characteristics of the electrical distribution system. Installing the DG units near to the load centers may solve the basic problems such as power losses and voltage drops. The overall system is simulated in MATLAB environment in which, the three-phase grid-connected photovoltaic system is integrated with the parallel loads. The outputs are being measured in the form of the power, voltage and current at various points in the existing system and are analyzed in both the normal and faulty conditions. At Point of Common Coupling (PCC), variations in voltage and current under different types of fault conditions, intentional islanding and nuisance tripping of load cases are analyzed without protection device.

Keywords Protection coordination · Distribution grid planning · Distributed generation · Renewable energy source · Voltage regulation

---

P. Harish Kumar · R. Mageshvaran (✉) · G. M. Baleboina
School of Electrical Engineering, Vellore Institute of Technology, Vellore, Tamil Nadu 632014, India
e-mail: rmageshvaran@vit.ac.in

K. Vasavi
Department of Electrical and Electronics Engineering, East Point College of Engineering, Bangalore, India

N. Zhou and S. Hemamalini (eds.), *Advances in Smart Grid Technology*, Lecture Notes in Electrical Engineering 688, https://doi.org/10.1007/978-981-15-7241-8_11

1 Introduction

In general, the Distributed Generation (DG) is defined to be a technique by which a little amount of electricity is produced by using non-conventional energy resources and later distributed to the customer's end. The methodology of DG is a primary way of using renewable energy. The distribution system (DS) network performance is influenced in several ways by DG. DG is connected with the renewable energy sources to create a less environmental effect on power generation, which provides enormous scope for installation of more number of DG in a system in the future. Installation of DG in the system can make the voltage profile improvement of the system, but in turn, it can affect the voltage regulation [1]. If a DG is installed or placed close/near to the load, losses generated in the DS can be minimized. Increasing DG in a system can have significant impacts on the network distribution system [2]. Installing DGs in the distribution system will change the utility system configuration. This assists in a unique challenge for controlling, planning, protection, and managing of the system condition [3]. General, DG, coupled with the DS, is depicted in Fig. 1 [4, 5]. (a) Shows DG coupled in parallel to the grid, (b) Shows DG coupled in parallel to the grid by a switch.

- When DG coupled in parallel to the main grid under the normal conditions, the distribution system and DG provide power to the load and DG is reserved in the standby mode, which can avoid the power interruption.
- DG coupled in parallel to the grid by using a switch when the main power interruption occurs at grid side, and the switch will open and, in this state, DG will supply the power to the load.

This paper provides a DG impact on distribution systems based on the technical point of view. An effect of DG has become a significant concept which is to be analysed carefully because of the new DG installation at the distribution level. To extract maximum power, the MPPT regulator begins to control PV voltage through a varying duty cycle [6]. This paper addresses the monitoring of output DG (Photovoltaics) voltage, power, and power injected into the grid both in normal case and faulty case

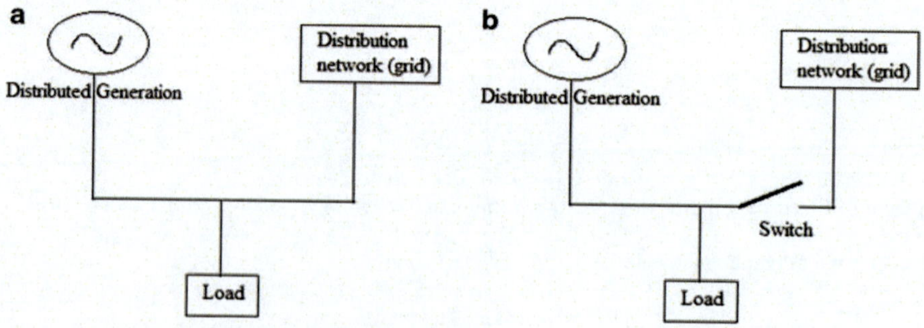

Fig. 1 **a** DG coupled in parallel to the grid. **b** DG coupled in parallel to grid by the switch

and also monitoring of fault current and voltage under different types of fault cases. Control of voltage and currents at PCC under the conditions of intentional islanding and nuisance tripping of DG units by adding and removing loads are analysed.

2 Types of DGs

DG is a mini power generating system, which provides a small amount of power to the customer loads through a utility distribution system [7]. The DG system has classified into two main groups; one is the inverter-based DG system and the second is a rotating machine DG system. DGs are classified as follows based on the active power injection and reactive absorption power in the DS [8]:

- DG injects only the active power into the distribution system (Photovoltaics, fuel cells, microturbines, and batteries).
- DG which injects only reactive power (Synchronous compensators).
- DG which injects the active power and absorbing the reactive power (wind turbine induction generators).
- DG injects both the active and reactive power into the distribution system (synchronous machines like combined heat and power generating units and gas turbines).

Reconfiguration of the distribution system with DGs mainly depends on critical factors as follows [9, 10]:

- The size and type of DGs connected to DS.
- DG units must be coupled based on voltage levels.
- Depending on previous generation levels, DG has to connect to the distribution system.

3 Impact of Distributed Generation Integration

DG uses, and integration can improve the power quality and power market reliability. Improvements in power quality and reliability are significant impacts of DG integration through the active distribution network. Installation of the DG in distribution system causes various effects on it, as presented in [11–13]. The installing and operating DG radially in the distribution system without any generation can have an impact on the voltage conditions and power flow both at customers and utility side [14, 15].

3.1 Impact of DG on Voltage Regulation

DS controls the voltage at substation with support of the load tap changing transformers. On the distribution feeders, voltage is being controlled by using the line regulators. By using the shunt, the capacitor voltage is controlled on feeders and along the transmission lines. The Voltage regulation relies mainly on the unidirectional power flow in the system, where the line regulators are fitted in a distribution system to compensate the line drop. The DG installation causes a change in the voltage profile along the feeder resulting in both the direction and magnitude of the active and reactive power flows in the system. The effect of DG installation on the voltage regulation factor can either be positive/negative based on the distribution system design, DG features and location.

3.2 Impact of DG on Power Loss

DG operating in a system shows an impact on the active power losses [15]. The power losses can be assessed for the low voltage distribution system and compared for different cases based on DG penetration and location. In minimizing the power losses in the distribution system, DG location and capacitor placement play a significant role. The difference between the position of DG and placement of capacitor is DG injects both active power and reactive power into the system, and capacitor banks only provide the system with reactive power [16]. In [17], the optimal power flow algorithms are used to analyse the impact of DG placement in the operation of the distribution system, taking into account the study of the voltage profile, power losses and system price.

3.3 Impact of DG on Harmonics

DG placement planning must be studied carefully for determining whether harmonics will be restricted within the DG position and fulfilling the IEEE-519 standard. In [18], the impact on system harmonics due to the installation of PV type in the distribution system has examined. In analyzing the impact of DG on system harmonics, the parameters considered are the level of penetration of DG insertion into the system, the number of DG units installed, and their location.

3.4 *Impact of DG on Short Circuit Levels of the Network*

Installing DG in the distribution system may increase protection issues, considering the network short circuit levels [19]. Variation in short circuit currents that are most subject to DGs network configuration, location and size. The short circuit level can increase the system fault currents when associated with the system normal operating conditions. The fault current impact from a small size, power generating unit is not large, but this will rise in the fault current at which it is located in the system. The effect of DG of faults current mainly is contingent on the factor like type, location and size of the DG from fault location.

3.5 *Islanding*

Islanding of the system occurs when single DG or group of DGs continually generate power to a part of the utility distribution system, even though grid power is not supplying [20]. Islanding can be harmful to utility workers, who may not know a circuit is still powered and may prevent automatic system re-connection. The solar panels will continue to deliver electricity in the event of a power outage as long as irradiance is adequate. The circuit disconnected by the disruption becomes an "Island". By operating upstream circuit breakers, fuses and automatic sectionalizing switch can separate the main utility system to form islanding cases. In [21], methods for finding the islanding of the DG through Multi-gene Genetic Programming has been proposed.

Islanding can be either unintentional islanding and intentional islanding. Intentional islanding means a state in which DG is continuously delivering the power to the loads when a disturbance occurs in the main utility system unfortunately to enhance the reliability of the system. The DG provides the power to required load demand until reconnection with the main utility system take place. Unintentional islanding is done during the maintenance of the system and/or failure of feeder's network, during these cases, DG is isolated and power to loads is supplied from the utility system [22].

4 Simulation Results

To measure the performance of the impact of PV type DG integration on the radial distribution system, consider a PV array with 66 parallel strings and series strings five characterized by a total output power of 100 kW simulated in MATLAB software. MATLAB model has depicted in Fig. 2 respectively. The following four cases are considered to identify the impact of PV type DG on the power system.

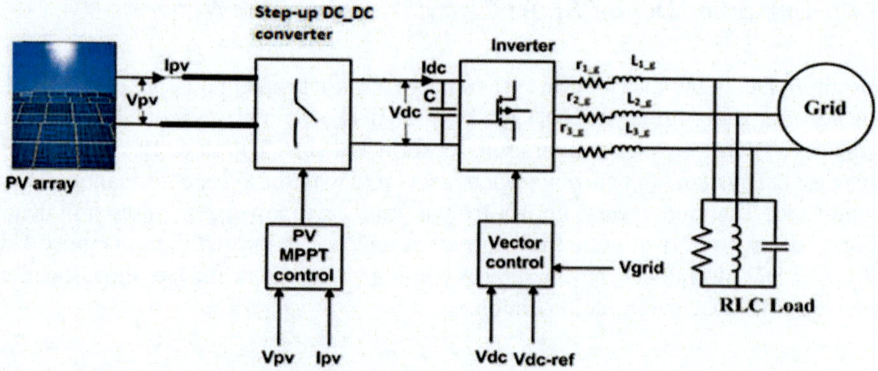

Fig. 2 Photovoltaics system connected to the grid in parallel with a load

4.1 Case 1—Performance Evaluation Under Normal Condition

The normal values of solar irradiation (1000 W/m^2) and ambient temperature (25 °C) are considered in this case study. In this case, in the normal operating conditions of the radial distribution system, PV type DG is coupled to the grid in parallel with the load. The active power is continually supplied to the load by the main utility system, and DG is maintained in standby mode to prevent power interruption. In Fig. 3 it can be seen, the waveforms of DG output power at an irradiance of 1000 W/m^2 and temperatures of 25 °C, boost converter input voltage, inverter input DC voltage, inverter output AC voltage and current, and power injected into the grid under normal operating condition. The results depict the amount of PV power generated at PCC is equal to active power, no impact on the system when DG is connected.

4.2 Case 2—Performance Evaluation Under Various Fault Conditions

In this case study, DG (PV) has coupled to the grid in parallel with the load is simulated by creating different types of short circuit faults within the system at PCC point. The following fault sequence is considered to the evaluation of the impact of PV type DG on the radial DS, with fault occur at $t = 2.1$ s and fault is cleared at $t = 2.8$ s having a fault resistance equal to 25 Ω. Under these faults occurring condition, the DG provides the power to the load. Figure 4 clearly shows the current and voltage waveforms at the PCC and fault voltage and current under the influence of LG, LLG, LLLG with fault occurs at $t = 2.1$ s and fault is cleared at $t = 2.8$ s. From the results, it is concluded that the effect of DG, due to faults depends mainly on the factors like the size of DG, the type of DG installed and the DG distance from the area of fault

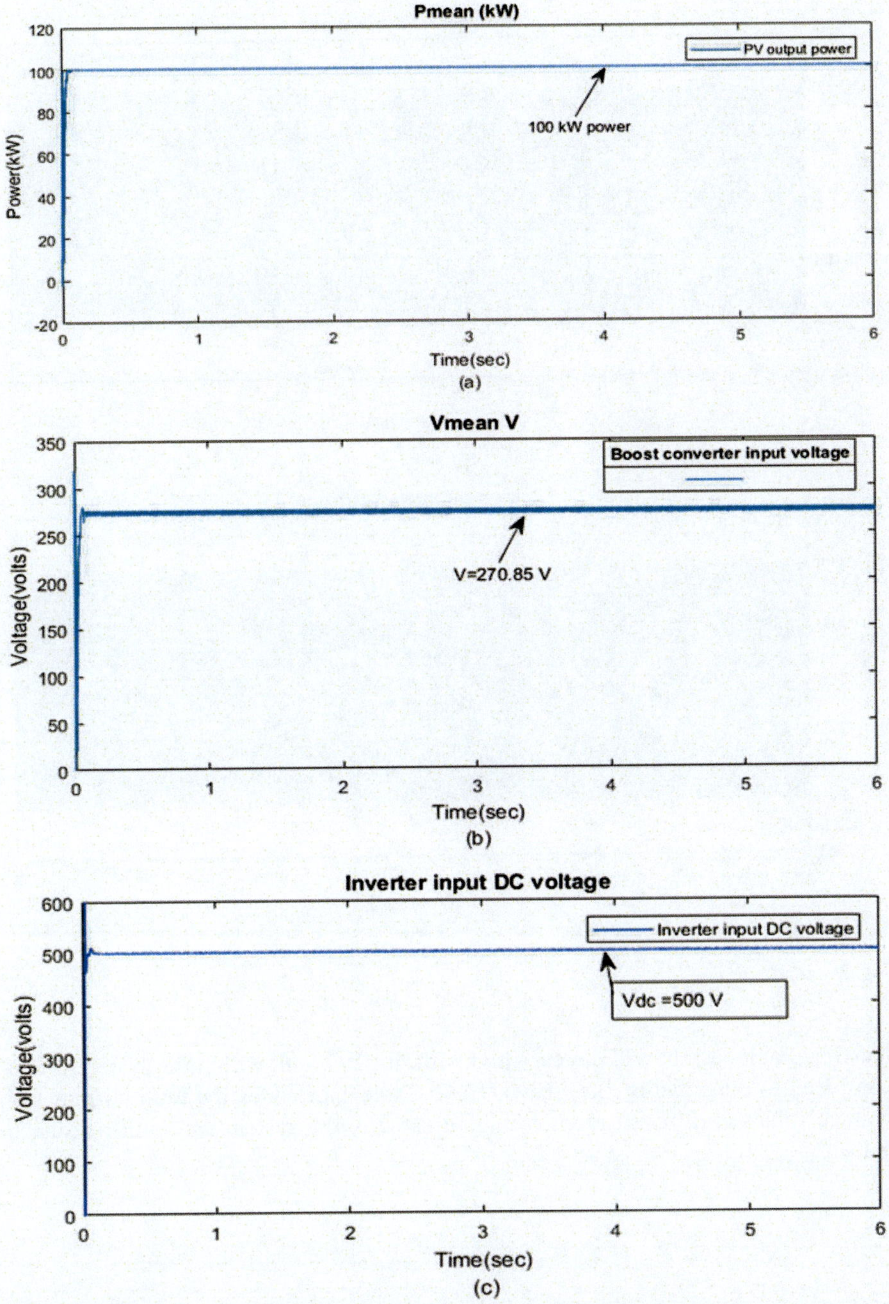

Fig. 3 Performance under normal conditions when DG (PV) has coupled to the grid in parallel with the load: **a** Generated PV output power. **b** Boost converter input voltage. **c** Inverter input DC voltage. **d** Output inverter voltage. **e** Output inverter current

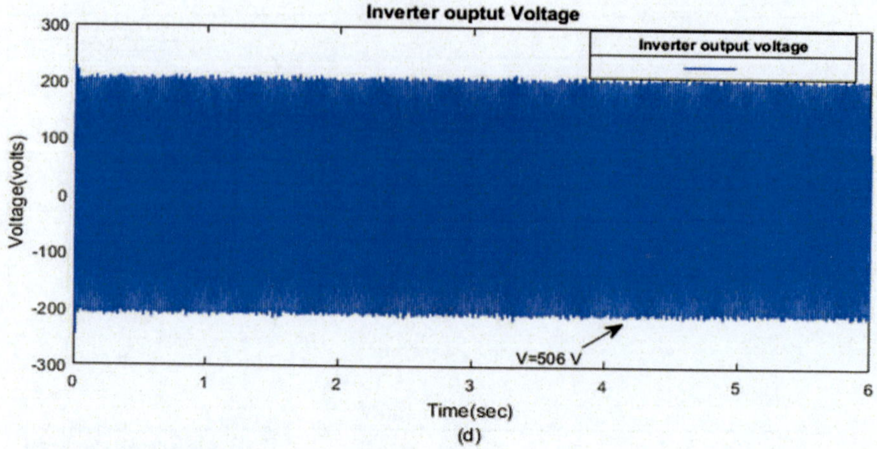

Fig. 3 (continued)

location. Fault current will increase highly during LG fault within the system, when compared to other faults like LLLG, LLG. Table 1 provides the fault current and voltage values at different types of fault created in the system and fault resistance values.

Fig. 4 Performance under various fault conditions when DG (PV) coupled to the grid in parallel with the load

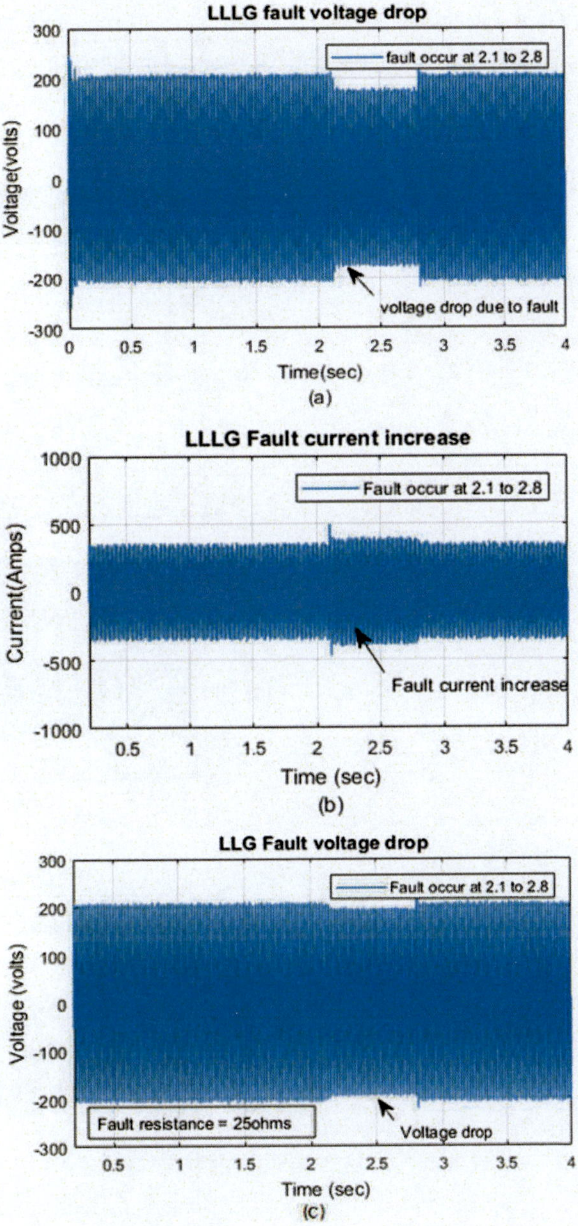

Fig. 4 (continued)

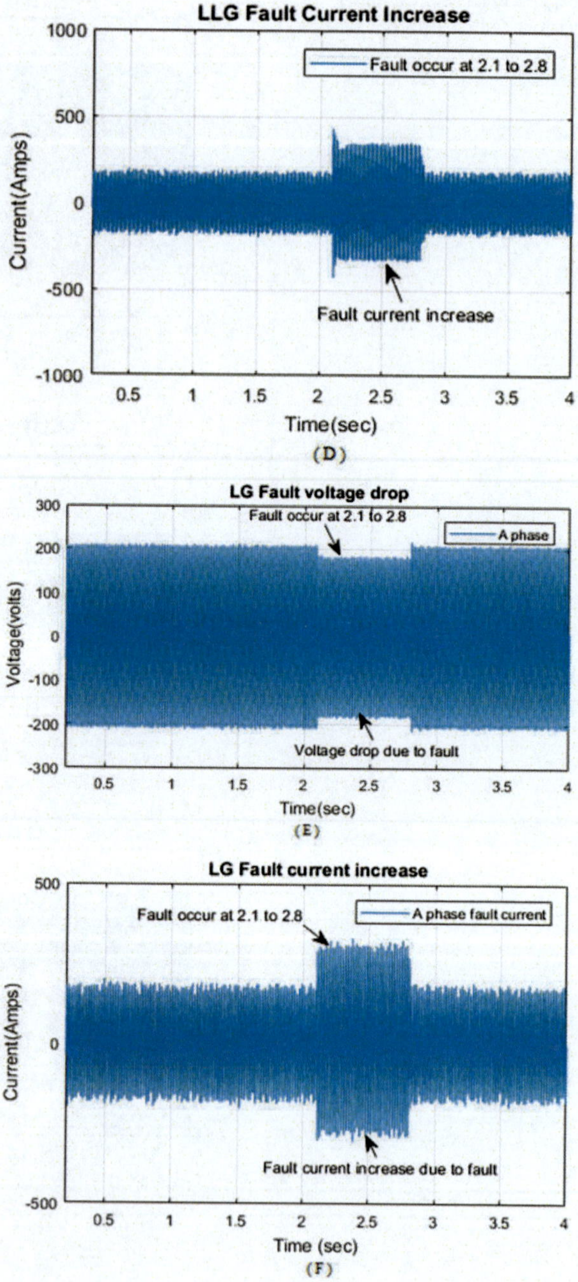

Table 1 Fault current and voltage values at different types of faults and fault resistance

Types of faults	Fault resistance value			
	$R = 25\ \Omega$		$R = 50\ \Omega$	
	Increased fault current value (A)	Dropped fault voltage (V)	Increased fault current value (A)	Dropped fault voltage (V)
LG	144.8	19.27	77.04	10.843
LLG	135.89	15.66	68.48	7.24
LLLG	94.27	36.13	59.92	18.07

4.3 *Case 3—Performance Evaluation Under Islanding Condition*

In this case study, PV type DG has coupled to the grid in parallel with the load, and intentional islanding has done. In this simulation condition, intentional islanding has performed by opening the DG at Pulse width = 20% and voltage, current at PCC have analysed. From results, it has concluded that after at 2 s no voltage and current is observed in the waveform at PCC point. In this state, the generating units should be able to sustain the load demand by maintaining appropriate voltage and the frequency level in the limit in the islanded system condition. Figure 5 shows clearly the results of the waveforms of current and the voltage at a PCC point under the intentional islanding case.

4.4 *Case 4—Performance Evaluation Under Nuisance Tripping of Loads*

In this case study, PV type DG has coupled to the grid in parallel with load, and a nuisance, tripping is created by adding and removing the load like the following sequence first load(20kW) is added at $t = 4.5$ s, again another load(40kW) is added at $t = 2.5$–3.5 s, and finally, yet the load(40 kW) is removed at $t = 3.5$–5 s, voltage and current at the PCC are measured. Figure 6 clearly depicts waveforms of voltage and current when the load is added and removed at PCC point. From the results, it has concluded that the voltage and current are suddenly increased with the injecting of load into the distribution system. Due to this DG power generation gets influenced, and the system goes into unbalances condition and reliability of the system also gets affected. Table 2 provides the PCC current and voltage values when load is added and removed suddenly. Table 3 provides description of simulink components used for modelling the system in Matlab.

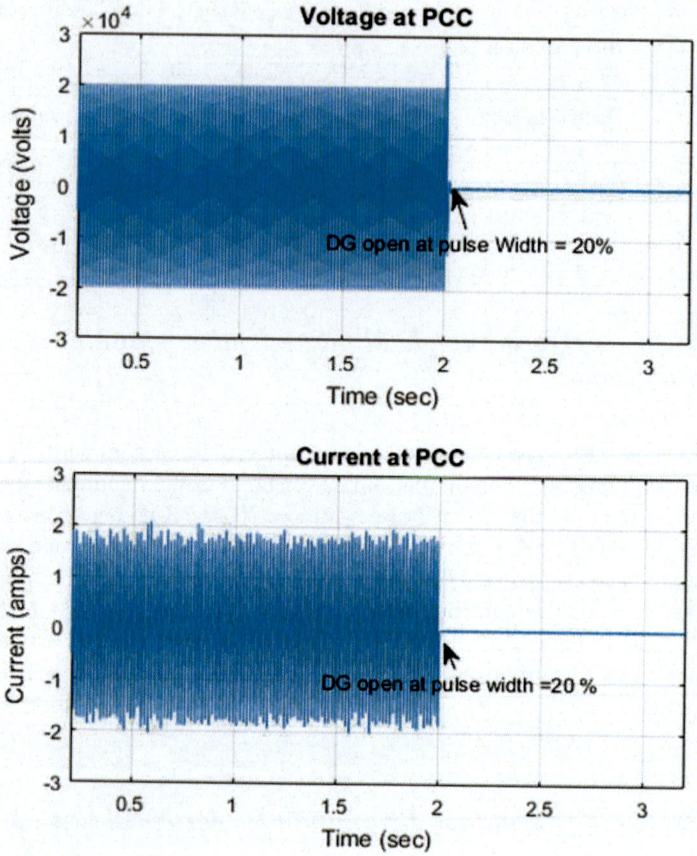

Fig. 5 Performance of the system under intentional islanding is created by opening the DG when DG (PV) coupled to the grid in parallel with the load

5 Conclusion

From the results obtained, it has been concluded that by interconnecting the renewable energy sources to the distribution systems, the overall current requirement to the load has to be shared both by the grid source as well as by the newly inserted DG. From the results obtained, the most severe voltage dip occurs near the PCC point is due to the occurrence of LLLG fault, which also raises the fault current level. The fault current rises in a system when an LG fault occurs at PCC point, which in turns cause a decrease in the voltage dip. The solution for the protection of the system at PCC point, due to insertion of DG or a sudden increase in load may increase fault current levels. This problem can be isolated by the insertion of a solid-state fault current limiter in series with the existing DG. The circuit breaker installed in the DS take a long time to detect the fault in the system. To overcome this delay, a fault

Fig. 6 Performance evaluation under nuisance tripping of loads (adding and removing) when DG (PV) has coupled to the grid in parallel with the load

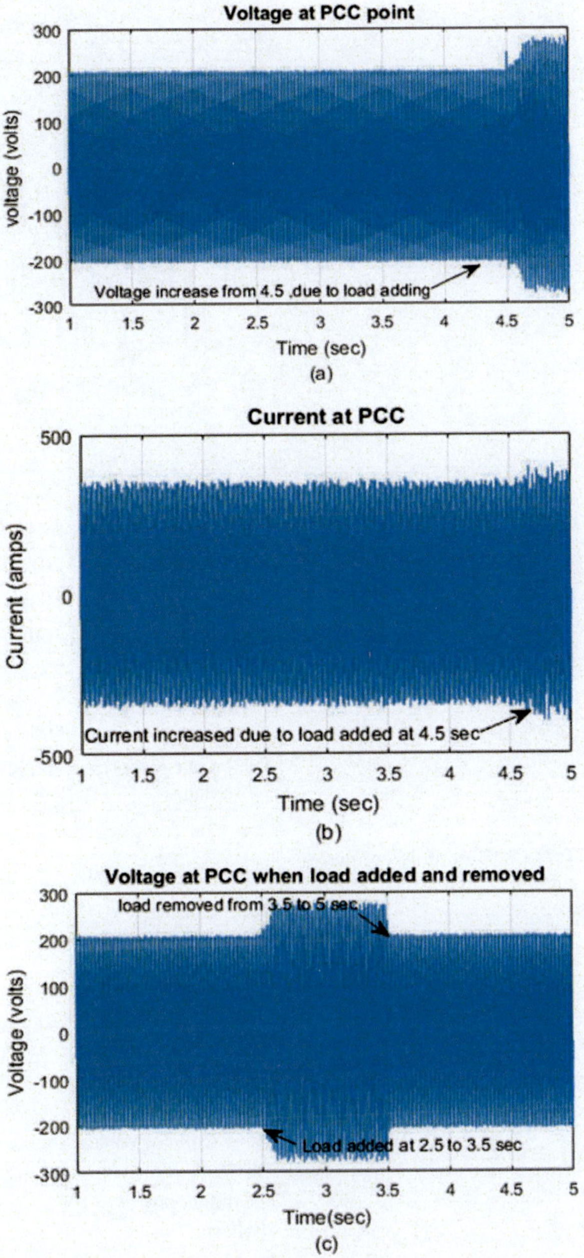

Fig. 6 (continued)

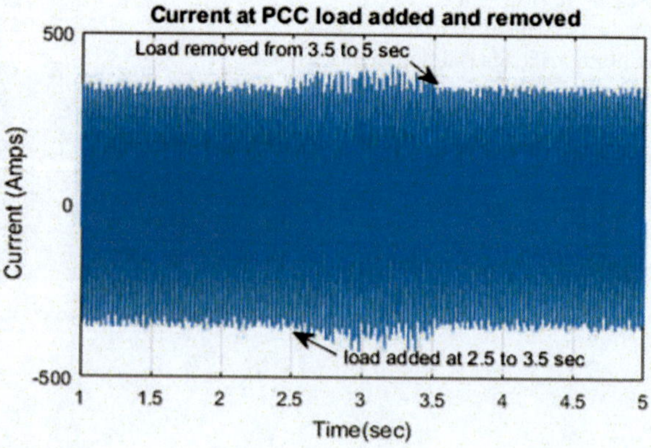

(d)

Table 2 PCC current and voltage values

Parameters	PCC Point voltage	PCC Point current
Load added	274.154 V	424.010 A
Load removed	201.154 V	334.241 A
Load value	60 kW	
PV power	100 kW	
Vdc (boost converter)	272.82 V	
Vdc inverter	500.15 V	
The output power of PV modules	100 KW, (66*5*305.226)	

Table 3 Description of the simulink components

Simulink components	Parameters
Boost converter	5 kHz DC-DC,273 V to 500 V DC
Filter values	$L = 5e{-3}$, $C = 100e{-6}$
Load value	50 kW
PV power	100 kW
Capacitor bank filtering harmonics	10 k var
3 level 3 phase voltage source converter	500.15 V to 260 V AC
The output power of PV Modules	100 kW (66 * 5 * 305.226)
Three-phase Coupling Transformer	100 kVA, 260/25 kV

current limiter known as superconductor fault current limiters have to be installed. The influence of DG current in the faulted area, miscoordination of circuit breakers have to be avoided, at the very first rise of short circuit current.

Acknowledgements The authors are extremely grateful to the Chancellor, Vice-Chancellor and Vice-Presidents, of VIT University, Vellore, for providing the excellent infrastructure facilities and encouragement which have made this research work possible.

References

1. Bhise DR, Kankale RS, Jadhao, S (2017) Impact of distributed generation on the protection of the power system. In: International conference on innovative mechanisms for industry applications (ICIMIA). Bangalore 2017, pp. 399–405
2. Quezada VHM, Abbad JR, Roman TGS (2006) Assessment of energy distribution losses for increasing penetration of distributed generation. IEEE Trans Power Syst 21(2):533–540
3. Sadeh J, Bashir M, Kamyab E (2010) Effect of distributed generation capacity on the coordination of the protection system of distribution network. In: IEEE/PES transmission and distribution conference and exposition, Latin America, Sao Paulo, pp. 110–115
4. Hou S, Gao Q (2011) Review of the impact of distributed generation on the distribution system. In: International conference on advanced power system automation and protection, Beijing, pp. 219–222
5. Cirrincione M, Pucci M, Vitale G (2008) A single-phase dg generation unit with shunt active power filter capability by adaptive neural filtering. IEEE Trans Ind Electron 55(5):2093–2110
6. De Brito MAG, Sampaio LP, Luigi G Jr, Guilherme A, e Melo, Canesin CA (2011) Comparative analysis of MPPT techniques for PV applications. In: 2011 International conference on clean electrical power (ICCEP), pp. 1–6
7. Adefarati T, Bansal R (2016) Integration of renewable distributed generators into the distribution system: a review. IET Renew Power Gener 10(7):873–884
8. Razavi SE, Rahimi E, Javadi MS et al (2019) Impact of distributed generation on protection and voltage regulation of distribution systems: a review. Renew Sustain Energy Rev 105(C):157–167
9. Roy NK, Pota HK (2015) Current status and issues of concern for the integration of distributed generation into electricity networks. IEEE Syst J 9:933–944
10. Prakash P, Khatod DK (2016) Optimal sizing and siting techniques for distributed generation in distribution systems: a review. Renew Sustain Energy Rev 57:111–30
11. Zayandehroodi H, Mohamed A, Shareef H, Mohammad Jafari M (2011) Impact of distributed generations on power system protection performance. Int J Phys Sci 6:3873–3881
12. Balamurugan K, Srinivasan D, Reindl T (2012) Impact of distributed generation on power distribution systems. Energy Procedia 25:93–100
13. Khani D, Yazdankhah AS, Kojabadi HM (2012) Impacts of distributed generations on power system transient and voltage stability. Int J Electri Power Energy Syst 43:488–500
14. Barker PP, de Mello RW (2000) Determining the impact of Distributed generation on power systems: part 1–radial distribution systems. In: Proceedings IEEE Power Engineering Society Summer Meeting, pp. 1645–1654
15. Chiandone M, Campaner R, Pavan AM, Sulligoi G, Piccoli G (2014) Impact of distributed generation on power losses on an actual distribution network. In: International conference on renewable energy research and application (ICRERA), Milwaukee, WI, pp. 1007–1011
16. Borges CLT, Falcao DM (2006) Optimal distributed generation allocation for reliability, losses, and voltage improvement. Int J Electri Power Energy Syst 28(6):413–420

17. Singh B, Gyanish BJ (2018) Impact assessment of DG in distribution systems from the minimisation of total real power loss viewpoint by using optimal power flow algorithms. Energ Rep 4:407–417
18. Saed JA, Quraan M, Samara Q, Zizzo G (2017) Impact of integrating photovoltaic based DG on distribution network harmonics. In: 2017 IEEE international conference on environment and electrical engineering, pp. 1–5
19. Uqaili MA, Sahito AA, Halepoto IA et al (2014) Impact of distributed generation on network short circuit level. In: 2014 4th International conference on wireless communications, vehicular technology, information theory, and aerospace and electronics systems (VITAE), Aalborg, pp. 1–5
20. Morsi WG, El-Hawary ME (2019) On the appropriate monitoring period for voltage flicker measurement in the presence of distributed generation. Electri Power Syst Res 79(4):557–561
21. Pedrino EC, Yamada T et al (2019) Islanding detection of distributed generation by using the multi-gene genetic programming-based classifier. Appl Soft Comput 74:206–215
22. Tran T, Nguyen DT, Fujita G (2018) Islanding detection method based on injecting perturbation signal and rate of change of output power in dc grid-connected photovoltaic system. Energies 11(5):1313–1319

Enhancement of Power Quality in a 3ph-3bus Distribution System with Unified Power Quality Conditioner

S. K. Abdul Pasha and N. Prema Kumar

Abstract The main objective towards electrical-distribution system is to experience the consumer's demands for energy later acquiring the huge electrical-energy from the transmission/substation. Various network compositions remain feasible to suit the needed supply reliability. Protection, control and supervising apparatus are implemented to empower the adequate process of distribution network. This effort mainly analyzes on 3phase, 3bus distribution system by introducing an UPQC with various load configurations. The outcomes are correlated as regards real-power, reactive power, voltage, current-THD. The Significance of the present work is to upgrade the quality of power for a 3-ph, 3-bus distribution network by introducing an UPQC with various load configurations.

Keywords Three phase three bus distribution system (TTBDS) · THD · DVR · AF · SVM

1 Introduction

Today, the commonly used parameter in case of 'power systems' is the sustention of quality of power. Later developing voltages, the power-engineers are feeling hard to transmit as well as distribute power to the load end, on account of various loads from the point of distribution to be more conscious to the interruptions in the voltage and harmonics. Bharat [1] introduced an ideal procedure for designing a 'unified power-quality conditioner' (UPQC) model along with least conceivable Volt Ampere rating depends on the 'compensation' requirements. A unique design procedure, analogous-algorithm were recommended to sort the extensive elements in an 'UPQC', such-like the 'series-inverter', 'shunt-inverter', and 'series-transformer' equivalent to the minimal feasible entire V-A rating. Sachin and Bhim [2] presented for designing

S. K. A. Pasha (✉)
Anurag Engineering College (A), Kodad, Telangana, India
e-mail: abdulpsh@gmail.com

N. P. Kumar
Andhra University College of Engineering (A), Vishakhapatnam, Andhra Pradesh, India

N. Zhou and S. Hemamalini (eds.), *Advances in Smart Grid Technology*, Lecture Notes in Electrical Engineering 688, https://doi.org/10.1007/978-981-15-7241-8_12

and outcome study of a 3phase one stage 'solar-photovoltaic integral unified power quality conditioner' (SPV-UPQC). The 'SPV-UPQC' consisted of a shunt-series voltage balancers tied end to end with usual 'dc-link'.

An innovative procedure to place, integrate and monitor of 'unified power quality conditioner' (U.P.Q.C.) in distributed-generation (D.G) depended grid tied/'autonomous micro grid/micro-generation (μG)-model will be represented by Shafiuzzaman K. Khadem [3]. Popular design and control implementation of UPQC-based on adjustable phase angle monitoring method was presented by Jian-Ye et al. [4]. For this, the 'optimal volt-ampere' (OVA) ratings of the converters in the U.P.Q.C. were examined due to system compensation needs. The 'phase angle control (P.A.C.)' technique was reviewed and adorned to have the aspect of altering the online-VA loading by accommodating the analogous 'displacement-angle'.

Sachin and Bhim [5] proposed an altered 'pq-theory' dependent control of a solar-photovoltaic (PV) array-integrated unified power quality conditioner (SPV-U.P.Q.C.). The model incorporated elegant energy production along with enhanced quality of power by that the increasing performance of the system. Jian Ye et al. [6] investigated the suitable measures of the U.P.Q.C. model depends on the compensation needs. A well defined approach was defined to revise the measures of the U.P.Q.C. model, which evaluates the primary ratings of the shunt-series converter, and series- transformer.

Shubh lakhsmi and Sanjib [7] presented the modeling, pointing approach for 'open-unified power quality conditioner' (O-U.P.Q.C.) integrated-photovoltaic (P-V) resemblance of the system in radial- distribution configurations to raise the energy-efficiency and 'PQ'. Flexible parallel filtering and monitoring of 'U.P.Q.C. for enhanced non linear-loads was presented by Saurav-Roy-Choudhury et al. [8]. The remaining capability power of U.P.Q.C. (later swell-sag compensations) was outlined for mitigation of harmonics.

Sachin and Bhim [9] proposed a monitoring process depends on 'second-order generalized integrator' (SOGI) and 'deferred signal cancellation' (DSC) for enhanced monitoring of one-phase 2-stage 'solar-photovoltaic-array-integrated-unified power quality conditioner'(SPV-U.P.Q.C.). Leonardo et al. [10] presented accurated study's affecting 'sizing, stability-analysis and power-flow via series, parallel power-converters in a multi-functional 3-phase Distributed-Generation (DG) model comprises of a one-stage photovoltaic (OPV) system incorporated into an 'unified power quality conditioner' (U.P.Q.C.).

Sergio and Fernando [11] presented the deployment of a usual 3-phase 4-wire (3P4W) distribution network utilizing a 1–3phase 'unified power quality conditioner (U.P.Q.C.) topology', called UPQC-1Phase-to-3Phase. A 1-phase transformer less unified power quality conditioner (OPTL-UPQC) was implemented by Victor et al. [12]. Away from having no isolation-transformer, the recommended structure uses 4 switching gadgets only, forming 2 half-bridge voltage-source inverters-1 connected in parallel with the load and another one linked in series with the ac-mains.

Shubh lakshmi presented an 'on-line operational optimization procedure' to find the optimal 'reactive-power/reactive volt-ampere (VAr) set points' for "open unified power quality conditioner (O-U.P.Q.C.)" by changing load-demand of a distribution-network. Shubh lakshmi presented a dual-objective planning scheme to optimal placement of 'open-unified-power-quality-conditioner (O-U.P.Q.C.) by concurrently synthesizing the "photo-voltaic-hosting-capacity (PVHC)" and energy-loss of 'distribution systems'. A non-linear 'discrete-time' model over an superlative controller utilising an 'unified power quality conditioner (U.P.Q.C.)' was proposed by Hamidreza Nazaripouya for weak/islanded grids.

2 Research Gap

The exceeding literature does not deals with the improvement of power-quality in a 3-ph, 3-bus distribution model. Hence, present work deals with the power quality enhancement for a three-phase, three-bus distribution network with the introduction of an SVM inverter based UPQC topology.

3 System Description

3.1 One Line Diagram of a Three Bus Scheme

The 1-line 'schematic-model' for three-bus-system will be delineated in Fig. 1. In this case bus 1, bus 2 will be the generator-buses where as bus '3' will be a load bus connected by linear, non linear loads.

3-bus model is considered, power flow is carried for finding the real-reactive powers, current THD's through the lines 2-1 and 1-3. When load flow runs between the lines 2-1 and 1-3, power flow is less in the line 2-3 and THD for the AC source current of non linear load is high.

Fig. 1 One-line-diagram for 3bus system

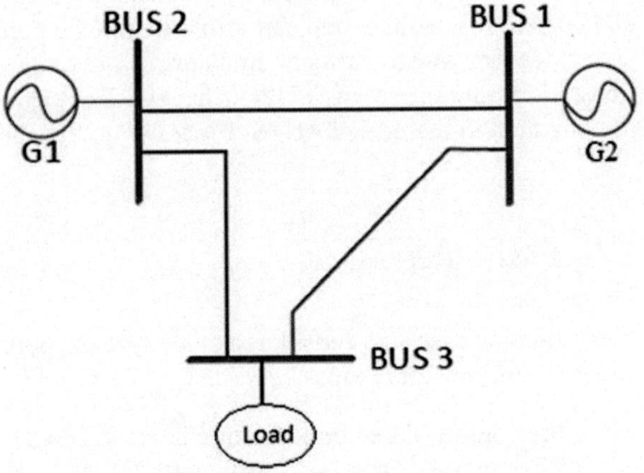

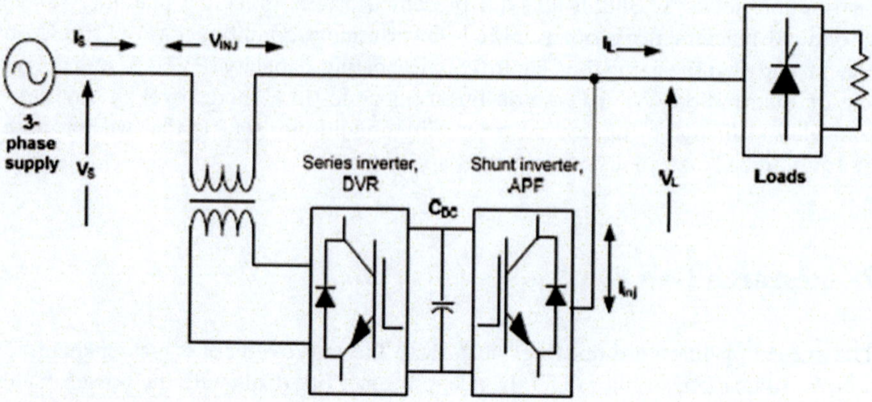

Fig. 2 Circuit diagram of UPQC

Table 1 UPQC-system parameters

Element	Value
Source	3-phase, 415 V, 50 Hz
Shunt active power filter	$L = 0.02$H; $C = 0.08$ μF
Series active power filter	$L = 0.09$H; $C = 10$ μF
Transformer capacity	1:1150kVA
DC-link capacitor	2000 μF

3.2 *Configuration of UPQC*

UPQC have been proposed in recent years for enhancement of power quality in the electrical distribution-systems. Development in manufacturing of power semiconducting devices have led to better characteristics such as higher voltage rating, current rating and at the same instant higher switching frequency. Besides to repress voltage and current distortions, one can also cope up with the power ability disputes such as voltage-sag, swell, currents, imbalances in voltages, 'flickers, frequency oscillations, interruptions, surges". UPQC has superior sag/swell compensation capability, as compared to the other FACTS devices (Fig. 2) (Table 1).

3.3 *Case Studies*

The following cases are considered to analyze the performance of a three bus system with various considerations.

(i) three phase, Three bus system without UPQC.
(ii) Three phase, Three bus system with UPQC.

(iii) Three phase Three bus system with SVM inverter based UPQC (Fig. 3).

(i) **Three phase, Three bus system without UPQC:** The basic representation for a 3-ph, Three-bus model is presented at Fig. 4 which is connected by two generators and load. This section mainly deals with modeling and simulation of Three-phase three bus-systems without UPQC. Linear and non-linear loads will be linked at load bus (Figs. 5, 6, 7) (Tables 2, 3, 4, 5).

Location of UPQC: Load flow indicates that less power flows through the line 2-3 when compared with lines 2-1 and 1-3. Hence location of UPQC is identified between the buses 2-3. Now, load flow is performed with UPQC between the buses 2 and 3.

(ii) **Three phase, Three bus system with UPQC:** The 'single-line' diagram for a three-bus model with UPQC be delineated (Figs. 8, 9, 10, 11, 12, 13) (Table 6, 7).

(iii) **Three phase, three bus system with SVM inverter based UPQC:** Circuit diagram of Three phase 3-bus model along with SVM-Inverter based UPQC is shown below. SVM- method has the benefit of an ideal-output and also decreases the harmonic-content of the receiving end voltage/current (Figs. 14, 15, 16, 17) (Table 8, 9, 10, 11).

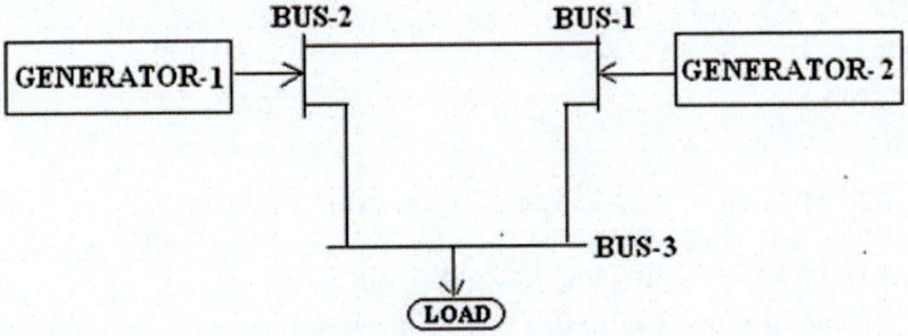

Fig. 3 Block-diagram for 3-phase, 3bus scheme

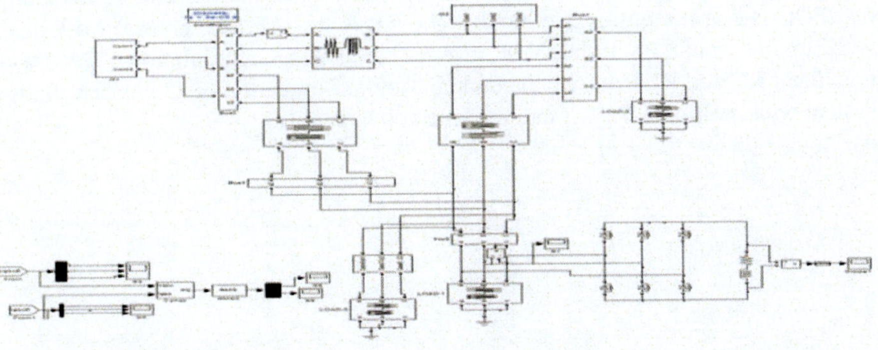

Fig. 4 Circuit diagram of three phase 3-bus system without UPQC

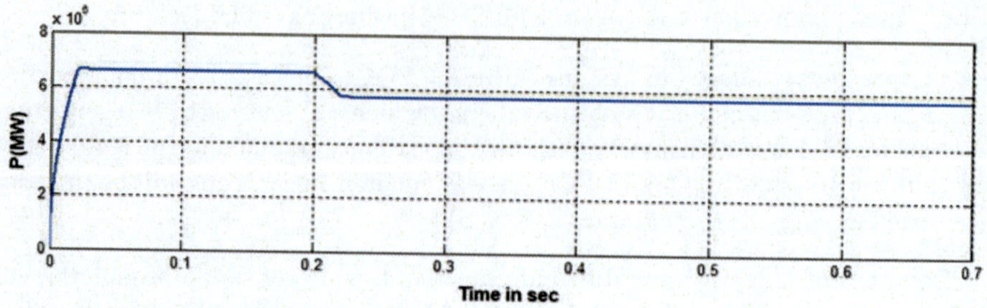

Fig. 5 Real-power in third bus

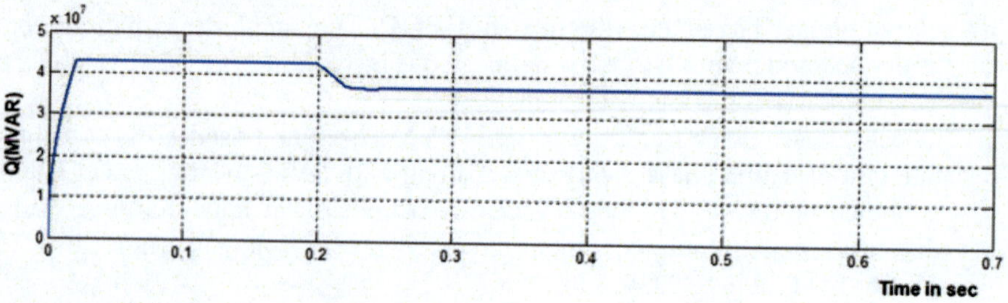

Fig. 6 Reactive-power in third bus

4 Conclusion

From the case studies, it is observed that the value of THD reduces from 12.59% Without UPQC to 2.86% of SVM-inverter based UPQC. The superior voltage response of 10.08 kV of with SVM inverter based UPQC is obtained as compared to all the cases. The improved values of real & reactive powers obtained as 6.830 MW and 52.31 MVAR respectively. The contribution of this work is to employ SVM-UPQC in T.B.S and reduce THD to acceptable levels. The advantages of UPQC are voltage-sag-compensation ability and harmonic attenuation. The disadvantage of UPQC is that it requires two blocks—i.e. DVR and AF. The present work deals with simulation of TBS with S.V.M. inverter based UPQC. Simulation of TBS with non linear RC and RL loads can be done in future. Simulation of 33 bus distribution system with multiple UPQC's can be done at a later date.

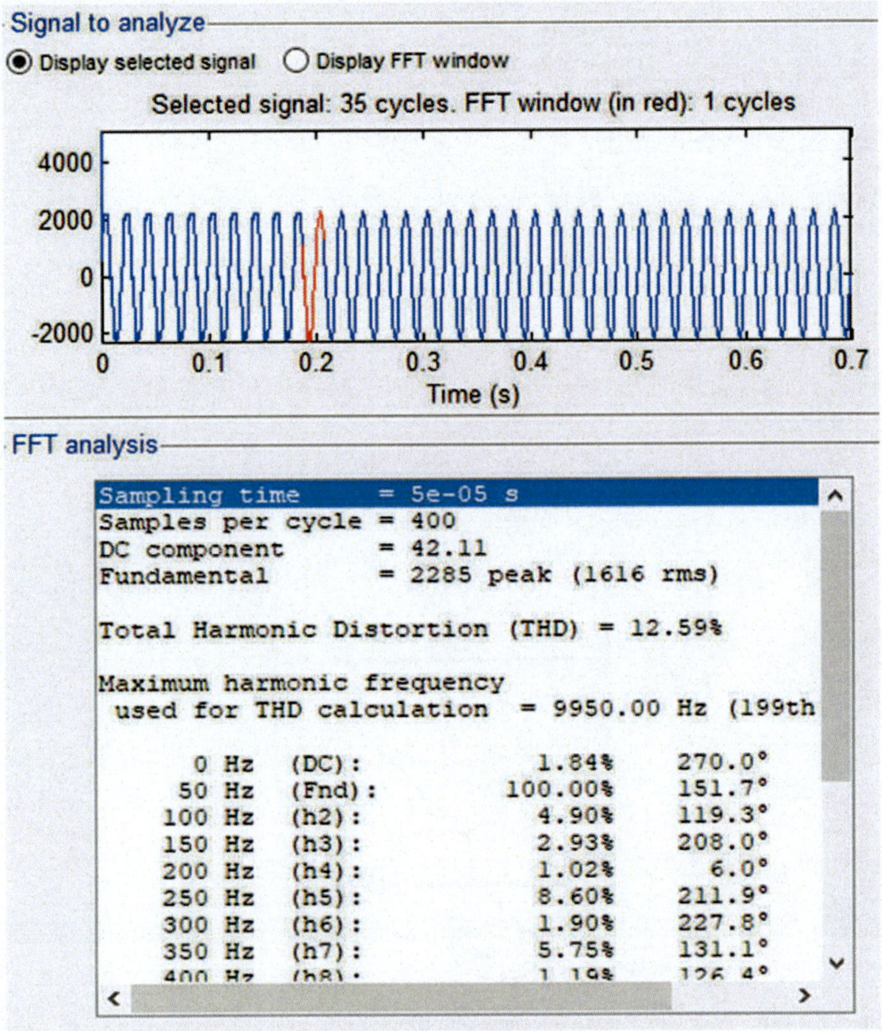

Fig. 7 Current-harmonics

Table 2 Line-data of a 3bus system

S. No.	Line	Line-resistance (Ω)	Line-inductance (H)	Line-reactance(Ω)
1.	1-3	0.023	0.0029	0.018
2.	1-2	0.0480	0.00058	0.182
3.	2-3	0.035	0.00037	0.116

Table 3 Load-data for 3-bus system

S. No.	Bus	Line-resistance (Ω)	Line-inductance (H)	Line-reactance (Ω)
1.	Bus-2	2.41	0.0192	6.028
2.	Bus-3	0.048	0.00255	0.800

Table 4 Harmonic analysis without UPQC

Harmonic	% HD without UPQC
I_3	2.93%
I_5	8.60%
I_7	5.75%

Table 5 Summary of voltage, real power and reactive power

Parameters	Bus-3
V (KV)	9.35
P (MW)	5.723
Q (MVAR)	38.23

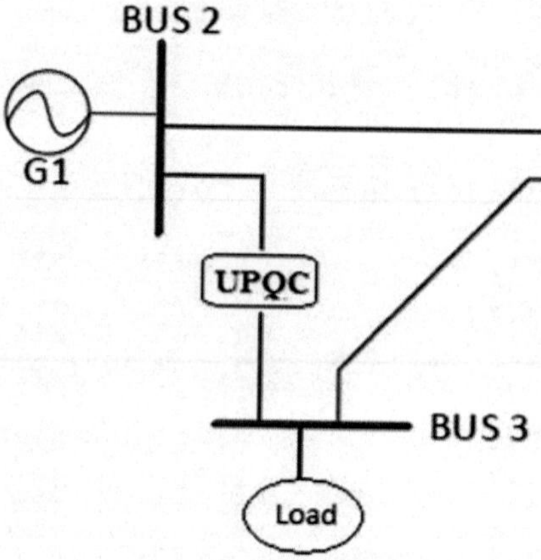

Fig. 8 "Single-line-diagram" for a three bus with UPQC

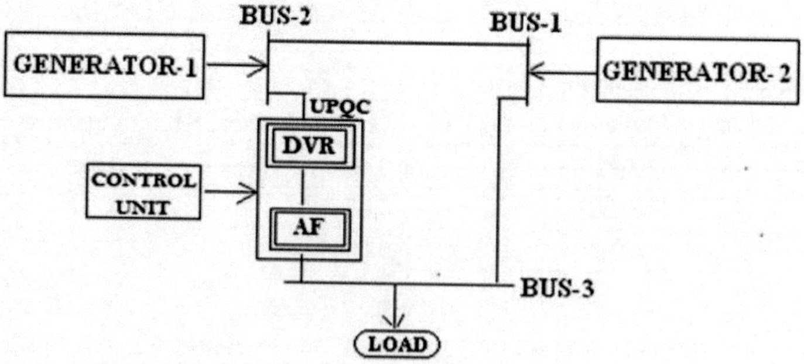

Fig. 9 Block diagram of 3-phase, 3bus system and UPQC

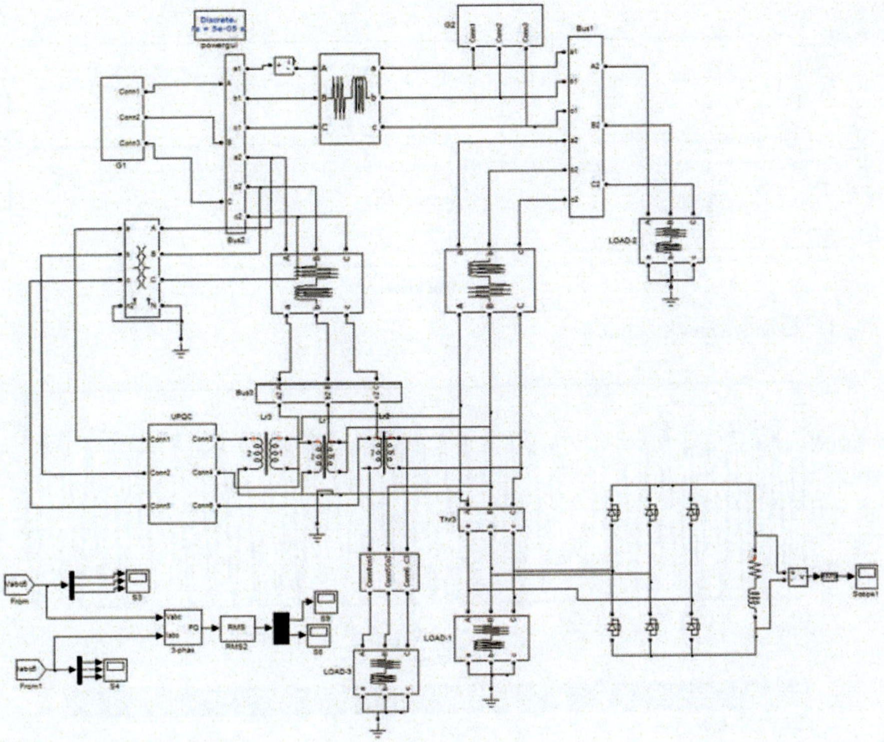

Fig. 10 Simulink-diagram of a three phase 3-bus with UPQC

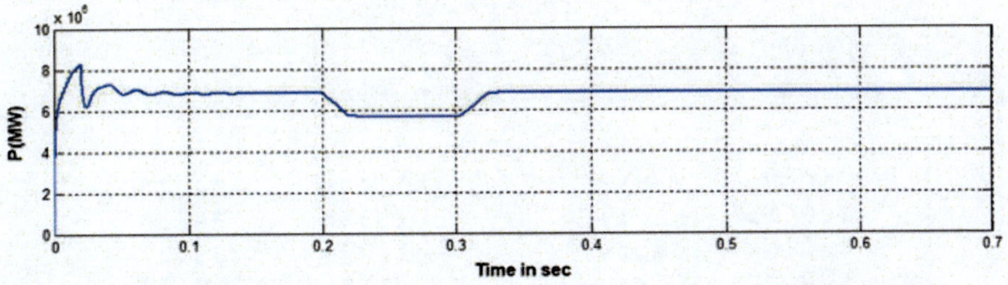

Fig. 11 Real-power in bus 3

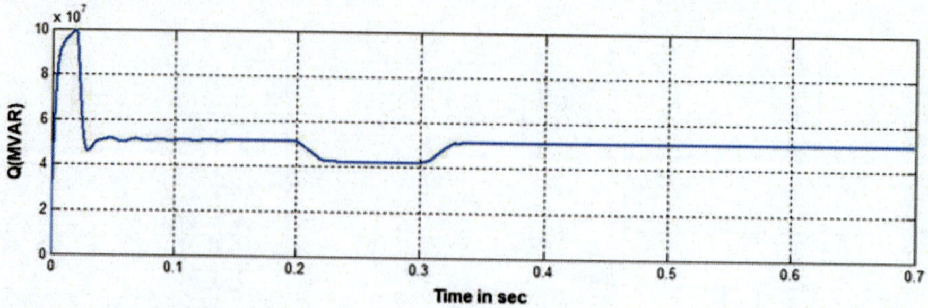

Fig. 12 Reactive power in bus-3

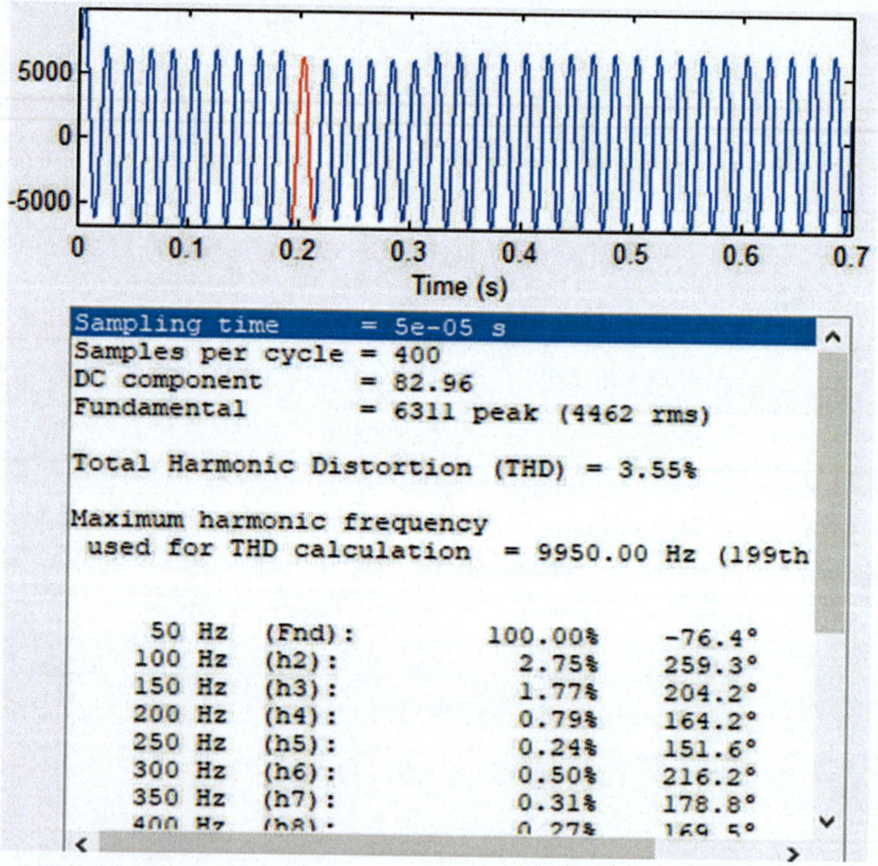

Fig. 13 Current THD in bus-3

Table 6 Harmonic analysis with UPQC

Harmonic	% HD with UPQC
I_3	1.77
I_5	0.24
I_7	0.31

Table 7 Real power, reactive power and voltage summary

Parameters	Bus-3
V (kV)	10.05
P (MW)	6.12
Q (MVAR)	51.14
V_0 (kV)(Non-linear load)	10.05

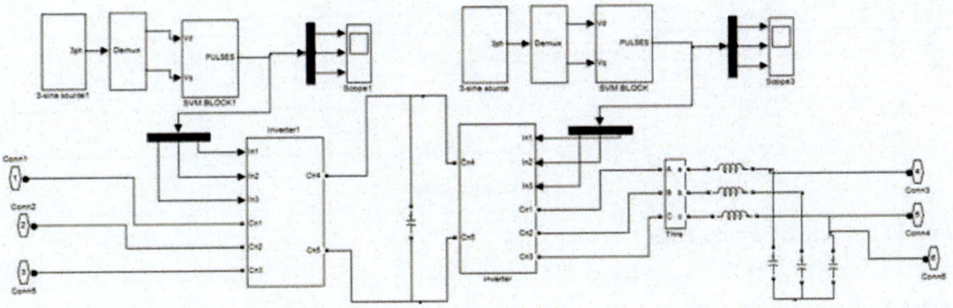

Fig. 14 Block diagram of SVM inverter

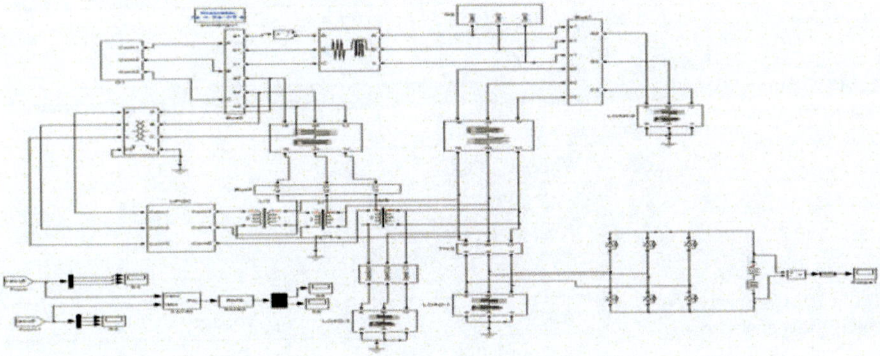

Fig. 15 Circuit diagram of SVM inverter based UPQC

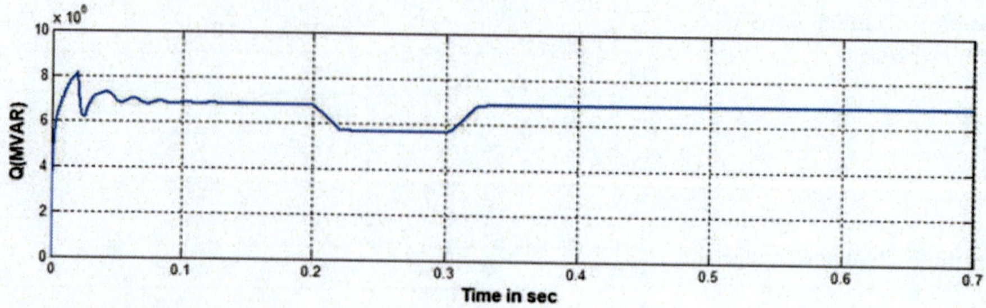

Fig. 16 Real-power in bus-3

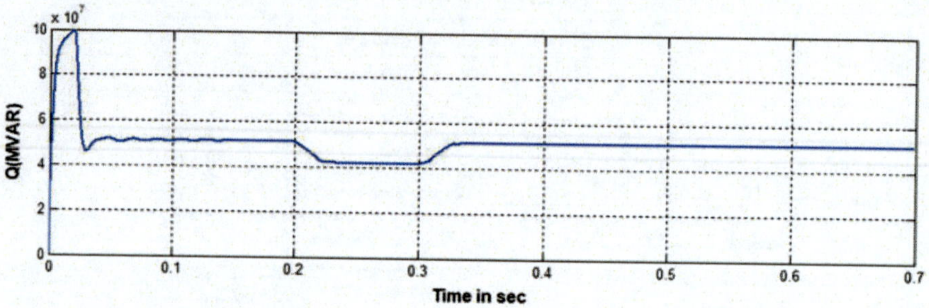

Fig. 17 Reactive-power in bus 3

Table 8 Harmonic analysis with UPQC-SVM inverter

Harmonic	% HD with SVM based UPQC
I_3	1.33
I_5	0.85
I_7	0.27

Table 9 Summary of real-powers and reactive-powers

Case	Real-power (MW)	Reactive-power (MVAR)
Without UPQC	5.723	38.23
With UPQC	6.12	51.14
SVM based UPQC	6.830	52.31

Table 10 Comparison of voltage responses

Case	Voltage (kV)
Without UPQC	9.00
With UPQC	10.05
SVM based UPQC	10.08

Table 11 Summary of current THD'S

Case	THD (%)
Without UPQC	12.59
With UPQC	3.55
SVM based UPQC	2.86

References

1. Ambati BB, Khadkikar V (2014) Optimal sizing of UPQC considering VA loading and maximum utilization of power-electronic converters. IEEE Trans Power Delivery 29(3)
2. Devassy S, Singh B (2018) Design and performance analysis of three-phase solar PV integrated UPQC. IEEE Trans Ind Appl 54(1)
3. Khadem SK, Basu M, Conlon MF (2015) Intelligent islanding and seamless reconnection technique for microgrid with UPQC. IEEE J Emerg Sel Top Power Electron 3(2)
4. Ye J, Gooi HB, Wu F (2018) Optimal design and control implementation of UPQC based on variable phase angle control method. IEEE Trans Ind Inf 14(7)
5. Devassy S, Singh B (2017) Modified pq-theory-based control of solar-PV-integrated UPQC-S. IEEE Trans Ind Appl 53(5)
6. Ye J, Gooi HB, Wu F (2018) Optimization of the size of UPQC system based on data-driven control design, IEEE Trans Smart Grid 9(4)
7. Lakshmi S, Ganguly S (2018) Modelling and allocation of open-UPQC-integrated PV generation system to improve the energy efficiency and power quality of radial distribution networks. IET Renew Power Gener 12(5)
8. Choudhury SR, Das A, Anand S, Tungare S, Sonawane Y (2019) Adaptive shunt filtering control of UPQC for increased nonlinear loads. IET Power Electron 12(2)
9. Devassy S, Singh B (2017) Control of solar photovoltaic integrated UPQC operating in polluted utility conditions. IET Power Electron 10(12)
10. Campanhol LBG, Da Silva SAO, De Oliveira AA, Bacon VD (2019) Power flow and stability analyses of a multifunctional distributed generation system integrating a photovoltaic system with unified power quality conditioner, IEEE Trans Power Electron 34(7)
11. Da Silva SAO, Negrão FA (2018) Single-phase to three-phase unified power quality conditioner applied in single-wire earth return electric power distribution grids. IEEE Trans Power Electron 33(5)
12. Cheung VS-P, Yeung RS-C, Chung HS-H, Lo AW-L, Wu W (2018) A transformer-less unified power quality conditioner with fast dynamic control. IEEE Trans Power Electron 33(5)

Genetic Algorithm and Graph Theory Approach to Select Protection Zone in Distribution System

S Ramana Kumar Joga, Pampa Sinha, and Manoj Kumar Maharana

Abstract In this paper, a Genetic Algorithm and Graph Theory-based approach has been proposed for the Protection Zone Selection for Distribution System. The Proposed method is designed and developed to split electrical distribution system into protection zones containing busses and protection relays or fault detectors and also to decrease the calculation burden in dealing with a large set of signal data. Genetic Algorithm based heuristic Search method is used to place fault detectors at optimal location, and it carried out in MATLAB. IEEE33 bus radial distribution system is tested for validating the proposed system.

Keywords Graph theory · Fault detection · Expert system · Protection zone · High impedance faults · Genetic algorithm

1 Introduction

Faults in Electrical Distribution System are unavoidable for many uncontrollable reasons like birds, weather-related reasons. These faults lead to the blackout including equipment failure, lightening at conductors in power system. In a power distribution system, many faults may occur like high impedance faults, short circuit faults, open circuit faults and ground faults etc.; early detection of fault ensure the reliability of the power system and also decrease the human and property loss, and it is more discussed in [1]. It is necessary to protect the distribution system from faults; the fault affected area must be kept isolated from the rest of the power system. This type of Protection Scheme is known as "Zone Protection". In this scheme, when a fault occurs in the particular zone, the circuit breakers within that zone only

S. R. K. Joga (✉) · P. Sinha (✉) · M. K. Maharana
School of Electrical Engineering, KIIT Deemed to Be University, Bhubaneswar, India
e-mail: sanset567@gmail.com

P. Sinha
e-mail: pampa.sinhafel@kiit.ac.in

M. K. Maharana
e-mail: mkmfel@kiit.ac.in

N. Zhou and S. Hemamalini (eds.), *Advances in Smart Grid Technology*, Lecture Notes in Electrical Engineering 688, https://doi.org/10.1007/978-981-15-7241-8_13

opened and it is isolated from the distribution system [2]. The primary consideration to select a protection zone is to reduce the fault effect on consumers by isolating them in the fault zone. Bus Bar Protection involves frame leakage, high impedance relays, medium impedance relays, and low impedance relays based upon centralised units, and all these are discussed in [3, 4]. Conventionally, electromechanical relays were the devices used to detect and isolate the faulty section by tripping the circuit breaker [5]. Subsequently, these electromechanical relays were substituted by static and numerical relays for greater accuracy, speed and reliability. To detect the type of fault occurs in the distribution system several machine learning and artificial intelligence approaches have been proposed by many researchers; most of them used multilayer perceptron. Sinha and Maharana proposed a fault classification method by using Artificial intelligence based classifiers such as back propagation neural network, Probabilistic Neural Network. Salat and Osowski proposed a fault detection technique by using support vector machine based machine learning classifier, it is discussed in [6]. A wavelet fuzzy based classification of faults are discussed in [7]. Reddy and Mohanta proposed an adaptive network based fuzzy inference system approach for locating faults in transmission system and it is discussed in [8]. The overall protection zone scheme for IEEE34 radial distribution bus system is discussed in [9]. A distance relay based protection zone scheme is developed in MATLAB Simulink by Farhana Ferdous and it is discussed in [10]. Kumar and Hansen proposed a numerical bus zone microprocessor based protection scheme, and it is given in [11]. A Transient based zone protection scheme using wavelet transform is discussed in [12]. The overall protection schemes for zone protection are reviewed by Singh and it is discussed in [13].

Most of the above mentioned researchers proposed zone protection schemes without considering any economic considerations. In this paper a genetic algorithm-graph theory approach based Protection Zone Selection is proposed for isolating fault zone from healthy zone in Electrical Distribution System. This methodology is proposed considering the economic consideration, by placing fault detectors at optimal location. The need of measurement devices and sensors are reduced, thereby the cost of protection scheme also get reduced. This proposed is used to separate zone in any fault conditions that occur in the Distribution System.

2 Proposed Methodology

2.1 Optimal Placement of Fault Detector

It is necessary to detect the fault in the distribution line; in order to detect the fault, Fault detectors should be placed to detect the fault. By reducing the number of fault detectors, the installation cost of protection system can be minimized. To minimize the number of fault detector and to find the optimal placement of a fault detector, a necessary optimization technique should be used. In this paper, a Genetic Algorithm based optimization technique is used to find the optimal location of the fault detector. A Genetic Algorithm is a heuristic search method used to find optimal solution for the optimization and search-related problems. This method is one of the Evolutionary Search method inspired by biological background named “Genetics”. Genetic Algorithm has its advantages compared to traditional computing methods, they are

a. It provides a list of optimal solutions, unlike traditional methods give only single solution.
b. Convergence to Optimal Solution is faster compared to traditional methods.
c. It is used for solving multi-objective problems
d. It doesn’t require any derivative information

In Basic Genetic Algorithm optimization technique, it involves four stages namely initialization, Selection, Crossover and Mutation. Selection is the process that determines which population is to be survived and should allow to crossover process in the search method. There are different techniques to implement the selection process in the genetic algorithm; they are Tournament Selection, Rank Selection, Roulette Wheel Selection and Stochastic Universal Sampling. Crossover Process exactly imitates the biological reproduction process. In this process crossover operator selects any two solution strings randomly from the population with high probability Pc, called the crossover probability and some portion of the strings are exchanged between the strings. It results new offspring’s from the parent’s string. These Child strings contain the selected good behavior of the parent string. There are different techniques to implement the crossover process in the genetic algorithm; they are Uniform Crossover, One-Point Crossover and Multipoint crossover. Mutation Process introduces the new features to the new strings formed from the parent strings by crossover process. This Process ensures the diversity in the population.

The Objective function of the optimal placement problem (OPP) is taken as

$$\sum_{i=1}^{n} x_i \tag{1}$$

where x denotes the fault detector, i denote the bus number. Sum of all fault detectors to be placed at each and every bus is taken as objective function. Here IEEE33 radial distribution bus is considered for locating the optimal placement of fault detectors, i.e. the optimization problem must minimize the fault detectors to reduce the cost. In this we are considering non-constraint objective function, shown in Eq. (1), as we are not considering the cost function of the objective function. The main objective of this optimization problem is to minimize the fault detectors and place them at optimal location. This optimization problem (OPP) is solved by an evolutionary optimization technique known as genetic algorithm [14]. The basic flow chart of the genetic algorithm optimization technique is shown in Fig. 1.

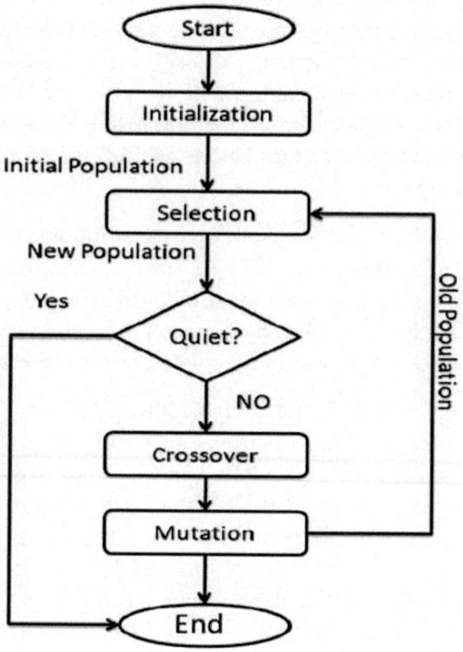

Fig. 1 Basic flow chart of genetic algorithm

The Specification considered for genetic algorithm optimization technique are

Population Size = 200
Crossover = uniform
Selection = Roulette wheel method
Fitness function = $\sum_{i=1}^{n} x_i$
Method = Single Objective function with no constraints

In this OPP, each fault detector at bus in objective function is considered as one gene, and all genes are combined to form a fitness function. Now it is required to find the fitness of each and every gene in the bus by checking the observability. It is considered that every bus must have one fault detector to check for the observability. From the measured values obtained from the measuring device, apply Kirchhoff and ohms law to measure the other variables. Check whether the bus is observable by this process. This analysis is repeated until all the busses are completely observable, if any bus is not observable, a pseudo-measurement current phasor can be assigned. The placement problem have following steps, they are

Step 1: from the given power system, assign chromosome
Step 2: generate initial population, let it be 200
Step 3: Calculate fitness function
Step 4: Apply Selection Operator
Step 5: Apply Crossover Operator
Step 6: Apply Mutation Crossover

Step 7: Repeat the process until G generations are completed, if not go to step 3

The genetic algorithm optimization is carried out in MATLAB considering the above specifications. After the Optimization of fitness function, Authors observed that 11 optimal locations are minimized out of 33 bus locations.

Number of Optimal Locations = 11

The 11 optimal bus locations are observed are 2, 3, 5, 9, 14, 17, 19, 22, 24, 26, 30.

By Considering these optimal measuring devices location, a new zone protection scheme should be designed. A new Zone Protection scheme for electrical distribution system is proposed which is based on Graph Theory.

3 Graph Theory-Based Protection Zone Selection

Graph theory based expert system search method is used to select the protection zone. In graph theory any complex power system can be simplified. The detailed explanation of expert system is discussed in [15]. It is one of the main reasons to choose graph theory based search method to choose protection zone. In this all power system topology is now called in graph theory topology like busses in distribution system are named as Edge in graph theory topology, similarly distribution line in power system are named as Vertex in graph theory topology. There are set of rules considered for the zone separation, they are

1. The Bus connected with an existed protection zone with a vertex will be combined to configure a new protection zone.
2. For the protection zones containing the same busses, only one of them is kept, and others should be eliminated.

According to these two rules of proposed protection zone selection, the following steps for IEEE4 bus system of Fig. 2 can be illustrated in Figs. 3, 4, 5.

a. Initially, each bus configures a new zone. For Example, there is a zone B1 for Bus 1;
b. For Each Protection Zone, the combinations of busses are connected to the nearest zone to form a new zone. For Example, Zone B1 evolve for zone B1ɵB2, while B1ɵB2 in Fig. 3 will evolve for zone B1ɵ B2ɵB3 and B1ɵB2ɵB4 in Fig. 4; in the same way, zone B1ɵB2ɵB3 and B1ɵB2ɵB4 evolve for zone B1ɵB2ɵB3ɵB4 in Fig. 5
c. The Graph Theory approach step search method will compare the new zone with one another, the zone with the same bus will be terminated; For Example in Fig. 5, there should be two-zone yield from B1ɵB2ɵB3 and B1ɵB2ɵB4, and the Identical one will be terminated according rule number 2.
d. If all the busses of a zone equal to number of busses in the distribution system, complete the search process. If not, Take step b for finding new zones.

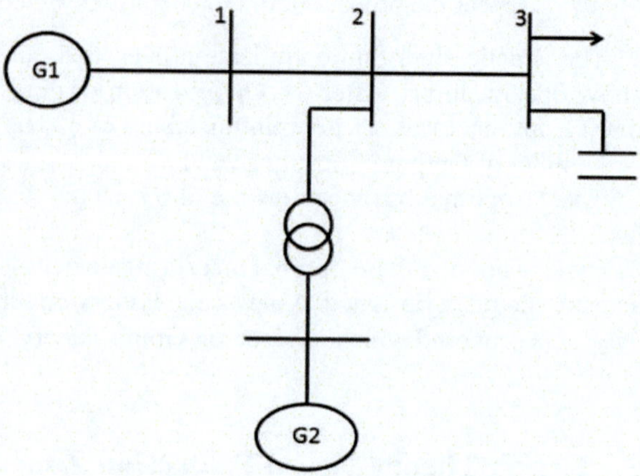

Fig. 2 Single line diagram of IEEE4 bus system

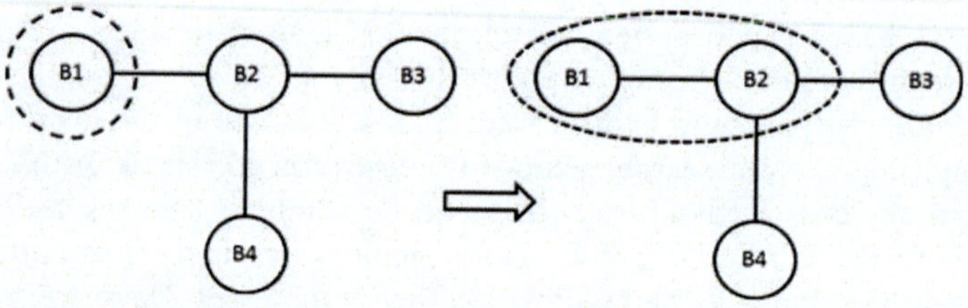

Fig. 3 Step b in search method

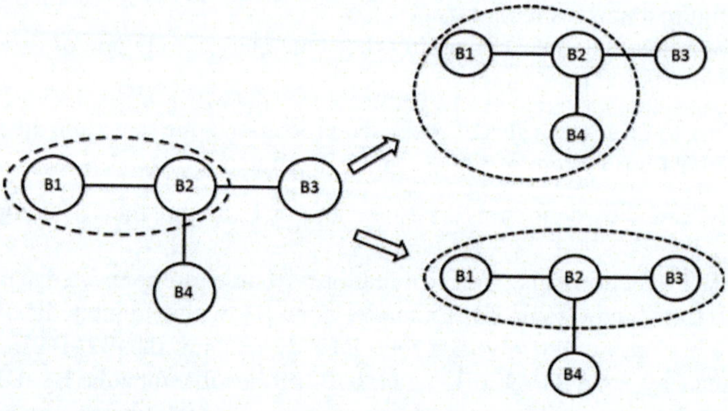

Fig. 4 Step b in search method

4 Results and Discussions

The Proposed Protection Zone Selection search method is tested on IEEE33 Radial Distribution System. The Single line diagram of the IEEE33 bus is shown in Fig. 6.

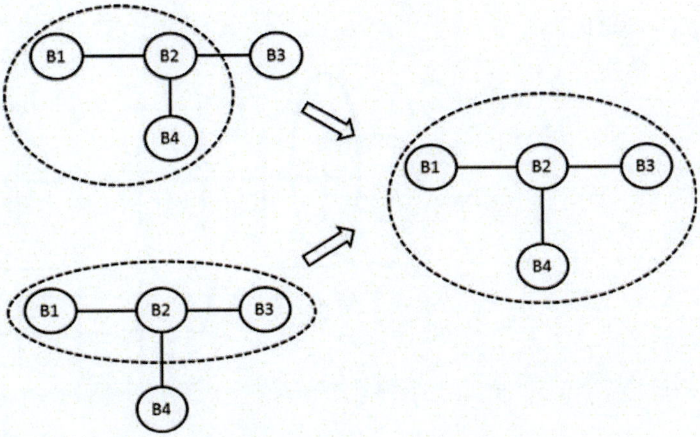

Fig. 5 Step c and d in search method

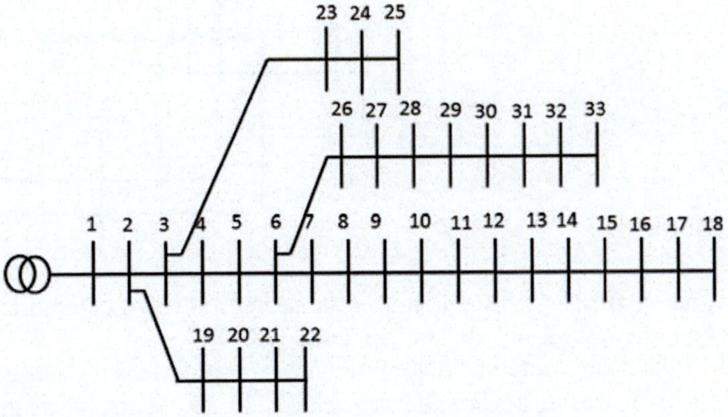

Fig. 6 Single line diagram of IEEE33 bus system

The step wise search procedure is developed in program; the following steps are applied to the IEEE33 bus in order to select a zone protection through the proposed method. The following steps are

Step1: Initially, Every basic bus configures a primary zone, for a case, bus 1 in the IEEE33 bus shown in Fig. 6 itself a Zone. This Initial zone is shown in Fig. 7

Step2: For Each Protection Zone, it will combine with any bus that connects with zone from the new zone.

Step 3: Now, the Search rule compares the new zone with one other, the zone which has same bus connection will be terminated. It is shown in Fig. 8.

Step 4: if all the busses of a zone equal to the number of busses of the distribution system, then complete the search process. If not, Take step 2.

Fig. 7 Step 1 in search method

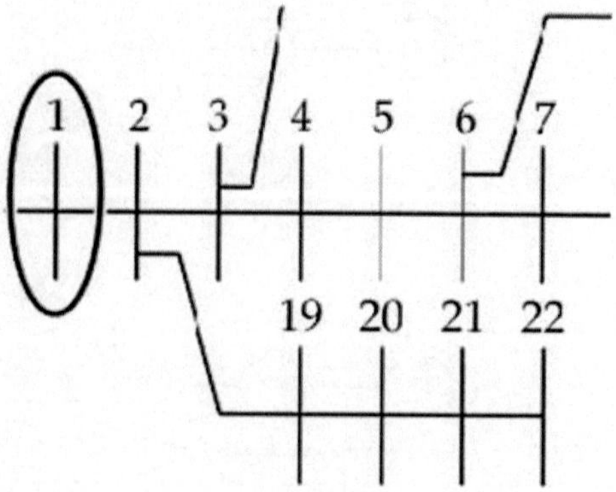

Fig. 8 Step 3 in search method

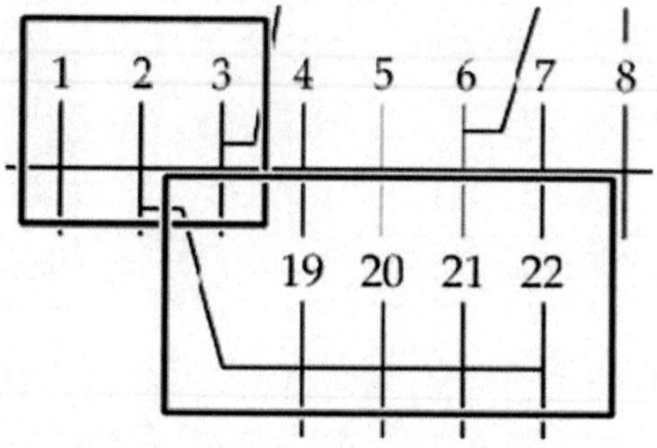

The results obtained after the step-4 in the search process is stored for further fault detection, the results are tabulated in Table 1.

In Table, it is clearly obtained that all the measuring devices are placed in each and every zone. Measuring devices that are bolded 2, 3, 19, 22 and 24 are grouped in zone 1, similarly measuring devices that are bolded 5, 26, 30 are grouped in zone 2 and measuring devices that are bolded 9, 14, 17 are grouped in zone 3. It assures that every zone is distributed with measuring devices with protective relays for gathering the fault signal data. It also ensures the smooth run of the power system and ensures the problem of power outages to the non-fault regions.

Table 1 Protection zone selection results for IEEE33 bus RDS

Zone	Bus numbers
Zone 1	1, **2**, **3**, **19**, 20, 21, **22**, 23, **24**, 25
Zone 2	4, **5**, **26**, 27, 28, 29, **30**, 31, 32, 33.
Zone 3	7, 8, **9**, 10, 11, 12, 13, **14**, 15, 16, **17**, 18

Bolded Measuring Devices are Optimal Measuring devices from 33 busses

5 Conclusions

In this paper from the obtained results, the following conclusions are made

a. Through genetic algorithm based optimization technique, measuring devices are placed at optimal locations successfully.
b. A Graph Theory based search method is proposed ensuring the measuring devices at each and every zone.
c. The Proposed method is successfully tested on IEEE33 bus, it satisfied the above conclusions.
d. Most of the previous zone selection methods either divided zone according to impedances values obtained from measuring devices or analyzing the fault signal through the signal processing analyzers, in this paper zone protection doesn't depend on fault type and condition.
e. This proposed zone protection is tested successfully and detected the location of short circuit faults in electrical distribution system by using Support Vector Machine classifier with more than 96% accuracy in performance.

6 Future Scope

In this paper only zone selection search method is proposed, there are some applications and future scope are mentioned below

a. Short Circuit Faults in Electrical Distributions are detected through Artificial Intelligence based classifiers and can be grouped into protection zones by this proposed method.
b. A microprocessor based Fault detector can be developed though this proposed search method and can apply to any fault zone identification process.

References

1. Sinha RKJP, Maharana MK (2019) Artificial intelligence in classifying high impedance faults in electrical power distribution system. SSRN Electron J 1–5
2. Qin BL, Guzman-Casillas A, Schweitzer EO (2000) A new method for protection zone selection in microprocessor-based bus relys. IEEE Trans Power Deliv 15(3):876–887
3. Guzmán A, Qin BL, Labuschagne C (2005) Reliable busbar protection with advanced zone selection. IEEE Trans Power Deliv 20(2I):625–629
4. Steenkamp L, Labuschagne C, Stokes-waller E (2017) Tutorial : complex busbar protection application. In: 34th western protective relay conference, pp. 1–11
5. Jana S, De A (2017) A novel zone division approach for power system fault detection using ANN-based pattern recognition technique détection de défaut des réseaux électriques utilisant la technique de reconnaissance de formes basée sur RNA. Can J Electr Comput Eng 40(4):275–283

6. Salat R, Osowski S (2004) Accurate fault location in the power transmission line using support vector machine approach. IEEE Trans Power Syst 19(2):979–986
7. Jayabharata Reddy M, Mohanta DK (2007) A wavelet-fuzzy combined approach for classification and location of transmission line faults. Int J Electr Power Energ Syst 29(9):669–678
8. Reddy MJB, Mohanta DK (2008) Performance evaluation of an adaptive-network-based fuzzy inference system approach for location of faults on transmission lines using Monte Carlo simulation. IEEE Trans Fuzzy Syst 16(4):909–919
9. Ashok SP (2014, May) Modeling and protection scheme for IEEE 34 radial distribution feeder with and without distributed generation
10. Farhana Ferdous R (2018) Zone protection system of transmission line by distance relay using Matlab/Simulink. In: International conference on advancement in electrical and electronic engineering, pp. 073–077
11. Kumar A, Hansen P (1993) Digital bus-zone protection. IEEE Comput Appl Power 06(04):29–34
12. Gafoor SA, Devi NR, Ramana Rao PV (2008) A transient current based bus zone protection scheme using wavelet transform
13. Singh M (2017) Protection coordination in distribution systems with and without distributed energy resources-a review. Prot Control Mod Power Syst 2(27)
14. Allagui B, Ben Aribia H, Hadj Abdallah H (2012) Optimal placement of Phasor Measurement Units by genetic algorithm. In: 2012 1st international conference on renewable energies and vehicular technology REVET, pp. 434–439
15. Su S, Liu J, Xiao H, Jiang D (2005) Expert system for wide area protection zone selection. Proc IEEE Power Eng Soc Trans Distrib Conf 2005:1–3

Analysis on DVR Based on the Classification of Converter Structure and Compensation Schemes

P. Priyadharshini and P. Usha Rani

Abstract Nowadays power quality issues are the greatest challenges to the upcoming development. Especially in case of using the sensitive loads the voltage disturbances will be high. Many research based solutions has been raised to overcome the PQ issues. One the best solution is power electronic switches based dynamic voltage restorer. The DVR becomes mature power quality device. The Converter structure and compensation schemes plays a major role in deciding the capability of the DVR. Hence in this paper, classification of DVR based on converter structure and compensation schemes have been analysed.

Keywords Power quality (PQ) · Dynamic voltage restorer (DVR) · Total harmonic distortion (THD) · Unscented kalman filter (UKF) · Improved synchronous reference frame (ISRF)

1 Introduction

During the past decades the usage of the critical loads has been increased, moreover the electronic devices has been used for inevitable applications. The applications like controlling, computerized applications and power conversions are done in more level. So the PQ problems also increased that leads to data loss, productivity reduction etc. The PQ mainly related with the voltage are voltage Sag, voltage Swell, Flicker, Voltage Fluctuations and unbalance. Generally voltage according to the definition of IEEE standard it is the decrease in the rms voltage between 0.1 and 0.9 p.u at normal frequency. Similarly the voltage swell is the increase in the rms voltage between 1.1 and 1.8 p.u. [1, 2]. These are the major problems in PQ which causes overheating, the sudden voltage dip affects the performance of the device.

P. Priyadharshini (✉) · P. Usha Rani
RMD Engineering College, Kavaraipettai, India
e-mail: priyadharshini.eee@rmd.ac.in; pur.eee@rmd.ac.in

N. Zhou and S. Hemamalini (eds.), *Advances in Smart Grid Technology*, Lecture Notes in Electrical Engineering 688, https://doi.org/10.1007/978-981-15-7241-8_14

2 Principle of Operation of DVR

The block diagram of DVR is shown in Fig. 1. The main components of the DVR are Converter, Injection Transformer, Filter, Energy Storage Device and By Pass Switch. There are three modes of operation namely Standby mode, Injection Mode and Protection Mode [3]. Whenever the voltage dip occurs DVR will injects the voltage in series with the supply voltage to safe guard the sensitive loads form the grid side disturbance.

Standby mode:

In this mode, the DVR will be in standby it will not supply any voltage since there is no voltage disturbance in the grid side voltage

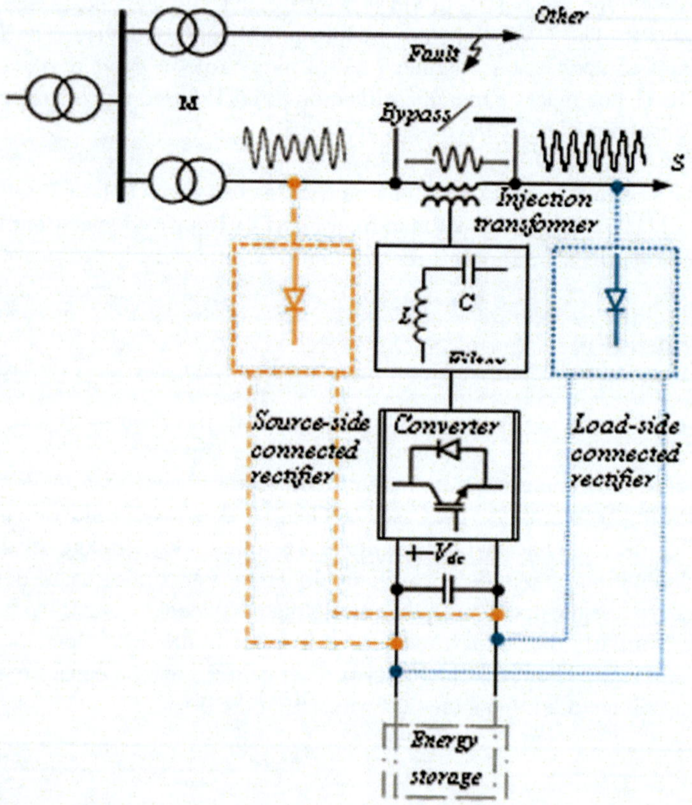

Fig. 1 Typical schematic diagram of DVR

Injection mode:

In this mode when the voltage disturbance is detected by the controller, then control signals will be generated which allows the bypass switch to open. Now the DVR will go into the injection mode and supply the voltage through the injection Transformer.

Bypass mode:

In this mode DVR will be protected from the fault current during the downstream. This mode is also called as protection mode. The current path should not be cut down because it will give arise to heavy voltage so an another way will be provided for the current to flow.

3 Comparison of DVR Based on Converter Topology [4–7]

In the DVR the important part is converter, hence based on the structure of converter it is classified mainly into

- AC–AC Converter
- DC–AC Converter

Figure 2 shows the classification of DVR based on the converter structure. Table 1 gives the comparison of the DVR with its capabilities and restrictions. Table 2 gives the comparison between switching cell based DVR and conventional DVR.

4 Evaluation of DVR Based on the Compensation Schemes

4.1 Commensurable Voltage Dip/Raise [8, 9]

Methodology: Positive and Negative Sequence Extraction by Using Multiple Reference Frame (Fig. 3).

Concepts: This technique used the multiple dq transformation used to extract the negative and positive sequence from the fundamental and harmonic waves to analyze and adjust the each component. This is easy to judge whether smooth dq transformation is possible.

Disadvantages: THD level is not reduced.

Advantages: Accuracy is very high. Dynamic Response time is meliorating without any voltage loss.

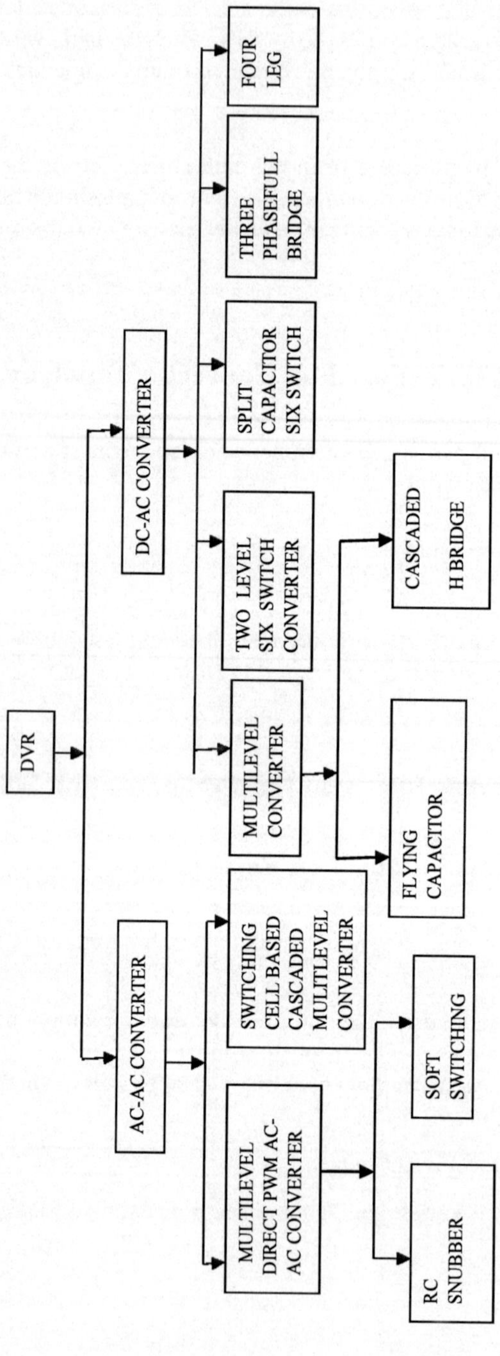

Fig. 2 Classification of DVR based on converter structure

Table 1 Comparison of DVR based on converter structure

S. No.	Topology	Capabilities	Restrictions
1	a. Flying capacitor based multi level converter	More reliability switching combination. Allowing more control in the voltage balance Using FC more than three level is more attractive	Increases the number of capacitors based on the output voltage Voltage balancing is quite challenging
	b. Cascaded H bridge multi level converter	Injection transformer is neglected portable and reliable	More isolated DC source for each H bridge
2	Two level six switch converter	Design is simple	Unable to compensate the voltage disturbance up to the level
3	Split capacitor-six switch inverter	Deign topology is easy based on the switches Voltage disturbance can be balanced	Output and input voltage ratio based on the capacitors voltage
4	Three full bridge	Independent phase control on the voltage disturbance	Construction is complex
5	Four leg	DC link is removed Compensation is done for longer time period	Switching frequency is high. Additional switches are used
6	AC–AC	Not employing the DC bulky capacitor Size of the converter is reduced Power loss is reduced Overall efficiency can be increased	Voltage compensation is not up to the depth. Voltage sag is not supported by the grid
7	Multilevel direct PWM AC–AC converter (RC Snubbed)	Used for scaling higher voltage higher power level	Commutation problem occurs due to bidirectional switches Converter efficiency also reduced Output is distorted because energy is dissipated in the RC circuit
8	Multilevel direct PWM (pulse width modulation) AC–AC converter (soft commutation)	Voltage level is improved Power level is increased	Reliable operation cannot be obtained Input/Output voltage and current waveforms are highly distorted
9	Switching cell based cascaded multilevel converter	Commutation problem is reduced Polarity of the voltage and current is not sensed Highly reliable robust	Control schemes are required for each phase

Data obtained from [1, 3]

Table 2 Comparison of conventional DVR and switching cell based DVR [1, 3, 19, 20]

Terminology	Conventional DVR	Switching cell based DVR
Structure	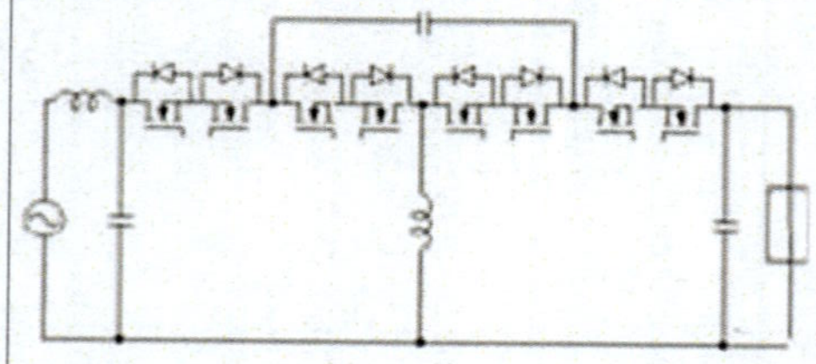	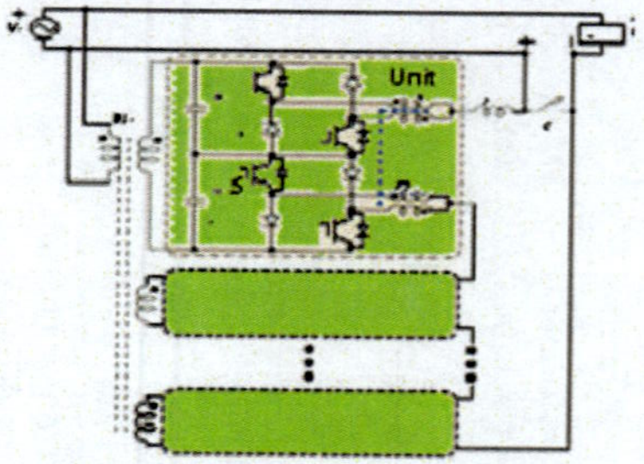
Switch	Number of switches are more MOSFET has more switching loss	Number of switches is less IGBT has less switching loss
Inductor	Discrete inductor	Magnetically integrated coupled inductor. [21–24] The size of the inductor is compact and it does not attain the saturation by the inductor current
Peak over shoot	Shoot Through Problems are more due to fault triggering	The shoot through problems is reduced due to minimal stray inductance
Commutation	Commutation problem can be reduced by sensing the polarity of voltage and current	Commutation problem can be reduced without sensing the Polarity of voltage and current
Reliable	Reliability is less during the highly distorted voltage waveform	Reliability is more even during the dead time, overlap time, during highly distorted voltage

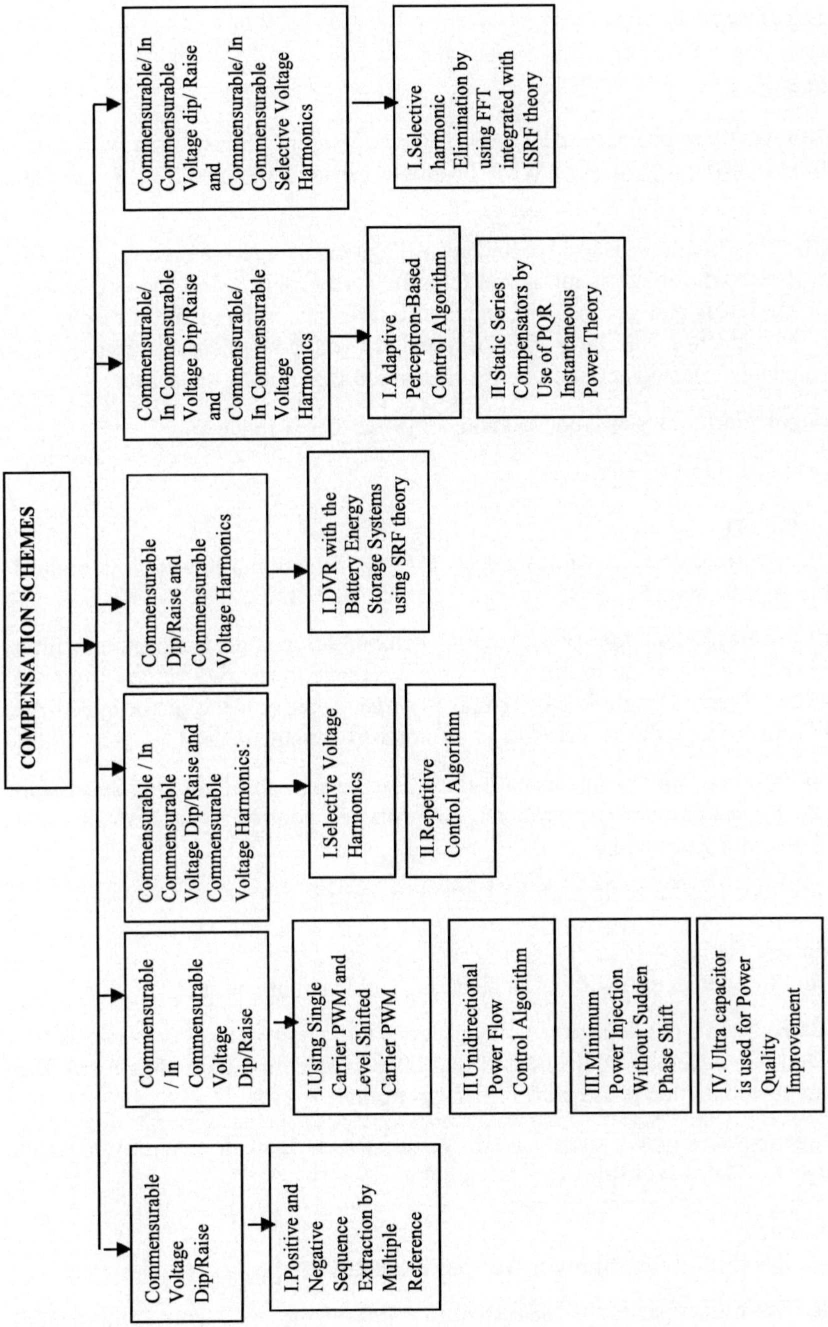

Fig. 3 Classification of DVR based on compensation schemes

4.2 Commensurable and in Commensurable Voltage Dip/Raise

Methodology I

DVR with two three phase Inverter with open end winding Transformer with either level-shifted carrier PWM (LSPWM) or single-carrier PWM (SCPWM) strategy [10].

Concepts: This technique is used to find out voltage sag using hysteresis control. The superior threshold voltage is measured then the system will select the two closest high voltage levels and get the difference with reference voltage and it discharges through the capacitor. Similarly inferior threshold voltage is measured two closest lower voltage is selected and the voltage is charged through the capacitor.

Advantages: Harmonics are reduced and converter loss is reduced.

Disadvantages: Cost is very high

Methodology II
Diode Bridge Rectified Transformer less DVR unidirectional power flow control algorithm is proposed [11].

Concept: In this method three phase rotating compensation angle is measured within the three phase compensation limit. With the reference to the two values proposed three phase reference angle is calculated. In order to prevent the distortion in the compensation voltage the progressive phase rotating scheme is used.

Advantages: Maximum Compensation limit, Progressive phase rotating speed and dc-link safety operation considerations, the proposed control algorithm can solve reverse power flow problems.

Disadvantages: Response time of the DVR is extended.

Methodology III
Minimum Power Injection without introducing sudden phase shift [12].

Concept: By using the injection voltage control strategy the angle between the voltage and the current is controlled so that DVR is injecting the voltage with the minimum power sudden phase shift is also controlled.

Advantage: Sudden phase jump can be avoided on the load side. With minimum power injection **Disadvantage**: It is not reliable:

Methodology IV
Ultracapacitor is used for improving the power quality [13].

Concept: The ultracapacitor is linked with dc link of the DVR.This combination providing stiff compensation and also it provides the compensation separately for sag and swell without depending on the grid voltage.

Advantage: Voltage Restoration has been improved. Independently compensates the voltage sag and swell without considering the grid voltage.

Disadvantage: Size is increased due to the use of capacitor.

4.3 Commensurable/ In Commensurable Voltage Dip/Raise and Commensurable Voltage Harmonics

Methodology I
Selective Voltage Harmonic Compensation Control in Medium Level Voltage [14].

Concept: Resonant Feedback based Controllers have been used to remove the selective voltage harmonics. It is applicable for both single and three phases.

Advantages: Effective Voltage Regulation on Harmonics is obtained.

Disadvantages: Cut off Frequency Variation has to be considered.

Methodology II
Repetitive Control Scheme is used to compensate the Voltage Disturbance [12]:

Concept: The entire voltage disturbance is considered with the bandwidth. The control structure is using the feedforward term which is developed to overcome the transient response. The feed backward term is used for zero voltage control.

Advantages: Very Simple, Single controller is used to control the voltage sag/swell and harmonics, Realistic, Excellent Voltage Regulation.

Disadvantages: For Sensitive Loads the compensation is not up to the level.

4.4 Commensurable Voltage Dip/Raise and Commensurable Voltage Harmonics

Methodology: DVR with the Battery Energy Storage Systems using SRF theory [15].

Concept: The compensation can be done by injecting or absorbing the real or Reactive power. When the injected voltage is in the quadrature with the current then DVR supplies the reactive power using the self supported DC bus.

Advantages: Compensation is done using battery source for the reduced rating voltage source converter.

Disadvantages: Voltage injection Capability and Energy optimization is less.

4.5 Commensurable/ in Commensurable Voltage Dip/Raise Commensurable/ in Commensurable Voltage Harmonics

Methodology I
Adaptive Perceptron-Based Control Algorithm [16].

Concept: This adaptively tracks and gets the selected voltage harmonics. The fundamental and harmonic component is extracted in a well versed manner. Flow chart sets the three bench marks if any disturbance is detected and it is compared with the top threshold value then the corresponding signal is produced to compensate the voltage dip

Advantages: Robust, Highly superior, Compensate the voltage swell even it occurs along with the voltage sag.

Disadvantages: Voltage harmonics is minimized but the percentage is less.

Methodology II
Static Series Compensators by Use of PQR Instantaneous Power Theory [17].

Concept: The three phase voltage is directly converted into pqr coordinates without the time delay, compensation voltages are produced by this coordinated systems separately for each phase without time delay. By using this theory harmonics is selected and compensated.

Advantages: Reference Wave Generator is used to find out information about the proper terminal voltage. By controlling one dc value compensation is done for all three phase.

Disadvantages: Transformer design is not considered.

4.6 Commensurable/ In Commensurable Voltage Dip/Raise Commensurable/ In Commensurable Selective Voltage Harmonics

Methodology
Selective harmonic Elimination by using FFT integrated with ISRF theory [18].

Concept: FFT is integrated with ISRF theory in this the grid voltage is used to find out the fundamental component and selective harmonics. The mathematical evaluation has been done to find out the reference voltage dip/arise and harmonics. The original signal is compared with the PWM signal. Based on this the inverter will get the signal and inject the control voltage.

Advantage: Very effectively compensates the voltage.

Table 3 Comparison of THD level [9, 16, 18]

Compensation schemes	THD level %
Unscented kalman filter based DVR	5.32
Adaptive perceptron	4.53
Repetitive csontroller	3.07
ISRF integrated with FFT	2

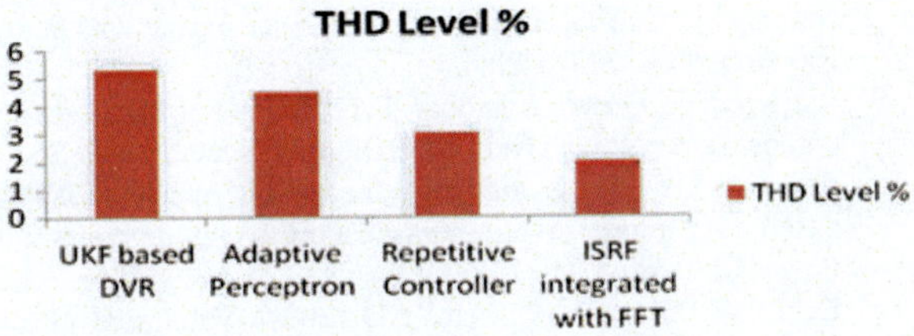

Fig. 4 Comparison of compensation schemes based on THD level

5 Comparison of the Compensation Schemes Based on the THD Level

THD values of different compensation schemes are given in the Table 3 and Fig. 4. By comparing the compensation schemes the FFT Integrated with ISRF technique gives the better THD level. From the inference the best scheme is ISRF with FFT technique which yields reduced THD level.

6 Conclusion

This paper provides the brief idea relating to the converter structure and compensation schemes of the DVR. Based on the converter structure switching cell based DVR is efficient since it has less switching loss, reliable, THD level is also less. Based on the compensation schemes FFT with ISRF provides maximum output with less THD. The combination of switching cell with FFT integrated with ISRF will provide the high output with less loss.

References

1. Cha H, Kim S, Kim H-G (2018) Dynamic voltage restorer using switching Cell structured multilevel AC-AC converter. IEEE Trans Power Electron
2. Kusko A, Thompson MT (2007) Power quality in electrical systems. McGraw Hill
3. Kangarlu MF, Babaei E, Blaabjerg F (2017) A comprehensive review of dynamic voltage restorers. IEEE Trans Ind Appl 92:136–155

4. Kaniewski J, Szczesniak P, Jarnut M, Benysek G (2015) Hybrid voltage sag/swell compensators: a review of hybrid ac/ac converters. IEEE Ind Electron Mag 9(4):37–48
5. Li L, Tang D (2012) Cascade three-level ac/ac direct converter. IEEE Trans Ind Appl 59(1):27–34
6. Divan D, Sastry J (2009) Control of multilevel direct ac converters. In: Conversion Congress and Exposition, pp 3077–3084
7. Kwon BH, Min BD, Kim JH (1996) Novel topologies of ac choppers. IEE Proc Electr Power Appl 143(4):323–330
8. Chen G, Zhu M, Cai X (2013) Medium voltage level dynamic voltage restorer compensations strategy by positive and negative sequence extractions in multiple reference frames. IET Power Electron 7(7):1747–1758
9. SasiKiran P, Gowri Manohar T (2018) UKF based estimation approach for dvr control to compensate voltage swell in distribution systems Ain Shams Eng J 9:441–453
10. Carlos GAA, Dos Santos EC, Jacobina CB, Mello JPRA (2016) Dynamic voltage restorer based three-phase inverters cascaded through an open-end winding transformer. IEEE Trans Power Electron 31(1):188–199
11. Lam C-S, Wong M-C, Han Y-D (2008) Voltage swell and overvoltage compensation with unidirectional power flow controlled dynamic voltage restorer. IEEE Trans Power Delivery 23(4):2513–2521
12. Li GJ, Zhang XP, Choi SS, Lie TT, Sun YZ (2007) Control strategy for dynamic voltage restorers to achieve minimum power injection without introducing sudden phase. IET Gener Transm Distrib 1(5):847–853
13. Somayajula D, Crow ML (2015) An integrated dynamic voltage restorer-ultra capacitor design for improving power quality of the distribution grid. IEEE Trans Sustain Ener
14. Newman MJ, Holmes DG, Nielsen JG, Blaabjerg F (2005) A dynamic voltage restorer (DVR) with selective harmonic compensation at medium voltage level. IEEE Trans Ind Appl 41(6):1744–1753
15. Jayaprakash P, Singh B, Kothari DP, Chandra A, Alhaddad K (2014) Control of reduced-rating dynamic voltage restorer with a battery energy storage system. IEEE Trans Ind Appl 50(2):1295–1303
16. Elnady A, Salama MMA (2005) Mitigation of voltage disturbances using adaptive perceptron-based control algorithm. IEEE Trans Power Delivery 20(1):309–318
17. Lee S-J, Kim H, Sul S-K, Blaabjerg F (2004) A novel control algorithm for static series compensators by use of PQR instantaneous power theory. IEEE Trans Power Electron 19(3):814–827
18. Adnan, Mustafa (2018) Multipurpose compensation scheme for voltage sag/swell and selective harmonics elimination in distribution systems. IEEE Trans Ind Appl 18:1
19. Kwon B-H, Jeong GY, Han S-H, Lee DH (2002) Novel line conditioner with voltage up/down capability. IEEE Trans Ind Electron 49(5):1110–1119
20. Choi N-S, Han B-M, Nho E-C, Cha H (2010) Dynamic voltage restorer using PWM AC-AC converter. In: Proceedings Power Electronics Conference (WEC) 2010, June 21–24, pp 2690–2695
21. Feng C, Liang J, Agilidis VG (2007) Modified phase-shifted PWM control for flying capacitor multilevel converters. IEEE Trans Power Electron 22(1):178–185
22. Peng FZ (2009) Revisit power conversion circuit topologies–recent advances and applications. In: Proceedings IEEE IPEMC, pp 188–192
23. Chapelsky C, Salmon J, Knight AM (2009) High-quality single-phase power conversion by reconsidering the magnetic components in the output stage - building a better half-bridge. IEEE Trans Ind Appl 45(6):2048–2055
24. Salmon J, Ewanchuk J, Knight AM (2009) PWM inverters using split-wound coupled inductors. IEEE Trans Ind Appl 45(6):2001–2009
25. Nielsen J, Blaabjerg F (2005) A detailed comparison of system topologies for dynamic voltage restorers. IEEE Trans Ind Appl 41(5):1272–1280

26. Aeoliza EC, Enjeti NP, Moran LA, Montero-Hernandez OC, Kim S (2003) Analysis and design of a novel voltage sag compensator for critical loads in electrical power distribution systems. IEEE Trans Ind Appl 39(4):1143–1150
27. Sankaran C (2002) Power quality. CRC Press LIC, United States of America
28. Usha Rani P (2014) Voltage swell compensation in an interline dynamic voltage restorer. J Sci Ind Res 73(1):29–32
29. Usha Rani P, Rama Reddy S (2011) Dynamic voltage restorer using space vector pwm control algorithm. Eur J Sci Res 56(4):462–470
30. Usha Rani P, Rama Reddy S (2011) Modelling and simulation of interline dynamic voltage restorer for voltage sag/swell compensation. J Electr Eng Romania 11(3):166–172(11.3.23) (ISSN 1582-4594)
31. Choi N-S, Han B-M, Nho E-C, Cha H (2010) Dynamic voltage restorer using PWM converter. In: Proceedings Power Electronics Conference (WEC) 2010 June 21–24, pp 2690–2695
32. Soeiro TB, Petry CA, Fagundes JCS, Barbi I (2011) Direct ac–ac converters using commercial power modules applied to voltage restorers. IEEE Trans Ind Electron 58(1):278–288
33. Park C-Y, Kwon J-M, Kwon B-H (2012) Automatic voltage regulator based on series voltage compensation with ac chopper. Proc. IET Power Electron. 5(6):719–725
34. Babaei E, Kangarlu MF (2012) Sensitive load voltage compensation against voltage sags/swells and harmonics in the grid voltage and limit downstream fault currents using DVR. Electric Power Syst Res 83:80–90
35. Divan D, Sastry J, Prasai A, Johal H (2008) Thin ac converters—a new approach for making existing grid assets smart and controllable. In: Proceedings IEEE power electronics specialist conference, pp 1695–1701 (PESC)
36. Enjeti PN, Choi S (1993) An approach to realize higher power PWM AC controller. In: Proceedings 8th annual applied power electronics conference and exposition (APEC'93), Mar 7–11, pp 323–327
37. Hyunhak S, Honnyong C, Heung-Geun K, Dong-Wook Y (2013) A novel single-phase PWM ac–ac converters without commutation problem. In: Proceedings IEEE energy conversion congress and exposition, pp 2355–2362
38. Shin H, Cha H, Kim HG, Yoo DW (2015) Novel single-phase PWM AC–AC converters solving commutation problem using switching cell structure and coupled inductor. IEEE Trans Power Electron 30(4):2137–2147
39. Khan AA, Cha H, Ahmed HF (2016) High-efficiency single-phase AC–AC converters without commutation problem. IEEE Trans Power Electron 31(8):5655–5665
40. Tolbert LM, Peng FZ, Khan FH, Li S (2009) Switching cells and their implications for power electronic circuits. In: Proceedings IEEE IPEMC, pp 773–779
41. Khan FH, Tolbert LM, Peng FZ (2006) Deriving new topologies of dc–dc converters featuring basic switching cells. In: Proceedings IEEE workshop computer power electronics, pp 328–332
42. Salmon J, Knight A, Ewanchuk J, Noor N (2008) Multi-level single phase boost rectifiers using coupled inductors. In: Proceedings IEEE International power electronics and motion conference, pp 3156–3163
43. Salmon J, Ewanchuk J, Knight A (2008) Single phase multi-level PWM inverter topologies using coupled inductor. In: Proceedings IEEE power electronics, pp 802–808
44. Galeshi S, Iman-Eini H (2016) Dynamic voltage restorer employing multilevel cascaded H-bridge inverter. IET Power Electron 9(11):2196–2204
45. Jiang F, Tu C, Shuai Z, Cheng M, Lan Z, Xiao F (2016) Multilevel cascaded-type dynamic voltage restorer with fault current-limiting function. IEEE Trans Power Deliv 31(3):1261–1269
46. Pires VF, Guerreiro M, Silva JF (2011) Dynamic voltage restorer using a multilevel converter with a novel cell structure. In: Proc EUROCON, pp. 1–4
47. Ajami A, Armaghan M (2009) A new concept of multilevel DVR based on mixed multi-cell cascaded topology. In: Proc ICMET. pp 398–402
48. Sadigh AK, Babaei E, Hosseini SH, Farasat M (2009) Dynamic voltage restorer based on stacked multi cell converter. In: Proc ISIEA, pp 419–424

49. Babaei E, Farhadi Kangarlu M, Sabahi M (2009) Dynamic voltage restorer based on multilevel inverter with adjustable dc-link voltage. IET Power Electron 7(3):576–590
50. Perez J, Cardenas V, Moran L, Nunez G (2009) Single-phase AC-AC converter operating as a dynamic voltage restorer (DVR). In: Proc IECON, pp. 1938–1943
51. Babaei E, Farhadi KM (2011) A new topology for dynamic voltage restorers without DC-link.. In: Proc ISIEA, pp. 1009–1014
52. Babaei E, Farhadi KM, Sabahi M (2010) Mitigation of voltage disturbances using dynamic voltage restorer based on direct converters. IEEE Trans Power Deliv 25(4):2676–2683
53. Lozano JM, Ramirez JM, Correa RE. A novel dynamic voltage restorer based on matrix converters. In: Proc MEPS, pp 1–7
54. Choi NS, Han BM, Nho EC, Cha H. Dynamic voltage restorer using PWM AC-AC converter. In: Proc IPEC, pp 2690–2695
55. Wang B (2010) A series compensator based on AC-AC power converters with virtual quadrature modulation. In: Proc PEDG, pp 438–443

Modeling and Simulation Analysis of Shunt Active Filter for Harmonic Mitigation in Islanded Microgrid

R. Zahira, A. Peer Fathima, D. Lakshmi, and S. Amirtharaj

Abstract The Microgrid is a combination of the distributed generation (DG) systems that deliver power to its local networks. The DG system constitutes non-conventional sources such as wind turbine, fuel cell, battery, and Solar. By connecting microgrid to the non-linear loads, it generates harmonic current into the system, which disturbs the sinusoidal waveform and power factor. The power quality compensator is used for improvement of the reliability and efficiency of the system. In this paper, first power quality issues are analyzed by connecting loads. As the system is connected to load it disturbs the sinusoidal waveform by introducing harmonics. To reduce these harmonics, passive filters are used. But due to the certain drawbacks such as complication in design and huge size of filter shunt active power filter (SAPF) is introduced in this system. These SAPF is controlled by an instantaneous power theory, which works as on line power quality analyzer and limits THD according to the IEEE 519-1992 standard. Depending on the power requirement of the nonlinear load, the proposed control for SAPF allows balanced line with near sinusoidal current and also maintains unity power factor. The presented model is simulated in the Matlab/simpower environment and results are validated.

Keywords Microgrid · Distributed generation system · Power quality · Shunt active power filter · Control strategies · Renewable energy · Harmonics · THD

R. Zahira (✉)
Department of Electrical and Electronics Engineering, BSA Crescent Institute of Science and Technology, Chennai, Tamilnadu, India
e-mail: zahirajaved@gmail.com

A. Peer Fathima
School of Electrical Engineering, Vellore Institute of Technology (VIT), Chennai, Tamilnadu, India
e-mail: peerfathima.a@vit.ac.in

D. Lakshmi · S. Amirtharaj
Department of Electrical and Electronics Engineering, Academy of Maritime Education and Training (AMET) (Deemed to be an University), Chennai, Tamilnadu, India
e-mail: lakshmiee@gmail.com

N. Zhou and S. Hemamalini (eds.), *Advances in Smart Grid Technology*, Lecture Notes in Electrical Engineering 688, https://doi.org/10.1007/978-981-15-7241-8_15

1 Introduction

To generate pollution free and environmental friendly power with Renewable Energy Source (RES), a large number of micro-producers will see the electrical grid in near future. To meet up the major power demand, it is necessary to utilize the RES such as wind energy, solar energy, biomass energy, etc. [1]. Photovoltaic (PV) energy and wind energy are increasing resources in distributed energy source (DES) [2]. The cluster of the DGs built in with biomass, solar energy and wind energy forms the Micro Grid (MG). Figure 1 presents the model of various DG's connected to non linear load. The MG's are emerging alternative source for developing green energy scenario and extending its support of utility grid [3]. The role of power electronic devices has become vital in distributed generation and integrating the RES with grid. Power electronic devices act as interface between RES and power grid. The power quality issues raises when the loads switches to the islanded mode from the grid connected mode. Extra harmonics are Injected in grid with the increased use of non-linear loads and sensitive equipment in the industrial and domestic area [4].

Harmonics distortion causes raise in temperature, malfunctioning of protective relays, mains and other control units. Therefore, it is essential to reduce the THD of the system within standard limit [5]. IEEE 519, IEEE 929-2000 and IEEE P-1547 are the standards available to monitor the limits of harmonics and other power quality issues. An enhanced awareness of power quality compensators has been developed among electricity consumers for enabling the working of the system within the restricted, standard limit. These compensators monitor the voltage and frequency in the line and when any deviation occurs, they compensate with proper control methods. Therefore, microgrid ensures better-quality power in the network [6]. The controller based on instantaneous power theory is connected at the point of common coupling (PCC) [7]. The conventional PI regulator along with feed forward compensation supports the controller in order to achieve a high dynamic response. The Clarke's transformation converts a–b–c coordinate of voltage and current into α–β–o coordinate, which helps to generate the reference signal for controlling Voltage Source Inverter (VSI) [8, 9].

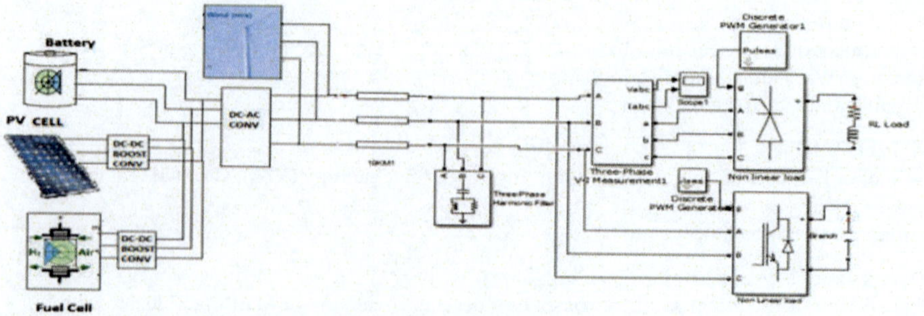

Fig. 1 Non-linear loads connected to distributed generation systems

The output of VSI is fed back to the line with inductor for compensation. The main objective of this work is to indicate improvement of the quality of power supply by attaining adequate voltage, current and frequency of the microgrid along with acceptable power sharing among the DG units [10, 11]. In this paper, Sect. 1 introduces the paper. Section 2, explains the modeling of various DG systems such as wind, PV cell, fuel cell, and battery. Section 3, deals with shunt active filter. Sections 4 and 5, validates and conclude the results.

A non-linear load is connected to the DG system through bus, which creates a disturbance in voltage and current magnitude. With the Fast Fourier transform (FFT), THD is measured, which is beyond the standard limit due to disturbances in the system. For lower rating, passive filters are implemented for compensating the disturbances. As the system size increases, the designing of passive filter becomes complicated. Moreover, the passive filters will be bulky in size for high rating system. To overcome this drawback, SAPF is used in this paper. In Section 3, SAPF design and its control method are discussed, in which simple and best control technique is used in this system to control VSI.

2 Modelling of Distributed Energy Sources

The RES such as wind energy, PV arrays, SOFC and battery are designed and coupled to the AC bus system through converters [9, 10]. Equations (1)–(3) give a complete sketch of the SOFC model [12]. Equations (4)–(10) show the design of wind turbine with several subsystems modeling [13, 14]. Equations (11)–(16) gives the design expression for the PV cells. Equations (17)–(22) gives the lead acid battery model, which acts as back up source for the system. The modelled PV cells and batteries are coupled to a load through the converter, with which it can supply to AC load. For improving the quality of power, the SAPF is connected at PCC. VSI is operated with current controlled technique for the Microgrid operating in grid connected. With this technique, it maintains unity power factor in AC bus. In addition, islanded Microgrid operates VSI in voltage-controlled technique, to maintain the AC bus voltage stability [15].

2.1 Modeling of Solid Oxide Fuel Cell (SOFC)

A SOFC is an electrochemical energy conversion system, where it converts chemical energy into heat and electrical energy [12]. Figure 2 shows SOFC simulated model. Equation (1) gives the chemical reaction, which is due to the material balance, is included with the change in the concentration of input and output flows into a volume of control as well as net production. Equations (2) and (3) expresses the voltage and current of the stack. Figure 3

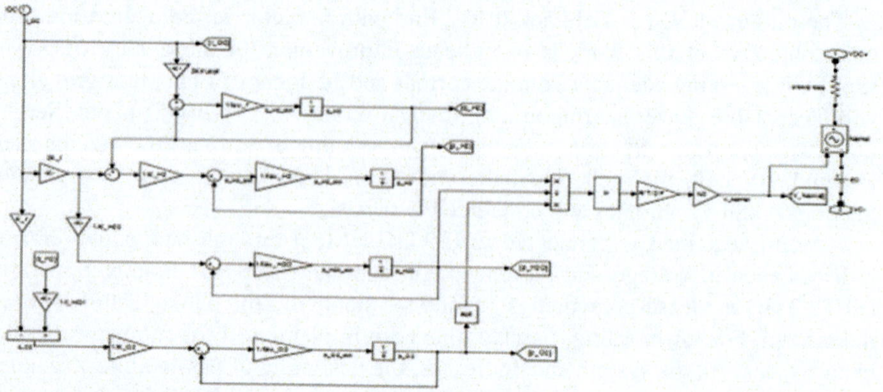

Fig. 2 Modelling of fuel cell

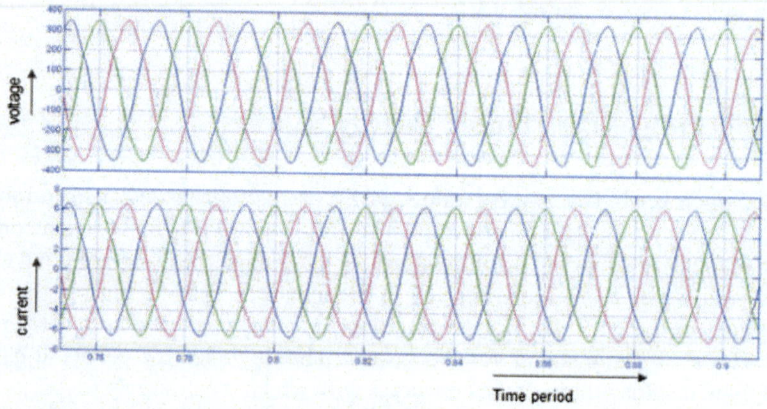

Fig. 3 SOFC output voltage and SOFC output current

shows the voltage and current output waveformafter connecting via an inverter that produces a voltage of about 320 V and 6.5A.

$$\left(\frac{P_r V_o}{R_{gas} T}\right)\dot{x} = N_i^{in} = N_i^r - N_i^o \tag{1}$$

where, V_o is the volume of the cell,
N_i^o are the flow levels at the cell input and output of the eight reactants respectively,
N_i^r is the reaction rate of the ith reactant. P_r is the pressure of the cells,
T is the temperature of the cells in K, and R_{Gass} is the constant of gas (8.31 J/ mole °K). V_{dc} is the number stack voltage and I_{dc} is the number stack current,

$$V_{dc} = C_n E_o + \left(\frac{C_n R_{gas} T}{2F}\right) \left[\ln \left(\frac{P_H \sqrt{P_{O_2}}}{P_{H_2O}}\right)\right] \tag{2}$$

$$I_{dc} = k_m \int \frac{(P_G - P_{ref})}{V} dt \tag{3}$$

Whereas, C_n is the number of cells in the stack associated with the series F is Faradays constant; K_m is the anode valve constant,
P_G is the power generator, and P_{ref} the reference power.

2.2 *Modelling of the Wind Turbine*

Kinetic energy is produced as the wind passed through the blades of the rotor. The energy derived from the rotor is double the wind speed, so the turbine must be planned to withstand the storms [13, 14]. Equation (4) gives the output power.

$$P_o = 0.5\, P_c V_\omega^3 A \sigma \tag{4}$$

where, P_o and P_c denotes Power and Power coefficient, V_ω gives the Velocity of the Wind, A is the rotor and density of air is denoted by σ. The P_c given in Eq. (3) is a kinetic energy extracted by the wind turbine. P_c is a function of the tip speed ratio δ and the Blade angle α which goes through furthermore analysis as expressed in Eqs. (5) and (6) [15]. Figure 4 shows the wind turbine modeling. With $\delta = 8.1$ and $\alpha = 0°$, Maximum value of Pc can be achieved.

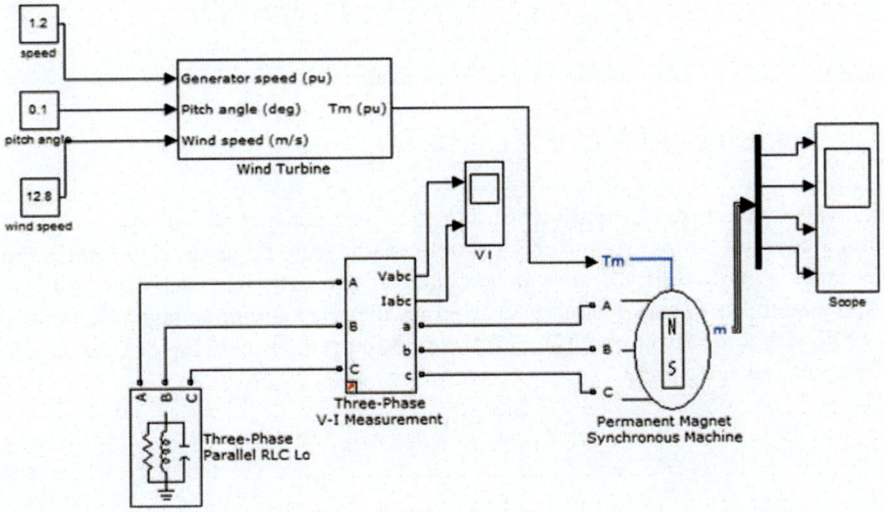

Fig. 4 Modeling of wind turbine

$$P_c(\delta, \alpha) = 0.5176 \left(\frac{116}{\delta_i} - 0.4\,\alpha - 5\right) \; e^{-21/\delta_i} + 0.0068\,\delta \tag{5}$$

$$\frac{1}{\delta_i} = \frac{1}{\delta + 0.08\alpha} - \frac{0.035}{\alpha^3 + 1} \tag{6}$$

Equations (7) and (8) give the expression of the voltage function of PMSM in the dq axes reference frame.

$$V_d = R_{i_d} + L_d \frac{\mathrm{d}i_\mathrm{d}}{\mathrm{d}t} - \omega_e L_q i_q \tag{7}$$

$$V_q = R_{iq} + L_q \frac{\mathrm{d}i_q}{\mathrm{d}t} - \omega_e L_d i_d + \omega_e \lambda_r \tag{8}$$

In the dq-axes reference frame, V_d and V_q are instantaneous stator voltages, i_d and i_q are instantaneous stator currents,
L_d and L_q are the inductances,
ω_e is the electrical angular speed of the rotor,
λ_r is the peak/maximum phase flux linkage due to the rotor.

The Active power stated as in Eqs. (9) and (10) derives the electromagnetic torque developed by a PMSM.

$$\mathrm{P_{em}} = \frac{3}{2}\,\omega_\mathrm{e}\,\left(\lambda_d\, i_q - \lambda_q\, i_d\right) \tag{9}$$

$$\mathrm{T_e} = \frac{3\mathrm{p}}{4}\left(\lambda_\mathrm{r} i_\mathrm{q} + (\mathrm{L}_d - \mathrm{L}_q)\; i_q i_d\right) \tag{10}$$

2.3 *Modeling of PV Cell*

Equivalent circuit model is shown in Fig. 5, which consists of a diode in parallel with a constant current source and a shunt resistor [11]. Equation (11) furnish the PV cell's output current (I_{pv}), which is equal to the light-generated current I_{lg} as expressed in Eq. (15). Equation (12) specifies the open circuit voltage (V_{oc}) of the pv cell with zero load current (I). I_d is the diode current given in Eq. (13), and I_{sh} is the shunt-leakage current.

$$I_{pv} = I_{lg} - I_d - I_{sh} \tag{11}$$

$$V_{oc} = V_t + I\,R_{se} \tag{12}$$

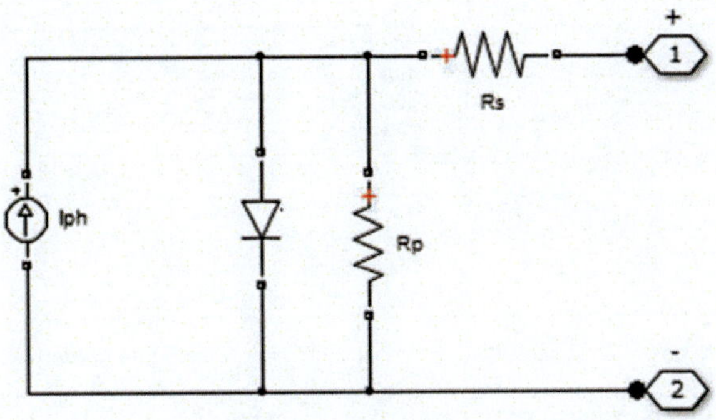

Fig. 5 PV module's equivalent circuit

$$I_d = I_s(qV_{oc}/\gamma_a\beta_z T) - 1 \tag{13}$$

where V_t is the *PV* cell terminal voltage and R_{se} and R_{sh} are series and shunt resistance, I_s is the diode saturation current, curve fitting constant given as γ_a, the electron charge ($1.6 * 10^{-19}$) in *C* is given as *q*, β_z is the Boltzmann constant ($1.38 * 10^{-23}$) J/K, *T* = temperature *K* [16]. Equation (14) gives the expression of load current (I_L), in which I_{sc} (cell reverse saturation current) is expressed in Eq. (16).

$$I_L = I_{lg} - I_{sc}\left\{e^{\left(\frac{qV_{oc}}{\gamma_a\beta_z T}\right)} - 1\right\}\frac{V_{oc}}{R_{sh}} \tag{14}$$

$$I_{lg} = \frac{S_r}{100}\{I_{sc} + K_I(T - 25)\} \tag{15}$$

$$I_{sc} = I_c\left(\frac{T^3}{T_r}\right)\cdot e^{\frac{qb_s}{Bk}\left(\frac{1}{T_r} - \frac{1}{T}\right)} \tag{16}$$

where, ideality factor of p-n junction diode is given as *B*, cell temperature [°C] as *T*, K_I is short circuit current temperature coefficient (0.0017 A/°C), S_r is solar irradiation in W/m^2, b_s is a band gap of silicon, T_r represents reference temperature (301.18 ± K), I_c represents cell saturation current.

2.4 Modeling of Lead-Acid Battery

To meet the additional peak demands, the batteries are coupled with the inverter in the system [11]. These batteries are used as spinning reserve during peak seasons. The chemical reactions that occur in a lead-acid battery represented by Eq. (17). Figure 6 shows simulation output voltage.

$$PbO_2 + Pb + 2H_2SO_2 \overset{\text{discharge/Charge}}{\longleftrightarrow} 2PbSO_4 + 2H_2O \tag{17}$$

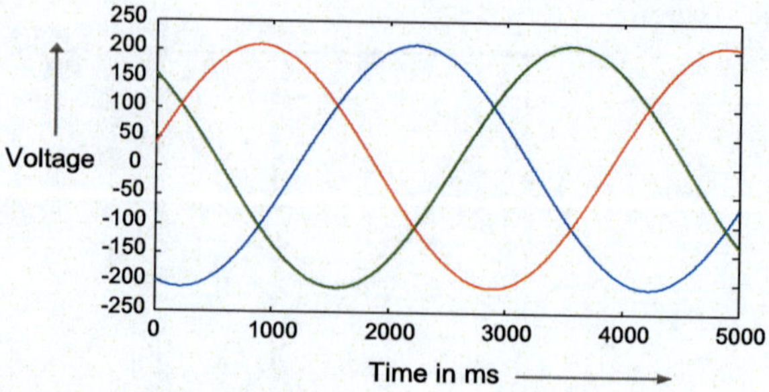

Fig. 6 Voltage waveform of battery source

Terminal voltage (V_t) is expressed as in Eqs. (18), and (19) to Eq. (22) gives charging and discharging voltage and resistance.

$$\mathrm{V}_t = \mathrm{V} + \mathrm{IR} \tag{18}$$

$$\mathrm{V_c} = [2 + 0.148\, soc] * \mathrm{N}_s \tag{19}$$

$$R_{\mathrm{ch}} = \frac{0.758 + \frac{0.1309}{[1.06 - soc]} \quad * N_s}{Q_{m\,c}} \tag{20}$$

where N_s is the number of 2 V battery cells in series; Q_{mc} is maximum battery capacity; *soc* is the current state of charge (Fig. 7).

$$V_{dcg} = [1.926 + 0.124 soc] * N_s \tag{21}$$

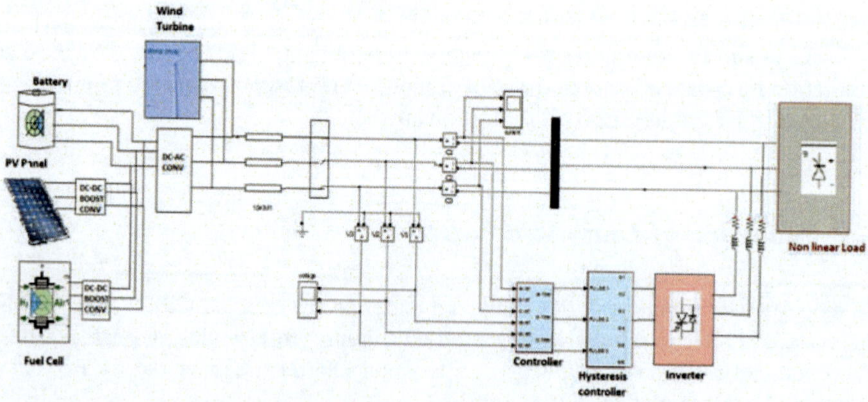

Fig. 7 Shunt active power filter with a voltage source inverter

$$R_{dcg} = \frac{0.19 + \frac{0.1037}{[soc - 0.14]} \quad * N_s}{Q_{mc}} \tag{22}$$

3 Shunt Active Filter

The shunt APF concept was first introduced by Gyugyi and strycula in 1976. APF cancels harmonics and compensates for reactive power. Akagi et al. in 1983 have suggested "The generalized theory of the instantaneous reactive power in three-phase circuits", also known as the instantaneous power theory, or p-q theory. Figure 8 shows the SAPF connection at PCC. Sensed Voltage and current of 'a–b–c' coordinate is converted to 'α-β-o' coordinate with Clarke's transformation as shown in Eq. (23) and (24). Real, Reactive power and Zero sequence power are calculated as in Eq. (25).

$$\begin{bmatrix} v_\alpha \\ v_\beta \\ v_o \end{bmatrix} = \sqrt{2/3} * \begin{bmatrix} 1/\sqrt{2} & 1/\sqrt{2} & 1/\sqrt{2} \\ 1 & 1/\sqrt{2} & -1/\sqrt{2} \\ 0 & \sqrt{3/2} & -\sqrt{3/2} \end{bmatrix} * \begin{bmatrix} v_a \\ v_b \\ v_c \end{bmatrix} \tag{23}$$

$$\begin{bmatrix} i_\alpha \\ i_\beta \\ i_o \end{bmatrix} = \sqrt{2/3} * \begin{bmatrix} 1/2 & 1/2 & 1/\sqrt{2} \\ 1 & 1/2 & -1/2 \\ 0 & \sqrt{3/2} & -\sqrt{3/2} \end{bmatrix} * \begin{bmatrix} i_a \\ i_b \\ i_c \end{bmatrix} \tag{24}$$

$$\begin{bmatrix} P_o \\ P \\ Q \end{bmatrix} = \begin{bmatrix} v_o & 0 & 0 \\ 0 & v_\alpha & v_\beta \\ 0 & -v_\beta & v_\alpha \end{bmatrix} * \begin{bmatrix} i_a \\ i_b \\ i_c \end{bmatrix} \tag{25}$$

$$\begin{bmatrix} i^*_{ca} \\ i^*_{cb} \\ i^*_{cc} \end{bmatrix} = \sqrt{\frac{2}{3}} * \begin{bmatrix} 1/\sqrt{2} & 1 & 1 \\ 1/\sqrt{2} & -1/\sqrt{2} & \sqrt{3/2} \\ 1/\sqrt{2} & -1/\sqrt{2} & \sqrt{-3/2} \end{bmatrix} * \begin{bmatrix} i^*_{co} \\ i^*_{c\alpha} \\ i^*_{c\beta} \end{bmatrix} \tag{26}$$

where, P_o is the instantaneous zero-sequence power; P the instantaneous real power; Q the instantaneous imaginary power. Equation (26) derives the 3Φ reference current,

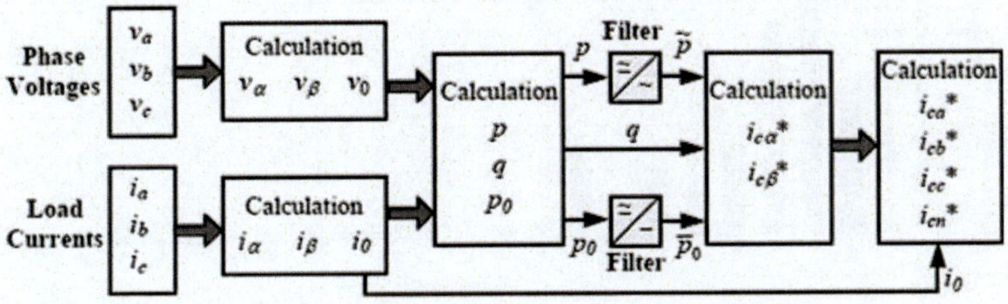

Fig. 8 Calculations for the constant instantaneous supply power control strategy

which is then subtracted with the load current and used to generate the corresponding PWM signal using hysteresis band. The hysteresis band produces six PWM signals for inverter. The output of the inverter is then transferred back to the line. As a result, the distortion is compensated and produces a near sine waveform [17, 18]. Figure 7 shows the overall block diagram of the system connected with shunt active filter. Figure 8 gives the detailed flow of calculating compensation current.

4 Simulation Analysis

Microgrid is modelled by multiple DG systems like PV cell, SOFC, wind and battery source in the MATLAB/Simulink 2014 package [19, 20]. The MATLAB/Simulink software helps to model, simulate, and analyze the system with linear and nonlinear loads in continuous time and sampled time. With this output of each source is analyzed and then connected to a linear and nonlinear load, to check the response of island grid [25–28]. Figure 9 shows the linear load response, in which it gives THD of 5.35% [21–23], which is acceptable limit as refereed to harmonics standard. When the system connected to non-linear loads, then the voltage and current waveform is distorted as shown in Figs. (10 and 11). In which the THD is measured as 16.19% for voltage waveform and 24.64% for current waveform, which is exceeding the limit of the harmonic standard shown in Table 1 [24]. Hence, to obtain the limits, compensation is necessary, which will satisfy the harmonic standards and to improves the power factor. SAPF with the hysteresis current controller is modelled and connected to the grid at PCC. The simulation is carried out using MATLAB/Simulink software in order to verify the proposed system. Figures 12a and b shows the real and reactive power of the system and to achieve balanced sinusoidal grid currents at unity power

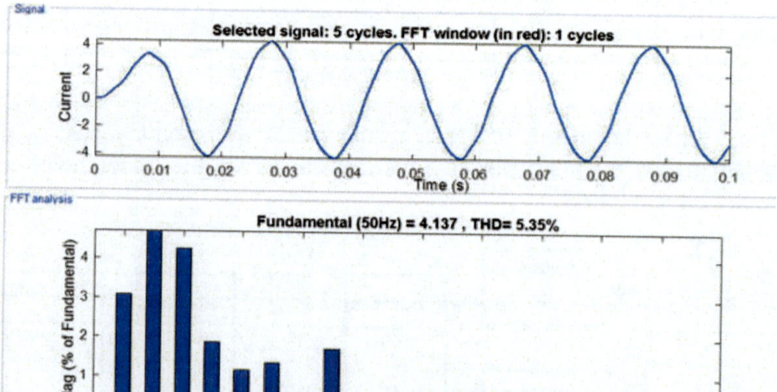

Fig. 9 Linear load Current waveform with 5.35% THD

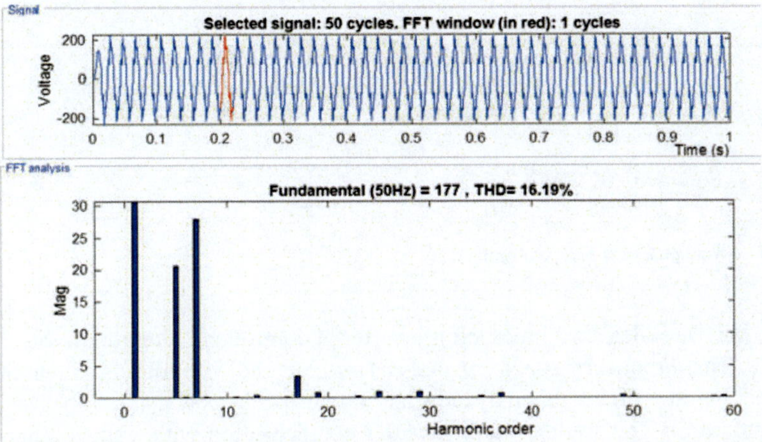

Fig. 10 Voltage distortion before filter with THD 16.19%

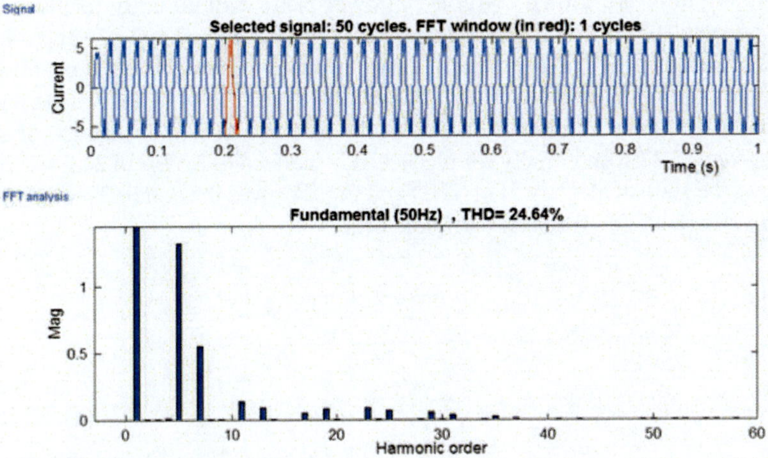

Fig. 11 Current THD before filtering

Table 1 Distortion limits as per IEEE 519_1992 harmonic standard

Maximum harmonic current distortion in percent of IL					
Individual harmonic order (odd harmonics)					
ISC/IL	<11	11 < h <17	17 < h <23	23 < h < 35	35 < h	THD
<20	4.0	2.0	1.5	0.6	0.3	5.0
20 < 50	7.0	3.5	2.5	1.0	0.5	8.0
50 < 100	10.0	4.5	4.0	1.5	0.7	12.0
100 < 1000	12.0	5.5	5.0	2.2	1.0	15.0
>1000	15.0	7.0	6.0	2.5	1.4	20.0

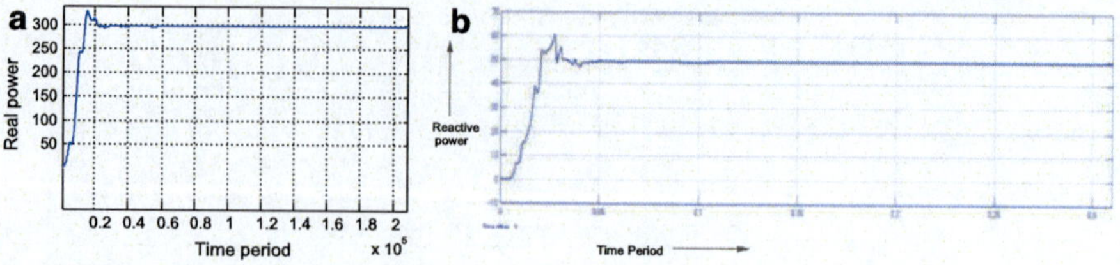

Fig. 12 **a** Real power. **b** Reactive power

factor, the Three-leg grid interfacing inverter is controlled using renewable energy sources. The Frequency spectrum analysis without and with filter is shown in Figs. (13 and 14) in which it is analyzed that compensating power the improves to 42 dB from 38 dB.The simulation result after compensation with reduced harmonic distortion of 2.57% is shown in Fig. 15.

Consolidated analysis of current THD and power factor of unbalanced load with and without filter are shown in Table 2. Power factor will be unity for linear load. As it connected to the non-linear load, power factor is reduced to 0.821 and it affects the system efficiency [26, 27]. These Power factor can be improved to near unity as 0.9925 through compensation. Table 3 gives a clear picture of harmonic reduction is each phase. In phase A, THD of 25.78% is reduced to 2.66%. For phase B, it attains 2.62% from 25.22% and finally for phase C it is reduced to 2.57% to 24.64%. Figure 16 shows the statistical analysis of THD and Fig. 17 gives the detailed representation of harmonic reduction for each harmonic order.

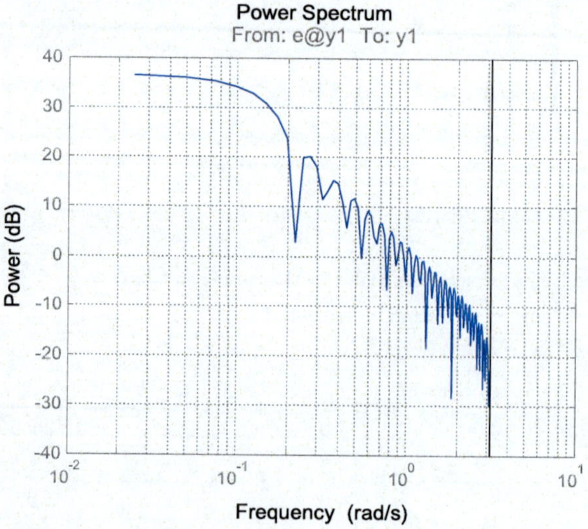

Fig. 13 Frequency spectrum analysis before filter

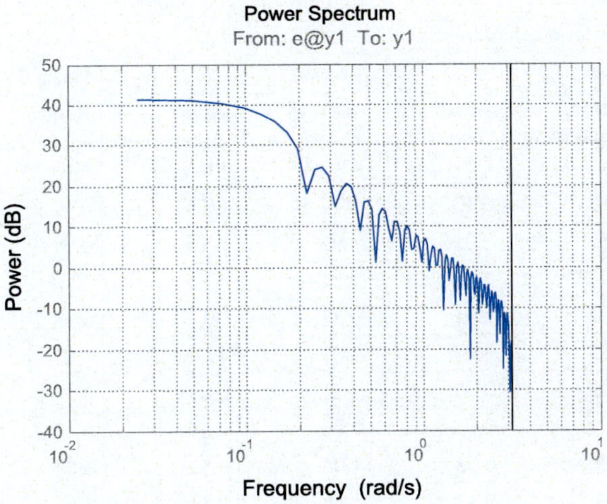

Fig. 14 Frequency spectrum analysis after filter

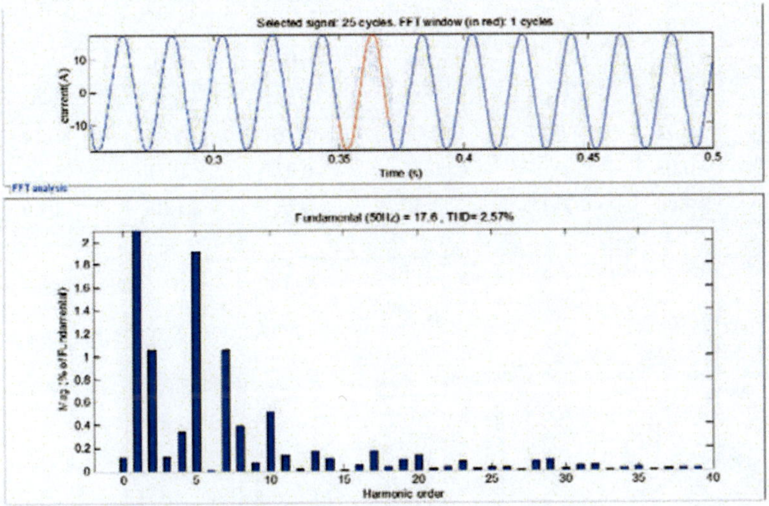

Fig. 15 Compensated current waveform with THD 2.57%

Table 2 Details of power factor and current THD

	Linear load output before connecting filter	Non-linear load output before connecting filter	After connecting filter
Power factor	1	0.821	0.9925
THD (%)	5.35	24.64	2.66

Table 3 Total harmonic distortion with harmonic order

Harmonics order	Phase A		Phase B		Phase C	
	Before connecting filter (%)	After connecting filter (%)	Before connecting filter (%)	After connecting filter (%)	Before connecting filter (%)	After connecting filter (%)
3	0.03	0.13	1.43	0.28	0.59	0.12
4	0.12	0.35	1.95	0.36	0.97	0.34
5	22.33	1.88	22.49	1.85	22.09	1.91
6	0.02	0.01	0.87	0.1	0.38	0.01
7	9.32	1.07	7.87	1.09	7.87	1.06
8	0.02	0.43	1.56	0.41	0.77	0.39
9	0.02	0.02	0.86	0.06	0.41	0.07
10	0.06	0.58	1.06	0.58	0.54	0.52
11	6.96	0.13	6.06	0.11	5.6	0.13
12	0.03	0.01	0.58	0.03	0.22	0.01
13	4.02	0.17	3	0.18	2.91	0.17
14	0.02	0.11	0.87	0.1	0.43	0.1
15	0.01	0	0.4	0.01	0.19	0
16	0.04	0.04	0.48	0.05	0.27	0.04
17	2.74	0.17	2.08	0.16	1.79	0.17
18	0.03	0.01	0.31	0.02	0.11	0.03
19	1.88	0.11	1.17	0.12	1.08	0.09

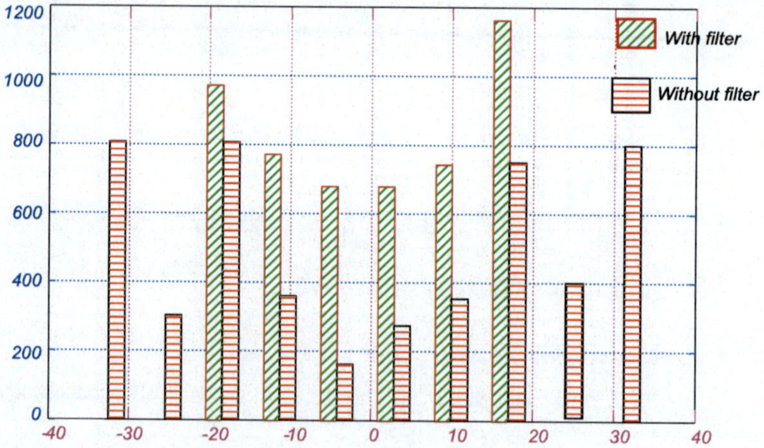

Fig. 16 Statistical graph of current THD—with and without filter

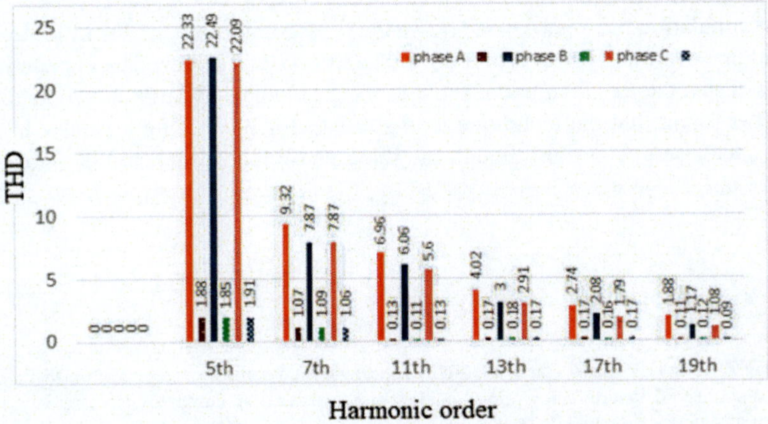

Fig. 17 Graphical representation of THD versus harmonic order for all phases

Table 3 shows the results before and after compensation in which the harmonic order 5th, 7th, 11th, 13th, and 17th causes the disturbances. After compensation, this level is within the limit according to IEEE harmonic standard 519 as in Table 1. Figure 17 represents the comparative graph of Total harmonic distortion with and without filter. For Phase A, the fifth harmonic reduces from 22.33% to 1.88%, 7th from 9.32% to 1.07%, 11th, 6.96% to 0.13%, 13th from 4.02% to 0.17% [27].

From Fig. 17 it is clear that the harmonic is reduced to the better level. The odd harmonics from 3rd to 19th range are mainly produces the large disturbances in the system. The THD levels of 5th harmonics are examined as 22.33% for phase A, 22.49% for phase B and 22.09%, for phase C. which is very well reduced to 1.88% for phase A, 1.85% for phase B and 1.91% for phase C respectively. Likewise, 7th harmonic THD level are reduced to 1.07, 1.09 and 1.06%.

5 Conclusion

In this paper, a dynamic simulation model including different micro sources like wind turbine, fuel cell, battery and PV cell has been presented in the Microgrid system. In addition, the quality of power is analyzed in the system when connected to linear and non-linear loads. The simulation results show that, the harmonic disturbances normally appear when connected to non-linear loads and the compensation is required for making the system reliable and efficient with neat power.

Traditionally, passive filters are used to eliminate line current harmonics and to increase the power factor. However, in some practical applications, the conventional solution becomes ineffective due to variation of magnitude and frequency, which creates random variation in the harmonic distortion. So, SAPF with pq theory is used for eliminating harmonics from the system and to improve the power factor.

The simulation has been carried out in the matlab environment. From Table 3, it is noticed that overall THD level of 2.66, 2.62 and 2.57% is achieved by connecting the proposed shunt active filter and it satisfies the harmonic standards shown in Table 1. Also it is found that the technique works satisfactorily yielding a sinusoidal unity power factor source current though the three-phase load connected is a non-linear reactive one.

References

1. Zeng Z, Yang H, Zhao R, Cheng C (2013) Topologies and control strategies of multi-functional grid-connected inverters for power quality enhancement: A comprehensive review. Renew Sustain Energy Rev 24:223–270
2. Singh B, Al-Haddad K, Chandra A (1999) A review of active filters for power quality improvement. IEEE Trans Industr Electron 46(5):960–971
3. Li Y, Vilathgamuwa DM, Loh PC (2005) Microgrid power quality enhancement using a three-phase four-wire grid-interfacing compensator. IEEE Trans Ind Appl 41(6):1707–1719
4. Xiaozhi G, Linchuan L, Wenyan C (2011) Power quality improvement for mircrogrid in islanded mode. Procedia Eng 23:174–179
5. Serban IOAN, Marinescu C (2008) Power quality issues in a stand-alone microgrid based on renewable energy. Rev Roum Sci Techn-Électrotechn et Énerg 53(3):285–293
6. Pepermans G, Driesen J, Haeseldonckx D, Belmans R, D'haeseleer W (2005) Distributed generation: definition, benefits and issues. Energy policy 33(6):787–798
7. Melício R, Mendes VMF, Catalão JPDS (2011) Comparative study of power converter topologies and control strategies for the harmonic performance of variable-speed wind turbine generator systems. Energy 36(1):520–529
8. Marei MI, El-Saadany EF, Salama MM (2004) A novel control algorithm for the DG interface to mitigate power quality problems. IEEE Trans Power Delivery 19(3):1384–1392
9. Kale M, ÖZDEMİR E (2015) A new hysteresis band current control technique for a shunt active filter. Turkish J Electr Eng Comput Sci 23(3):654–665
10. Li X, Song YJ, Han SB (2007, July). Study on power quality control in multiple renewable energy hybrid microgrid system. In: 2007 IEEE lausanne power tech, IEEE, pp 2000–2005
11. Temma K, Kono Y, Shimomura M, Kataoka M, Goda T, Uesaka S (2006) Proposal and development of power quality improvement method under islanding operation in a micro-grid. IEEJ Trans Power Energy 126:1032–1038
12. Das S, Das D, Patra A (2014) Operation of solid oxide fuel cell based distributed generation. Energy Procedia 54:439–447
13. Kesraoui M, Chaib A, Meziane A, Boulezaz A (2014) Using a DFIG based wind turbine for grid current harmonics filtering. Energy Convers Manag 78:968–975
14. Hoseinpour A, Barakati SM, Ghazi R (2012) Harmonic reduction in wind turbine generators using a Shunt Active Filter based on the proposed modulation technique. Int J Electr Power Energy Syst 43(1):1401–1412
15. Hassan AA, Fahmy FH, Nafeh AESA, El-Sayed MA (2010) Modeling and simulation of a single phase grid connected photovoltaic system. WSEAS Trans Syst Control 5(1):16–25
16. Cobben JFG, Kling WL, Myrzik JMA (2005, November) Power quality aspects of a future micro grid. In: 2005 International conference on future power systems, IEEE, pp 5
17. Zahira R, Fathima AP, Muthu R (2014) Hardware and simulation modelling of shunt active filter controlled by sliding mode. Int J Appl Eng Res 9(24):8229–8235
18. Zahira R, Lakshmi D (Aug, 2011) Control techniques for improving quality of power-a review. In: The greatness of any research work is only through its wider applicability

19. Rustemli S, Cengiz MS (2015) Active filter solutions in energy systems. Turkish J Electr Eng Comput Sci 23(6):1587–1607
20. Barote L, Marinescu C (2014) Software method for harmonic content evaluation of grid connected converters from distributed power generation systems. Energy 66:401–412
21. Zahira R, Fathima AP, Muthu R (2014, January) SPWM technique for reducing harmonics in three-phase non-linear load. In: 2014 IEEE 2nd international conference on electrical energy systems (ICEES), IEEE, pp 56–60
22. Zahira R, Fathima AP, Muthu R (2016) Harmonic reduction in wind power generating system using shunt active filter with SPWM technique. Circuits Syst 7(04):157
23. Mohamed F (2006) Microgrid modelling and simulation. Helsinki University of Technology, Finland
24. Duffey CK, Stratford RP (1989) Update of harmonic standard IEEE-519: IEEE recommended practices and requirements for harmonic control in electric power systems. IEEE Trans Ind Appl 25(6):1025–1034
25. Zahira R, Lakshmi D, Ravi CN, Sasikala S (2018) Power quality enhancement using grid connected PV inverter. J Adv Res Dyn Control Syst 10(05):309–317
26. Ramya N, Sasikala, Zahira R (2018) A noval multi input grid connected bidirectional converter for hybrid systems. Int J Pure Appl Math 119(12):2993–3004
27. Zahira R (2018) Design and performance analysis of a shunt active filter for power quality improvement
28. Chishti F, Murshid S, Singh B (2019) Development of wind and solar based AC microgrid with power quality improvement for local nonlinear load using MLMS. IEEE Trans Ind Appl 55(6):7134–7145

Zahira Rahiman received B.E (Electrical and Electronics Engineering) from University of Madras in 2004, M.E (Power Systems Engineering) from B.S.A.Crescent Engineering College, Chennai in 2006 and the doctoral degree from Anna University in 2018. Having 14 years of experience, She started her teaching carrier as Lecturer in 2006 Promoted as Assistant professor, Senior assistant professor, and Associate Professor (2018) in Tagore Engineering College. She is Currently working as Assistant Professor in BSA Crescent Institute of Science and Technology. She has published over 19 national and international journals and nearly 15 National and international conferences Proceedings. She has Received Young Educator and Scholar Award in 10th National Teachers' Day Awards 2019 from NFED Association (Coimbatore), Women Researcher Award, Outstanding scientist Award and Innovative Technological Researcher & Dedicated Academician Award (Electrical Engineering). She has guided over 20 UG and PG candidates. She was life Member in 6 Professional Bodies, and acting as reviewer for reputed journals. Her area of interest are Power quality, harmonic suppression, Active filter control techniques, renewable energy systems, Microgrid, electric vehicle charging systems.

Dr. A. Peer Fathima has obtained her B.E degree in Electrical &Electronics Engineering from Madurai Kamaraj University, M.E degree in High voltage Engineering from Anna University,Chennai, M.S degree in Electronics and Control from BITS Pilani. She received her Ph.D degree from Anna university, Chennai.She has been in the teaching profession for the past 28 years .She has published over 80 papers in International journals and national conferences.She has guided several P.G and U.G projects. She produced three Ph.Ds and guiding six Ph.D scholars at VIT. Her main teaching and research interest encompasses power system operation and control in deregulated powersystems,renewable energy applications in power systems.

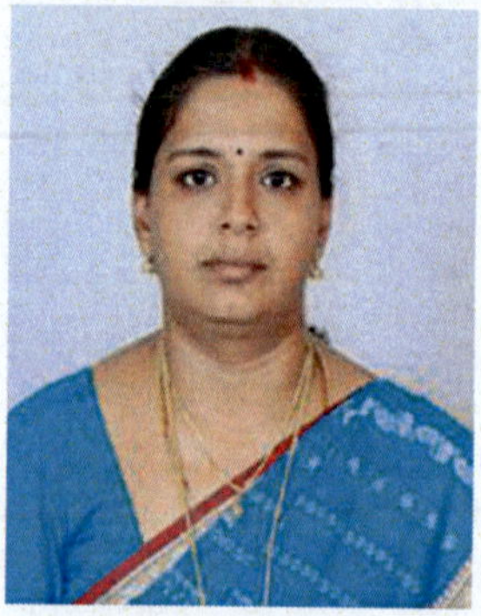

D. Lakshmi received her undergraduate degree in Electrical and Electronics Engineering from Madras University in 1999 and postgraduate degree in Power Systems from Anna University in 2006. She finished her Ph.D. in 2018 from Anna University, Chennai at SSN College of Engineering, India working as an Associate Professor in Department of Electrical & Electronics Engineering at Academy of Maritime Education and Training (Deemed to be an University), Chennai. She has been teaching field for more than 18 years. She has published 21 papers in International Journals and more number of papers in Conference Proceedings and has supervised many undergraduate and postgraduate projects. Her research interests include deregulated power system, power generation and operation, and power system control.

S. Amirtharaj received B.E. in Electrical and Electronics from Coimbatore Institute of Technology in 1998, M.E. from Thiagarajar College of Engnieering, Madurai in 2006. Currently, he is pursuing Ph.D. in Anna University, Chennai. He has an experience of 16 years of experience in the teaching field and presently he is working as Assistant Professor in in Department of Electrical and Electronics Engineering at Academy of Maritime Education and Training (Deemed to be an University), Chennai. He has received received many awards for his academic excellence. His research area includes Design and analysis of high efficiency power converter for low power photovoltaic generation system.

Fault Diagnosis of Self-aligning Conveyor Idler in Coal Handling Belt Conveyor System by Statistical Features Using Random Forest Algorithm

S. Ravikumar, V. Muralidharan, P. Ramesh, and Cheran Pandian

Abstract A coal handling system equipment is a bulk material handling system; it plays an important role in key mechanical industries. It holds the important aspects of a country economy in mining, smelter plants, thermal power plants, process industries, etc. Considering operation attributes of the coal handling belt conveyors, various parameters have to be taken into account while designing, as it has to convey materials from one location to another continuously for most part of the year. In broad spectrum, the flat belt coal conveyor has to be with maximum load handling ability, for conveying over long distance in a single stroke. Hence, it has to be steadfast in design, with easy operation and maintenance and high dependability in function. Self-aligning conveyor roller (SACR) is an important element in coal belt conveyor. It is placed between the carrying conveyor idlers to vary the sideways dislocation caused by imbalance loading which is difficult to avoid in harsh loading conditions. When the coal conveyor belt moves against the carrying rollers, there is a difference in frictional force between two sides, which will make the top strand of the coal belt conveyor to twist toward the center. Further, the crisscross movement, offset from the center line, and damage of coal conveyor belt were competently prevented by self-aligning conveyor roller. As SACR is found to be critical in coal belt conveyor systems, it becomes compulsory to supervise its smooth and continuous functioning. To make sure this certain, condition monitoring of self-aligning conveyor roller (SACR) should be done periodically which principally creates a classification or categorization problem. Self-aligning conveyor roller is made of vital elements like groove ball bearing, main central shaft, and the external shell. In this case, it is categorized with the below mentioned cases such as coal handling belt conveyor with SACR running in no-fault condition (NFC), with groove ball bearing fault condition (BBFC), with main shaft fault (MSF), with combined ball bearing fault condition and main shaft fault (BFC & MSF). A model investigational arrangement has been made as per the actual coal handling belt conveyor operating conditions and research

S. Ravikumar (✉) · V. Muralidharan
B.S Abdur Rahman Crescent Institute of Science and Technology, Chennai, India
e-mail: ravi78usa@gmail.com

P. Ramesh · C. Pandian
Acadmey of Martime Education and Training (AMET), Chennai, India

N. Zhou and S. Hemamalini (eds.), *Advances in Smart Grid Technology*, Lecture Notes in Electrical Engineering 688, https://doi.org/10.1007/978-981-15-7241-8_16

requirement. Followed by the fabrication of SACR setup, the vibration signatures were obtained from the model for frequently occurring fault discussed earlier. These vibration signatures are fed to digital convertor and transformed to digital signals. From the digital vibration signals, statistical features were calculated. Then, effective statistical features were extracted and provided as input to random forest algorithm, followed by categorization, which was performed by random forest algorithm. In the current work, the random forest algorithm achieved 90.2% categorization accuracy, which summarizes the algorithm correctness in fault prediction self-aligning conveyor roller failure and life assessment.

Keywords Self-aligning conveyor roller (SACR) · Coal belt conveyor system (CBCS) · Random forest · Confusion matrix

1 Introduction

Self-aligning conveyor roller is an inevitable module of (CBCS) coal belt conveyor system; it will become damaged due to certain unavoidable operating conditions such as incoherent stresses, contemptible oil usage in lubricating in site conditions, wearable rubber seals, multidimensional angle loading of coal from ship unloader, traveling hoppers and from transfer towers, and continuous operation of coal handling conveyors due to demanding conditions of operation. The vital elements of SACR which fail often are groove ball bearing which balances the outer roller revolution and main shaft which holds the entire load of the SACR. Any malfunction of these vital SACR elements will consequently develop poor operating conditions of the coal belt conveyor, which subsequently will trail to failure of coal belt conveyor system. In this regard, huge economical and fatal damages have been reported in various coal handling belt conveyor thermal power plants in TNEB as well as across the world. This unquestionably insists the ultimate need for malfunction prediction system in SACR, which is the main focus of this research work and purpose of exploration of this industrial work. The assorted faults measured for this research work were SACR operating in no-fault condition (NFC), with groove ball bearing fault condition (BBFC), with main shaft fault (MSF), with ball bearing fault condition, and main shaft fault (BBFC & MSF). The damage caused to any of these vital elements in SACR influences the operation and effectiveness of the coal belt conveyor system. The traditional used technique uses the oil residue and component wear identification, which are not as effective in predication of the faults in advance. These traditional monitoring methods detect the faults after the fault was developed in SACR and create a failure in the system. As a change, fast Fourier transform (FFT) technique dependence was employed for the type of continuous operating elements. In FFT-type system, the vibration signature data signals are evaluated in frequency domain. The focal issue with FFT-based technique is it is very influential and effectual for the stationary vibration signature data, (i.e.) for the signal data for which the attribute frequency does not vary with respect to time. But the SACR considered for research and fault exploration rotates continuously, while the coal belt conveyor system is in operation. Hence, it becomes obvious that the attribute frequency will vary insignificantly with respect to time domain. In this regard, it is impossible to sideline with FFT-based techniques for the fatal damage causing component of SACR. Thus, it can be confirmed that vibration signature data alone will overcome this signal exploration that will lead to augmented results with vibration signature. Considering the vibration signature acquisition, the components in a coal belt conveyor system interact each other a making vibration signature acquiring a bit tedious. But the variation signatures are acquired with advanced sensors and the various faults are categorized.

1.1 Random Forest Algorithm

Random forest algorithm (RF) generally referred as a universal solution provider in algorithm study. It is a versatile and smart machine learning method that can perform both classifications and regression tasks and can also perform dimensional reduction methods, outlier values, treat missing values, and other steps of data exploration as well. The reliability of RF lies on developing real-time solution for most of the technical problem as it is known for the ensemble way of learning as a group of weak models are combined to form a powerful model [1]. Advanced hybrid fuzzy decision tree in random forest algorithm can be used to identify the disease by analyzing the medical records of the patient in this research work. An expert system was built that can diagnose disease using the random forest algorithm. The use of medical information for diagnosing an illness a sickness is on the growth, and the tool developed in this research work helps to improve the analytic accurateness of diagnosing. [2] The random forest algorithm can be for estimation of traffic based on the traffic congestion level occurs due to social media usage. It also deals with applying sentiment analysis for the tweets to classify it either into positive or negative or neutral tweets. Priority is given to the traffic level, which is tweeted by the majority of people. A system is developed to predict traffic based on date, day and holiday or not data along with three hours of traffic data prior to the targeted hour, and the results were astonishing as the data traffic measure was very precise. This algorithm is more useful as it is applicable to both classification and regression (prediction). This algorithm has given a great accuracy of about 86% when applied to real-time traffic [3]. An improved random forest algorithm with multiple weight-based majority voting has been that presented for imbalanced credit score datasets. The proposed method is compared with the traditional random forest and single weight-based random forest classifiers. Further, the misclassification rate of minority class instances is highly reduced using the proposed method [4]. The Apache Spark Random forest algorithm model was used in the breast cancer diagnosis in an effective manner. The algorithm with the provided dataset predicts the cancer faster and provided accurate results. Random forest algorithm's baseline and pipeline features was implemented and compared in terms of accuracy, error, and performance and found to be good with real time diagnosis [5]. A comparative study between Naïve Bayes classifier and Bayes net classifier has been made to find fault and diagnosis of monoblock centrifugal pump. The algorithms are based on conditional probability, and literatures related to applications of these algorithms are only few. The comparative study concludes that the Bayes net algorithms perform better than its counterpart [6]. A rough set-based rule learning and fuzzy classification of wavelet features for fault diagnosis of monoblock centrifugal pump has been evolved, and the results were good when compared to traditional algorithms [7]. Monoblock centrifugal pump fault was diagnosed using discrete wavelets, and J48 decision tree algorithm was helpful in identification of effective statistical features; the discrete wavelet-based technique provided a much better accuracy when compared with similar algorithm in the data field [8]. This research work explores the fault identification of self-aligning troughing rollers in belt conveyor system using k star algorithm in an elucidated new approach for condition monitoring [9]. Further, the condition monitoring of self-aligning carrying idler (SAI) in coal handling belt conveyor system uses critical statistical features like mean, standard deviation, and kurtosis. Decision tree algorithm was effectively utilized in fault diagnosis of SAI critical elements [10]. In this research, the initial errors seem to progress gradually with respect to time, and there is a swop of typical operation to error operation, using of non-continuous decision function provide essential backing in establishing the progress and intensity of the error. It identifies the proper decision function in identifying the fault through diagnosis of a continuous binary mixture distillation column and the result proved it is practically achievable [11]. The effectiveness of the research lies in effectual utilization of the lazy algorithms and its variants. This ultimately proved that the k star lazy algorithm was efficient in categorization and data mining province, which can further be explored. Subsequently, the authors have effectively found special classification methods as unique fusion system for fault identification. It utilizes wavelet-dependent categorization technique which can be implemented with multi-gradient attributes for error identification in revolving machines [12]; the gear failure causes were found out using vibration signatures acquired under various conditions of the gear damage and wavelet theory utilized generated very good results [13]. In this work, the researchers have explored the errors of reciprocating compressors where the piston clearance increase causes failure of the system and performance issues; support vector machine (SVM) was utilized for this error categorization with big data analysis [14]; further, the wavelet features were used in transmission line fault identification using discrete methods uniquely developed for this application [15]. It also explains the spectral attributes of higher order which is an excellent choice of categorization problem with efficient output dependent [16]. The wind turbine fault was categorized using texture analysis, and the categorization accuracy used in this fault diagnosis was found to be more than 80% which was accepted as a better output efficiency in this field [17]. The algorithm developed using K star was able to reduce the maintenance cost by 15–20%, which was an ultimate achievement in this domain [18]. The authors studied about the vibration signature capturing method for disorder finding in journal bearing on multiscale permutation entropy (MPE) when evaluated with similar classifiers (ISVM-BT). The random forest algorithm was implied in categorization of tool used for plain turning and milling as the algorithm was able to categorize the faults causing the tool failure. Even though random forest has been utilized in researched here, and\there has not been a proper application of this universal algorithm in effective manner in many research areas in spite of its wonder applicability in the mechanical component applications

1.2 Bulk Material Handling Systems

Power utilization and requirement surge everyday in a country like India the necessitate for power generation is ultimatum and the methods are highly questionable irrespective of the difficulties in the operation and maintenance methods. Coal-based thermal power generating stations are prominently installed throughout the country due to abundant availability of coal in the country, which is mined and conveyed through various huge earth moving equipments and unimaginable methods. When it comes to conveying millions of tons of coal per annum, huge equipments like stacker and reclaimer are prominently used in most of the thermal power stations depending upon the generation capacity and boiler fuel need. In mining sites, high-capacity long-distance overland conveyors, primary crushers with high material crushing capacity, and secondary crushers with good pulverizing ability are the best choices. In Mettur thermal power station, wagon tipplers are used for unloading coal from railway wagons conveyed via Chennai port and a dedicated railway line for uninterrupted coal supply to the 1050 MW unit. Such huge demand of 40–50 thousand tons of coal per day is managed by long-distance overland conveyors in most of the coal-based thermal power generating stations. The NCTPS (North Chennai thermal power station) has 16 km of internal conveyor throughout the plant.

Belt conveyor system
One of the important aspects and means of convening coal throughout the plant continuously in an uninterrupted manner is coal belt conveyor system. It has important components like head pulley which has the driving unit arrangement in the order of ht motor, fluid coupling, rigid coupling, and helical gar box which is in turn connected to the drive pulley. Over the drive pulley, a fiber multilayered belt of having 4 or more plays is laid and connected like an endless belt through a tail pulley on the other end. The important aspects of belt conveyor system are that it can transfer coal continuously for most time in any type of terrain with ease without much human intervention. Further, the operation and maintenance of coal handling belt conveyor system is much more obliged when compared to other means of material handling system under testing conditions. Hence, for conveying such bulk quantity of materials, belt conveyors are preferred in most of the thermal power unit's throughout the world, which seems to be wise choice until now.

1.3 Self-Aligning Conveyor Roller (SACR)

Self-aligning conveying roller (SACR) is a vital element for conveyor operation without any hazardous manner. It helps to transmit the power of conveyor of the conveyor in a silky manner without any component damage and failure. Hence it is subjected to heavy forces and high amount of strain in the internal elements. The SACR is very essential in any coal handling belt conveyor system, which can adjust itself automatically to the type of load because of its free swiveling nature. The SACR frame is fixed over the conveyor supports in between the coal carrying rollers. It has important groove ball bearing and main shaft which is a key in handling the heavy conveyor belt. Failure in any of these key elements has to be addressed immediately; if delayed, it may cause grave danger to the entire conveying system. It has caused fatal fire accidents in TNEB-based thermal power plants. Hence, the design and fabrication of SACR is given utmost importance. But still because of the grave operating circumstances, these components tend to fail which can cause huge financial loss and fatal.

In this situation, to keep the conveyor under proper operation, there is a need for fault preventive system, which is essential for a bulk material handling equipment which supplies the basic need of power generating stations. So ultimately a high-level fault and failure deterrent system is high priority. A high-end technique considering the basic onsite and field conditions has to be utilized in this scenario, since it has to operate under rough environment. Taking such critical factors into consideration, an investigational setup was made after taking relevant precautionary needs while deigning and fabricating the setup. The set was integrated with sensors and apt vibration signature data acquiring devices. The raw vibration signatures obtained from the integrated setup of sensors; NDC devices were converted to digital data.

The raw data collected from the investigational setup after conversion into digital ones were analyzed using LabVIEW software and converted into basic MS excel files for simple usage for further processing. The data in xls format were utilized for calculating the mathematical statistical features like median mode, range, variance, etc. Once the statistical features are calculated using relevant macro programs which is shown in Fig. 1. Consequently only statistical feature which had high influential over the error and failure which is very critical as utilization of all the calculated statistical features cannot provide an effectual output. Hence, selection of vital and influential statistical attributes is topmost priority since this can substantially have an impact on the processing speed and time taken.

Once the assortment of important statistical attributes is completed, a better categorization algorithm has to be chosen for further error and failure detection. It has been established that random forest categorization has been a universal tool under this kind of classification. Hence, it is good choice to choose random forest algorithm tool for detection of SACR faults. Apart from this, a good-quality fault diagnosing tool has to be utilized for classification, random forest classifier has been chosen for classification of faults

```
        ''Macro11Macro
        For i = 2 To 250
    Workbooks.Open Filename:="C:\ C:\Users\Elcot\Desktop\Phd\ bearing fault \a_0" & i & ".xlsx"
    ActiveCell.Offset(2, 1).Range("A1:A13").Select     Selection.Copy
    Windows("a_001.xlsx").Activate
    ActiveCell.Offset(1, 0).Range("A1").Select
    Selection.PasteSpecial Paste:=xlPasteValues, Operation:=xlNone, SkipBlanks _:=False, Transpose:=True
    Application.CutCopyMode = False
    Windows("a_0" & i & ".xlsx").Activate
    ActiveWindow.Close
    ActiveCell.Offset(2, 1).Range("A1:A13").Select     Selection.Copy
    Windows("a_001.xlsx").Activate
    ActiveCell.Offset(1, 0).Range("A1").Select
    Selection.PasteSpecial Paste:=xlPasteValues, Operation:=xlNone, SkipBlanks _          :=False,
Transpose:=True
    Application.CutCopyMode = False
    Windows("a_0" & i & ".xlsx").Activate
    ActiveCell.Offset(2, 1).Range("A1:A13").Select     Selection.Copy
    Windows("a_001.xlsx").Activate
    ActiveCell.Offset(1, 0).Range("A1").Select
    Selection.PasteSpecial Paste:=xlPasteValues, Operation:=xlNone, SkipBlanks _          :=False,
Transpose:=True
    Application.CutCopyMode = False
    Windows("a_" & i & ".xlsx").Activate
    ActiveWindow.Close .
```

Fig. 1 Macro programs

Further the raw data collected, the conversion process and environment of data obtained were quite complex, which may reduce the efficiency and effectiveness of the entire condition monitoring process; hence, selection of categorization algorithm is quite critical. As random forest algorithm is entropy-dependent algorithm which will be able to cope up with the complexity nature of the vibration signature articulated, random forest algorithm has been chosen for this exploration work.

2 Investigational Exploration and Vibration Signals Acquisition

A sequential process explained in stages can be clearly seen in the flowchart shown in Fig. 2. The investigational integrated setup fabricated is meant for collecting of all different faults to be analyzed.

In view of this, all the four categories of fault which occur frequently are studied in this research work: SACR operates under no-fault condition (NFC), SACR has groove

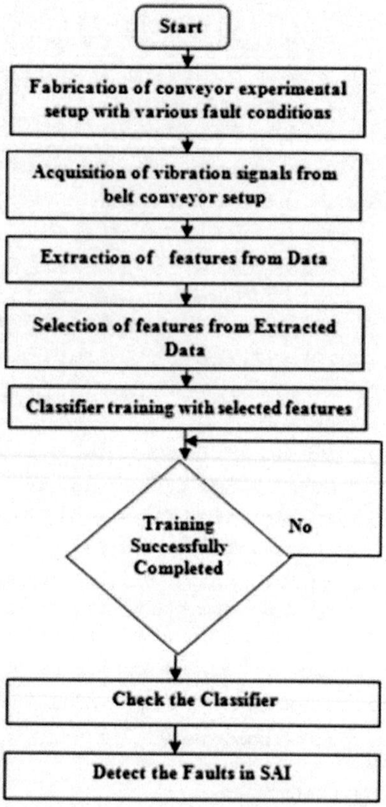

Fig. 2 Flowchart for fault diagnosis of using random forest

ball bearing fault condition (BBFC). The external ring thickness 8.5 mm—actual measured thickness 9 mm), SACR having main shaft fault condition (MSFC), (main central shaft ground to 11.8 mm—actual diameter 13 mm), SACR having main shaft and groove ball bearing fault (BBFC & MSFC). The investigational arrangement has a drive unit which is powered by an AC motor of 2 HP; it has nylon 3 ply conveyor belt similar to that of used in coal belt conveyor. It has SACR installed as per the different conditions to be analyzed. The AC motor of 2 Hp is integrated with a rigid coupling to a "v" belt speed reducer for lessening the rotational rpm of the investigational arrangement to the specified rotational rpm. Subsequently, self-aligning conveyor roller was installed at correct transition distance which was followed in coal belt conveyor design, fabrication, and installation. SACR fault identification is carried out in the subsequent stage by stage process (a) design, fabrication and installation of SACR investigational setup with various error conditions, (b) raw vibration signature data acquisition, (c) mathematical statistical feature calculation, and (d) mathematical statistical feature categorization. Initially, the belt conveyor model was allowed.

The vibration signatures were collected from various SACR elements which were considered as critical for failure as discussed earlier. Based on the data collected from

Table 1 Main shaft internal diameter readings

S. No.	Dia of shaft before grinding in mm	Dia of shaft after grinding in mm	Side
1	13.01	9.88	Left side 1
2	13.00	9.90	
3	13.00	9.98	Right side 2
4	12.99	9.99	

Table 2 Groove ball bearing external ring thickness readings

S. No.	Dia of roller bearing before grinding in mm	Dia of roller bearing after grinding in mm
1	6.50	4.0
2	6.51	3.90

coal belt conveyor fields, the groove ball bearing and the main shaft with relevant wear and tear were prefabricated as per Tables 1 and 2, and the vibration signature readings are obtained.

Various conditions of SACR having groove ball bearing wear and main shaft wear and tear set one by in the investigational arrangement and corresponding vibration signature data were collected. A sample of 250 nos of reading were collected for each conditions from the investigational setup for analysis.

.The investigational arrangement can be viewed in Fig. 3. A smart accelerometer sensor (Sl. No. 3055B1) was installed over the vibration area in the SCAR frame to obtain the vibration data (Fig. 4).

The signal received must be properly trained before being utilized for fault analysis. So a signal conditioning (ADC—analog-to-digital converter) shown in Fig. 5 unit is integrated with the investigational setup. The vibration signatures obtained from the training unit was fed to the relevant software for further doling out. The LabView software was used to record the vibration signals in the digital form and store in the computer hard disk memory.

Fig. 3 Dyatron accelerometer sensor

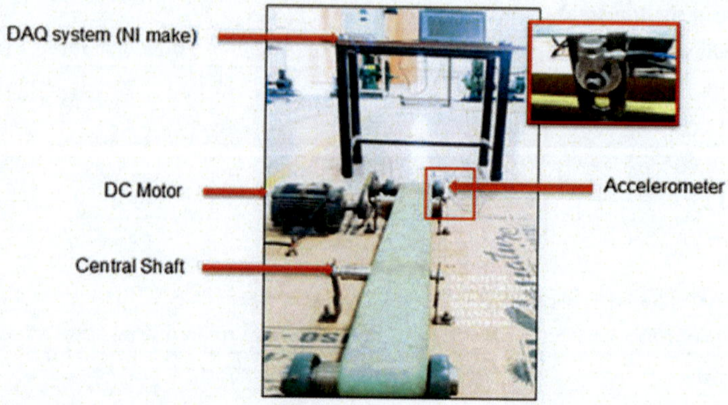

Fig. 4 Experimental setup OF SACR

Fig. 5 Data acquisition NSIC device

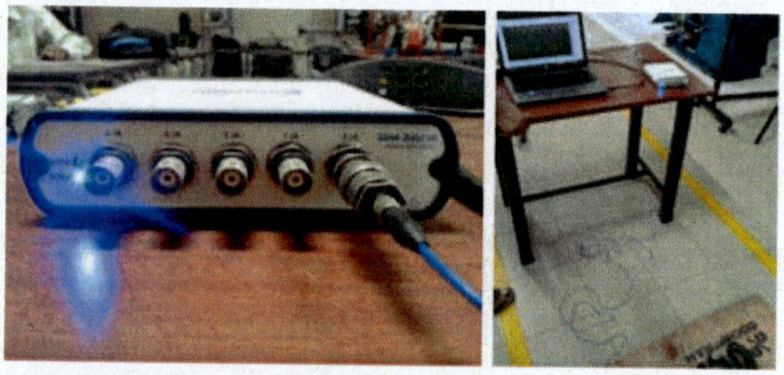

3 Statistical Features Extraction and Selection

Feature Extraction
Mathematical statistical features were extracted from the vibration digital signature. The various parameters are mean, median, mode, standard error, standard deviation, kurtosis, skewness, minimum value, maximum value, sample variance, and range.

Selection of Important Features
All the extracted features might not be used for classification. Only statistical feature which had high influential over the error and failure which is very critical is utilized, and the inefficient ones are eliminated. Hence, selection of vital and influential statistical attributes is topmost priority since this can substantially have an impact on the processing speed and time taken. Certain features such as standard deviation, kurtosis, range, and mean were supportive for categorizing various fault conditions. Hence, other mathematical statistic features could be purged from the set of feature to improve the effectiveness of the random forest algorithm and decrease the processing time.

4 Result and Discussion

The result of the 1000 samples fed into the random classifier has been given below, and it is discussed in detail in this section. The grouping of fault with random classifier is discussed as follows:

i. Categorization using random forest classifier
ii. Random forest performance evaluation.

Relation: fo1
(Combined faults)-
weka filters unsupervised attribute Remove-R3, 9-12
Instances: 1000
Attributes: 8
Mean, Standard Error, Standard Deviation, Sample
Variance, Kurtosis
Skewness, Range, Class
Test mode: 10-fold cross-validation
KStar options: -B 20 -M a
Classifier model (full training set)
Random Forest
Bagging with 100 iterations and base learner weka.
Classifiers. Random Forest Tree
Time taken to build model: 0.31 s

Correctly categorized Instances	902.00	90.20%
Incorrectly categorized Instances	98.00	9.80%
Kappa statistic	0.86	
Mean absolute error	0.07	
Root mean squared error	0.19	
Relative absolute error	18.65%	
Root relative squared error	44.42%	
Total Number of Instances	1000.00	

The final result received from 1000 no. of samples was analyzed with the help of confusion matrix generated along the random forest output. From the output received, it is known that the no fault-free condition (NFC) is mentioned in the first row, continued by main shaft fault condition (MSFC). Then, the groove ball bearing fault condition (BBFC) and the combined main shaft fault (MSFC) and groove ball bearing fault condition (BBFC) represented in the third and fourth rows in the Table 3. One can clearly understood from the confusion matrix (Table 3) that out of the 250 sample Nos, a portion of the data were considered for training and the remaining data is classified for different error conditions of the self-aligning conveyor roller. In the confusion matrix, the diagonal elements are used to indicate the number of correctly categorized vibration signal and the incorrectly categorized

Table 3 Confusion matrix of random forest algorithm with selected features

	NFC	MSFC	BBFC	MSF& BBFC
FFC	212	38	0	0
CSF	33	202	0	15
(BFC)	0	0	250	0
CSF& BBFC	0	12	0	238

data signals are located as non- diagonal elements. This is how the categorization accuracy is represented in the confusion matrix. In SACR case (Table 3), inferring the result of the algorithm, out of 1000 trial data 212 were no-fault condition (FFC) data signals have been exactly categorized and the balance thirty eight points infer they were main shaft fault (MSFC). Similarly, 202 data points of main shaft fault (MSFC) have been correctly categorized and 32 were incorrectly categorized as no-fault condition. Thus, the confusion matrix is inferred and categorization accuracy was 90.20%. These results achieved are for a sample data for the signals acquired in particular site conditions so the categorization accuracy of 90.20% does guarantee a parallel efficiency for all such feature vibration signal data. Based on the result, it can be accomplished that 90.2% accuracy achieved by RF is best in this category

5 Conclusion

An investigational arrangement was designed, fabricated, and installed for the self-aligning conveyor roller. It was fabricated to replicate different faults and error that would occur in actual coal handling belt conveyors. The vibration signal data were acquired using ADC. The collected vibration signal in analog form has been processed by data mining tools to convert to digital format and further processed to extract the mathematical statistical features like mean, median, etc. All the extracted features might not be used for classification. Only statistical features which had high influential over the fault classification are utilized, and the inefficient ones are eliminated. The filtered processed digital data were fed into random forest algorithm for further categorization. The result achieved is for a set of sample data and signals obtained in particular site conditions. And a categorization accuracy of 90.20% was achieved, which does guarantee a similar efficiency for all such feature vibration signature data. Hence, it can be concluded that random forest algorithm is better in fault diagnosis of SACR among the existing classifiers.

References

1. Dhanalakshimi R, Geetha C, Sethukarasi T (2019) Monitoring and detecting disease in human adults using fuzzy decision tree and random forest algorithm. Int J Recent Technol Eng 7:2277–3878
2. Sridevi K, Ganesan T, Samrat BVS, Srihari S (2019) Traffic analysis by using random forest algorithm considering social media platforms. Int. J. Recent Technol. Eng. 7:2277–3878
3. Pannir Selvam RR, Saleem IAM, Alenezi A (2019) Classification of imbalanced class distribution using random forest with multiple weight based majority voting for credit scoring. Int J Recent Technol Eng 7:2277–3878
4. Krishna TH, Rajabhushanam C (2019) Breast cancer prognosis with apache spark random spark random forest forest pipeline. Int J Recent Technol Eng 7:2277–3878
5. Sugumaran V, Muralidharan V (2012) A comparative study of Naïve Bayes classifier and Bayes net classifier for fault diagnosis of monoblock centrifugal pump using wavelet analysis. Appl Soft Comput 12:2023–2029
6. Muralidharan V, Sugumaran V (2013) Selection of discrete wavelets for fault diagnosis of monoblock centrifugal pump using the J48 algorithm. Appl Artif Intell 27:1–19
7. Sugumaran V, Muralidharan V (2013) Rough set based rule learning and fuzzy classification of wavelet features for fault diagnosis of monoblock centrifugal pump. Measurement 46:3057–3063
8. Ravikumar S, Kanagasabapathy H, Muralidharan V (2018) Fault diagnosis of self-aligning troughing rollers in belt conveyor system using k-star algorithm. Measurement 133:341–349
9. Kanagasabapathy H, Muralidharan V, Ravikumar S (2014) Condition monitoring of self aligning carrying idler (SAI) in belt-conveyor system using statistical features and decision tree algorithm. Measurement 112:221–224
10. Demetgul M, Yildiz K, Taskin S, Tansel IN, Yazicioglu O (2014) Fault diagnosis on material handling system using feature selection and data mining techniques. Measurement 55:15–24
11. Jiang F, Liu G, Du J, Sui Y (2016) Initialization of K-modes clustering using outlier detection techniques. Inform Sci 332:167–183
12. Ito T, Myojin M, Hirai Y, Kaneko K (2014) Fault-tolerant routing in (n, k)—Star Graphs. In: 15th international conference on parallel and distributed computing. Appl Technol 223:113–118
13. Shukla SNS (2014) A review on k-means data clustering approach. Int J Inform Comput Technol 4:1847–1860
14. Saini I, Singh D, Khosla A (2013) QRS detection using K-nearest neighbor algorithm (KNN) and evaluation on standard ECG databases. J Adv Res 4:331–344
15. Li P, Kong F, He Q, Liu Y (2013) Multiscale slope feature extraction for rotating machinery fault diagnosis using wavelet analysis. Measurement 46:497–505
16. Jena DP, Panigrahi SN, Kumar R (2013) Gear fault identification and localization using analytic wavelet transform of vibration signal. Measurement 46:1115–1124
17. Qi G, Zhu Z, Erqinhu K, Chen Y, Chai Y, Sun J (2018) Fault-diagnosis for reciprocating compressors using big data and machine learning. Simul Model Practice Theory 80:104–127
18. Li Y, Xu M, Yu W, Huang W (2016) A new rolling bearing fault diagnosis method based on multiscale permutation entropy and improved support vector machine based binary tree. Measurement 77:80–94

Levy Interior Search Algorithm-Based Multi-objective Optimal Reactive Power Dispatch for Voltage Stability Enhancement

N. Karthik, A. K. Parvathy, R. Arul, and K. Padmanathan

Abstract Reactive power resource management is a crucial and vital step in order to have a safe and cost-effective power system option when it comes to voltage stability. The optimal reactive power dispatch (ORPD) has a key aim to find out the appropriate control variable values, for example, shunt VAR compensator settings, generator bus voltages, and tap settings of on-load tap change (OLTC) transformers. This objective is designed, by taking the constraints into account, and to ensure that the objective function is reduced. In the current research work, the researchers proposed a levy interior search algorithm (LISA) in order to elucidate multi-objective optimal reactive power dispatch challenges that mitigate the real power loss, and at the same time, it also saves the voltage quality. In this research article, the researchers considered real power loss and voltage stability index as objective functions. The paper has a primary objective, i.e., by taking large-scale power system into account, the researchers proposed a multi-objective interior search algorithm in order to get rid of multi-objective ORPD problem and to yield the best results from the secure and cost-effective operation of electric power systems. The researchers simulated the setup under IEEE 57-bus and large-scale IEEE 300-bus system to emphasize the efficacy of LISA. From the results, it can be confirmed that the proposed approach yielded much better results than the existing algorithms.

Keywords Optimal reactive power dispatch · Interior search algorithm (ISA) · Levy interior search algorithm (LISA) · Real power loss · Voltage stability index · Meta-heuristic optimization

N. Karthik (✉) · A. K. Parvathy
Hindustan Institute of Technology and Science, Chennai, India
e-mail: nkarthik@hindustanuniv.ac.in

R. Arul
Vellore Institute of Technology, Chennai Campus, Chennai, Tamil Nadu 600127, India

K. Padmanathan
Agni College of Technology, Thalambur, Chennai, India

N. Zhou and S. Hemamalini (eds.), *Advances in Smart Grid Technology*, Lecture Notes in Electrical Engineering 688, https://doi.org/10.1007/978-981-15-7241-8_17

1 Introduction

In today's electric power systems, the optimal reactive power dispatch (ORPD) remains a prominent challenge that needs to be addressed. The key aim of the conventional ORPD problem is to mitigate the loss of real power and simultaneously meeting different requirements [1]. In the researcher conducted earlier [2], the researchers proposed a method to find out the voltage stability index. The study conducted by Raghunatha et al. [3] proposed a numerically stable and sequential primal–dual LP algorithm with the aim to enhance the system static voltage stability index. The literature overcomes the multi-objective reactive power flow challenge through the adaptive immune algorithm proposed in one of the studies [4]. In this study, the researchers selected voltage deviation at load bus, real power loss as well as voltage stability index as the objective functions. In case of few real applications, it becomes mildly not enough to consider a single target of power loss. In the recent times, the ORPD problem is rechristened as multi-objective optimization problem (MOP). The study [5] proposed the enhanced particle swarm optimization (PSO) in order to eliminate the multi-objective optimal power flow problem with the help of four various objective functions.

According to the studies [6, 7], there is a reduction in the power loss in addition to L-index as well, in case of multi-objective ORPD (MOORPD). The MOORPD is one of the parts in OPF challenge that consists of various elements like deviation in voltage stability index or voltage, quantity of reactive power supplied by parallel compensators in order to mitigate the real power loss, determination of optimal adjustment of magnitudes of generator voltages, and the position of transformer taps. In continuation to these, it also meets various physical as well as operational challenges brought by the power system itself. The MOORPD is nothing but a generally known large-scale, multifaceted, multi-constrained, highly nonlinear and mixed integer programming problem created by various entities such as nonlinear power flow equations, discrete variables, and many such constraints [8]. For the past ten years, there is a tremendous increase experienced in electric power systems, thanks to basic alterations that are required in both supply as well as demand ends and this improvement tends to impact the security concerns in power system and may result in increased losses. In line with this, these enhancements mandate the transmission system operations to be updated in terms of their optimization strategies so as to ensure the economic scheme as well as to overcome all security concerns. Such optimization strategies usually have parallelly working, two or three minimizing or maximizing (i.e., conflicting) objectives which are subjected under different constraints [9, 10].

The researchers shifted their interest toward ORPD problem in the recent years and that can be understood through the flourishing research in this area to ensure a stable, safe, and secure operation of the electric power systems. Currently, a number of algorithms have been proposed by various researchers as a solution for MOORPD problem. For instance, in the study [11], a modified NSGA-II was proposed by the researchers based on the strategies such as dynamic crowding distance and controlled elitism. The purpose was to enhance the range of non-dominated solutions. At the time of validating the technique, the voltage deviation and the large-scale test systems were not considered and the method was implemented in medium-sized test systems, for instance, IEEE-30-bus and IEEE 118-bus. In the study [12], the researchers proposed an improved version of multi-objective PSO so as to mitigate the power losses and at the same time reduce the voltage deviations in power systems. But the control variables were reported in the exterior end of the acceptable limits due to the non-induction of constraint handling method. In the literature [13], the researchers proposed an updated version of the MOPSO to overcome MOORPD, though this study did not focus on the control variables. A chaotic improved PSO-based multi-objective optimization was proposed by the researchers and implemented in medium-sized systems, and in this research, Chen et al. [14] aimed at reducing voltage stability index as well as power losses. The results yielded for the control variables were observed only outside the imposed limits.

The study conducted by Ghasemi et al. [15] proposed a novel multi-objective algorithm, named as chaotic parallel vector evaluated interactive honey bee mating optimization (CPVEIHBMO) algorithm as a solution to tackle MOORPD problem. However, this study remained unfeasible for the dependent variables. The firefly algorithm was modified and proposed in the study [16] to overcome the multi-objective active/reactive ORPD problem that accompanies load as well as wind power generation uncertainties in a standard IEEE 30-bus system. Though this was an insufficient test system, this experimentation was conducted to assure the evenness of the proposed algorithm. As few other studies, this study too missed to discuss the optimal control variables. As a result, the prominent drawback of these algorithms is that it cannot dominate to converge to the Pareto frontier at the time when maintaining enough diversity. Further, the complex nature of the objective functions utilized (for instance, ORPD problem), it lasts to converge toward the local optimum. In other terms, it is interesting to determine the optimal solution or near optimal solution of such problem simultaneous to providing the assurance for the safety concerns and guarantee for the admissible limits of the control variables.

In the study [17], quasi-oppositional differential evolution was proposed to elucidate the ORPD problem. Various optimization algorithms such as hybrid PSO and imperialist competitive algorithm [18], Gaussian bare-bones water cycle algorithm [19], and ant lion optimization algorithm (ALO) [20] were successfully experimented to elucidate the ORPD problem. As per the literature [20], active power loss and voltage stability index were chosen as objective functions. Simulation was carried out on IEEE 30-bus, IEEE 118-bus and large-scale IEEE 300-bus systems. In the study conducted by Souhil et al. [21], a multi-objective ALO algorithm was proposed, taking three objective functions into consideration, to get rid of multi-objective ORPD problem. Arul et al. [22] conducted a study in which the researchers proposed chaotic self-adaptive differential harmony search algorithm to handle the optimal power flow problem. In the literature [23], optimal VAR dispatch problem was successfully solved using chaos-embedded krill herd algorithm. Rajan et al., [24] developed weighted elitism-based ALO algorithm to elucidate the optimal VAR problem. In the recent years, various optimization techniques, for instance, whale optimization algorithm [25], moth-flame optimization algorithm [26], differential search algorithm [27], and fuzzy adaptive PSO [28] were proposed to overcome the ORPD problem. The study conducted by Gandomi [29] proposed art-inspired optimization algorithm titled interior search algorithm (ISA) which had interior design and decoration as its base, and the algorithm is aimed at determining the global optimum solution. In [30], ISA was implemented to address the multi-objective economic emission dispatch problem. In case of conventional ISA, new variables were generated using rand functions which resulted in the local optima. To enhance the search capability of ISA, the levy flight was integrated with ISA to generate better optimal solution. In LISA, the levy flight method was used to generate random variables. In this paper, a novel multi-objective optimization algorithm named levy interior search algorithm (LISA), based on ISA optimization algorithm, was proposed to elucidate the MOORPD problem in large-scale power system.

Further, LISA algorithm has not been applied so far to handle MOORPD problem, especially in large-scale power systems. The effectiveness of the proposed LISA algorithm was already validated in both medium-sized IEEE 57-bus system and large-scale IEEE 300-bus system. Furthermore, simulation is carried out using the LISA algorithm employing two different strategies and the ISA algorithms with regard to computation efficiency and solution quality.

2 Problem Formulation

The ORPD problem formulation was modeled with two different objective functions such as real power loss and voltage stability index.

As portrayed in any other optimization problem, the traditional ORPD problem can be formulated as (1)–(3) [6].

2.1 *Real Power Loss Mitigation*

Then, the transmission line loss can be determined as

$$F_1(x, u) = P_{\text{loss}} = \sum_{n=1}^{\text{NL}} g_n\left[V_j^2 + V_k^2 - 2V_j V_k \cos\left(\delta_j - \delta_k\right)\right] \tag{1}$$

where g_n denotes the conductance of nth line that associates bus j to bus k, whereas the NL denotes the number of transmission lines.

2.2 Minimization of Voltage Stability Index

Since the operations of state-of-the-art power systems are nearby the stability limits, the investigation of voltage stability has become an increasing phenomenon in power system analysis. So, the voltage stability index was inducted as an objective function in the current study in order to enhance the security of electric power systems and to forecast the voltage collapse. Kessel and Glavitsch [2] set the operating range of the indicator L between 0 and 1. So, one can define the above-mentioned objective as.

$$F_2(x, u) = \min \text{VSI} = \text{Minimize}\big(\max\big(L_j\big)\big) \quad (2)$$

where L_j of the jth bus is characterized by the following expression:

$$L_j = \left| 1 - \sum_{i=1}^{N_{PV}} F_{ji} X \frac{V_i}{V_j} \{\theta_{ji} + (\delta_i - \delta_j)\} \right| \quad j = 1, 2, \ldots, N_{PQ}$$

With

$$F_{ji} = |F_{ji}|\theta_{ji},\ V_i = |V_i|\delta_i,\ V_j = |V_j|\delta_j F_{ji} = -[Y_1]^{-1}[Y_2]$$

2.3 Equality and Inequality Constraints

Transmission line losses must be taken into consideration with the objective of accomplishing proper economic load dispatch. Transmission line losses can be found out by the following Newton–Raphson and B-coefficients approaches. The optimal power flow solution, attained by employing Newton–Raphson, was used to decide the real power loss P_{loss}. The active power loss subjected to equality constraints can be expressed as:

$$P_{gj} - P_{Dj} - V_j \sum_{j=1}^{NB} V_k\big[G_{jk}\cos(\delta_j - \delta_k) + B_{jk}\sin(\delta_j - \delta_k)\big] = 0 \quad (3)$$

$$Q_{gj} - Q_{Dj} - V_j \sum_{j=1}^{NB} V_k\big[G_{jk}\sin(\delta_j - \delta_k) + B_{jk}\cos(\delta_j - \delta_k)\big] = 0 \quad (4)$$

where $j = 1,2, \ldots$ NB; NB denotes the total number of buses; V_j and V_k represent the magnitude of the voltage at bus j and bus k, respectively; Q_{gj} denotes the reactive power output at jth bus; δ_j and δ_k are voltage angles at bus j and bus k, respectively; B_{jk} and G_{jk} are the transfer susceptance and conductance between bus j and bus k, respectively; P_{Dj} and Q_{Dj} are the jth bus active and reactive power load, respectively.

Every generating unit's real power generation output must be confined by maximum as well as minimum boundaries as follows.

$$P_{gi\min} \le P_{gi} \le P_{gi\max} \tag{5}$$

P_{gi} represents the real power output of ith generating unit, whereas $P_{gi\max}$ and $P_{gi\min}$ are the maximum and minimum real power outputs of ith generating unit correspondingly.

$$Q_{gi\min} \le Q_{gi} \le Q_{gi\max} \tag{6}$$

$$V_{gi\min} \le V_{gi} \le V_{gi\max} \tag{7}$$

Equation (8) is related to tap changer of transformers which are used with the intention of regulating the magnitude of the voltage. Equation (9) is associated with output of all the switchable shunt components, for instance, capacitor banks. As a final point, (10) & (11) are security limitations that comprise of loading of transmission lines' constraints and magnitude of the load voltage.

$$T_{\mathrm{r}}^{\min} \le T_{\mathrm{r}} \le T_{\mathrm{r}}^{\max} \tag{8}$$

$$Q_{\mathrm{C_s}}^{\min} \le Q_{\mathrm{C_s}} \le Q_{\mathrm{C_s}}^{\max} \tag{9}$$

$$V_{\mathrm{L_k}}^{\min} \le V_{\mathrm{L_k}} \le L_{\mathrm{L_k}}^{\max} \tag{10}$$

$$\left|S_{\mathrm{l_t}}\right| \le S_{\mathrm{l_t}}^{\max} \tag{11}$$

3 Fuzzy Logic-Based Assortment of Finest Compromise Solution

The selection of the finest compromise solution in decision making method from the set of available optimal solutions is an important task. The researchers deployed fuzzy membership approach [27] to identify the finest compromise solution. The jth objective function f_j of individual k is classified by a membership function μ_j^k due to indeterminate feature of the decision maker's conclusion defined as

$$\mu_j^k = \begin{cases} 1 & f_j \le f_j^{\min} \\ \frac{f_j^{\max} - f_j}{f_j^{\max} - f_j^{\min}} & f_j^{\min} < f_j < f_j^{\max} \\ 0 & f_j \ge f_j^{\max} \end{cases} \tag{12}$$

where $f_j^{\max}$ and $f_j^{\min}$ denote the minimum and maximum value of jth fitness function available in the non-dominated solutions, whereas the normalized membership function μ^k is also calculated similarly for every non-dominated result k as

$$\mu^k = \frac{\sum_{j=1}^{N} \mu_j^k}{\sum_{k=1}^{r} \sum_{j=1}^{N} \mu_j^k} \tag{13}$$

where r denotes the whole non-dominated solutions. The finest compromise solution consists of the maximum value, μ^k.

4 Levy Interior Search Algorithm

ISA is an aesthetic optimization algorithm that replicates the attractive approaches used in embellishing and designing a particular interior space [29]. ISA incorporates two indispensable features to determine the solution in search space. In order to improve the optimization capability of ISA, the levy interior search algorithm is generated through the implementation of levy flights principle, a flight pattern that can be observed in birds, whereas the proposed algorithm is termed as levy interior search algorithm (LISA).

(i) Composition stage to address exploration

The objective of this feature is to find out an appropriate composition for components that create an attractive background so as to meet the requirements of the customer. In this stage, the composition of elements gets altered with the intention of obtaining a better aesthetic view.

(ii) Mirror inspection approach to address exploitation

In order to create an attractive decoration, the mirror inspection work is modeled. In this stage, a mirror is located in the vicinity of the global best to establish an appropriate outlook.

The primary objective of using LISA is to formulate a design problem in such a way that it overcomes all constraints. Figure 1 represents the flowchart for LISA.

The steps followed in the implementation of LISA algorithm is summarized as:

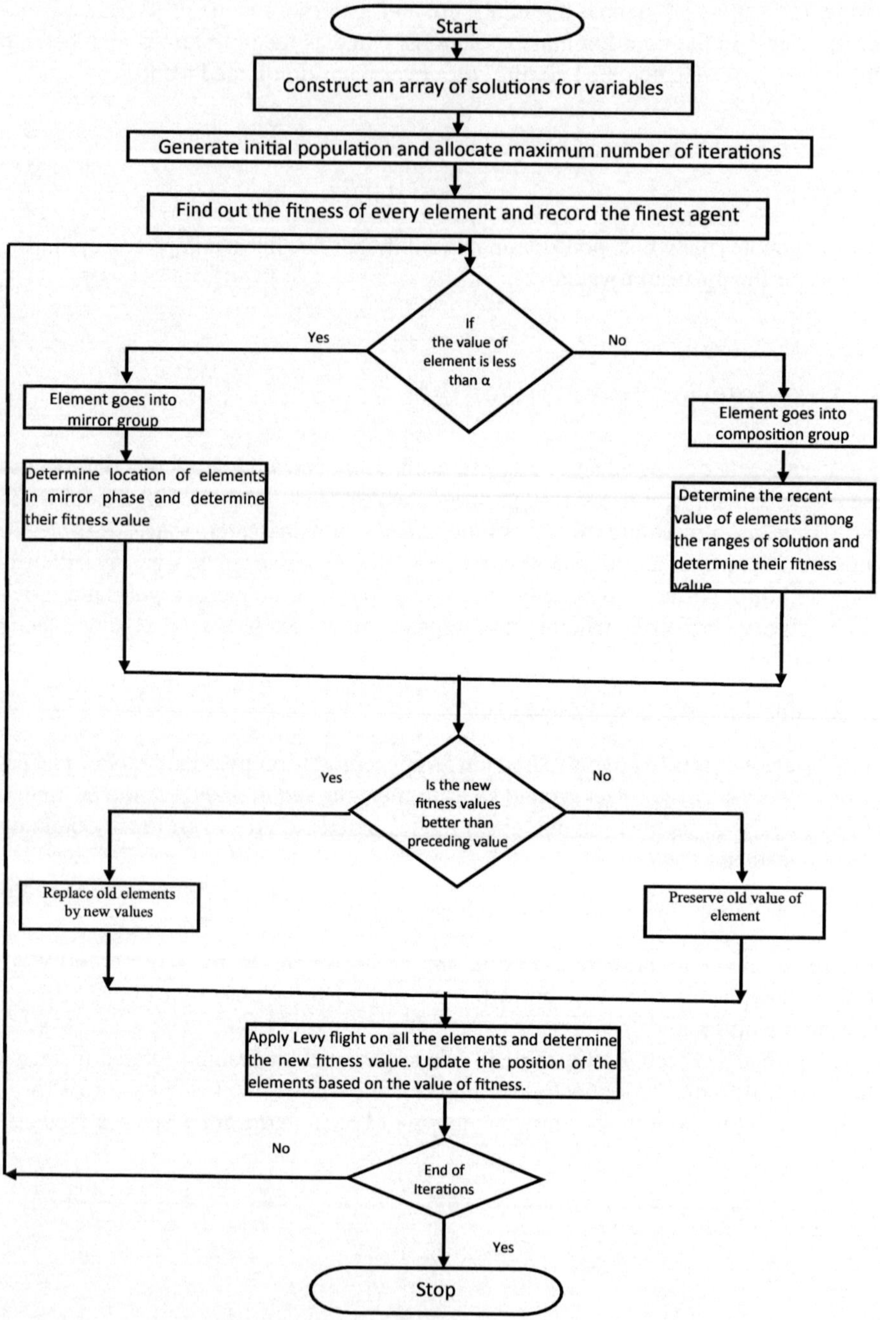

Fig. 1 Flowchart for Levy interior search algorithm

Step 1: Characterize the search space and indiscriminately construct a population of elements x_k for $k = 1, 2, 3 \ldots n$ between lower boundaries (LB) and upper boundaries (UB).

Step 2: The objective f_k should be assessed for complete elements. The finest fitness element is showcased as the global best, which otherwise denoted as x_{gb}.

Step 3: Indiscriminately categorize the other elements into two groups based on the probability of α, composition stage, and mirror inspection work. If the value of r_3 is less than α, the element is placed into the mirror group or else it falls under the composition group. Here r_3 is a random number between 0 and 1.

In a conventional ISA, α has been considered as a predetermined value [4]. It can be observed that when there are huge number of elements in composition phase, it enhances the ability to conduct global search. With the purpose of improving the convergence speed, the current study utilized a dynamic α that differs linearly up to 80% of the iterations, yet it is retained as a constant close to 1 after that. When the α is kept near 1 at the final stages of the iterations, it leads to larger mirror group that can probably help in the process of local search. To conclude, the variable α results in faster convergence.

Step 4: Every element needs to be modified in the composition phase in an arbitrary fashion enclosed by the narrow search space.

$$x_j^k = \text{LB}^k + \left(\text{UB}^k - \text{LB}^k\right)r_2 \tag{14}$$

where x_j^k corresponds to jth solution in kth run and r_2 is an indiscriminate value between 0 and 1.

Step 5: In the mirror phase, a mirror is positioned expansively flanked by each element and the best element. The position of the mirror can be determined as:

$$x_{m,j}^k = r_3 x_j^{k-1} + (1 - r_3)x_{\text{gb}}^k \tag{15}$$

where r_3 is an indiscriminate value ranging from 0 to 1. The image's altered position is denoted via

$$x_i^k = 2x_{m,i}^k - x_i^{k-1} \tag{16}$$

Step 6: The location of the global finest component is to some extent altered by employing levy flight approach. This levy flight method is employed for exploration process which is related to confined search.

$$x_{\text{gb}}^j = x_{\text{gb}}^{j-1} + \delta * \text{Levy}(\lambda) \tag{17}$$

where δ is referred as a scale factor which is assigned in proportion to dimension of the exploration space. Here δ is fixed as 1. Levy(λ) characterizes the step length which is incorporated by the levy distribution with infinite values of variance and mean with $1 < \lambda < 3$.

Step 7: The fitness value of the altered locations of element as well as the images should be calculated followed by careful preservation of the global best element when the fitness value at the updated position outperforms the preceding position.

Step 8: The steps 2–7 mentioned above need to be repeated until the time the maximum number of iterations is reached.

5 Simulation Results

In order to substantiate the capability of LISA in elucidating the ORPD problem, LISA was implemented in IEEE 57-bus and IEEE 300-bus test systems. The simulation was carried out using MATLAB software for 50 independent test runs using LISA and ISA algorithm. As per the literature [31, 32], the number of control variables as well as the active power losses in the initial conditions for IEEE 57-bus and 300-bus system were considered.

5.1 *Test System 1 (IEEE 57-Bus Test System)*

In the standard IEEE 57-bus test system, there exists a total of 7 generating units at buses 1, 2, 3, 6, 8, 9, and 12. In this, seventeen transformers had an off-nominal tap ratio, whereas three parallel compensators were in the buses 18, 25, and 53. The power demand was considered to be 12.508 p.u. In order to overcome the ORPD problem, the proposed LISA algorithm was incorporated whereas the application results obtained were compared and analyzed with the results obtained when applying MOCIPSO, MOIPSO, MOPSO, MOEPSO, and ISA optimization techniques. From the comparison of simulation results, it can be observed that LISA provides better efficiency than MOCIPSO, MOIPSO, MOPSO, MOALO, and ISA in terms of superiority of solution and computation time. The convergence characteristics attained using ISA, LISA strategy-I, and strategy-II are presented in Figs. 2,3, and 4, respectively. The trade-off characteristics between active power loss and voltage stability index attained when using ISA, LISA strategy-I, and LISA strategy-II are shown in Figs. 5, 6, and 7, respectively. It can be concluded from the application results that the proposed LISA algorithm revealed a very modest performance when compared to ISA, MOCIPSO, MOIPSO, MOPSO, and MOALO. Moreover, the computational time for LISA was less when compared to ISA, MOALO, MOPSO, and MOEPSO algorithms. The application results amply substantiate the ability of LISA to overcome the equality and inequality constraints brought by the ORPD problem (Tables 1 and 2).

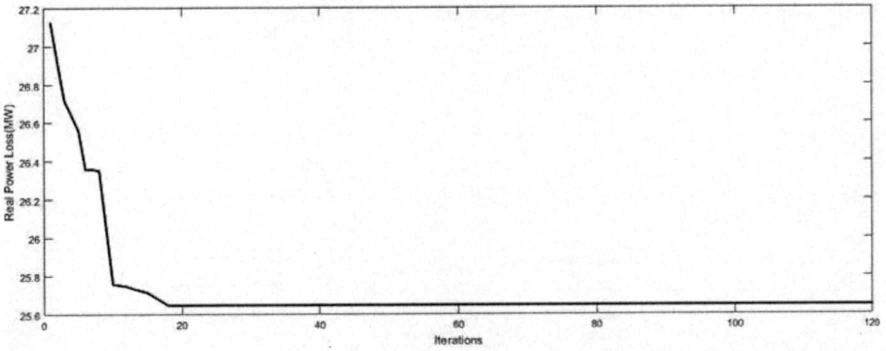

Fig. 2 Convergence characteristic for IEEE 57-bus system using ISA

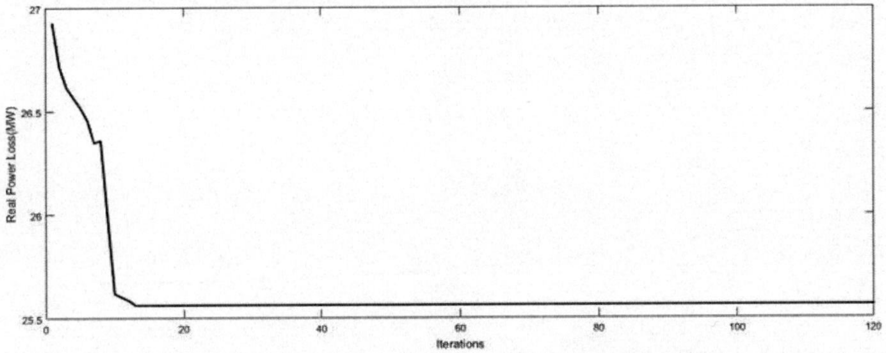

Fig. 3 Convergence characteristic for IEEE 57-bus system using LISA Strategy-I

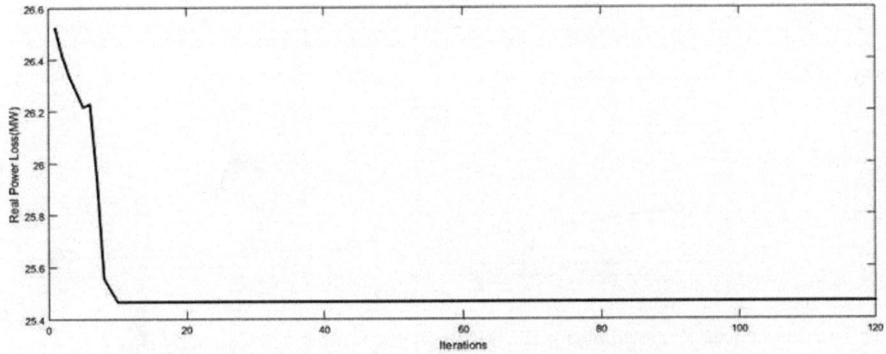

Fig. 4 Convergence characteristic for IEEE 57-bus system using LISA Strategy-II

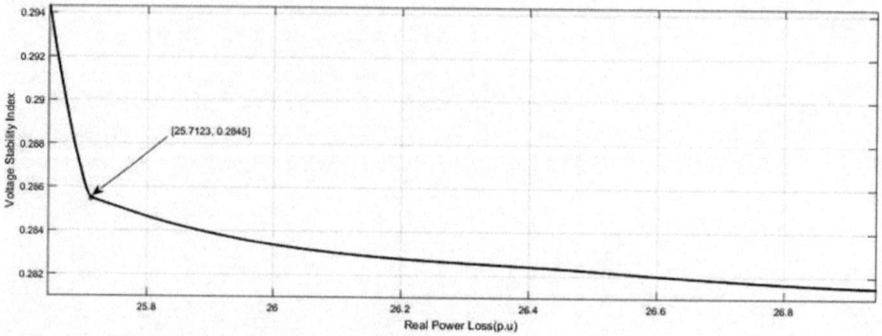

Fig. 5 Best-obtained Pareto-fronts for IEEE 57-bus system using ISA

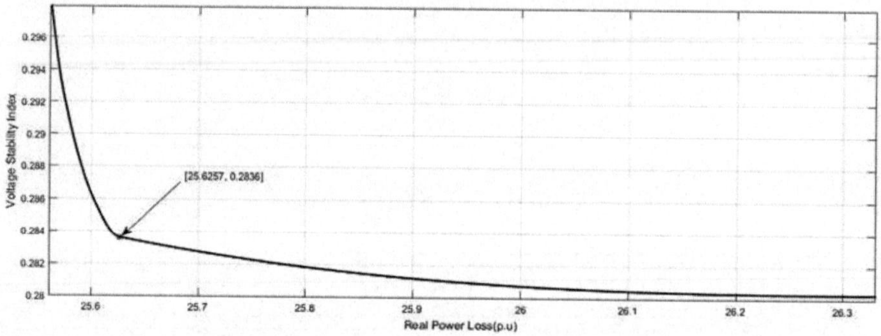

Fig. 6 Best-obtained Pareto-fronts for IEEE 57-bus system using LISA Strategy-I

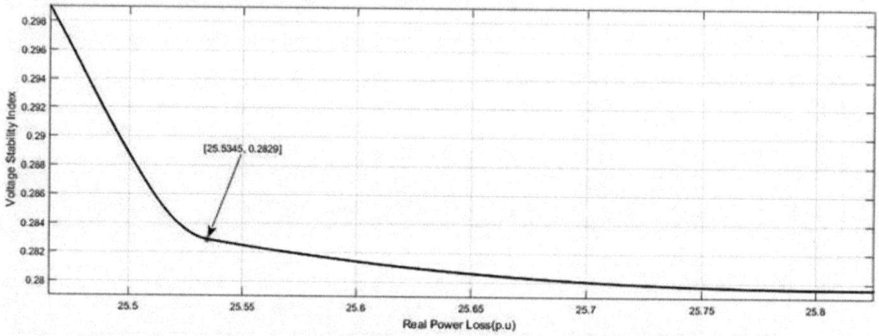

Fig. 7 Best-obtained Pareto-fronts for IEEE 57-bus system using LISA Strategy-II

Table 1 Comparison of simulation results for the minimization of real power loss and L-Index for IEEE 57-bus system

MOCIPSO [14]				MOIPSO [14]			MOPSO [14]		
Variables	Min PLoss	Min L-Index	BCS	Min PLoss	Min L-Index	BCS	Min PLoss	Min L-Index	BCS
V_{G1} (p.u.)	0.90000	0.90000	1.10000	0.90000	0.90000	0.90000	1.10000	1.10000	1.10000
V_{G2} (p.u.)	1.04996	0.90000	0.93520	1.10000	1.10000	1.10000	0.90000	0.90000	0.90000
V_{G3} (p.u.)	1.10000	0.93820	0.90000	0.90000	0.90000	0.90000	1.10000	1.10000	1.10000
V_{G6} (p.u.)	0.99769	0.90831	0.90000	0.90000	0.90000	0.90000	1.10000	0.99771	1.05681
V_{G8} (p.u.)	0.90000	1.10000	1.07345	0.90000	0.90000	0.90000	0.90000	0.90000	0.90000
V_{G9} (p.u.)	0.90000	1.10000	0.90000	1.10000	1.10000	1.10000	1.08959	0.96893	1.05494
V_{G12} (p.u.)	1.01203	1.08472	0.96240	0.90000	0.90000	0.90000	1.10000	1.10000	1.10000
$T_{4\text{-}18}$	0.96000	0.96000	0.96000	1.10000	1.10000	1.10000	0.90000	0.90000	0.90000
$T_{4\text{-}18}$	0.90000	0.90000	0.90000	0.90000	0.90000	0.90000	0.97000	1.00000	1.01000
$T_{21\text{-}20}$	1.00000	1.00000	1.01000	1.02000	1.00000	0.99000	1.01000	1.10000	1.04000
$T_{24\text{-}25}$	1.10000	1.10000	1.10000	0.90000	0.90000	0.90000	0.90000	0.90000	0.90000
$T_{24\text{-}25}$	1.10000	1.10000	1.10000	1.10000	1.10000	1.10000	1.10000	1.10000	1.10000
$T_{24\text{-}26}$	1.00000	1.02000	1.01000	0.90000	0.90000	0.90000	1.10000	1.10000	1.10000
$T_{7\text{-}29}$	0.93000	0.93000	0.93000	0.93000	1.01000	0.97000	0.91000	0.91000	0.91000
$T_{34\text{-}32}$	0.98000	0.90000	0.93000	0.90000	0.90000	0.90000	0.90000	0.90000	0.90000
$T_{11\text{-}41}$	0.97000	0.94000	0.96000	0.97000	0.96000	0.96000	0.90000	0.90000	0.90000
$T_{15\text{-}45}$	0.94000	0.94000	0.94000	0.95000	0.93000	0.96000	0.92000	0.92000	0.92000

(continued)

Table 1 (continued)

MOCIPSO [14]				MOIPSO [14]			MOPSO [14]		
Variables	Min PLoss	Min L-Index	BCS	Min PLoss	Min L-Index	BCS	Min PLoss	Min L-Index	BCS
$T_{14\text{-}46}$	0.92000	0.92000	0.92000	0.90000	0.90000	0.90000	0.90000	0.90000	0.90000
$T_{10\text{-}51}$	0.93000	0.93000	0.93000	0.97000	0.95000	0.97000	0.90000	0.90000	0.90000
$T_{13\text{-}49}$	0.90000	0.90000	0.90000	0.90000	0.90000	0.90000	1.10000	1.10000	1.10000
$T_{11\text{-}43}$	0.90000	0.90000	0.90000	0.90000	0.90000	0.90000	0.91000	0.91000	0.91000
$T_{40\text{-}56}$	1.08000	1.10000	1.08000	1.10000	1.10000	1.10000	1.10000	1.10000	1.10000
$T_{39\text{-}57}$	1.00000	1.05000	1.02000	0.90000	0.90000	0.90000	0.97000	1.03000	1.01000
$T_{9\text{-}55}$	0.92000	0.92000	0.92000	0.93000	0.98000	0.95000	0.92000	0.92000	0.92000
Q_{C18} (p.u.)	0.00000	0.00000	0.00000	0.00000	0.00000	0.03500	0.07500	0.18500	0.17500
Q_{C25} (p.u.)	0.18000	0.18000	0.18000	0.01800	0.18000	0.15000	0.00000	0.00000	0.00000
Q_{C53} (p.u.)	0.04200	0.05400	0.04800	0.00000	0.00000	0.00000	0.00000	0.00000	0.00000
P_{loss} (p.u.)	0.27075	0.27254	0.27122	0.28449	0.29501	0.28799	0.31424	0.31595	0.31462
L-Index	0.24274	0.23291	0.23695	0.29632	0.24087	0.24770	0.29123	0.28834	0.28952
MOALO [21]				ISA			LISA Strategy-I		
Variables	Min PLoss	Min L-Index	BCS	Min PLoss	Min L-Index	BCS	Min PLoss	Min L-Index	BCS
V_{G1} (p.u.)	1.0638	1.0456	1.0550	1.0635	1.0454	1.0548	1.0637	1.0455	1.0547
V_{G2} (p.u.)	1.0486	1.0298	1.0381	1.0484	1.0297	1.0379	1.0481	1.0296	1.0377
V_{G3} (p.u.)	1.0510	1.0297	1.0383	1.0513	1.0299	1.0381	1.0517	1.0291	1.0382
V_{G6} (p.u.)	1.0570	1.0336	1.0424	1.0569	1.0334	1.0427	1.0569	1.0337	1.0426
V_{G8} (p.u.)	1.0967	1.0764	1.0830	1.0965	1.0767	1.0832	1.0965	1.0763	1.0827
V_{G9} (p.u.)	1.0666	1.0462	1.0552	1.0663	1.0459	1.0555	1.0664	1.0459	1.0551

(continued)

Table 1 (continued)

MOALO [21]				ISA			LISA Strategy-I		
Variables	Min PLoss	Min L-Index	BCS	Min PLoss	Min L-Index	BCS	Min PLoss	Min L-Index	BCS
V_{G12} (p.u.)	1.0805	1.0642	1.0701	1.0801	1.0639	1.0698	1.0805	1.0639	1.0694
$T_{4\text{-}18}$	1.0195	0.9978	1.0053	1.0192	0.9979	1.0049	1.0192	0.9969	1.0047
$T_{4\text{-}18}$	1.0340	1.0112	1.0210	1.0337	1.0110	1.0207	1.0335	1.0104	1.0206
$T_{21\text{-}20}$	1.0990	1.0952	1.0983	1.0987	1.0949	1.0981	1.0987	1.0951	1.0976
$T_{24\text{-}25}$	0.9909	0.9642	0.9735	0.9904	0.9639	0.9732	0.9904	0.9639	0.9731
$T_{24\text{-}25}$	1.0128	0.9844	0.9934	1.0126	0.9841	0.9937	1.0121	0.9839	0.9932
$T_{24\text{-}26}$	1.0998	1.0947	1.0977	1.0996	1.0952	1.0974	1.0992	1.0946	1.0973
$T_{7\text{-}29}$	0.9973	0.9693	0.9799	0.9971	0.9697	0.9795	0.9971	0.9692	0.9793
$T_{34\text{-}32}$	1.0305	1.0029	1.0136	1.0301	1.0031	1.0134	1.0302	1.0029	1.0135
$T_{11\text{-}41}$	1.0260	1.0057	1.0138	1.0258	1.0060	1.0135	1.0256	1.0059	1.0132
$T_{15\text{-}45}$	0.9904	0.9642	0.9737	0.9901	0.9639	0.9741	0.9901	0.9636	0.9737
$T_{14\text{-}46}$	0.9828	0.9579	0.9672	0.9831	0.9582	0.9671	0.9824	0.9579	0.9666
$T_{10\text{-}51}$	0.9963	0.9759	0.9853	0.9961	0.9760	0.9857	0.9958	0.9761	0.9855
$T_{13\text{-}49}$	1.0084	0.9902	0.9988	1.0081	0.9901	0.9991	1.0079	0.9902	0.9988
$T_{11\text{-}43}$	0.9987	0.9738	0.9825	0.9985	0.9741	0.9827	0.9982	0.9738	0.9822
$T_{40\text{-}56}$	1.0291	1.0038	1.0140	1.0288	1.0041	1.0137	1.0287	1.0034	1.0137
$T_{39\text{-}57}$	0.9995	0.9765	0.9864	0.9992	0.9767	0.9862	0.9991	0.9770	0.9867
$T_{9\text{-}55}$	1.0650	1.0421	1.0525	1.0647	1.0423	1.0523	1.0648	1.0421	1.0525
Q_{C18} (p.u.)	5.8080	5.8077	5.8632	5.8078	5.8075	5.8635	5.8075	5.8071	5.8632
Q_{C25} (p.u.)	8.6317	8.3751	8.4914	8.6313	8.3748	8.4917	8.6314	8.3747	8.4914

(continued)

Table 1 (continued)

MOALO [21]				ISA			LISA Strategy-I		
Variables	Min PLoss	Min L-Index	BCS	Min PLoss	Min L-Index	BCS	Min PLoss	Min L-Index	BCS
Q_{C53} (p.u.)	13.5386	13.3489	13.4402	13.5383	13.3487	13.4399	13.5385	13.3488	13.4394
P_{loss} (p.u.)	26.3952	27.2634	26.7477	25.6456	26.9451	25.7123	25.5613	26.3224	25.6257
L-Index	0.2994	0.2854	0.2885	0.2942	0.2815	0.2855	0.2978	0.2803	0.2836

LISA Strategy-II

Min P_{Loss}	Min L-Index	BCS
1.0636	1.0452	1.0552
1.0482	1.0295	1.0382
1.0515	1.0294	1.0385
1.0572	1.0338	1.0425
1.0969	1.0766	1.0829
1.0667	1.0463	1.0554
1.0807	1.0641	1.0697
1.0194	0.9971	1.0051
1.0338	1.0107	1.0208
1.0991	1.0953	1.0979
0.9906	0.9643	0.9734
1.0125	0.9842	0.9935
1.0995	1.0949	1.0976
0.9974	0.9694	0.9796
1.0304	1.0031	1.0137

(continued)

Table 1 (continued)

LISA Strategy-II		
Min P_{Loss}	Min L-Index	BCS
1.0257	1.0061	1.0134
0.9902	0.9638	0.9739
0.9827	0.9581	0.9669
0.9959	0.9762	0.9854
1.0082	0.9904	0.9990
0.9984	0.9740	0.9823
1.0289	1.0035	1.0139
0.9994	0.9769	0.9865
1.0652	1.0424	1.0527
5.8077	5.8074	5.8634
8.6316	8.3749	8.4916
13.5387	13.3491	13.4397
25.4662	25.8218	25.5345
0.2987	0.2796	0.2829

Table 2 Statistical results for IEEE 57-bus system

Objective functions	Obtained results	ISA	LISA Strategy-I	LISA Strategy-II
Active power losses (MW)	Best solution	25.6456	25.5613	25.4762
	Average value	26.1564	26.0948	26.0015
	Worst solution	27.2156	27.1346	27.0154
	Standard deviation	0.1356	0.1294	0.1210
L-Index	Best solution	0.2815	0.2801	0.2798
	Average value	0.2886	0.2864	0.2845
	Worst solution	0.2993	0.2947	0.2905
	Standard deviation	0.0119	0.0112	0.0108

5.2 *Test System 2 (IEEE 300-Bus Test System)*

With the purpose of testing the capability of LISA simulation in large-scale power systems, the researchers utilized IEEE 300-bus system [32] which had a total of 69 generating units along with 411 transmission lines from which 107 branches protrude with off-nominal tap ratios. Further, an additional 14 parallel compensators were also considered as per the suggestions in the literature [14]. The total load demand considered was 235.258 p.u. The minimum and maximum value of the control variables were chosen from [20]. Figures 8, 9, and 10 portray the convergence characteristics attained using ISA, LISA strategy-I, and strategy-II, respectively. The trade-off characteristics between real power loss and voltage stability index attained when using ISA, LISA strategy-I, and LISA strategy-II are presented in Figs. 11, 12, and 13, respectively. Table 3 presents the best solution obtained for real power loss and voltage deviation Index. The simulation results revealed that the magnitudes of the voltages at all load buses were inside the enacted boundaries. It can be concluded from the application results that the proposed LISA algorithm yielded better solution

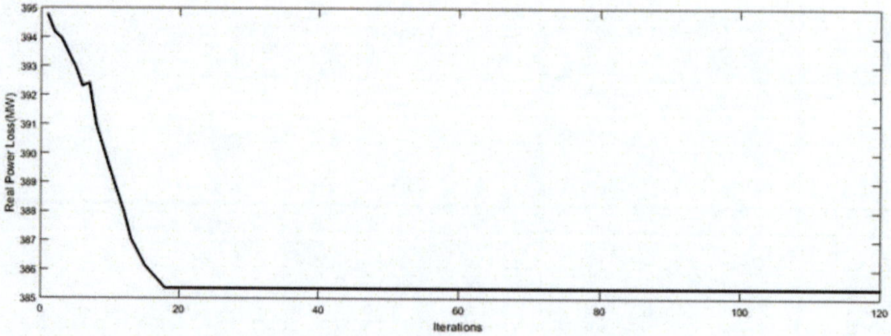

Fig. 8 Convergence characteristic for IEEE 300-bus system using ISA

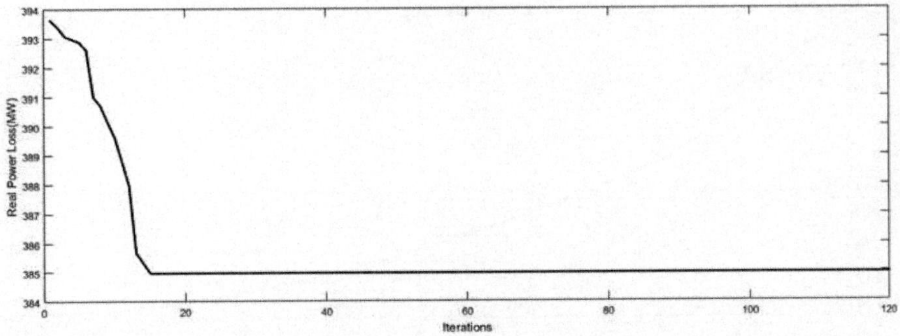

Fig. 9 Convergence characteristic for IEEE 300-bus system using LISA Strategy-I

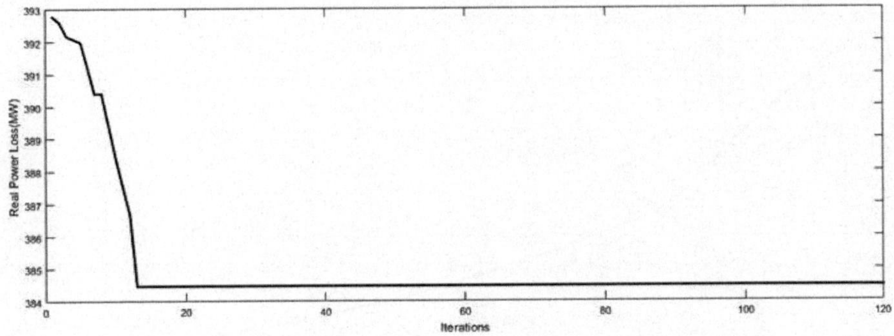

Fig. 10 Convergence characteristic for IEEE 300-bus system using LISA Strategy-II

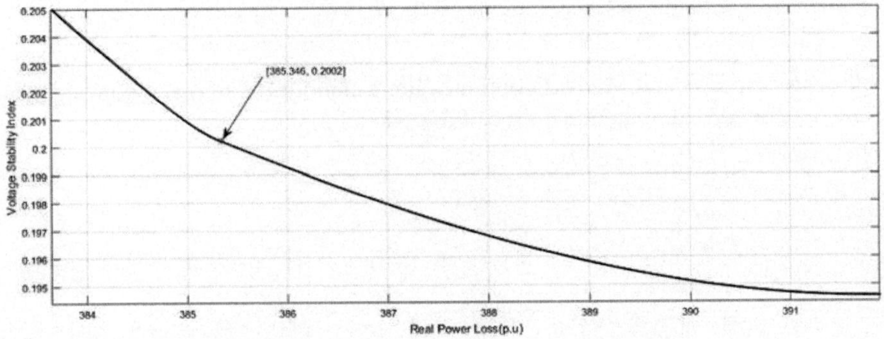

Fig. 11 Best-obtained Pareto-fronts for IEEE 300-bus system using ISA

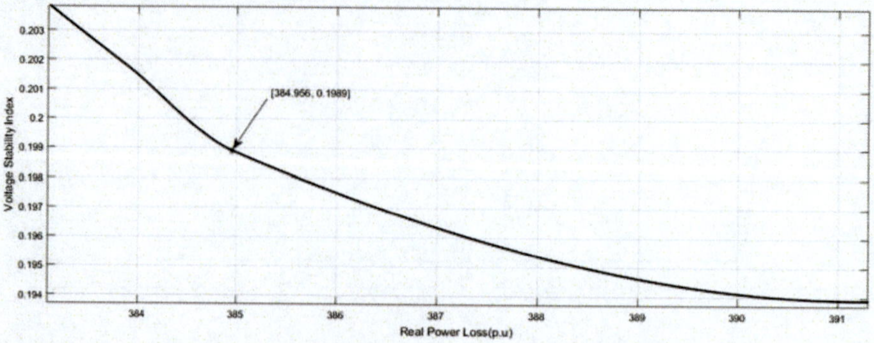

Fig. 12 Best-obtained Pareto-fronts for IEEE 300-bus system using LISA Strategy-I

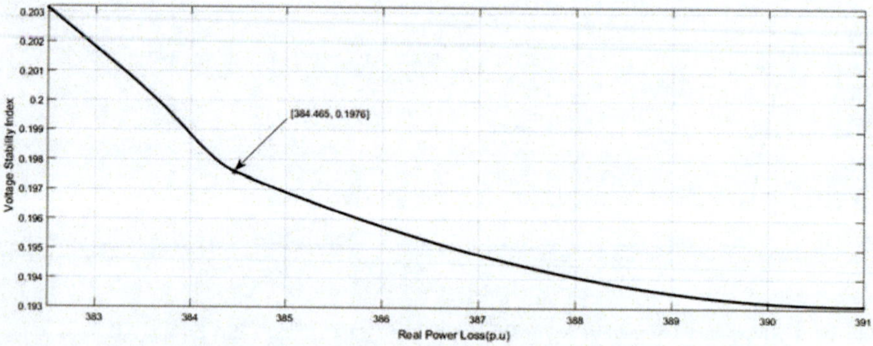

Fig. 13 Best-obtained Pareto-fronts for IEEE 300-bus system using LISA Strategy-II

with respect to ISA and MOALO optimization techniques in terms of convergence characteristic and computational time (Tables 4 and 5).

Table 3 Comparison of optimal results for the minimization of real power loss and voltage deviation for IEEE 300-bus system

	Limits		MOALO [21]			ISA		
Variables	Q_{min}	Q_{max}	Min P_{Loss}	Min VSI	BCS	Min P_{Loss}	Min VSI	BCS
Q_{96}	0	4.5	1.991	1.915	1.962	1.993	1.914	1.958
Q_{99}	0	0.59	0.509	0.500	0.506	0.510	0.498	0.508
Q_{133}	0	0.59	0.444	0.433	0.440	0.439	0.432	0.438
Q_{143}	4.5	0	−2.574	−2.621	−2.59	−2.572	−2.619	−2.60
Q_{145}	4.5	0	−3.377	−3.428	−3.95	−3.375	−3.431	−3.94
Q_{152}	0	0.59	0.1721	0.1629	0.1685	0.1719	0.1630	0.1679
Q_{158}	0	0.59	0.5325	0.5163	0.5274	0.5324	0.5159	0.5276
Q_{169}	2.5	0	−2.231	−2.262	−2.238	−2.229	−2.263	−2.237
Q_{210}	4.5	0	−4.194	−4.214	−4.201	−4.196	−4.215	−4.203
Q_{217}	4.5	0	−3.305	−3.386	−3.328	−3.306	−3.387	−3.329
Q_{219}	1.5	0	−0.8038	−0.8135	−0.807	−0.8037	−0.8136	−0.809
Q_{227}	0	0.59	0.375	0.362	0.369	0.377	0.363	0.371
Q_{268}	0	0.15	0.1180	0.1157	0.1170	0.1182	0.1153	0.1169
Q_{283}	0	0.15	0.0847	0.0825	0.0839	0.0850	0.0823	0.0841
P_{Loss}			384.5104	392.7140	386.241	383.6948	391.8125	385.346
L-Index			0.2058	0.1983	0.2010	0.2049	0.1946	0.2002

Variables	LISA Strategy-I			LISA Strategy-II		
	Min P_{Loss}	Min VSI	BCS	Min P_{Loss}	Min VSI	BCS
Q_{96}	1.994	1.916	1.959	1.992	1.917	1.963
Q_{99}	0.509	0.497	0.507	0.512	0.501	0.509
Q_{133}	0.436	0.431	0.437	0.438	0.429	0.439
Q_{143}	−2.569	−2.617	−2.599	−2.570	−2.620	−2.597
Q_{145}	−3.372	−3.427	−3.941	−3.373	−3.429	−3.938
Q_{152}	0.1716	0.1625	0.1677	0.1718	0.1627	0.1678
Q_{158}	0.5325	0.5160	0.5274	0.5327	0.5161	0.5275
Q_{169}	−2.227	−2.259	−2.236	−2.231	−2.262	−2.233
Q_{210}	-4.194	−4.213	−4.204	−4.197	−4.214	−4.202
Q_{217}	-3.305	−3.388	−3.322	−3.308	−3.391	−3.317
Q_{219}	-0.8035	−0.8136	−0.812	−0.8039	−0.8138	−0.811
Q_{227}	0.374	0.361	0.369	0.376	0.364	0.373
Q_{268}	0.1179	0.1148	0.1170	0.1181	0.1152	0.1172
Q_{283}	0.0847	0.0824	0.0843	0.0849	0.0826	0.0842
P_{Loss}	**383.1634**	**391.2156**	**384.956**	**382.7892**	**390.9154**	**384.465**
L-Index	**0.2037**	**0.1939**	**0.1989**	**0.2024**	**0.1931**	**0.1976**

Table 4 Statistical results for IEEE 300-bus system

Objective functions	Obtained results	ISA	LISA Strategy-I	LISA Strategy-II
Active power losses (MW)	Best solution	385.346	384.956	**384.465**
	Average value	386.514	385.965	385.376
	Worst solution	389.469	388.946	388.617
	Standard deviation	0.1638	0.1694	0.1564
L-Index	Best solution	0.2002	0.1989	0.1976
	Average value	0.2024	0.1998	0.1992
	Worst solution	0.2057	0.2053	0.2049
	Standard deviation	0.0106	0.0101	0.0096

Table 5 Comparison of CPU time (s) of various methods in test system 1 and test system 2

Algorithms	IEEE 57-bus system	IEEE 300-bus system
MOCIPSO [14]	432.71375	NA
MOIPSO [14]	431.29943	NA
MOPSO [14]	424.59392	NA
MOALO [21]	151.55	NA
ALO [20]	NA	4022.9
ISA	147.34	4017.32
LISA Strategy-I	142.17	4015.47
LISA Strategy-II	140.78	4013.68

6 Conclusion

This paper presented the application of LISA to elucidate the ORPD problem by considering real power loss and voltage stability index. This research article assesses the capability of LISA algorithm to overcome the large-scale multi-objective ORPD problem so as to enhance the performance of power systems operation and simultaneously reduce the real power losses through precise fine-tuning of control variables. Being an intricate multi-objective optimization problem, the ORPD has both independent decision variables and multiple conflicting objectives. The proposed algorithm was successfully validated on standard IEEE 57-bus and IEEE 300-bus test systems. The simulation results obtained using LISA were compared with other meta-heuristic optimization techniques such as ISA, MOALO, MOPSO, and MOEPSO. The application results revealed that the LISA was highly effective than ISA, MOALO, MOPSO, and MOEPSO in terms of the superiority in attaining the optimum solutions and computational time. Moreover, the application results evidently substantiated the effectiveness of LISA to create a set of Pareto optimal solutions. The proposed LISA algorithm subsequently revealed an upright capability to even handle large-scale power systems when it comes to elucidating the sophisticated multi-objective optimization problems. Thus, the future researchers can consider LISA algorithm as an efficient tool as a solution for various complex engineering optimization problems. With regard to the future scope of the current research, the authors would like to explore the areas of reduction in active power loss and voltage stability index in view of the fact that there are uncertainties in system load and wind power generations in a wind-integrated power system. LISA is robust and consumes higher execution time compared to other optimization algorithms. Furthermore, LISA seeks for the global optimal solution by hitting a fine steadiness among exploration and exploitation progressions. LISA can be employed to elucidate large-scale power system optimization problems.

References

1. Yan W, Liu F, Chung CY, Wong KP (2006) A hybrid genetic algorithm-interior point method for optimal reactive power flow. IEEE Trans Power Syst 21(3):1163–1169
2. Kessel P, Glavitsch H (1986) Estimating the voltage stability of a power system. IEEE Trans Power Delivery 1(3):346–354
3. Raghunatha R, Ramanujam R, Parthasarathy K, Thukaram D (1999) Optimal static voltage stability improvement using a numerically stable SLP algorithm, for real time applications. Int J Electr Power Energy Syst 21(4):289–297
4. Xiong H, Cheng H, Li H (2008) Optimal reactive power flow incorporating static voltage stability based on multi-objective adaptive immune algorithm. Energy Convers Manag 49(5):1175–1181
5. Niknam T, Narimani MR, Aghaei J, Azizpanah-Abarghooee R (2012) Improved particle swarm optimisation for multi-objective optimal power flow considering the cost, loss, emission and voltage stability index. Gener Transmiss Distrib 6:515–527
6. Roy PK, Mandal B, Bhattacharya K (2012) Gravitational search algorithm based optimal reactive power dispatch for voltage stability enhancement. Electr Power Components Syst 40:956–976
7. Saraswat A, Saini A (2013) Multi-objective optimal reactive power dispatch considering voltage stability in power systems using HFMOEA. Eng Appl Artif Intell 26:390–404
8. Sayah S, Zehar K (2008) Modified differential evolution algorithm for optimal power flow with non-smooth cost functions. Energy Convers Manage 49(11):3036–2042
9. Abido MA, Bakhashwain JM (2005) Optimal VAR dispatch using a multi-objective evolutionary algorithm. Int J Electr Power Energy Syst 27(1):13–20
10. Khorsandi A, Hosseinian SH, Ghazanfari A (2013) Modified artificial bee colony algorithm based on fuzzy multi-objective technique for optimal power flow problem. Electr Power Syst Res 95:206–213
11. Jeyadevi S, Baskar S, Babulal CK, Iruthayarajan MW (2011) Electrical power and energy systems solving multiobjective optimal reactive power dispatch using modified NSGA-II. Int J Electr Power Energy Syst 33(2):219–228
12. Chen G, Liu L, Yanyan Guo SH (2016) Multi-objective enhanced PSO algorithm for optimizing power losses and voltage deviation in power systems. Compel—Int J Comput Math Electr Electron Eng 35(1):350–372
13. Zeng Y, Sun Y (2014) Solving multi-objective optimal reactive power dispatch using improved multiobjective particle swarm optimization. In: Control and decision conference (2014 CCDC), China, 31 May–2 June 2014. IEEE, pp 1010–1015
14. Chen G, Liu L, Song P, Dua Y (2014) Chaotic improved PSO-based multi-objective optimization for minimization of power losses and L index in power systems. Energy Convers Manage 86:548–560

15. Ghasemi A, Valipour K, Tohidi A (2014) Multi objective optimal reactive power dispatch using a new multi objective strategy. Int J Electr Power Energy Syst 57:318–334
16. Liang R, Wang J, Chen Y, Tseng W (2015) An enhanced firefly algorithm to multi-objective optimal active/reactive power dispatch with uncertainties consideration. Int J Electri Power Energy Syst 64:1088–1097
17. Basu M (2016) Quasi-oppositional differential evolution for optimal reactive power dispatch. Electr Power Energy Syst 78:29–40
18. Mehdinejad M, Mohammadi-Ivatloo B, Dadashzadeh-Bonab R, Zare K (2016) Solution of optimal reactive power dispatch of power systems using hybrid particle swarm optimization and imperialist competitive algorithms. Electr Power Energy Syst 83:104–116
19. Jordehi R (2017) Gaussian bare-bones water cycle algorithm for optimal reactive power dispatch in electrical power systems. Appl Soft Comput J 57:657–671
20. Mouassa S, Bouktir T, Salhi A (2017) Ant lion optimizer for solving optimal reactive power dispatch problem in power systems. Eng Sci Technol 20(3)
21. Mouassa S, Bouktir T (2018) Multi-objective ant lion optimization algorithm to solve large-scale multi-objective optimal reactive power dispatch problem. COMPEL: Int J Comput Math Electr Electron Eng 35(1):350–372
22. Arul R, Ravi G, Velusami S (2013) Solving optimal power flow problems using chaotic self-adaptive differential harmony search algorithm. Electr Power Components Syst 48:782–805
23. Mukherjee A, Mukherjee V (2016) Chaos embedded krill herd algorithm for optimal VAR dispatch problem of power system. Electr Power Energy Syst 82:37–48
24. Rajan A, Jeevan K, Malakar T (2017) Weighted elitism based ant lion optimizer to solve optimum VAr planning problem. Appl Soft Comput J 55:352–370
25. ben oualid Medani K, Sayah S, Bekrar A (2017) Whale optimization algorithm based optimal reactive power dispatch: a case study of the Algerian power system. Electr Power Syst Res
26. Ng R, Mei S, Herwan M, Mustaffa Z, Daniyal H (2017) Optimal reactive power dispatch solution by loss minimization using moth-flame optimization technique. Appl Soft Comput J 59:210–222
27. Abaci K, Yamacli V (2017) Optimal reactive-power dispatch using differential search algorithm. Electr Eng 99
28. Naderi E, Narimani H, Fathi M, Narimani MR (2017) A novel fuzzy adaptive configuration of particle swarm optimization to solve large-scale optimal reactive power dispatch. Appl Soft Comput
29. Gandomi AH (2014) Interior search algorithm (ISA): a novel approach for global optimization. ISA Trans
30. Karthik N, Krishnamoorthy Parvathy A, Arul R (2019) Multi-objective economic emission dispatch using interior search algorithm. Int Trans Electr Energy Syst 29(1)
31. The IEEE 57-Bus Test System [online]. Available at http://www.eewashington.edu/research/pstca/pf57/pg_tca57bus.htm
32. Zimmerman RD, Murillo Sänchez CE, Thomas RJ (2011) MATPOWER: steady-state operations, planning, and analysis tools for power systems research and education, power systems. IEEE Trans Power Syst 26(1):12–19

Feasible Settlement Process for Primary Market Using Distributed Slack Power Flow Strategy

F. Ruby Vincy Roy, A. Peer Fathima, and Arunachalam Sundaram

Abstract The paper proposes feasible and acceptable distributed slack power flow (DSPF) formulation for the settlement process of active power market in deregulated system with single-sided bidding. In general market clearing is done using single slack bus model but whenever marginal bus changes loss price varies drastically. In order to prevent the sudden changes in loss price, optimization using distributed slack model is suggested in this paper. The loss factor approach is used in determining the feasible participation factor. The proposed strategy is verified using PJM 5 bus system.

Keywords Locational marginal pricing · OPF · DSPF

1 Introduction

For decades, vertically integrated electric utilities monopolized the generation, transmission and distribution services to the customers. To implement competition, these monopolies have to unbundle their services so as to bring privatization into the picture, thereby proposed restructured models that brought open access environment for the customers to choose their service provider for electric energy. The consumer has to pay back the generation, transmission and distribution companies for supplying the expected demand.

Henceforth tariffs are charged for the customers based on their usage. Different methods are available for the price calculation. But there is need for giving a feasible strategy for the usage of supply for the customers. This strategy needs to address the

F. Ruby Vincy Roy (✉) · A. Peer Fathima (✉)
School of Electrical Engineering, Vellore Institute of Technology, Chennai 600127, India
e-mail: rubyvincy.royf2019@vitstudent.ac.in

A. Peer Fathima
e-mail: peerfathima.a@vit.ac.in

A. Sundaram
Jubail Industrial College, Al Jubail 31961, Kingdom of Saudi Arabia
e-mail: sundaram_a@jic.edu.sa

N. Zhou and S. Hemamalini (eds.), *Advances in Smart Grid Technology*, Lecture Notes in Electrical Engineering 688, https://doi.org/10.1007/978-981-15-7241-8_18

contribution of each generating company in the market structure thereby acceptable prices can be charged for their contribution and for the usage by the customers. The Independent system operator (ISO) provides the freedom to the customers to choose the suppliers. This increases the mode of competition in the restructured environment. The paper carries out a simple technique of charging the prices of the energy consumed by the distribution company.

2 Literature Survey

For primary transaction, clearing and settlement in a deregulated power system for the active power market is posed as optimal power flow (OPF) problem. The paper [1] proposes ACOPF model for the purpose of producing optimal power flows for the participants of market clearing and analyzes locational marginal pricing methodology. It has been a dominant approach in power markets because the losses are included and produce accurate results. Due to robustness and high computational speed the DCOPF model is most widely used than ACOPF in all conventional methodology [2]. The resistance of the lines and the transformers are neglected in DCOPF formulation.

The major challenge in DCOPF methodology is to address the loss allocation and calculation of loss price [3]. Energy transactions are done using location marginal pricing strategy and LMP is decomposed into reference price at the slack bus, the marginal price for loss in transmission, and the limiting price of constraints in power with single slack bus.

Similar LMP decomposition is obtained when distributed slack bus concept is used in the power-flow formulation [4]. But selection of participation factor plays a very important role in proper loss allocation. This paper suggests the importance of participation factors which will uniquely allocate loss according to the loss contributed by each generator.

3 Distributed Slack Bus

In balanced transmission systems, distributed slack buses were introduced in order to overcome the limitations of the single slack bus concept. System losses during the clearing and settlement process are to be addressed by the generating companies using participation factors. The concept of distributed slack bus was introduced in order to overcome the following drawbacks:

- The main drawback of single slack bus approach is that whenever the marginal bus changes, there is a large variation in the loss component of the LMP.

- All the existing methods use marginal bus as the reference bus for calculating LMPs. LMPs change drastically when the marginal bus changes for every hour in the hour-ahead market (HAM).
- Major impact took place in pricing for the contribution made by the generating companies when single slack bus was used.

4 OPF Formulation Using DSPF Model

The objective is to calculate the optimized schedules of generation and marginal prices so as to produce feasible results for the market clearing problem.

4.1 Assumptions

The major assumptions are:

- Each bus has at most one resource.
- Bus voltage are constant at 1 p.u
- PV buses ordered from 1 to $N - 1$ where N is the total no. of buses.
- Reference bus is Nth bus with phase angle $\theta_N = 0$
- A distributed slack bus (imaginary bus at the pseudo node) whose injection δ (the mismatch) is dispersed to all the buses by participation factors α_i

4.2 Mathematical Model

Mathematical model for the market clearing problem with loss is given below [1],

To Determine: Optimum schedule of GENCOs P_{G}
To minimize objective function:

$$\mathrm{Min} \sum_{i=1}^{\mathrm{NG}} C_{\mathrm{G}i} \cdot P_{\mathrm{G}i} \tag{1}$$

Sub. to:

Equality constraint: Zero difference (Initial mismatch at imaginary node)

$$\delta = 0 \tag{2}$$

Equality constraint:

$$\underline{P_i}(\theta') - (P_{gi} - P_{di}) - \alpha_i\delta = 0, \quad i = 1, 2 \ldots N \tag{3}$$

Inequality constraint: Line flow constraint,

$$\underline{g}(\theta') - g^{\mathrm{Max}} \leq 0 \tag{4}$$

Limits:

$$P_{\mathrm{Gi}_{\min}} \leq P_{\mathrm{Gi}} \leq P_{\mathrm{Gi}_{\max}}, \quad i = 1, 2 \ldots .NG \tag{5}$$

where

$\theta' = [\theta_1, \theta_2, \ldots, \theta_{N-1}]^T$ represents the voltage phase angle and $\theta_N = 0$
$\underline{P_i}(\theta')$ nonlinear function of active power injections at bus i

$$P_i(\theta) = V_i \sum_{i=1}^{N} V_j \left[G_{ij} \cos(\theta_i - \theta_j) + B_{ij} \sin(\theta_i - \theta_j)\right] \tag{6}$$

$g(\theta')$ active power flows carried by transmission lines
g^{Max} transmission line rating
P_{Gi} active power generation at the ith Genco
α_i is the participation factor at ith bus
δ is the injection at the pseudo node where distributed slack bus is located
NG is the total no. of GENCOs.

4.3 Optimization for Lossy Model

The constrained optimization is carried out using the linear programming solver in MATLAB which solves the nonlinear equations and produces optimized results of the generation schedules and the locational marginal prices (LMPs) at each Generating company (GENCo). The initial value for the problem is taken from the lossless dispatch schedules (obtained from aggregated supply offer curve) [5].

4.4 Segregation of LMPs

Each LMP is decomposed into three components:

(1) Marginal price at the distributed slack bus
(2) The marginal price of the transmission losses and
(3) The marginal price of the network constraint.

The LMP decomposition is mainly meant for settlement process and to introduce financial price risk hedging instruments. For the decomposition, DSPF analysis is carried out.

The Lagrangian function:

$$L(P_{G1} \ldots P_{GN,}\theta_1 \ldots \theta_N, \delta, \lambda, \mu) = \sum_{i=1}^{NG} C_{Gi}.P_{Gi} + \lambda_o\delta$$
$$+ \sum_{i=1}^{N} \lambda_i[P_i(\theta') - (P_{gi} - P_{di}) - \alpha_i\delta$$
$$+ \sum_{k=1}^{NL} \mu_K[g_K(\theta') - g_K^{Max}] \quad (7)$$

where

$C_{Gi} \cdot P_{Gi}$	Energy bid function of *i*th bus
λ_o	LMP at distributed slack bus
λ_i	LMP of real power at bus i
μ_k	Transmission constraint cost
NL	Total no. of line.

5 Distributed Slack Bus Formulation

The DSPF is carried out using the power flow equation given in (3).The first-order Taylor series expansion of the power flow equation is given below;

$$\begin{bmatrix} \frac{\partial P'(\theta')}{\partial\theta'} & -\alpha' \\ \frac{\partial P_N(\theta')}{\partial\theta'} & -\alpha_N \end{bmatrix} \begin{bmatrix} \partial\theta' \\ \partial\delta \end{bmatrix} = \begin{bmatrix} \Delta P'(X^{(h)}) \\ \Delta P_N(X^h) \end{bmatrix} \quad (8)$$

Transmission losses of the system in terms of active power in given by:

$$P_{Loss} = \sum_{i=1}^{N} P_i + \delta \quad (9)$$

The incremental loss factor is given by:

$$\frac{\partial P_{Loss}}{\partial P} = [11\ldots1]^T + \frac{\partial\delta}{\partial P} \quad (10)$$

Calculated by Jacobian matrix inversion as,

$$\begin{bmatrix} \frac{\partial \theta'}{\partial P} \\ \frac{\partial \delta}{\partial P} \end{bmatrix} = \begin{bmatrix} \frac{\partial P'(\theta')}{\partial \theta'} & -\alpha' \\ \frac{\partial P_N(\theta')}{\partial \theta'} & -\alpha_N \end{bmatrix}^{-1} \tag{11}$$

where

$$P' = [P_1, P_2 \ldots P_{N-1}], \quad \alpha' = [\alpha_1, \alpha_2 \ldots \alpha_{N-1}] \tag{12}$$

By referring to the KKT condition in Yong and Zuyi, 2010 the equation for the bus prices are given below;

$$\begin{bmatrix} \lambda' \\ \lambda^N \end{bmatrix} = \begin{bmatrix} 1 \\ 1 \\ \cdot \\ \cdot \\ 1 \end{bmatrix} \lambda^o - \left(\frac{\partial P_{\text{Loss}}}{\partial P} \right)^{\text{T}} - T^{\text{T}} \mu \tag{13}$$

where $\lambda' = [\lambda_1, \lambda_2, \ldots \lambda_{N-1}]$.

The Power Transfer distribution Matrix (PTDF) or T Matrix is given by,

$$T = \frac{\partial g}{\partial P} = \frac{\partial g}{\partial \theta} \frac{\partial \theta}{\partial P} \tag{14}$$

can be obtained from Eq. (11).

6 Participation Factor

The concept of participant-based pricing used in deregulated power system is discussed in [6, 7]. In general, there are three methods to calculate the participation factors. They are (1) slack bus variation, (2) generator power adjustment (GPA) variation, and (3) load power adjustment (LPA) variation [4].

The existing system uses one of the above methods for calculating the participation factor for the various buses. Using this participation factor the market clearing problem is done using linear programming and the optimized schedules and the LMP at various buses are obtained [9]. For the LMP decomposition, DSPF analysis is carried out.

But selection of participation factor plays a very important role in proper loss allocation to the participants. The proposed method uses **Loss Factor Approach (LFA)** for calculating the participation factors whose values will be proportional to the incremental transmission loss, thus resulting in feasible and optimal LMPs at various buses [10].

The loss factor (LF_i) at the ith bus is given as

$$\mathrm{LF}_i = \frac{\partial P_{\mathrm{Loss}}}{\partial P} * P_i \tag{15}$$

where

$\frac{\partial P_{\mathrm{Loss}}}{\partial P}$ is the incremental loss factor
P_i is the power injection at the ith bus.

The participation factor α_i is expressed as

$$\alpha_i = \frac{\mathrm{LF}_i}{\sum_{i=1}^{N} \mathrm{LF}_i} \tag{16}$$

7 Solution Procedure for Single-Side Auction Power Loss Market

1. Collect the bus-wise real power demand data
2. Collect the supply offer- price and quantity from GENCOs
3. Clear the lossless real power market
4. With the cleared quantities as the initial value and participation factors obtained from the any one of the three methods, perform the constrained optimization in section C using MATLAB linear programming solver.
5. Using the schedules obtained from the linear programming; perform the distributed slack power flow analysis for Eq. (3) and determine the LMP components [8]
6. Repeat step 4 and 5 using Participation factors obtained from LFA
7. Compare the price results obtained from existing and proposed method
8. Carry out the settlement procedure for the GENCO.

8 Results and Discussions

A sample PJM five-bus system is considered for the market clearing of single side auction primary market. The single-line diagram is shown in Fig. 1.

The lossless dispatch for a single side auction primary market is performed using the demand data and GENCOs bid data as shown in Tables 1 and 2, respectively.

Schedules of successful GENCOs obtained from lossless dispatch are given in Table 3. It is obtained from the aggregated supply offer curve given in Fig. 2.

Using the schedules (Table 3) and the participation factor obtained from **LPA** (Table 4), the market clearing problem is carried out. The results of the optimized schedules and the LMPs for various buses are displayed in Tables 5 and 6, respectively.

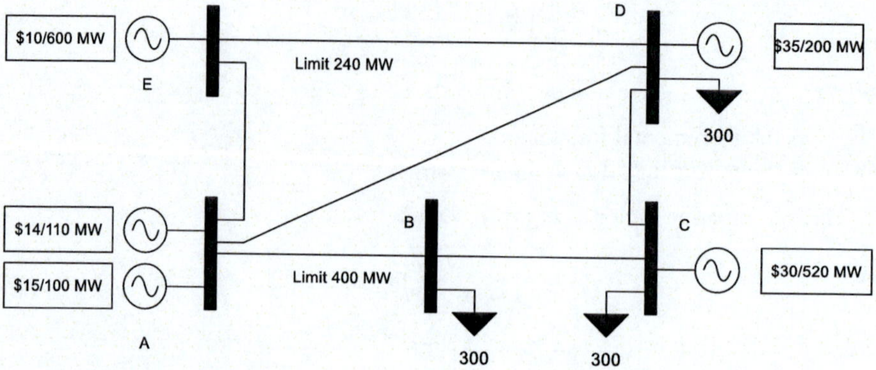

Fig. 1 Sample PJM 5 bus system

Table 1 Demand data for PJM five bus system

BUS	B	C	D
DISCOs	D1	D2	D3
$P_{D_{Max}}$ (MW)	300	300	300

Table 2 GENCOs offer data for PJM five bus system

BUS	A	A	C	D	E
GENCOs	Gen 1	Gen 2	Gen 3	Gen 4	Gen 5
$P_{G_{Max}}$ (MW)	110	100	520	200	600
Bid price ($/Mwh)	14	15	30	35	10

Table 3 Optimal schedules of the GENCOs

Bus No.	GENCO NO	Quantity (MW)
1	Gen 1	110
	Gen 2	100
2	–	–
3	Gen 3	90
4	–	–
5	Gen 5	600
	Total generation	900

From the optimized schedules obtained in Table 5 and the total demand the generation loss is estimated as 8.84 MW. On carrying out the DSPF formulation, the total line loss is estimated as 9.041 MW and the LMP components are obtained and given in Table 7

Repeating the steps of optimization using participation factors obtained from **LFA** (given in Table 4), the LMP components are listed in Table 8. The total loss obtained

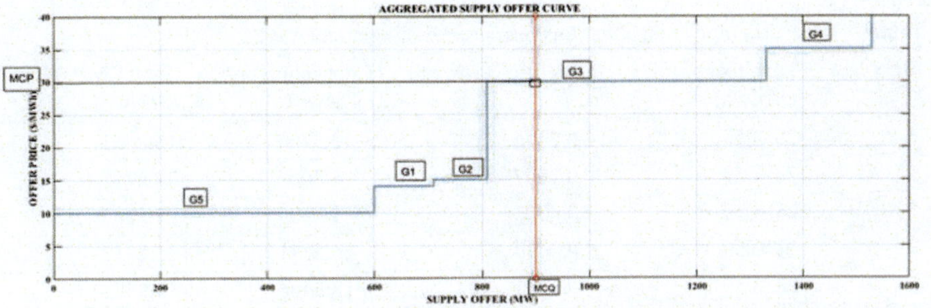

Fig. 2 Lossless dispatch of the aggregated supply offer curve

Table 4 Participation factors using LPA

Bus No.	P.F from LPA	P.F. from LFA
1	0	0.097
2	0.333	0.2216
3	0.333	0.2514
4	0.333	0.01738
5	0	0.4123

Table 5 Optimized schedules using LPA

Bus No./GENCO No.	Schedules (MW) P_G	Phase angle θ
1 (G1)	110	−0.11488
1 (G2)	100	−0.12337
3 (G3)	0	−0.05822
4 (G4)	116.75	0.74053
5(G5)	582.09	−0.008566

Table 6 Bus prices obtained using LPA

Bus No.	LMP at Bus ($/Mwh)
1	15.77
2	24.121
3	27.182
4	35.0000
5	10.0000

from LMP decomposition for the proposed method after performing distributed slack optimal power flow studies is 8.933 MW.

On comparing the results in Tables 7 and 8, we can determine that reduced and feasible values of pure energy price, loss price, and congestion price can be obtained

Table 7 LMP decomposition at various buses for existing method

BUS No.	Pure energy price ($/Mwh)	Loss price ($/Mwh)	Congestion price ($/Mwh)	Total LMP price ($/Mwh)
1	28.4059	−0.2395	−12.5764	15.8461
2	28.4059	0.3807	−4.8060	24.2415
3	28.4059	0.4306	−1.7709	27.3283
4	28.4059	0.0494	6.5771	35.2912
5	28.4059	−0.3631	−18.3466	9.9532

Table 8 LMP decomposition for the proposed method

Bus No.	Pure energy price ($/Mwh)	Loss price ($/Mwh)	Congestion price ($/Mwh)	Total LMP price ($/Mwh)
1	18.4747	−0.1558	−2.4703	15.8486
2	18.4747	0.2476	5.4734	24.1957
3	18.4747	0.2801	8.5078	27.2626
4	18.4747	0.0322	16.6761	35.1829
5	18.4747	−0.2362	−8.2519	9.9867

from the proposed method. It helps in producing accurate LMP values in less computational time. These accurate prices can be used in settlement process and to reduce the hedging risks. The difference in results from both the methods proves that participation factor have a high impact on the LMP decomposition. Hence, the proposed method would be beneficial for the participants to perform market clearing and settlement process.

8.1 *Settlement Process*

Table 9 gives the comparison of settlement process for both the existing and the proposed method. The surplus amount in the proposed method is less when compared to the existing method, and thus, it is financially reasonable for the customers.

Table 9 Comparison of settlement process for existing & proposed method

	Net MW of DISCOs	Net MW of GENCOs	Net payment by DISCOs ($)	Net payment to GENCOs ($)	Surplus
Existing method	900	909.047	25,823.55	12,955.18	12,868.36
Proposed method	900	908.933	25,984.92	13,278.20	12,706.72

9 Conclusion

This paper presents a detailed study on market clearing and settlement for the primary market with transmission loss using DSPF. The locational marginal prices are estimated and the segregation is done using distributed load flow analysis. A new approach in determining the participation factor is discussed in this paper. Thus, optimal and feasible prices can be calculated from this approach.

Appendix

The line data of PJM 5 bus system in given below:

From Bus (NL)	To BUS (NR)	*R* (P.U)	*X* (P.U)	1/2 B (P.U)	X'MER TAP (A)	LINE LIMIT (MVAR)
1	2	0.00281	0.0281	0.01	1	999
1	4	0.00304	0.0304	0.01	1	999
1	5	0.00064	0.0064	0.01	1	999
2	3	0.00108	0.0108	0.01	1	999
3	4	0.00297	0.0297	0.01	1	999
4	5	0.00297	0.0297	0.01	1	240

References

1. Fu Y, Li Z (2006) Different models and properties on LMP calculations. IEEE Power and Energy Society General Meeting, PES, 11pp
2. Orfanogianni T, Gross G (2007) A general formulation for LMP evaluation. IEEE Trans Power Syst 22(3)
3. Schweppe FC, Caramanis M, Tabors RD, Bohn RE (1988) Spot pricing of electricity. Kluwer, Norwell, MA
4. Meisel J (1993) System incremental cost calculations using the participation factor load-flow formulation. IEEE Trans Power Syst 8(1):357–363
5. Arunachalam S, Abdullah Khan M (2013) Linear programming formulation for market clearing of primary market incorporating tracking based loss factors. Eur J Sci Res 99:520–630
6. Sundaram A, Khan MA (2018) Electric power components and systems market clearing and settlement using participant based distributed slack optimal power flow model for a double sided electricity auction market—part II based distributed slack optimal power flow model for a double S. Electr Power Components Syst 1–11 (2018). https://doi.org/10.1080/15325008.2018.1460640
7. Sundaram A, Abdullah Khan M (2018) Market clearing and settlement using participant based distributed slack optimal power flow model for a double sided electricity auction market—part I. Electr Power Components Syst 5008:1–12. https://doi.org/10.1080/15325008.2018.1460639

8. Panigrahi BK (2012) Locational marginal pricing (LMP) in deregulated electricity market. Int J Res Advent Technol 4(2231):101–105
9. Litvinov E, Zheng T, Rosenwald G, Shamsollahi P (2004) Marginal loss in LMP calculation. IEEE Trans Power Syst 19(2):880–888
10. River M, Pérez-Arriaga IJ (1993) Computation and decomposition of spot prices for transmission pricing. In: Proceedings of PSCC, Avignon, France, pp 371–378

Optimal Placement of DG and Controlled Impedance FCL Sizing Using Salp Swarm Algorithm

C. Vasavi and T. Gowri Manohar

Abstract Smart grids have become one of the most emerging technologies in power systems by its fashionable design, reliability, and efficiency. Smart grid is the integration of multiple DGs. In this context, DG (Distributed generation) plays a significant role in power system. When DG is connected to the distribution network, the impact of fault currents is high which can cause protection coordination failure in smart grids. In this paper, three phase fault current which is the highest fault current in distribution network and its effect on DG is discussed. Controlled Impedance Fault Current Limiter (CI FCL) combination of both resistive and reactive component is used to maintain fault currents. The problem is formulated as integer, non-linear constrained problem. Optimal location of DG and CI FCL sizing is determined by using Salp Swarm Algorithm (SSA) and it is tested on IEEE 33 and 69 standard distribution bus systems in MATLAB2017. A comparison between the developed SSA and PSO is executed in terms of complexity, No. of iterations, and computational time. The results show that the proposed method is effective.

Keywords Distributed generation · Fault current limit (FCL) · Salp swarm algorithm (SSA) · FCL sizing · Optimal DG location

1 Introduction

Traditional grids are the interconnection of all transmission lines, sub transmission network, distribution substation. Transmitting and distribution of power causes many losses and power outages in the system. This results in failure in meeting the demands of the consumers. Now, the recent trend is smart grids. It provides more reliable and efficient two way of communication between utility and grid. The major advantage of smart grid is adjustment to differ situations by its fashionable design. Smart grids is the combination (or) integration of all DGs in the network in order to have clean environment DGs plays a vital role by meeting the demands of the consumer. While

C. Vasavi (✉) · T. Gowri Manohar
Department of EEE, Sri Venkateswara University, Tirupati, India
e-mail: cvasavi11788@gmail.com

N. Zhou and S. Hemamalini (eds.), *Advances in Smart Grid Technology*, Lecture Notes in Electrical Engineering 688, https://doi.org/10.1007/978-981-15-7241-8_19

using DGs in the power distribution network we are enjoying many benefits such as power quality, voltage profile improvement and reduction in power losses also[1]. Mohammadi et al. has discussed the various effects of distributed generation in terms of voltage profile, fault currents and voltage stability.

However, besides its merits the main problem is when DG is placed in the distribution network its fault currents increases. This causes the protection coordination problem in the network. Traditionally Relays and C.Bs are used in order to trip the unhealthy part of the system from distribution network. Whereas, the amplitudes of the fault currents in the system is very high that fuses, Relays, C.Bs cannot withstand Hence in order to limit the fault current a new approach has been found, i.e., Fault Current Limiter(FCL) [2]. Behazadirafi.et al. explained the problems due to overcurrent's in power system and uses a new a model fault current limiter. This helps in preventing the increase of failure rate and increases reliability.

FCL is a semiconductor device which suppresses the fault current to maintain constant. There are different types of FCLs and research is going on to utilize these FCLS in the distribution network. In [3] Bayati et al., proposed a method based on hybrid Genetic Algorithm is used to find optimal placement of FCL and restore the coordination of direct overcurrent relays and reduce the operating time of DOCL. This method is also used to reduce the size of the FCL and which in turn minimizes the cost of FCL. Resistive type of FCL is used due to its lower cost compared to other types of FCL.

In [4] Yasin et al. proposed the concept of installing small scale energy sources namely DG over the distribution network which has gained a great interest. DG is defined as the limited power resource applied less than 10 MW that connected either to the substation or a distribution feeder (or) load levels. In this paper, different types of faults are analyzed in the distribution network while connecting to DG and find the optimal location and sizing of DG with minimum power losses.

In [5] HTS FCL is considered. The optimization problem is solved by using PSCAD considering angular and voltage stability issues instead of conventional FCL. The basic type of HTS–FCL contains a short resistor in parallel with the HTS element. In this case the HTS element will provide a very fast fault limitation and due to its high resistivity in comparison with the shunt element for the remaining fault duration which lowers the stresses on the HTS wire and boosts the recovery time. In [6], Cuckoo optimization and PSO algorithm were adopted to coordinate overcurrent relays of micro grid. It was tested both for grid connected mode and island mode using resistive and inductive FCLs. In [7] Inductive and resistive FCLs are examined to limit the current drawn by DG anywhere in the looped power delivery systems to restore the original relay coordination.

Fault current limiters can be either passive or active. Passive type of FCL causes extra impedance and results in voltage drop during normal operation because they are placed constantly in the power system circuit. Whereas the active type of FCL provide low impedance during normal operation and high impedance during fault condition. Active type of FCL would be possible to increase or decrease the impedance under appropriate conditions. Figure 1 presents the active type of FCL.

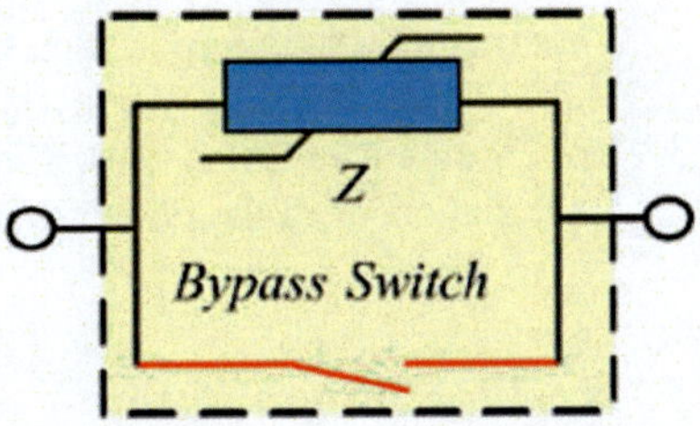

Fig. 1 Active type of FCL

In [8, 9] presents the controlled impedance (CI) which is active type of FCL used for the protection of distribution system. In [10, 11] explained that the protection coordination is important for islanded mode of operation and PSCAD is implemented in practical network to overcome the problems of using overcurrent relays, the application of FCL type and size is proposed [12]. In [13] Hoshyarzadeh Genetic algorithm (GA) is implemented for optimal placement of DG and FCL sizing. GA and other optimization algorithms have been used for various engineering problems [14–17]. The previous studies from the literature clearly shows the revolution of emerging meta-heuristic based algorithms such as GA [3], PSO, Cuckoo search algorithm [6] which are evolutionary [19] and swarm intelligence [20]. Therefore a new Salp Swarm Algorithm [18] is employed to deal the objective function.

SSA is heuristic swarm based approach which is inspired by the social behavior of salps. The salps which are similar to jelly fishes have translucent—cylinder shaped body. The interesting behavior of salps is they form as salp chains for food foraging.SSA has been applied to most of the power system problems [21–23] because of its merits

- It is able to solve non-linear problems as most of the power system problems are non-linear in nature.
- It gives best fitness values.
- Requires less no. of. Iterations.
- Less computational time when compared to GA and PSO.
- Fine tuning performance can be obtained with less no of adjustments.

In this paper SSA is adopted for both optimal placement of DG and CI FCL sizing. The power system network is grid connected and synchronous based type of DG is considered.

2 Contributions of the Paper

1. Fault current levels in DG connected mode and non-DG connected mode differs from each other. In order to maintain constant current magnitudes we are using CI FCL and optimal location of DG by using Salp swarm algorithm (SSA).

2. This method gives importance in finding the correct buses for placing the DGs and FCL size. FCL may be Resistive, Inductive or both.
3. This problem is tested on IEEE 33, 69 bus systems for faults like three phase fault that occur in the power system network and the results are verified.

3 Problem Formulation

Distributed generation is one of the most important aspects discussed in smart grids and micro grids because DG provides power supply at the customer site. However, using DG in the power system network changes the short-circuit current. Under normal configuration there is no change in the basic current but when DG is connected to the power system network there is change in the directions of short-circuit currents and the problems are taken into consideration and is solved by using SSA optimization by taking IEEE 33 and 69 bus standard distribution systems in two different configurations

Configuration A: Grid connected without DG Connection
Configuration B: Grid connected with DG connection.

In Figs. 6 and 9 we are comparing three phase fault currents without DG and with DG connected mode at buses 2, 3, 19 for 33 bus 3 DG operation and at buses 2, 8 for 2 DG operation. In 69 bus system DG is connected at buses 18, 32, 49 for 3DG and 5, 17 for 2 DG operations. The difference between the fault currents is calculated and we are installing FCL in the distribution systems in series with DG at buses where DG's are connected in order to minimize the change in the impedance matrices of these configurations.

For both IEEE 33 and 69 bus systems the total demand is about 4.5 MW and this is provided by three 1.5 MW DGS or two 2 MW DGS for 3 DG and 2 DG operations. The impedance matrix in Configuration A is without DG connected to the grid, in Configuration B is DG connected to the grid and here both the optimal placement of DG and FCL sizing is considered as the objective function.

$$Z_{\mathrm{B}} = F(\mathrm{Ci}_r, L_r, P_r, \mathrm{Ci}_s) \quad \forall r = 1{:}m \tag{1}$$

Z_{B} represents the impedance matrix of the Configuration B,
m is the no of DGs in the system, r represents DG between m DG'S.
Ci_r Controlled Impedance in series with the DG number r
L_r Location of DG number
P_r Power generated by DG number r
Ci_s Source Impedance at the utility.

FCL source impedance is connected to bus 1. FCL is always placed in series with DG to reduce the fault current magnitude. Both real and imaginary parts are considered. It is given as follows

$$\mathrm{Zci}_r = R_r + jX_r \quad \forall r = 1{:}m \tag{2}$$

$$\mathrm{Zci}_s = R_s + jX_s \tag{3}$$

where R_r and R_s are resistive part and X_r and X_s are for reactance.

The ith diagonal element (Z_{ii}) of the matrix represents Thevenin's impedance the fault current of this bus is calculated as follows. Consider a circuit connected to no. of series and parallel resistors and a load resistance and convert it into Thevenin's equivalent circuit (Fig. 2).

Remove all loads at the output find open circuit voltage V_{th} and Thevenin's equivalent resistance R_{th} by making all the voltages sources short-circuit and current sources open circuit and calculate resistance seen from the output (Fig. 3).

Then find the short-circuit current at the fault bus

$$I_{\mathrm{short}} = V_{\mathrm{th}}/R_{\mathrm{th}}$$

In this way, calculate fault current at bus i, assuming the voltage at all the buses as 1 P.u, V = 1 P.u

$$I_f \text{bus } i = 1/Z_{ii} \tag{4}$$

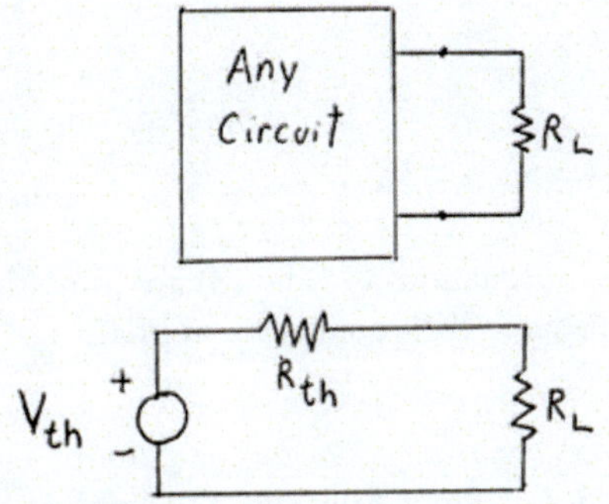

Fig. 2 Thevenin's equivalent circuit

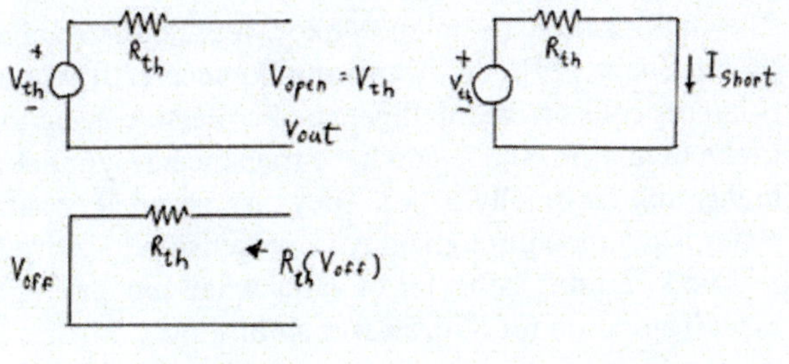

Fig. 3 Circuit for R_{th} and I_{sh}

Z_{ij} is the fault current between bus i and bus j, As mentioned earlier the objective function is expressed as

$$\sum_{i=1}^{m} \sum_{j=1}^{m} [Z_{A_{i,j}} - Z_{B_{i,j}}] \tag{5}$$

Therefore fault current levels can be maintained by minimizing the difference in two matrices as low. By running the optimization the location of DG and FCL size are obtained

No. of variables for m DG'S is given as

$$\text{No. of variables} = m_{\text{FCL}} + m_{\text{DG}} = m_{\text{FCL}(R+jX)} + m_{\text{DG}} \tag{6}$$

FCL consists of both resistive and reactive components and the limitations of controlled impedance (FCL) is given as follows

$$0 \leq R_r \leq R_{\text{max}_r} \tag{7}$$

$$X_{\text{min}_r} \leq X_r \leq X_{\text{max}_r} \tag{8}$$

$$0 \leq R_s \leq R_{\text{max}_s} \tag{9}$$

$$X_{\text{min}_s} \leq X_s \leq X_{\text{max}_s} \tag{10}$$

Equations (7) and (8) are limitations of CI FCL, Eqs. (9) and (10) boundaries of Source FCL connected at PCC.

3.1 Salp Swarm Algorithm

The optimization techniques like Meta-heuristic algorithms are categorized into Evolutionary [19] and swarm intelligence [20] whereas swarm intelligence algorithm depends on the intelligence of collective behavior of creatures. SALPS belong to the family of salpide and have translucent-cylinder shaped body, their tissues are highly similar to jelly fishes. They also move very similar to jelly fish, in which the water is pumped through body as propulsion to move forward (Fig. 4) [24].

The swarming behavior of salps when navigating and foraging in oceans is the main inspiration for Salp swarm algorithm.

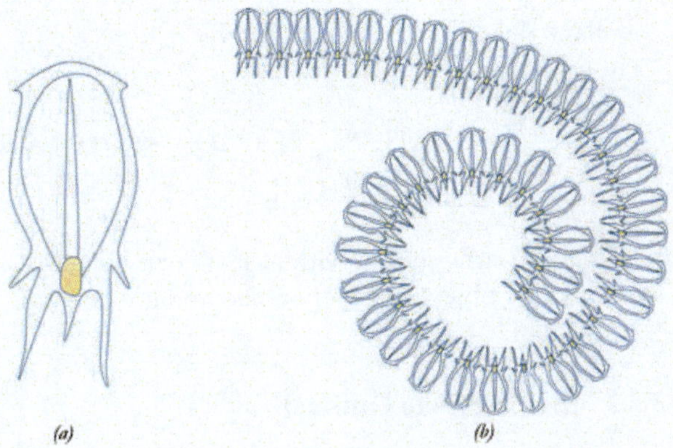

Fig. 4 **a** Individual salp, **b** salp chain

3.2 General Form

Let us consider a function

$$f_i(x) = \{ f1(x), f2(x) \dots fn(x)\} \tag{11}$$

Subject to

$$g_k(x) \geq 0, \quad i = 1, 2 \dots a \tag{12}$$

$$h_i(x) = 0, \quad i = 1, 2 \dots b \tag{13}$$

$$\text{lb}_i \leq x_i \leq \text{ub}_i \quad i = 1, 2 \dots c \tag{14}$$

where

n no. of objectives
a no. of inequality constraints
b no. of equality constraints
lb_i lower bound of the ith variable
ub_i upper bound of the ith variable.

The three main challenges for a multiobjective optimization to find a Pareto optimal front, convergence and distribution of solutions.

3.2.1 Def 1: Pareto Governing

Assume two vectors $Y = (y_1, y_2 \dots y_k)$.

And $Z = (z_1, z_2 \dots z_k)$

Vector Y dominates Z then $(Y < Z)$
If and only if

$$\forall i \in \{1, 2 \ldots k\}, [fi(Y) \leq fi(z)] \wedge \exists i \in \{1, 2 \ldots t\}$$
$$[fi(Y) < fi(Z)]$$

Pareto governs means with at least one objective function return a better value with no other objective function becoming worse-off.

3.2.2 Def 2: Pareto Optimal

Assuming $Y \in V$, Y is a Pareto optimal solution if and only if

$$\{\exists Z \in V | Z < Y\}$$

In every Multi objective problem, there is a set of feasible solutions are Pareto optimal where no further improvement can be made is called Pareto optimal solution.

3.2.3 Def 3: Pareto Optimal Set

The set that includes Pareto optimal solutions

$$Ps = \{Y, Z \in V | \exists Z < Y\}$$

3.2.4 Def 4: Pareto Optimal Front

The Pareto optimal solutions are called ineffective solution lies on the optimal front

$$\forall i \in \{1, 2 \ldots t\}$$

$$Pf = \{fi(Y) | Y \in Ps\}$$

From the four definitions, solutions can be easily justified for solving the multiobjective situations. During Optimization, each Salp is compared against all the vault residents using the Pareto governing operators

- If a Salp overcomes a solution in the stockpile they have to be swapped
- If a Salp governs a set of solution in the storehouse, they all should be removed from the repository and the Salp should be added in the stockpile.

For mathematical model of the salp chains, the salp population is divided into leader and followers. The leader is front of the chain and guides the remaining salps directly or indirectly. Similar to other Swarm Optimization methods; the position of salps is defined in an *n*-dimensional search space where n is the no of variables of a given problem. The position of all salps are stored in a two dimensional matrix '*X*' it is also assumed that there is a food source called '*F*' in the search space as the swarms target

3.3 Algorithm by Using Salp Swarm Optimization

Step 1: Initialize
Read the line data, load data
No.of search agents
No. of population of Salps
No. of iterations
The search agents here is the DG location and CI–FCL sizing.
Let Salp population

$$X_i = (i = 1, 2, \ldots n) \text{ Considering } U_b \text{ and } l_b \tag{15}$$

Step 2.: Calculate the fitness of each search agent and determine the ineffective Salps.
Step3: Update the stockpile considering the obtained ineffective Salps
Step 4: If the stockpile becomes full, call the repository maintenance procedure to remove one repository resident add the ineffective Salp to the storehouse.
Step 5: Choose a source of food from repository F
Step 6: Update r_1 by Eq. 16

$$r_1 = 2e - (4t/M_t)2 \tag{16}$$

It balances the exploration and exploitation

t is the current iteration
M_t is the maximum no of iterations.

Step 7: For each Salp (X_i)
If ($i == 1$)
Update the position of leading Salp by

$$X_j^1 = \{F_j + r_1((\mathrm{Ub}_j - \mathrm{lb}_j)\, r_2 + \mathrm{lb}_j)\; r_3 \geq 0 \\ F_j - r_1((Ub_j - lb_j)\, r_2 + lb_j)\; r_3 \leq 0\} \tag{17}$$

X_j^1 shows position of first Salp (leader) in the *j*th dimension

Table 1 Optimization variables constraints for SSA in 33 and 69 bus systems

Variable	Minimum	Maximum
33 bus system		
X_r & X_s	−10	10
R_r & R_s	0	10
m	2	33
69 bus sytem		
X_r & X_s	−10	10
R_r & R_s	0	10
M	2	69

Table 2 SSA Implementation data to 33 and 69 bus systems

Search agents	30
Max. no. iterations	50
Upper boundary	80.60
Lower boundary	42.20

Table 3 Simulation data

33 bus and 69 bus source and DG system data	
Utility data	$MVA_{sc} = 100$ MVA, $X/R = 6$
DG transformer reactance	$X^+ = X^- = 5\%$ Y-grounded
DG reactance	$X^+ = X^- = 9.67\%$
Base KV	12.47
Base MVA	100

F_j Position of food source
r_1, r_2, r_3 random numbers

Else
Update the position of follower Salp (Tables 1, 2 and 3).

4 Results and Discussion

Simulation is carried out in MATLAB2017 for both IEEE 33 & 69 bus standard distribution systems in 3 DG and 2DG operations and the results obtained are compared with PSO optimization. MOSSA is used to find the best location for DG and FCL sizing. In traditional power systems we use fuses, relays and C.B's whenever fault occurs but in modern power systems like smart grid and micro grids the protection coordination failure mainly occurs when the power system network is connected to DG. Hence, therefore CIFCL(controlled impedance Fault current limiter) is used

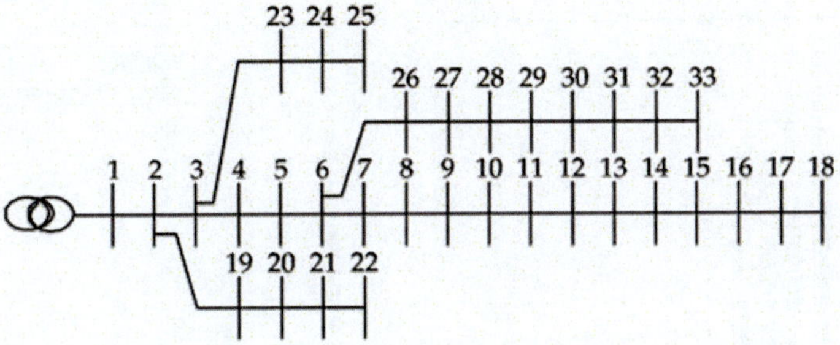

Fig. 5 Single-line diagram of IEE33 bus sytem

for controlling the tremendous amount of fault currents occurring at the buses and reduce the magnitude of fault current. This active type of FCL is connected in series with DG acts at high impedance during fault condition and less impedance during normal operation.FCL source is connected at the PCC.

4.1 33 Bus System

Figure 5 shows the single-line diagram of IEEE 33 bus system. It consists of one main lateral (1–18) and three sub laterals (19–22, 23–25, 26–33) the total real and reactive power loads is P = 3715 kW, Q = 2300 kVAr The DGS are connected to buses 2, 3, and 19 for 3 DG (three 1.5 MW DG's) at buses 2, 8 for 2 DG operation(2 MW DG's) and Fig. 6 shows the comparison of three phase fault current during normal operation and DG connected mode. The buses which is connected near to the substation has high fault current whereas the buses far away from the substation has low fault current. Figure 7 shows comparison of three phase fault current with the DG connected and FCL sizing.

4.2 69 Bus System

Figure 8 shows the single-line diagram of IEEE 69 bus system it consists of one main lateral (1–27) and seven sub laterals(28–35, 36–46, 47–50, 51–52, 53–65,66–67, 68–69) and the total real and reactive power loads is $P = 3{,}801.9$ kW, $Q =$ 2694.1 kVAr and the DGS are connected to 18,32,49 buses for 3 DG and at buses 5,17 for 2 DG operations. Figure 9 shows comparison of fault current in normal mode and DG Connected mode. Figure 10 shows comparison of fault current with DG and FCL size. Tables 4 and 5 gives the optimal location for DG, FCL sizing and fitness values for 33 & 69 bus systems in 3DG and 2DG operations by using Salp swarm optimization and the results are compared with PSO.SSA gives the best optimal size

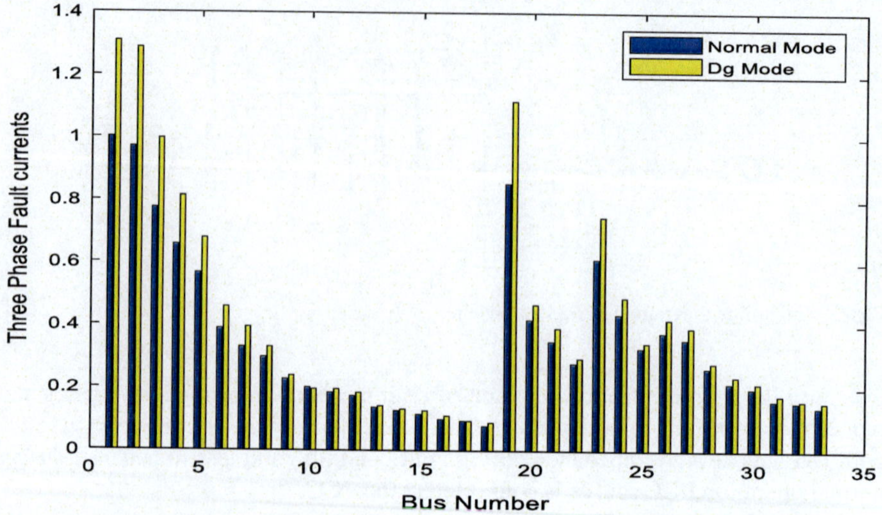

Fig. 6 Comparison of three phase fault current in normal mode and DG connected mode

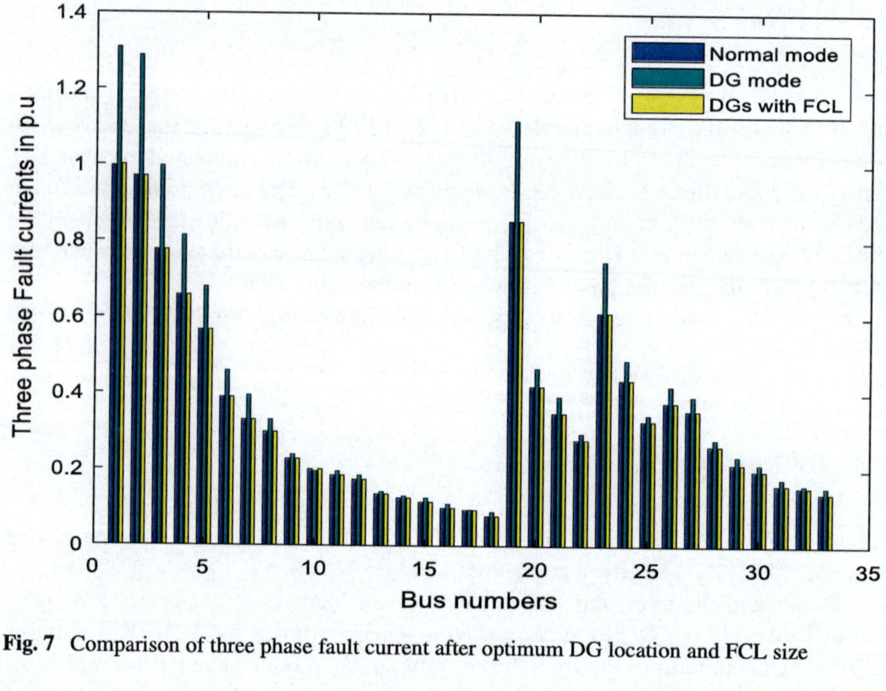

Fig. 7 Comparison of three phase fault current after optimum DG location and FCL size

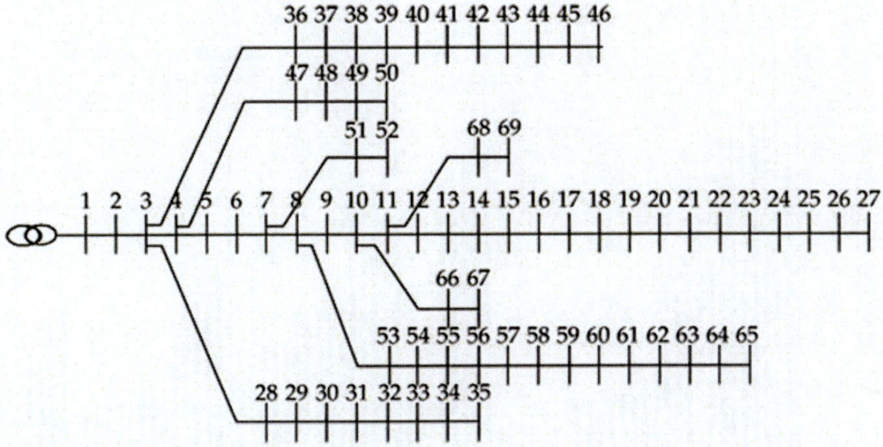

Fig. 8 Single line diagram of IEEE 69 bus system

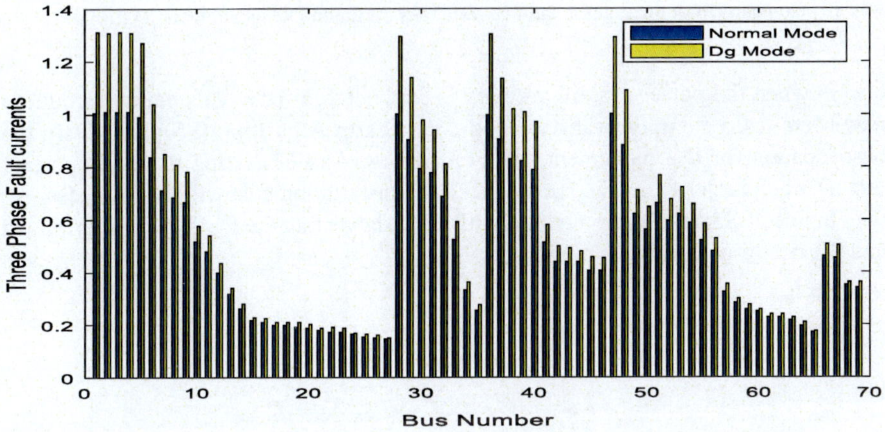

Fig. 9 Comparison of three phase fault current in normal mode and DG connected mode

of fault current limiter so that the fault current magnitude is reduced and maintained constant and also results shows that no.of.iterations required for PSO is more when compared to MOSSA this reduces the complexity and time for optimization.

5 Conclusion

This paper discuss about the distribution network connected to DG and its impact on short-circuit currents. During normal operation of the system the fault currents may not cause much damage to the system whereas protection coordination failure

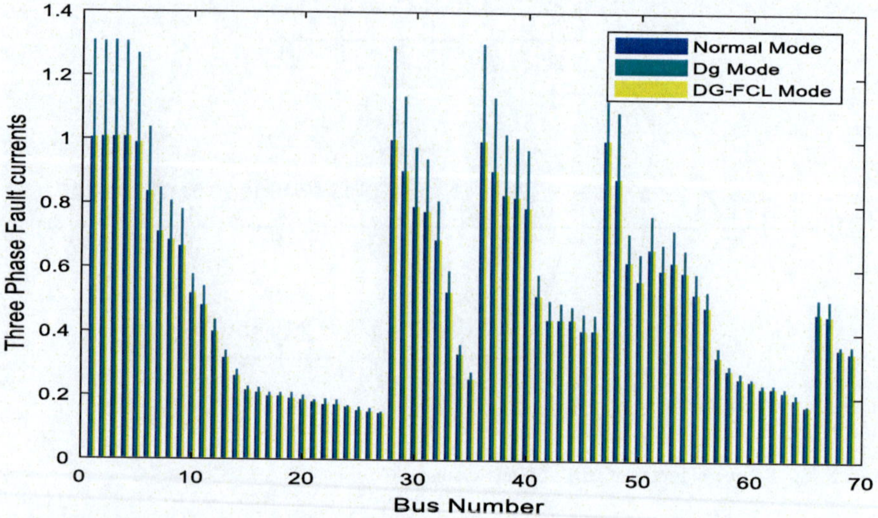

Fig. 10 Comparison of three phase fault current after optimum DG location and FCL size

occurs when the power system network is connected to DG. This paper introduces new Meta-heuristic optimization called Salp swarm Algorithm (SSA) to obtain the best location for DG placement and FCL sizing (As FCL sizing is also important factor) when larger FCL size is connected the installation cost will increase. Hence, the obtained SSA results are verified with PSO shows that the FCL size is minimized and the best fitness value is achieved.

Table 4 PSO and SSA results in 3 DG and 2 DG operation for 33 bus distribution system

PSO results for 3 DG					SSA results for 3 DG				
DG number	Location	FCL size	FCL source	Best fitness	DG number	Location	FCL size	FCL source	Best fitness
DG1 (1.5 MW)	2	1.8464 + 1.994j	0.11566 + 0.36305i	9.9140	DG1 (1.5 MW)	2	1.7842 + 1.299j	0.1019 + 0.2324i	9.9824
DG2 (1.5 MW)	3	6.3021 + 4.2043j			DG2 (1.5 MW)	3	7.3021 + 4.2040j		
DG3 (1.5 MW)	19	3.89 + 2.35j			DG3 (1.5 MW)	19	0.1230 + 4.3820j		
PSO results for 2 DG operation					SSA results for 2 DG operation				
DG number	Location	FCL size	FCL source	Best fitness	DG number	Location	FCL size	FCL source	Best fitness
DG1 (2 MW)	3	9.9882 + 4.2141j	0.012514 + 0.21371i	9.7	DG1 (2 MW)	2	1.952 + 3.27j	0.0110 +0.124i	9.49824
DG2 (2 MW)	19	6.0836 + 6.3737j			DG2 (2 MW)	8	1.4021 + 3.204j		

Table 5 PSO and SSA results in 3 DG and 2 DG operation for 69 bus distribution system

PSO results for 3 DG					SSA results for 3 DG				
DG number	Location	FCL size	FCL source	Best fitness	DG number	Location	FCL size	FCL source	Best fitness
DG1 (1.5 MW)	4	4.0267 + 0.29311j	0.01046 +0.30675i	1.0615	DG1 (1.5 MW)	18	3.827 + 0.293j	0.0103 +0.305i	1.0432
DG2 (1.5 MW)	28	2.2563 + 2.0257j			DG2 (1.5 MW)	32	2.65 + 2.025j		
DG3 (1.5 MW)	36	3.3053 + 6.2336j			DG3 (1.5 MW)	49	3.23 + 5.233j		
PSO results for 2 DG operation					SSA results for 2 DG operation				
DG number	Location	FCL size	FCL source	Best fitness	DG number	Location	FCL size	FCL source	Best fitness
DG1 (2 MW)	2	1.8318 + 4.5763j	0.029363 +0.19119i	5.4484	DG1 (2 MW)	5	8.524 + 5.214j	0.0128 +0.3105i	8.0432
DG2 (2 MW)	5	2.7356 + 5.1776j			DG2 (2 MW)	17	7.254 + 5.964j		

References

1. Behzadirafi S, Aravindan S (2018) Impact of power system protection and fault current limiters on component failure rates
2. Mohammadi P, El-Kishykyl H, Abdel-Akher M, Abdel-Salam M (2014) The impacts of distributed generation on fault detection and voltage profile in power distribution networks, pp 191–196
3. Bayati N, Hossein S, Sadeghi H, Hosseini A (2017) Optimal placement and sizing of fault current limiter in distributed generation systems using a hybrid genetic algorithm. Eng Technol Appl Sci Res 7(1):1329–1333
4. Yasin ZM, Sam'on IM, Aminudin N, Salim NA, Mohamad H (2017) Impact of distributed generation on the fault current in power distribution system. TELKOMNIKA Indones J Electr Eng 6(2):357–367
5. Surour Alaraifi M, El Moursi S, Zeineldin HH (2013) Optimal allocation of HTS-FCL for power system security and stability enhancement. IEEE Trans. Power Syst
6. Ahmarinejad A, Hasanpour SM, Babaei M, Tabrizian M (2016) Optimal overcurrent relays coordination in microgrid using Cuckoo Algorithm. In: 3rd international conference on power and energy systems engineering (CPESE 2016), 8–12 Sept 2016
7. El-Khattam W, Sidhu TS (2006) Restoration of directional overcurrent relay coordination in distributed generation systems utilizing fault current limiter
8. Boribun B, Kulworawanichpong T (2008) Comparative study on a fault current limiter with thyristor-controlled impedances. In: 13th international conference on harmonics and quality of power, pp 1–5
9. Salama MMA, Temraz H, Chikhani AY, Bayoumi MA (1993) Fault-current limiter with thyristor-controlled impedance. IEEE Trans Power Delivery 8(3):1518–1528
10. Laaksonen HJ (2010) Protection principles for future microgrids. IEEE Trans Power Electron 25(12):2910–2918
11. Amjady N, Keynia F, Zareipour H (2010) Short term load forecast of microgrids by a new bilevel prediction strategy. Trans Smart Grid 1(3):286–294
12. Agheli A, Abyaneh HA, Chabanloo RM, Dezaki HH (2010) Reducing the impact of DG in distribution networks protection using fault current limiters. In: 4th international power engineering and optimization conference (PEOCO), pp 298–303
13. Hoshyarzadeh AS, Zaker B, Khodaddost Arani AA, Gharchpetian GB (2018) Optimal DG allocation and Thyristor-FCL controlled impedeance sizing for smart distribution systems using genetic algorithm. Int J Electr Electron Eng 14(3)
14. Cvetkovski G, Petkovska L (2016) Genetic algorithm as a tool for multi-objective optimization of permanent magnet disc motor. Arch Electr Eng 65(2):285–294
15. Tamilselvan V, Jayabarathi T (2016) Multi objective Flower Pollination Algorithm for solving capacitor placement in radial distribution system using data structure load flow analysis. Arch Electr Eng 65(2):203–220
16. Bhattacharyya B, Rani S, Vais RI, Bharti IP (2016) GA based optimal planning of VAR sources using Fast Voltage Stability Index method. Arch Electr Eng 65(4):789–802
17. Farshad M, Sadeh J (2013) Fault locating in high voltage transmission lines based on harmonic components of one-end voltage using random forests. Iran J Electr Electron Eng 9(3):158–166
18. Mirjalili S et al (2017) Salp swarm algorithm: a bio-inspired optimizer for engineering design problems. Adv. Eng. Software
19. Bäck T (1996) Evolutionary algorithms in theory and practice: evolution strategies, evolutionary programming, genetic algorithms. Oxford University Press, Oxford
20. Blum C, Li X (2008) Swarm intelligence in optimization. Springer, Berlin
21. El-Fergany AA (2018) Extracting optimal parameters of PEM fuel cells using salp swarm optimizer. Renew Energy 119:641–648
22. Abdel-mawgoud H, Kamel S, Yu J et al (2019) Hybrid Salp Swarm Algorithm for integrating renewable distributed energy resources in distribution systems considering annual load growth. J King Saud Univ Comput Inform Sci

23. Hasanien HM, El-Fergany AA (2019) Salp swarm algorithm-based optimal load frequency control of hybridrenewable power systems with communication delay and excitation crosscoupling effect. Electr Power Syst Res 176:105938
24. Madin L (1990) Aspects of jet propulsion in salps. Can J Zool 68:765–777

Delicate Flower Pollination Algorithm for Optimal Power Flow

S. Dhivya, R. Arul, and K. Padmanathan

Abstract This paper has been emphasized to develop a new methodology for optimal power flow (OPF) using delicate flower pollination algorithm (DFPA). Being an essential key factor in the ocean of power sector, its operational features and controlling attributes have been making the available power resources to flow in a fair manner. The DFPA, a nature intriguing algorithm, has originated from the pollinating characteristics of flowers. The proposed method has sounded very louder on the global optimum solution using DFPA through propagating the exploitation phase of optimization by considering two test-case studies such as shortened fuel cost and real power loss decrement. Simulation results on IEEE_Standard 30 bus test system have clearly exhibited that the proposed method outperforms the existing numerous strategies.

Keywords Optimal power flow · Delicate flower pollination algorithm · Global optimum

Nomenclature

d	Delicate factor
DE	Differential evolution
DFPA	Delicate flower pollination algorithm
FPA	Flower pollination algorithm
$G_{ij} + B_{ij}$	Complex bus admittance matrix corresponding to i, j
g_{ij}	Line conductance between buses$-i$ and j
g^*	Current best solution

S. Dhivya (✉) · R. Arul
School of Electrical and Electronics Engineering, Vellore Institute of Technology, Chennai Campus, Chennai, Tamil Nadu 600127, India
e-mail: dhivya.s2019@vitstudent.ac.in

K. Padmanathan
Department of Electrical and Electronics Engineering, Agni College of Technology, Thalambur, Chennai, India

N. Zhou and S. Hemamalini (eds.), *Advances in Smart Grid Technology*, Lecture Notes in Electrical Engineering 688, https://doi.org/10.1007/978-981-15-7241-8_20

$iter^{max}$	Maximum number of iterations
$J(x, u)$	Objective function
k	Iteration counter
L	Levy distribution
NF	Number of flowers
NFC	Net fuel cost
NR	Newton--Raphson
nef	Number of elite flowers
ng	Number of generators
nt	Number of transformers
OPF	Optimal power flow
PV bus	Generator bus other than slack bus
P_{Gi}	Real power generation at bus-i
P_{Di}	Real power demand at bus-i
P_{Gs}^{limit}	Limit of real power generation at slack generator
p	Switch probability
Q_{Gi}	Reactive power generation at bus-i
Q_{Di}	Reactive power demand at bus-i
Q_{Ck}	Reactive power supplied by kth shunt compensator
Q_{Cnc}	Reactive power supplied by nth shunt compensator
Q_{Gi}^{limit}	Reactive power limit at ith PV bus
RPL	Real power loss
R_L	Line resistance
TTS	Transformer tap settings
Tt	Tap settings of tth transformer
Tnt	Tap settings of n-th transformer
V_i	Voltage at ith bus
V_{Ls}^{limit}	Voltage limit violation at ith load bus
V_{Gi}	Voltage magnitude at ith generator bus
V_{Gng}	Voltage magnitude at nth generator bus
V_{Li}	Voltage magnitude at ith load bus
X_L	Line reactance
X	Vector of dependent variables
x_i^t	Pollen of flower-l at tth iteration
λ_V, λ_Q	Penalty factors
δij	Phase angle between buses-i and j
ε	Uniform probability distribution
Γ	Standard gamma function
γ	Scaling factor
R	A set of load buses, whose voltages violate either the lower or upper limits
Z	A set of generator buses, whose Q_G violate either the lower or upper limits
φ	A set of transmission lines
Ω	A set of generator buses
ψ	Augmented objective function to be minimized
min	Lower limit

max Upper limit

1 Introduction

An elite real and reactive power transmission analysis have been detecting the best control parameters within its threshold limits through the objective functions for given loads. Since the common objective of optimal programming of power flow is cutting down the fuel cost, there have also been prevailing certain goals like minimal level of loss, profile assessment, and stability enrichment of voltage. As per the bus terminology, the known quantities of generator bus such as real power, magnitude of voltage, and similarly, reactive power at load bus and shunt compensators could be considered as control parameters. Moreover, the transformer tap settings are also included along with reactive power limits and voltage limits even though those of which fall under the unknown quantities at generator bus and load bus. The Newton--Raphson equations of power flow programming have been modelled under the equality constraints while the above said quantities on the bounds of inequality constraints [1–3].

Neither convexity nor linearity, an OPF has been portrayed as a highly constrained problem in the presence of uninterrupted and distinct control parameters. The minimization function has served its purpose for obtaining the best solution [4–6].

On looking back the last few decades, various methodologies have been forged with the vision over fine-optimal point for evaluating the OPF scenario, based on the mathematical tools, gradient search [3], linear programming [7, 8], nonlinear programming [9, 10], interior point method [11–13], and quadratic programming [14]. Eventhough most of the above methods had good rate of converging behaviors, they had the shortcomings such as local region of convergence and some illusions on constraint handling discontinuous variables.

The complexities arising in performing OPF analysis could be diminished by new-age heuristic algorithms like genetic algorithm (GA) [15–18], evolutionary programming [19–21], particle swarm optimization (PSO) [22], differential evolution (DE) [23], gravitational search [24], teaching-learning [25, 26], and harmony-search [27, 28]. Among all above-seen algorithms, the population vector comprising of several control variables had been framed through some key-processing tools which might be varied between one another. The volatile parameters could be picked up and carried into next stage of evolution through appropriate principle. However, the global departure of solution space has been attained at the end point in the task of OPF.

Now, flower pollination algorithm (FPA) have been authorized for multimodal optimization problems [29, 30]. This paper aims to suggest a new modified version of this algorithm known as a delicate flower pollination algorithm (DFPA)-based method for solving OPF in order to obtain the elite solution globally that could be validated on IEEE standard_30 bus test system for studying the performance.

2 Formulation of OPF Problem

The optimal analysis of power flow problem could be addressed as a minimization function of one or more objectives being optimized by manipulating number of constraints entitled with an overall representation as given below.

2.1 OBJECTIVE-1: Net Fuel Cost Minimization

The fuel cost of thermal power plants has been considered which follows the quadratic equation of second order. A net fuel cost (NFC) could be formulated as,

$$\text{Minimize} \quad J_1(x,u) = \sum_{i\in\Omega} F_i(P_{\text{G}i}) \tag{1}$$

where

$$F_i(P_{\text{G}i}) = a_i P_{\text{G}i}^2 + b_i P_{\text{G}i} + c_i \tag{2}$$

2.2 OBJECTIVE-2: Real Power Loss Reduction

The RPL of the system can be expressed as

$$\text{Minimize} \quad J_2(x,u) = P_{\text{L}} \tag{3}$$

where

$$P_{\text{L}} = \sum_{k\in\phi} G_{ij}\left(|V_i|^2 + \left|V_j\right|^2 - 2|V_i|\left|V_j\right|\cos\delta_{ij}\right) \tag{4}$$

The N-R power flow equations have been employed as equality constraints revolving around the power system.

$$P_{\text{G}i} - P_{\text{D}i} - V_i \sum_{j=1'}^{nb} V_j\left(G_{ij}\cos\delta_{ij} + B_{ij}\sin\delta_{ij}\right) = 0 \tag{5}$$

$$Q_{\text{G}i} - Q_{\text{D}i} - V_i \sum_{j=1'}^{nb} V_j\left(G_{ij}\sin\delta_{ij} - B_{ij}\cos\delta_{ij}\right) = 0 \tag{6}$$

x could be indicated as dependent vector containing unknown quantities such as magnitudes of voltage at load bus, reactive power supplied at PV buses, and active power produced at slack bus.

u could be expressed as independent tunable vector consisting magnitudes of voltages at generator bus, transformer tap settings, and reactive power injected by shunt compensators.

$$P_{\mathrm{G}i}^{\min} \leq P_{\mathrm{G}i} \leq P_{\mathrm{G}i}^{\max} \tag{7}$$

$$Q_{\mathrm{G}i}^{\min} \leq Q_{\mathrm{G}i} \leq Q_{\mathrm{G}i}^{\max} \tag{8}$$

$$Q_{\mathrm{C}i}^{\min} \leq Q_{\mathrm{C}i} \leq Q_{\mathrm{C}i}^{\max} \tag{9}$$

$$T_i^{\min} \leq T_i \leq T_i^{\max} \tag{10}$$

$$V_{\mathrm{G}i}^{\min} \leq V_{\mathrm{G}i} \leq V_{\mathrm{G}i}^{\max} \tag{11}$$

$$V_{\mathrm{L}i}^{\min} \leq V_{\mathrm{L}i} \leq V_{\mathrm{L}i}^{\max} \tag{12}$$

3 Delicate Flower Pollination Algorithm

The flowering plants have been evolving for more than billions of years, and flowers had strategic insight in evolution thereby spreading the fragrance around the plant world. As the ultimate aim of a particular flower is being able to reproduce new pollen gamete from which new flower could be flourished. The pollen from angiosperms is attached with gymnosperms through some pollinators in the form of small insects, wind, water, air, animals, and birds. Different kinds of flowers could make successive pollination through attraction over pollinator and yield new pollen with some delicacy.

First, pollination could be differentiated in two attributes: cross and self. On second hand, pollination could be classified based on its location either dependent or independent. Cross-pollination is the process of transferring pollen from one flower to other through some pollinating agents such as bees, whereas self-pollination does not require any pollinating agents. The transfer of pollen gamete could be happened by diffusion of wind and water. The fertilizing capability of a single flower has been made possible only in a way to get the pollen produced by another plant species which defines the dependent pollination. The self-reproduction of a plant could be possible from pollen of individual flower or group of flowers in a plant which might be called as independent pollination. Grass in land might perform self-pollination [29,

30]. Bees could have been entitled flower constancy in a way that these pollinators tend to contact the delicate flower species to have more pollen exchange through the scented fragrance. The strength of reproduction tends to attain the maximum range thereby nectar transportation smoothly entangled in a way to yield more number of flowering species.

The dependent, cross-pollination could be happened at indefinite distance by which the pollinating agents like birds and animals travel over the large areas could represent a whole reproduction around the global level. This indefinite modelling of distance has been calculated by Levy probability distribution in terms of varying step sizes. All random motion of pollen activities have been taken into account as constant reproductive nature of flower patches.

The phenomenon of flowering reproduction gives rise to a new supremacy called flower pollination algorithm whose objective must be the survival among the fittest. The elite flower could be observed which might replicate number of births to new younger species. All the above factors have been put together to form the rule base for this algorithm:

Rule 1: Global optimization must be represented in cross and dependent pollination actions, because pollinators have been training Levy flight.
Rule 2: Local optimization must be represented in self and independent pollination in which normal probability distribution takes place.
Rule 3: Flower delicacy should be increased by insects in relation with a reproduction capability that is in proportion to the resemblance of two different flowers.
Rule 4: The interface boundary between local and global optimization should be supervised by a switching probability $p \in [0,1]$.

Curiously, in real world, uncountable flowers could be rising from different plant species, and rate of producing gametes might be progressive in nature. In order to make the overall computation easier, one flower has been taken to yield one pollen gamete from which another one offspring is being protruding out. For easy understanding, a pollen x_i has been considered as a flower vector or solution vector.

As per the four rules, a pollen-reproduction-based algorithm, namely delicate flower pollination algorithm (DFPA), could be formulated with the two key factors envisioning on global and local optimization. In case of global optimization, pollens could be taken and diffused by pollinating agents which run over a long distance thereby reproducing the best pollen or flower (g^*). In order to elongate the global zone of optimization, a delicate factor has been introduced in FPA. Then, it could be called as delicate flower pollination algorithm (DFPA). Hence, global optimization and flower delicacy step along with a levy flight can be expressed as,

$$x_l^{t+1} = x_l^t + \gamma\ L(\lambda)\ (x_l^t - g*)d \tag{13}$$

where x_l^t indicates the pollen of flower-l at tth iteration and $g*$ tends to be the current best solution found among all solutions at the current generation. Here, γ is a scaling factor controlling the step size, and d is delicate factor.

Usually, $L(\lambda)$ the Levy flights-based index corresponds to the pollinating tendency. As long indefinite distances could be crossed using various distance steps, a Levy distribution might be used to enhance this random activity efficiently. Levy distribution confirms that L should not be void ($L > 0$).

$$L \approx \frac{\lambda \Gamma(\lambda) \sin(\pi \lambda/2)}{\pi} \frac{1}{s^{1+\lambda}}, \quad (s \gg s_o > 0) \tag{14}$$

Here, $\Gamma(\lambda)$ is the standard gamma function, and this distribution is valid for large steps $s > 0$. For the local optimization, both rule 2 and rule 3 could be represented as

$$x_l^{t+1} = x_l^t + \varepsilon(x_n^t - x_p^t) \tag{15}$$

where x_n^t and x_p^t denotes the pollen from different flowers of the same plant species enrolling the flower delicacy in a limited neighborhood. For a local random motion, x_n^t and x_p^t comes from the same species, then ε is drawn from a uniform distribution as [0, 1]. The exploitation characteristics of delicate factor are implemented in DFPA to depart at global optimum zone.

As per the rule base, flower optimization could take place at all scales, both local optimization and global optimization. The optimization, in the sense, pollination could be carried out randomly which might be pointed out by switching probability in the extensive search space of all species. The preliminary parametric studies showed that $p = 0.9$ might work better for our solution.

3.1 The DFPA Pseudocode Is Given Below

1. Select the variables such as population size, maximum number of iterations for termination, switch probability *p*, and delicate factor *d*.
2. Allot a population of NF flower vectors with random solutions.
3. Compute the objective function of each flower vector in the entire population.
4. Choose the current best solution and carry the corresponding solution vector.
5. For each flower vector in the population, generate a random number in a range of (0, 1). If this random number is greater than p ($p = 0.9$), perform local optimization, else do global optimization using Levy flight with delicate factor.
6. Output the best solution if the termination criteria is fulfilled
7. Otherwise update the iteration with the preceding solution vector. Proceed with step (3).
8. Stop.

4 Delicate Flower Pollination Algorithm Based OPF Program

Being a large-scale and complex optimization problem with a nonlinear objective function and related operating constraints, a fascinating effort is taken to develop an elegant strategy, involving the DFPA with its victorious vision in exploring the search space and obtain the elite solution for OPF. In general, the optimization task elongates over a volatile solutions for a given problem. Moreover, the pollen vectors of flowers are initialized by random storage of values enclosed within the upper and lower boundaries of the variables. The proposed DFPA-based solution method (DFPA) has been devised for representing the pollen vector and modelling the cost-augmented function.

4.1 Pollen Representing Control Vector

The control vectors have taken the forms of active power generations and magnitude of voltages at PV buses, tap settings of transformer, and reactive power supplied by shunt compensators. The position matrix of each pollen in the DFPA is formulated a vector form to contain these control variables as

$$\text{flower} = x = \left[P_{\text{G2}}, \ldots, P_{\text{G}j}, V_{\text{G1}}, V_{\text{G2}}, \ldots, V_{\text{Gng}}, T_1, T_2, \ldots, T_{\text{nt}}, Q_{\text{C1}}, Q_{\text{C2}}, \ldots, Q_{\text{Cnc}}\right] \tag{16}$$

4.2 Cost-Augmented Function

The algorithm suspects for optimality in its solution by minimizing a cost-augmented function, which could be framed from the objective function and the penalty terms containing the limit violation of the dependent variables such as reactive power generation at PV buses and voltage magnitude at PQ buses. The cost-augmented function is written as

$$\begin{aligned}\text{Minimize } \Psi = J(x,u) + \lambda_V \sum_{i \in \Re} \left(V_{Li} - V_{Li}^{\text{limit}}\right)^2 + \lambda_Q \sum_{i \in Z} \left(Q_{Gi} - Q_{Gi}^{\text{limit}}\right)^2 \\ + \lambda_P \left(P_{Gs} - P_{Gs}^{\text{limit}} \right)^2\end{aligned} \tag{17}$$

where

$$V_{\text{L}i}^{\text{limit}} = \begin{cases} V_{\text{L}i}^{\min} & \text{if } V_{\text{L}i} < V_{\text{L}i}^{\min} \\ V_{\text{L}i}^{\max} & \text{if } V_{\text{L}i} > V_{\text{L}i}^{\max} \end{cases} \tag{18}$$

$$Q_{Gi}^{\text{limit}} = \begin{cases} Q_{Gi}^{\min} & \text{if } Q_{Gi} < Q_{Gi}^{\min} \\ Q_{Gi}^{\max} & \text{if } Q_{Gi} > Q_{Gi}^{\max} \end{cases} \quad (19)$$

$$P_{Gs}^{\text{limit}} = \begin{cases} P_{Gs}^{\min} & \text{if } P_{Gs} < P_{Gs}^{\min} \\ P_{Gs}^{\max} & \text{if } P_{Gs} > P_{Gs}^{\max} \end{cases} \quad (20)$$

4.3 Strategy Direction

The random process of creating a newer pollen vector of flowers through major exploration and minor exploitation might be considered as one iterative procedure. It might be continued by taking the flowers obtained in the previous step as the initial pollen vector for next step. The position vector of flowers having the best cost function value has been present in memory along with its objective function at all steps. An entire DFPA iterative process could terminate its loop after reaching the fixed number of iterative steps. The solution process is depicted in the flowchart of Fig. 1.

5 Numerical Results

The two minimal objectives of net fuel cost and real power loss have been considered in this problem as mentioned earlier in Eqs. (1) and (3). The DFPA-based OPF have been simulated on IEEE standard_30 bus test system, whose data is being retrieved from Ref. [1]. The m-codes are developed in MATLAB 8.0 and executed on a 2 GHz i3 core processor-based computer. NR technique [31] has been incorporated to perform the load flow at iterative process. An optimal solution vector obtained by DFPA have been put forth to demonstrate the effectiveness with respect to the base case as given in Table 1.

The optimal functions such as NFC of case-1 and RPL of test-2 derived by the DFPA for IEEE_Standard 30 bus system have been distinguished between those techniques, namely ABC [32], enhanced and adapted GA (EGA) [17] and (AGA) [18], DE [23], and FPA [33] in Table 2.

It is obvious from Table 2 that the DFPA tends for shortening the NFC to the lowest value of 799.1532 $/h, and corresponding real power loss is noted as 8.615 MW. Since, our objective is fuel cost, loss has been neglected in case-1. Similarly, when we have considered the real power loss reduction as the main objective in case-2, we have yielded it as 4.3724 MW, and its respective fuel cost is 900.9635 $/hr. The fuel cost has been neglected in case-2. The percentage savings of fuel cost could be calculated by taking the difference between base case solution and the final solution and then multiplied by 100 for percentage and then divided by base case for several

Fig. 1 Flowchart of the DFPA based OPF

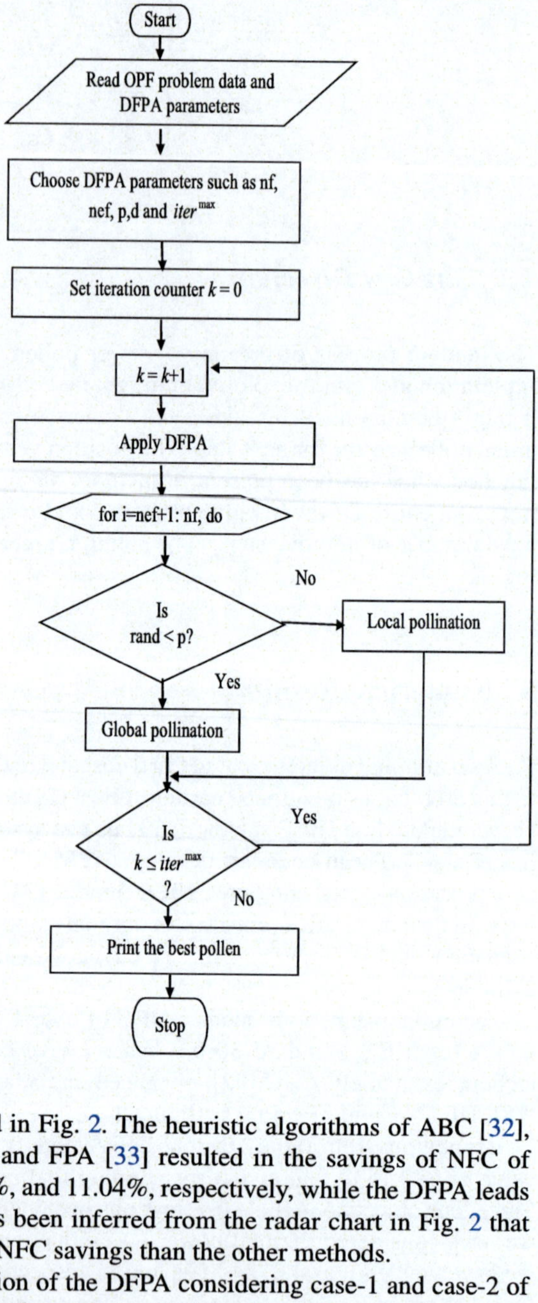

algorithms that have been plotted in Fig. 2. The heuristic algorithms of ABC [32], AGA [18], EGA [17], DE [23], and FPA [33] resulted in the savings of NFC of 11.22%, 11.31%, 11.35%, 11.38%, and 11.04%, respectively, while the DFPA leads to 11.40% savings of NFC. It has been inferred from the radar chart in Fig. 2 that the DFPA departs at incremental NFC savings than the other methods.

The converging curve orientation of the DFPA considering case-1 and case-2 of IEEE Standard_30 bus system has been shown in Figs. 3 and 4, respectively. It is very clear from Fig. 3 that the DFPA converges at iteration number 43 to the final

Table 1 Numerical results of DFPA-OPF

Control variables	Base case	Case-1: NFC optimization	Case-2: RPL optimization
P_{G1}	99.217	178.2138	90.2676
P_{G2}	80	47.9426	55.4874
P_{G5}	50	21.3346	47.1564
P_{G8}	20	20.5437	32.6934
P_{G11}	20	11.5232	23.6416
P_{G13}	20	12.5378	38.5259
V_{G1}	1.05	1.10000	1.0993
V_{G2}	1.04	1.0864	1.0900
V_{G5}	1.01	1.0577	1.0367
V_{G8}	1.01	1.0600	1.0612
V_{G11}	1.05	1.0569	1.0490
V_{G13}	1.05	1.0816	1.0929
$T_{6–9}$	1.078	0.9811	0.9831
$T_{6–10}$	1.069	0.9204	1.0581
$T_{4–12}$	1.032	0.9418	1.0154
$T_{28–27}$	1.068	0.9580	1.0935
Q_{C10}	0.0	0.0499	0.0472
Q_{C12}	0.0	0.0483	0.0500
Q_{C15}	0.0	0.0478	0.0500
Q_{C17}	0.0	0.0479	0.0500
Q_{C20}	0.0	0.0479	0.0341
Q_{C21}	0.0	0.0473	0.0500
Q_{C23}	0.0	0.0328	0.0376
Q_{C24}	0.0	0.0394	0.0500
Q_{C29}	0.0	0.0132	0.0076

Table 2 Distinguishable solutions of DFPA with other algorithms for all two tests of IEEE_Standard 30 bus OPF

	Base case	NFC ($/h)	RPL (MW)
		901.9354	5.812
	ABC [32]	800.66	9.0328
	AGA [18]	799.8441	8.9166
	EGA [17]	799.56	8.697
	DE [23]	799.2891	8.6150
	FPA [33]	802.3491	9.527
Case-1	DFPA	**799.1532**	8.615
Case-2	DFPA	900.9635	**4.3724**

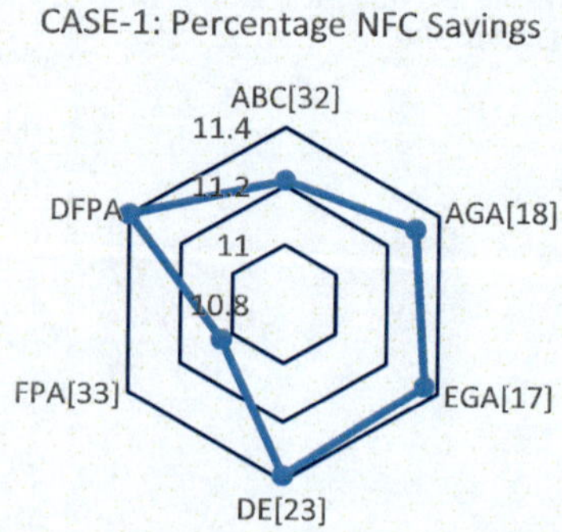

Fig. 2 Percentage NFC savings of DFPA for case-1 of IEEE_Standard 30 bus OPF

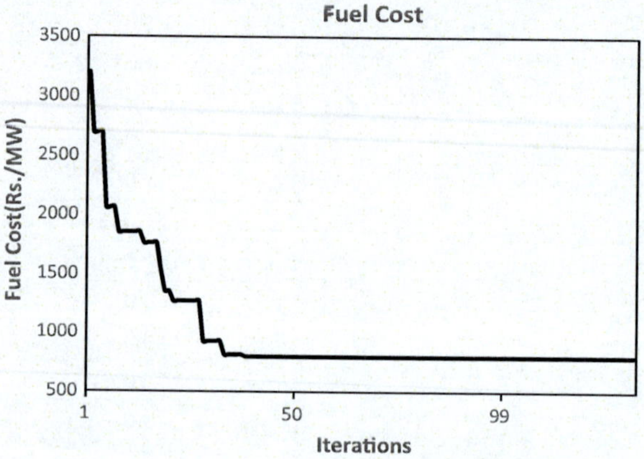

Fig. 3 Converging curve period of the DFPA for test-1 OPF of IEEE_Standard 30 bus

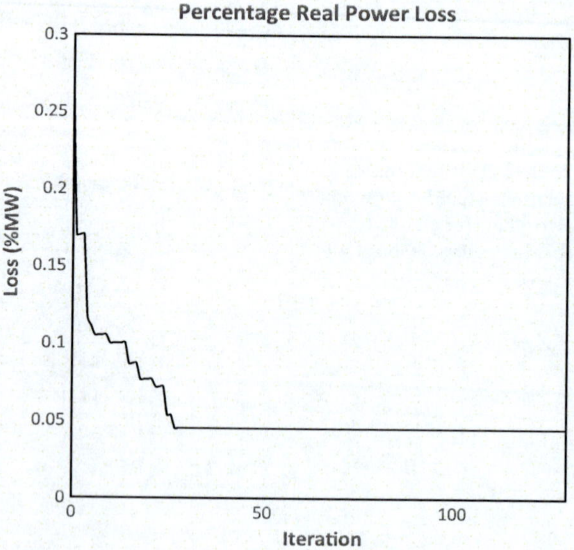

Fig. 4 Converging curve period of the DFPA for test-2 OPF of IEEE_Standard 30 bus

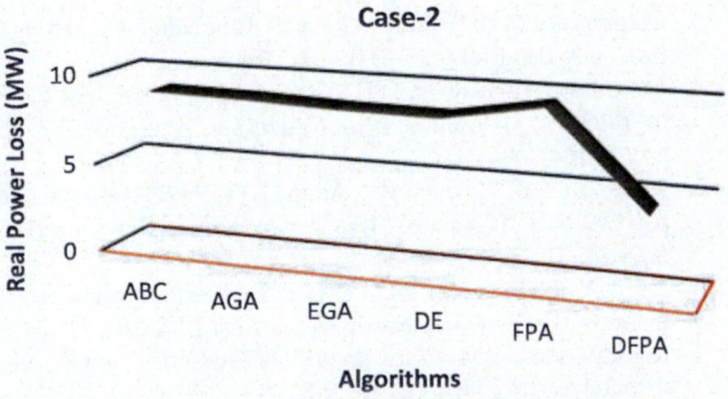

Fig. 5 Distinguishing features of RPL decrement for test-2 in IEEE_Standard 30 bus system

solution of cheaper fuel cost 799.1532 $/h, whereas in Fig. 4, DFPA converges faster at iteration number 25 to the final solution of loss reduction as 0.043724%.

It has been noted from Fig. 5 that the DFPA is relatively yielding the better result such that the real power loss gets reduced than the other approaches. The global optimum could be attained in the embedded pollen at the exponential rate of convergence. It could reveal that the DFPA has been landed at global zone of optimization.

6 Conclusion

An innovative DFPA strategy for resolving OPF analysis which has been incorporating a complexity and non-convexity in optimal characteristics has been elucidated in this paper. This method has been gliding with DFPA, inspired from the pollinating behaviors of flowers, foreseen an optimal solution spectating in unique optimization problems through the exploration and exploitation sequences. Its effectiveness in terms of global best solution, computational efficiency, and robustness has been demonstrated through the improved results on IEEE Standard_30 bus test system. The computational efforts are being found to be simple.

In forthcoming period, the flowers could be finely extended to multiple angiosperms for multiple objectives of OPF problems.

Acknowledgements The authors are being grateful to acknowledge the authorities of VIT University for the facilities provided to carry out this work.

References

1. Wood AJ, Wollenberg BF (1996) Power generation, operation and control. Wiley, New York
2. Elgerd OI (1999) Electric energy systems: theory—an introduction. Tata Mc Graw Hill, New Delhi

3. Carpentier J (1962) Contribution a l'Etude du Dispatching Economique'. Bulletin de la Societe Francaise des Electriciens 3:431–474
4. Momoh JA, El-Hawary ME, Adapa R (1999) A review of selected optimal power flow literature to 1993 Part I: nonlinear and quadratic programming approaches. IEEE Trans Power Syst 14:96–104
5. Momoh JA, El-Hawary ME, Adapa R ((1999) A review of selected optimal power flow literature to 1993 Part II: Newton, linear programming and interior point methods. IEEE Trans Power Syst 14:105–111
6. Lee KY, Park YM, Oritz JL (1985) A united approach to optimal real and reactive power dispatch. IEEE Trans Power Appar Syst PAS-104:1147–1153
7. Lee KY, Park YM, Oritz JL (1984) Fuel-cost minimization for both real and reactive power dispatches. IEE Proc C 131(3):85–93
8. Mangoli K, Lee KY (1993) Optimal real and reactive power control using linear programming. Electr Power Sys Res 26(1):1–10
9. Dommel H, Tinny W (1968) Optimal power flow solution. IEEE Trans Power Appar Syst PAS-87(10):1866–1876
10. Alsac O, Scott B (1974) Optimal load flow with steady state security. IEEE Trans Power Appar Syst PAS: 745–751
11. Wei H, Sasaki H, Kubokawa J, Yokoyama R (1998) An interior point nonlinear programming for optimal power flow problems with a novel structure data. IEEE Trans Power Syst 13:870–877
12. Yan X, Quantana VH (1999) Improving an interior point based OPF by dynamic adjustments of step sizes and tolerances. IEEE Trans Power Syst 14(2):709–717
13. Momoh JA, Zhu JZ (1999) Improved interior point method for OPF problems. IEEE Trans Power Syst 14(3):1114–1120
14. Burchett RC, Happ HH, Vierath DR (1984) Quadratically convergent optimal power flow. IEEE Trans Power Appar Syst 103:3267–3276
15. Lai LL, Ma JT (1997) Improved genetic algorithms for optimal power flow under both normal and contingent operation states. Int J Elercr Power Energy Syst 19(5):287–292
16. Bakirtzis AG, Biskas PN, Zoumas CE, Petridis V (2002) Optimal power flow by enhanced genetic algorithm. IEEE Trans Power Syst 17(2):229–236
17. Sailaja Kumari M, Maheswarapu S (2010) Enhanced genetic algorithm based computation technique for multi-objective optimal power flow. Int J Electr Power Energy Syst 32(6):736–742
18. Attia A-F, Al-Turki YA, Abusorrah AM (2012) Optimal power flow using adapted genetic algorithm with adjusting population size. Electric Power Components Syst 40:1285–1299
19. Yuryevich J, Wong KP (1999) Evolutionary based optimal power flow algorithm. IEEE Trans Power Syst 14(4):1245–1250
20. Ongsakul W, Tantimaporn T (2006) Optimal powers flow by improved evolutionary programming. Electr Power Comp Syst 34(1):79–95
21. Sood YR (2007) Evolutionary programming based optimal power flow and its validation for deregulated power system analysis. Int J Electr Power Energy Syst 29(1):65–75
22. Abido MA (2002) Optimal power flow using particle swarm optimization. Proc Int J Electr Power Energy Syst 24(7):563–571
23. Abou El Ela AA, Abido MA (2010) Optimal power flow using differential evolution algorithm. Electr Power Syst Res 80(7):878–885
24. Duman S, Güvenç U, Sönmez Y, Yörükeren N (2012) Optimal power flow using gravitational search algorithm. Energy Convers Manage 59:86–95
25. Shabanpour-Haghighi A, Seifi AR, Niknam T (2014) A modified teaching-learning based optimization for multi-objective optimal power flow problem. Energy Convers Manage 27:597–607
26. Ghasemi M, Ghavidel S, Gitizadeh M, Akbari E (2015) An improved teaching-learning-based optimization algorithm using Levy mutation strategy for non-smooth optimal power flow. Electr Power Energy Syst 65:375–384
27. Arul R, Ravi G, Velusami S (2013) Solving optimal power flow problems using chaotic self-adaptive differential harmony search algorithm. Electr Power Components Syst 41(8):782–805

28. Rajagopalan A, Sengoden V, Govindasamy R (2014) Solving economic load dispatch problems using chaotic self adaptive differential harmony search algorithm. Int Trans Electr Energy Syst. https://doi.org/10.1002/etep.1877(2014)
29. Yang XS (2012) Flower pollination algorithm for global optimization, unconventional computation and natural computation. Lect Notes Comput Sci 7445:240–249
30. Yang XS, Karamanoglu M, He X (2013) Multi-objective flower algorithm for optimization. Procedia Comput Sci 18:61–68
31. Tinney WF, Hart CE (1967) Power flow solution by Newton's method. IEEE Trans Power Appar Syst PAS-86(11):1449–1460
32. El-Fergany AA, Abdelaziz AY (2014) Artificial bee colony algorithm to allocate fixed and switched static shunt capacitors in radial distribution networks. Electr Power Components Syst 42(5):427–438
33. Prathiba R, Balasingh Moses M, Sakthivel S (2014) Flower pollination algorithm applied for different economic load dispatch problems. Int J Eng Technol 1009–1016 (2014)

Optimization of Electric Field Distribution Along a 400-kV Composite Insulator

C. Archana and K. Usha

Abstract Polymeric insulators are being widely used over ceramic insulators due to their tremendous merits. However, due to absence of the intermediate metal part, electric field and potential distribution along these insulators are non-uniform, which can be minimized by using suitable corona ring and grading ring. The aim of this work is to provide an optimum design of corona ring and grading ring for a 400-kV suspension-type polymeric insulator. Finite element method-based software is used for simulation purposes. This paper presents the results of 3D finite element calculations of electric field distribution along a 400-kV polymeric insulator. An optimum dimensions are predicted using multi-objective genetic algorithm. The predicted electric fields are compared with the actual values.

Keywords Corona ring · Grading ring · Electric field · Finite element method · Polymeric insulator · Genetic algorithm

1 Introduction

An uninterrupted power supply is the motive of all the power systems across the world. For efficient use of the power generated, the transmission and distribution lines play a vital role in improving the transmission capacity, thereby meeting the demand. High-voltage insulators take a significant part in transmission line for their factors like electrical insulation, corona discharge, mechanical support which affects the main motive of uninterrupted power supply. The reliable operation of the insulator directly affects the reliability of the grid [1, 2]. Polymeric insulators or non-ceramic insulators are now being widely used compared to ceramic insulators due to their advantages of surface hydrophobicity, pollution flashover performance, mechanical strength, less weight, low cost, resistance to vandalism (damage due to gun shots), high wet

C. Archana (✉) · K. Usha
SSN College of Engineering, Kalavakkam, Chennai, India
e-mail: archanac@ssn.edu.in

K. Usha
e-mail: ushak@ssn.edu.in

N. Zhou and S. Hemamalini (eds.), *Advances in Smart Grid Technology*, Lecture Notes in Electrical Engineering 688, https://doi.org/10.1007/978-981-15-7241-8_21

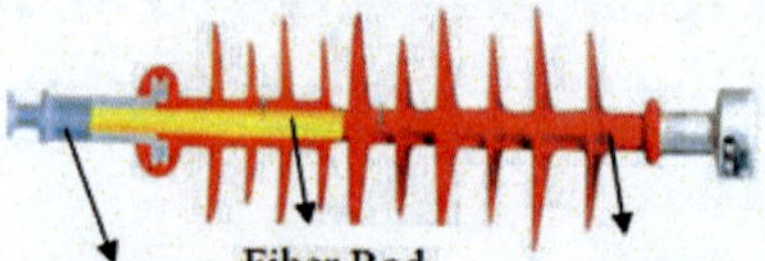

Fig. 1 Structure of a polymeric insulator

flashover and flashover voltage, withstand voltage is better than ceramic insulators. In addition to these features, polymeric insulators have higher mechanical strength-to-weight ratio [3]. But the surface field distribution and potential distribution for polymeric insulator is uneven due to the composite insulator structure characteristics. The intermediate metal parts in ceramic insulator which provided proper grading are absent in polymeric insulator. High surface electric field produces corona discharge or partial discharge which deteriorates the insulator surface leading to conduction inside the insulator. The enormous field not only produces corona discharge but also affects the transmission line electromagnetic environment, produces audible sound pollution, and causes electric erosion which leads to insulator ageing. The structure of a polymeric insulator is shown in Fig. 1. It consists of fibre reinforced plastic (FRP), silicon rubber (SiR) which provides sheading (housing) for the fibre rod to avoid contamination of the rod. It also provides leakage distance within a limited insulator length under contaminated and wet conditions. The metal end fittings are attached to the either end of the rod, where one end is attached to the line and other end to the tower.

The electric field is generally high in three main regions of the polymeric insulator which are at the triple junction point (interface of polymer, air, and metal fittings), around the grading ring and corona ring. [4, 5]. The electric strength in the vicinity of the critical regions should be below the critical limits. As per the literature, the critical field limits in these critical regions on dry clean polymeric insulators are

(1) On the surface of the shed, surrounding the end fitting and air surrounding (Triple Junction)—0.64 kV/mm
(2) Surrounding the corona ring—2.97 kV/mm
(3) Surrounding the grading ring—2.97 kV/mm.

Generally, the electric field is much higher in the high-voltage side compared to other parts of the insulator. When the electric field exceeds the critical values, corona discharge takes place which reduces the life span of the insulator. Installation of corona ring is the solution to minimize the field below the mentioned critical values. Corona ring transfers the maximum field density from the insulator surface to the ring surface, thus reducing the electrical stress on the insulator. Corona ring transfers the maximum field density between the end fittings and the first shed to the outer side of the ring, whereas grading ring is used to make voltage and electric field more uniform over the entire shed [9].

As per the literature, it is understood that there are no proper standards for design parameters and placement of corona ring and grading ring for all voltage ratings. The placement of corona ring has a major role as improper placement inflates the

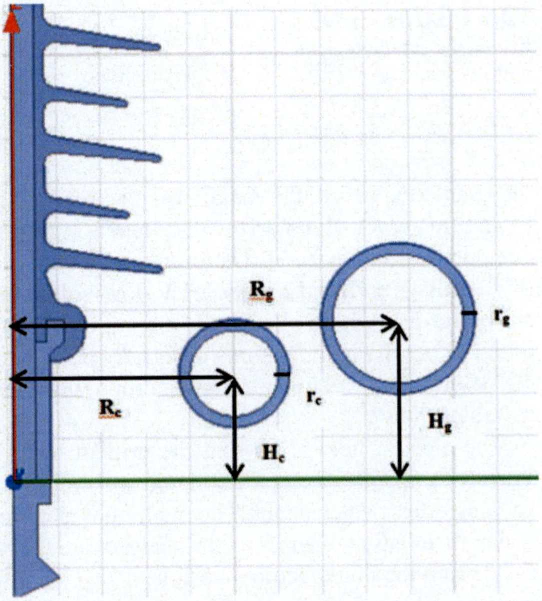

Fig. 2 Different design parameters used for the study

electric field rather minimizing it. Therefore, there is a need in designing the corona ring and grading ring to the optimum values for uniform electric field distribution on the surface of the insulator. Different design parameters of the ring such as corona ring radius (R_c), corona ring thickness (r_c), corona ring height (H_c), grading ring height (H_g), grading ring radius (R_g), and grading ring thickness (r_g) are varied, and an optimum solution is suggested. The parameters which are varied are shown in Fig. 2.

2 Simulation Studies

A 400-kV suspension-type insulator is taken for study. The length of the insulator is about 3400 mm long with 121 sheds with two metallic ends, namely ball and socket. The dimensions of ball and socket are as per IEC 120. One of the metallic ends is fitted to the tower, and the other end carries bundled conductor with two subconductors. The length of the conductor is 1.5 times the total length of the insulator. The dimensions of the housing are as per IEC 60815-3. The ball end (high voltage) is energized with a voltage of $400 \times \frac{\sqrt{2}}{\sqrt{3}} = 326$ kV and the socket (Ground) is energized with 0 V. Finite element method-based (FEM) software (Maxwell) is used. Finite element method is a numerical method which reduces complexity and produces accurate results. Simplified 3D model of a 400-kV insulator is designed, and the results are analysed. The simulation is computed with the relative permittivity of 4.3 for silicone rubber and 5.0 for fibre rod. The insulator is assumed to be under clean and dry condition. From

Table 1 Maximum electric field in the critical regions with corona ring

Critical regions	Example
Triple junction	1.7389
Corona ring	3.466

the simulation result, the maximum electric field in the critical regions is analysed. Literature survey summarizes that increasing number of sheds would increase the computational time [6, 7]. Therefore, number of sheds is decreased to 5 sheds at HV (high-voltage end) and 2 sheds at LV (low-voltage end). The forthcoming simulations are performed with 5 sheds at HV and 2 sheds at LV to reduce the computational time. As per the manufactures design $R_c = 100$ mm, $H_c = 50$ mm, $r_c = 20$ mm. A 400-kV insulator is simulated with corona ring alone, and the results are summarized in Table 1.

The electric field is beyond the critical limits in spite of placing corona ring. Therefore, the parameters are varied to analyse the field. One parameter is varied keeping other two constant. The maximum electric field is beyond the critical limits under all the cases. Therefore, the analyses are further continued by installing grading ring. From the manufactures design $R_g = 175$ mm, $r_g = 30$ mm and $H_g = 75$ mm is opted. Figure 3 shows the 3D simulation of a 400-kV insulator with tower under study.

The enlarged section of the high-voltage end (HV) of the insulator with the conductor, corona ring, and grading ring is shown in Fig. 4. The equipotential plot of 400-kV insulator is shown in Fig. 5.

From Table 2, it is clearly understood that the field is beyond the critical limits in the triple junction in spite of installation of the corona ring and grading ring. Figure 6 shows the maximum electric field at triple junction.

It can be inferred from the analyses that the location and the dimensions of corona ring and grading ring play a vital role in influencing the electric field in the critical regions. In spite of the electric field around corona ring and grading ring within the limits, the analyses are still studied even at the rings in order to ensure that the

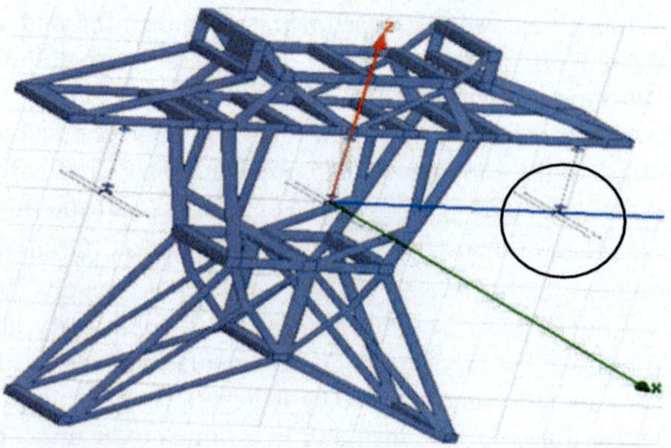

Fig. 3 3D simulation of a 400-kV insulator and tower in Maxwell

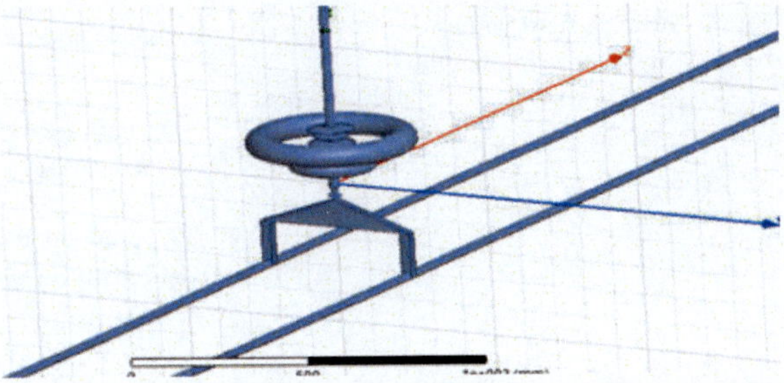

Fig. 4 Enlarged view of high-voltage end of the insulator

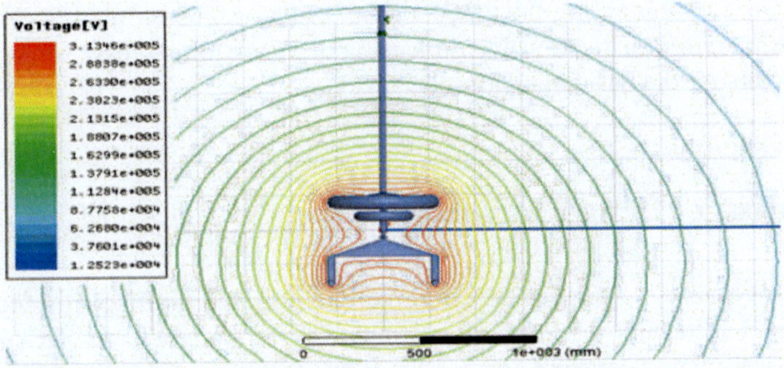

Fig. 5 Equipotential plot of a 400-kV insulator

Table 2 Maximum electric field in the critical region with both rings

Critical regions	Example
Triple junction	0.797
Corona ring	1.002
Grading ring	1.925

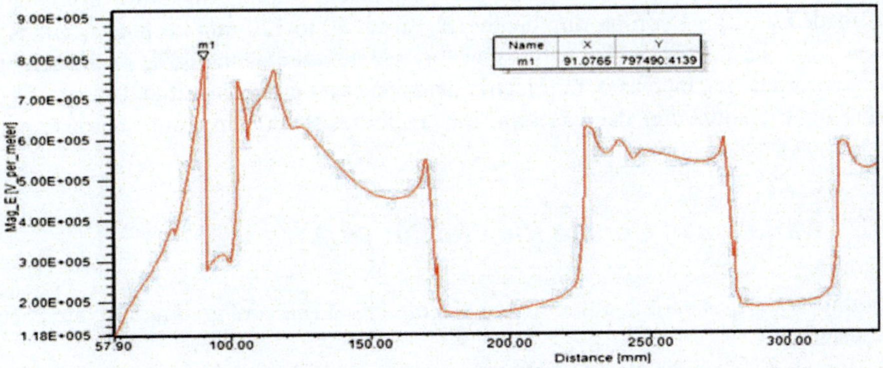

Fig. 6 Maximum electric field value at triple junction

Fig. 7 Electric field values by varying corona ring height

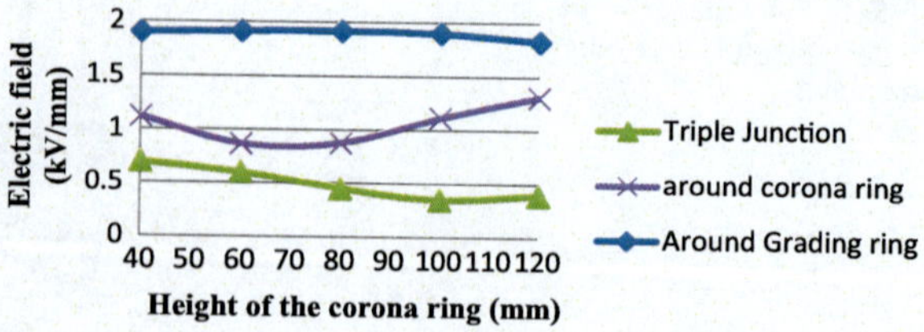

field is within the limits at all cases. Also, the location of rings reduces the dry acing distance by increasing the length of the insulator. Therefore, it is necessary to determine an optimum location and dimension to ensure longest dry arching distance without reducing the basic insulation level [8].

3 Analysis of Design Parameters

In order to obtain an optimum design of corona ring and grading ring for uniform electric field distribution in the critical regions, one parameter is varied keeping other five parameters constant as per manufactures design. When the corona ring is analysed, the grading ring parameters are kept constant and vice versa.

3.1 *Variation of Corona Ring Height* (H_c)

In this case, H_c is varied by keeping other parameters constant and the electric field is analysed in the triple junction region, corona ring, and grading ring. Now, an attempt is made by varying corona ring height (H_c) from 40 to 120 mm keeping r_c and R_c constant. The electric field at various locations is represented in Fig. 7. As the height of the corona ring increases, the electric field decreases gradually around the grading ring and triple junction, decreases and then increases at 100 mm around corona ring.

3.2 *Variation of Corona Ring Radius* (R_c)

In this case, R_c is varied; other parameters of corona ring and grading ring are kept constant.

Figure 8 shows the electric field values at the critical areas by varying R_c. It can be observed that the as the radius of the ring is increased, the field around the grading ring reduces. But the field around the corona ring increases. Similarly field around the triple junction also increases.

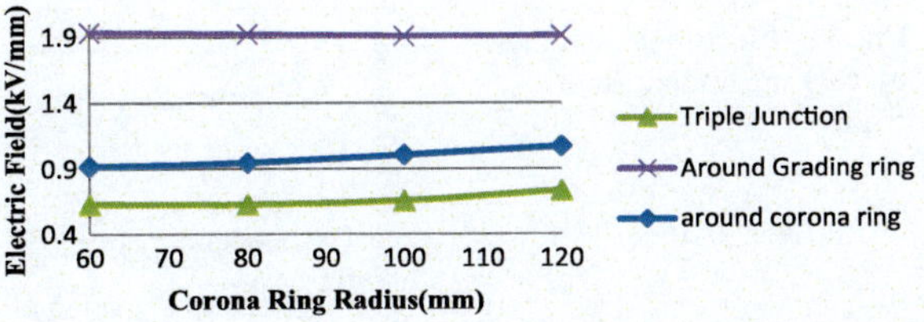

Fig. 8 Electric field values by varying corona ring radius

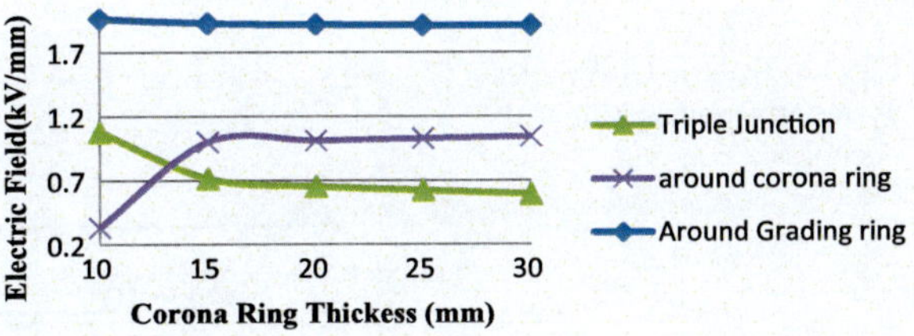

Fig. 9 Electric field values by varying corona ring thickness

3.3 Variation of Corona Ring Thickness (R_c)

The analyses are carried out by varying from 10 to 30 mm, and the other parameters are kept constant. Figure 9 shows the electric field values at the critical areas by varying $r_{c.}$

When the thickness of the corona ring increases, the field around the grading ring and triple junction decreases. But the field near the corona ring increases.

3.4 Variation of Grading Ring Height (H_g)

The simulation is further extended by varying H_g from 40 to 95 mm keeping the remaining parameters constant.

Figure 10 shows the electric field at the critical regions when H_g is varied. Electric field near the triple junction and the grading ring decreases, while field near the corona ring increases.

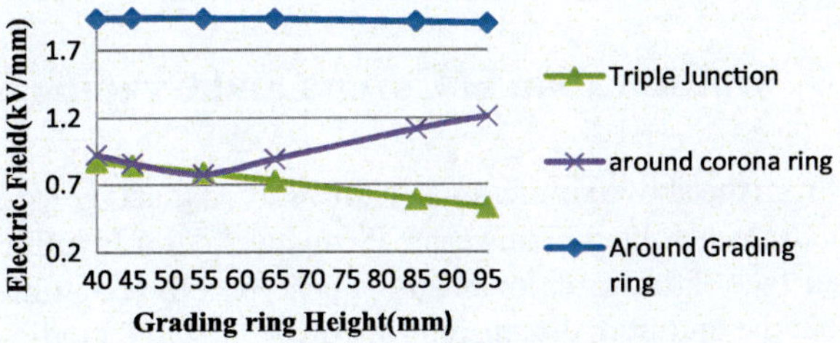

Fig. 10 Electric field values by varying grading ring height

Fig. 11 Electric field values by varying grading ring radius

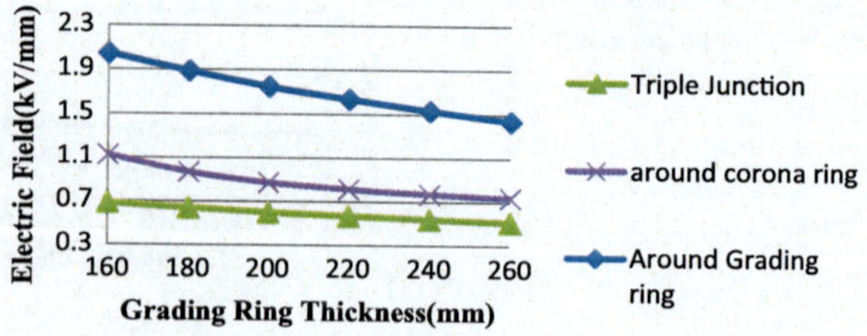

Fig. 12 Electric field values by varying grading ring thickness

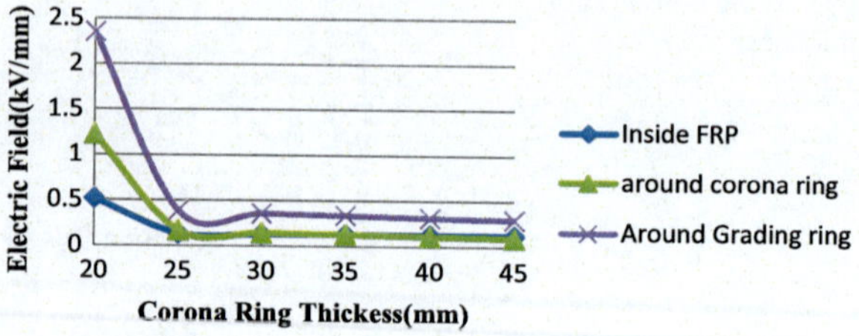

3.5 Variation of Grading Ring Radius (R_g)

In this case, r_g is varied and the other parameters are kept constant. It is noticed that the field decreases gradually when the grading ring radius is increased. Figure 11 shows the variation of electric field at the critical areas when the grading ring radius is varied.

3.6 Variation of Grading Ring Thickness (r_g)

The thickness of the grading ring is varied from 20 to 45 mm where H_g and R_g are kept constant. The field at all critical regions is observed to be at the maximum when the thickness of the grading ring is of least value. And as the ring size is increased, electric field reduces gradually. Figure 12 shows the variation of electric field at the critical regions when the thickness of the grading rings is varied.

4 Optimization of Corona and Grading Ring

From the above simulations, it is observed that there are many parameters and combinations involved in this case; it can considerably influence the maximum field in the surface of the insulator and the rings. Therefore, optimization technique is used to find out the optimum dimensions of corona ring and grading ring using OPTIMIZATION tool in MATLAB software.

The analyses are extended by,

1. Finding a mathematical relation between the maximum electric field and the ring parameters.
2. Applying multi-objective genetic algorithm as an optimization technique to find out the optimized parameters.

Nonlinear curve fitting is performed using curve fitting software, which gives the best fit equations for the above six considered cases where one parameter is varied keeping other five parameters constant. Equations (1), (2), and (3) correspond to the mean electric field equations at the triple junction, around the corona ring, around the grading ring, respectively. E is the function of corona ring height, corona ring radius, corona ring thickness, grading ring height, grading ring radius, and grading ring thickness.

$$\begin{aligned} E\left(H_c, R_c, r_c, H_g, R_g, r_g\right) = {} & 2.52 + 2.5 \times 10^{-7} x_1^3 - 9.6 \times 10^{-3} x_1 \\ & + 3.24 \times 10^{-7} x_2^3 - 3.5 \times 10^5 x_2^2 + 3.17 \times 10^{-7} x_3^3 \\ & - 0.06 x_3 + 7.05 \times 10^{-7} x_4^3 - 1.56 \times 10^{-4} x_4^2 \\ & + 4.756 \times 10^{-3} x_4 - 2.615 \times 10^8 x_5^3 + 2.27 \times 10^{-5} x_5^2 \\ & - 7.68 \times 10^{-3} x_5 - 2.15 \times 10^{-4} x_6^3 + 0.02 x_6 - 0.811 x_6 \end{aligned} \quad (1)$$

$$\begin{aligned} E\left(H_c, R_c, r_c, H_g, R_g, r_g\right) = {} & 3.9165 - 3.1 \times 10^{-6} x_1^3 + 9.48 \times 10^{-4} x_1^2 \\ & - 8.4 \times 10^{-2} x_1 + 0.0001 x_2^2 - 0.001381 x_2 \\ & + 4.36 \times 10^{-4} x_3^3 - 3.03 \times 10^{-2} x_3^2 + 0.66 x_3 \\ & - 9.106 \times 10^{-6} x_4^3 + 2.06 \times 10^{-3} x_4^2 - 0.142 x_4 \\ & - 3.4 \times 10_{-7} x_5^3 + 2.59 \times 10^{-4} x_5^2 - 6.3 \times 10^{-2} x_5 \\ & - 3.85 \times 10^{-4} x_6^3 + 4.14 \times 10^{-2} x_6^2 - 1.453 x_6 \end{aligned} \quad (2)$$

$$\begin{aligned} \left(H_c, R_c, r_c, H_g, R_g, r_g\right) = {} & 7.2987 - 6.45 \times 10^{-8} x_1^3 - 3.23 \times 10^{-5} x_1^2 \\ & + 3.5 \times 10^{-3} x_1 - 0.72 \times 10^{-2} x_2^5 + 3.16 \times 10^{-3} x_2 \\ & + 9.6 \times 10^{-6} x_3^3 - 3.51 \times 10^{-4} x_3^2 - 1.25 \times 10^{-8} x_4^4 \\ & + 3.57 \times 10^{-6} x_4^3 - 3.9 \times 10^{-4} x_4^2 + 1.83 \times 10^{-2} x_4 \\ & - 1.14 \times 10^{-7} x_6^3 + 9.3 \times 10^{-5} x_6^2 - 2.9 \times 10^{-2} x^3 \\ & - 7.18 \times 10^{-4} x_6^3 + 7.7 \times 10^{-2} x_6^2 - 2.701 x_6 \end{aligned} \quad (3)$$

As electric field (E) is the function of more than one variable, multi-objective genetic algorithm is used using OPTIM tool in MATLAB. These equations are given as an input to multi-objective genetic algorithm optimization technique. The objective functions are

Triple junction min E $(H_c, R_c, r_c, H_g, R_g, r_g) \leq 0.64$

Table 3 Constraints of the rings

Critical Regions	Constraints
Triple junction	$\text{lb}(50) \le H_c \le \text{ub}(100)$ $\text{lb}(80) \le R_c \le \text{ub}(120)$ $\text{lb}(20) \le r_c \le \text{ub}(30)$ $\text{lb}(50) \le H_g \le \text{ub}(100)$ $\text{lb}(175) \le R_g \le \text{ub}(225)$ $\text{lb}(30) \le r_g \le \text{ub}(45)$
Corona ring and grading ring	$\text{lb}(40) \le H_c \le \text{ub}(120)$ $\text{lb}(60) \le R_c \le \text{ub}(120)$ $\text{lb}(10) \le r_c \le \text{ub}(30)$ $\text{lb}(40) \le H_g \le \text{ub}(95)$ $\text{lb}(160) \le R_g \le \text{ub}(260)$ $\text{lb}(20) \le r_g \le \text{ub}(45)$

Corona ring $\quad \min \text{E}\,(H_c, R_c, r_c, H_g, R_g, r_g) \le 2.97$
Grading ring $\quad \min \text{E}\,(H_c, R_c, r_c, H_g, R_g, r_g) \le 2.97.$

The constraints with maximum bound and the minimum bound for all the critical cases (triple junction, corona ring, and grading ring) are listed in Table 3.

The bounds are increased in steps initially and then stopped when the critical limit is met. The electric field values at corona ring and grading ring are always within the limit (i.e. less than 2.97 kV/mm), whereas the field values at the triple junction are not within the limits. Therefore, the electric field at the triple junction region is of major concern and the analysis has been carried out at the triple junction region. The optimized dimensions of corona ring and grading ring are predicted and are listed in Table 4.

The simulation of the 400-kV insulator with the optimized dimensions of corona and grading ring is performed using MAXWELL, and the obtained electric field values are compared with the predicted values as shown in Table 5.

Table 4 Optimized dimensions of the rings

H_c (mm)	R_c (mm)	r_c (mm)	H_g (mm)	R_g (mm)	r_g (mm)
80.3	89.7	22	84	193.2	35

Table 5 Percentage difference between the predicted and actual electric field at various locations

Location	Predicted electric field	Actual electric field	% difference in electric field
Triple junction	0.6345	0.6176	−2.736
Corona ring	0.685	0.669	−2.391
Grading ring	1.69	1.64	−2.86

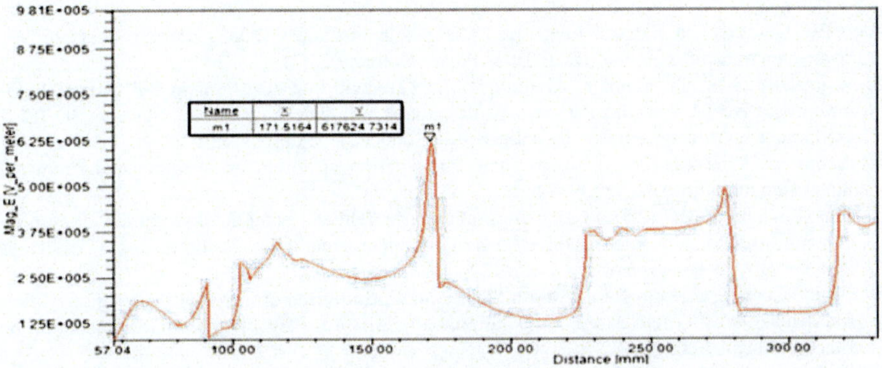

Fig. 13 Actual electric field plot for optimized dimensions

It is observed that the error is less than 3%. Hence, the predicted dimensions are the optimized dimensions where the field is well below the critical limits (Fig. 13).

5 Conclusion

Corona ring plays a vital role as they increase the life span of the insulator by decreasing the electric field below their critical limits. By installing corona ring alone, the field in the critical regions remains high. The design parameters are varied, and a mathematical relation between the electric field and the ring parameters (corona and grading) is formulated using curve fitting and fed as an input to multi-objective genetic algorithm optimization technique. Genetic algorithm predicts the optimized design parameters which are compared with the actual value. The proposed method optimizes corona ring and grading ring dimensions systematically rather than performing an old methodology of trial-and-error method. As a future work, it is proposed to use neural network as a replacement for curve fitting.

References

1. Huang D, Ruan J, Liu S (2010) Potentila distribution along UHV AC transmission line composite insulator and electric field distribution on surface of grading ring. High Voltage Eng 36(6):1442–1447
2. Huang D, Zheng Z, Huang Z, Guoli W, Ruan J, Li P (2013) Calculation model simplification study for porcelian insulator string potential and grading ring surface electric field distribution of UHV AC transmission line. In: Annual Report Conference on Electrical Insulation and Dielectric Phenomena, Oct 2013
3. Muniraj Ch, Chandrasekar S (2012) Finite element modelling for electric field and volatge distribution along the polluted polymeric insulator. World J Model Simul 8(4):310–320

4. IEEE Taskforce on Electric Fields and Composite Insulators (2008) Electric fields on AC composite transmission line. IEEE Trans Power Delivery 23(2)
5. Kanyakumari M, Shivakumara Aradhya RS (2012) Control of electric field and voltage distribution of a 765 kV system polymeric insulator used in Indian Transmission Systems. In: IEEE conference on the properties and applications of dielectric materials, July 2012
6. Anbarasan R, Usa S (2015) Electric field computation of polymeric insulator using reduced dimension modeling. 22(2):739–746
7. Zhao T, Comber MG (2000) Calculation of electric field and potential distribution along non-ceramic insulators considering the effects of conductors and transmission towers. IEEE Trans Power Deliv 15(1):313–318
8. Espino Cotes S, Chemey EA, Jayaram SH (2004) Optimization of Corona ring design for long-rod insulators using fem based computational analysis. In: IEEE International Symposium on Electrical insulation, pp 480–483
9. Anbarasan R (2015) Design optimization of polymeric insulators using reduced dimension modelling. Thesis,, Anna University

Vector Control Scheme for the PMSG-Based WPS Under Various Grid Disturbances

R. Vijaya Priya and R. Elavarasi

Abstract This paper proposes a simple modified vector control scheme for restraining the impacts on permanent magnet synchronous generator (PMSG)-based wind power system (WPS) integration under various grid disturbances. Voltage sag and harmonics are the major causes of grid voltage disturbances. The percentage of voltage sag for which WPS remains to be connected with the grid is dictated by fault ride-through (FRT) characteristics. Moreover, voltage sag occurred due to short-circuit fault has its associated phase-angle jump. Hence, it is necessary to map the voltage sag with phase-angle jump to meet the broad range of FRT requirements. A new analysis with major types of faults that can occur at the point of grid integration is presented in this paper. In order to satisfy the grid code compliances and reduced harmonic distortions, the modified vector control scheme is modelled with additional voltage and current oscillation cancellation blocks without the need for dual vector control loop. Extensive analytical simulation has been carried out in PSCAD/EMTDC to validate the superiority of proposed control scheme over the conventional schemes in terms of transient overshoots and oscillations that appear in dc-link voltage and real and reactive power of grid when PMSG is subjected to various disturbances.

Keywords WPS · FRT · PMSG

1 Introduction

Power quality improvement at the point of grid interface with the high-level integration of wind power systems (WPS) is the most important aspect to be given attention. Modern grid codes are emerging with a substantial rise in the integration of WPS

R. Vijaya Priya (✉)
School of Electrical Engineering, Vellore Institute of Technology, Vellore, Taml Nadu 632014, India
e-mail: vijayapriiyaa@yahoo.co.in

R. Elavarasi
AMET University, Chennai, India

N. Zhou and S. Hemamalini (eds.), *Advances in Smart Grid Technology*, Lecture Notes in Electrical Engineering 688, https://doi.org/10.1007/978-981-15-7241-8_22

into the transmission and distribution grid. Essentially, these grid codes firmly frame the fault ride-through (FRT) requirements at the interconnection point of WPSs to the grid, commonly referred as point of common coupling (PCC). FRT is explicitly defined with voltage-time profile, for which the wind farm must remain connected to the grid during grid disturbances. Figure 1 shows the FRT profile framed by the Indian Electricity Grid Code (IEGC) for WPS connected at the voltage level of 66 kV and above [1]. It is noted from Fig. 1 that a wind farm shall remain connected to the power network under the sag conditions when the voltage measured at PCC is within the shaded region for various grid fault conditions. A voltage sag of 85% is the most severe fault condition, as dictated by IEGC.

Most of the commercially available new MW scale wind turbines are based on variable-speed generators of either a doubly fed induction generator (DFIG) or a PMSG [2]. The DFIG based wind turbines are completely exposed to the grid faults, and special control measures must be incorporated to safeguard it against grid fault conditions. However, PMSG-based wind turbine is fully decoupled from the grid through power converters employed for interfacing and power control. Hence, with more fringing grid codes, PMSG-based WPSs are becoming more promising in ease of accomplishing FRT rather than DFIG. In addition to that, PMSG also offers several advantages of gearless construction [3, 4], full controllability of speed for evacuating maximum power, maximum reactive power support to the grid.

The operation of PMSG-based WPSs under various grid fault conditions has been periodically investigated in the literature. Initial works were mainly aimed to develop a scheme to use a dc-link braking chopper with and without an energy storage system to enhance the FRT requirement [5–7]. However, an additional hardware requirement increases the overall cost of the system.

The recent works have been focusing more and more on the controlling aspects of the WPS power converters to inject the generated wind power into the grid even under grid fault conditions. Hence, to meet up with the grid code compliance, several configurations of WPSs are employed [8–12]: voltage source converter (VSC)-based back–back topology, neutral point clamped (NPC) back–back converts, uncontrolled rectifier with DC chopper. The primary goal of these configurations is to suppress

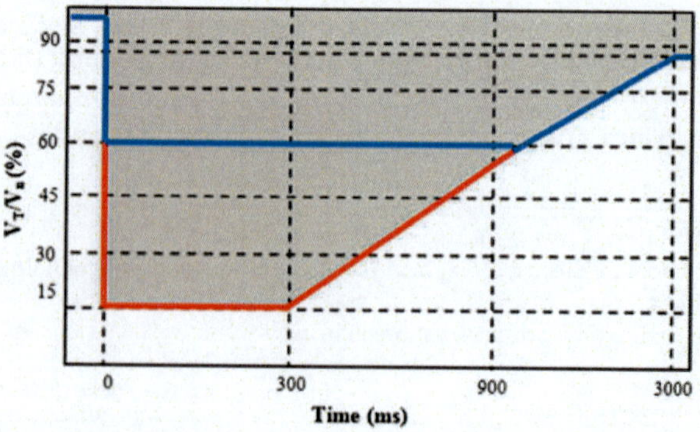

Fig. 1 FRT requirements of IEGC

the oscillation in the dc-link voltage and real and reactive power of the gird under symmetrical and unsymmetrical voltage sag. Specifically, the VSC-based back–back topology is gaining utmost importance with new innovative control techniques in recent decades. With the conventional dc-link voltage control on the grid side converter (GSC) a dual vector control scheme has been implemented in [13–15] for the positive and negative sequence of inner current loop. However, during grid fault, the GSC goes out of control in maintaining the dc-link voltage. As the machine side converter (MSC) is not directly connected to the grid, an alternative method has been suggested in [16] to employ the dc-link voltage control strategies on the MSC. The GSC dictates the amount of power to be injected into the grid. Hence, under fault conditions the speed of the generator is increased, thereby reducing the power injection into the grid to maintain the dc-link voltage for the given wind velocity. This controlling technique has also been addressed with hybrid adaptive proportional–integral (PI) controller in [17] and with the feedback linearization controller in [18]. Though the power of PMSG is controlled by GSC, still there is a need for dual current controllers for the positive and negative-sequence reference frames. Also, the reported works examine very limited grid disturbances with only the voltage sag magnitude without considering the corresponding phase jump.

This paper aims to develop a novel control scheme for PMSG-based WPS, which possibly addresses all kinds of abnormal gird conditions of symmetrical and unsymmetrical voltage sags, phase-angle jump and harmonics. Section 2 presents the *d–q* modelling of MSC and GSC. This section also examines the classification and characterization of various gird disturbances that occur at PCC. Section 3 proposes a simple modified vector control scheme for the GSC, which will be used to validate the simulation results over the conventional schemes under various grid faults. The control scheme is implemented in the positive synchronous frame (PSF) with the additional voltage and current oscillations compensation blocks, without the need for dual vector control loops. Finally, Sect. 4 simulates the analytical modelling of WPS in PSCAD/EMTDC and is subjected to various grid disturbances to investigate the transient overshoots and steady-state oscillations that appear in the dc-link voltage and real and reactive power of the grid.

2 A System Modelling

Figure 2 depicts the overall configuration of PMSG-based grid-integrated WPS. A two-level VSC topology is used for both the MSC and GSC. The MSC is needed to be controlled for maximum wind power evacuation for the given wind velocity. Also, in addition to the regular power flow control (dc-link voltage and grid reactive power control), the GSC is needed to be controlled for oscillation suppression in the system parameters during grid disturbances. Hence, for the proper modelling of a controller structure a complete analysis from modelling of wind turbine to the characterization of grid disturbances is required.

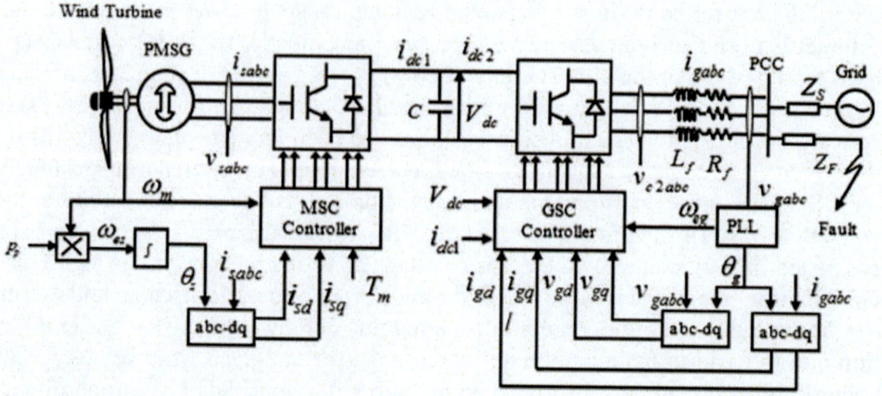

Fig. 2 PMSG-based grid-integrated WPSs

2.1 Modelling of Wind Turbine

The mechanical output power P_{m} of a wind turbine is formulated as given in (1) [19]

$$P_{\mathrm{m}} = \frac{1}{2}\rho A C_{\mathrm{p}} \nu^3 \tag{1}$$

Since the air density (ρ) and the swept area of the turbine (A) are constant, for the given wind velocity (ν), the output power of the turbine can be maximized by maximizing the power conversion coefficient Cp. Without pitch angle control mechanism, Cp will be the function of only tip-speed ratio λ, which is defined as

$$\lambda = \frac{\omega_{\mathrm{m}} r}{\nu} \tag{2}$$

The Cp–λ characteristics is constant for a given wind turbine. Hence by fixing λ for the maximum coefficient of Cp, the optimum speed is determined for a given wind velocity, which corresponds to the speed of the maximum output power of the turbine.

2.2 Modelling of PMSG

Figure 3 shows the space vector representation of PMSG, the permanent magnet; i.e., the flux of the machine is aligned along the *d*-axis, and *q*-axis is in front of axis d by an electrical degree of 900. For sinusoidal flux distributions of permanent magnet AC machine [26], the stator voltage in *d*–*q* axis in generator mode is modelled as

Fig. 3 Space vector representation of PMSG

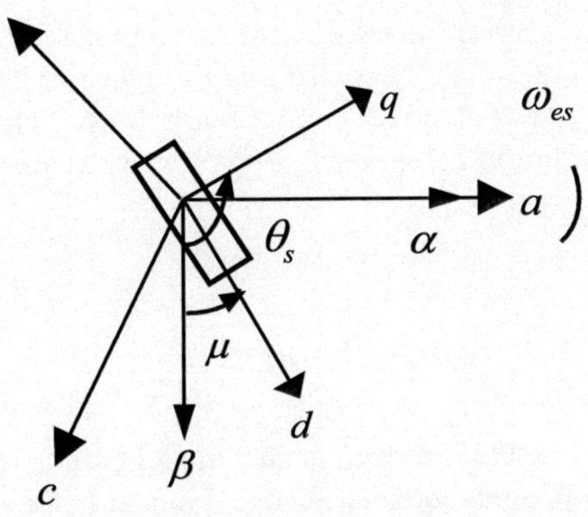

$$v_{sd} = L_{sd}\frac{di_{sd}}{dt} + R_s i_{sd} - L_{sq} i_{sq} \omega_{es} \tag{3}$$

$$v_{sq} = L_{sq}\frac{di_{sq}}{dt} + R_s i_{sq} + L_{sd} i_{sd} \omega_{es} + \psi_{PM} \omega_{es} \tag{4}$$

with

$$\psi_q = L_{sq} i_{sq} \tag{5}$$

$$\psi_d = L_{sd} i_{sd} + \psi_{PM} \tag{6}$$

Also, the expression for the generator electrical torque T_e, stator real and reactive power in *d–q* model is derived to be

$$T_e = 1.5 p_p \left(\psi_{PM} i_{sq} + \left(L_{sd} - L_{sq}\right) i_{sd} i_{sq}\right) \tag{7}$$

$$P_s = 1.5\left(v_{sq} i_{sq} + v_{sd} i_{sd}\right) \tag{8}$$

$$Q_s = 1.5\left(v_{sq} i_{sd} - v_{sd} i_{sq}\right) \tag{9}$$

The dynamic one mass model of the mechanical system is described by

$$T_m - T_e = J\frac{d\omega_m}{dt} + B\omega_m \tag{10}$$

The generator torque T_e for the surface mounted PMSG is obtained by equating $L_{sq} = L_{sd}$ in (7), which results in

$$T_e = 1.5 p_p \left(\psi_{PM} i_{sq}\right) \tag{11}$$

Since speed control is employed for maximum power extraction Te can be controlled by controlling the speed of the generator, which in turn controls the real power of the generator. Hence, from (11) it is clear that isq corresponds to a real power current component. However, the reactive power of the generator is controlled by controlling isd.

2.3 *Modelling of GSC*

The GSC control is also modelled in *d*–*q* frame by aligning the grid phase voltage along the *q*-axis. The GSC model in the *d*–*q* frame is given as

$$L_{\mathrm{f}}\frac{\mathrm{d}i_{\mathrm{gd}}}{\mathrm{d}t} = \omega_{\mathrm{g}} i_{\mathrm{gq}} L_{\mathrm{f}} + v_{\mathrm{gd}} - R_{\mathrm{f}} i_{\mathrm{gd}} - v_{\mathrm{c2d}} \tag{12}$$

$$L_{\mathrm{f}}\frac{\mathrm{d}i_{\mathrm{gq}}}{\mathrm{d}t} = -\omega_{\mathrm{g}} L_{\mathrm{f}} i_{\mathrm{gd}} + v_{\mathrm{gq}} - R_{\mathrm{f}} i_{\mathrm{gq}} - v_{\mathrm{c2q}} \tag{13}$$

$$C\frac{\mathrm{d}V_{\mathrm{dc}}}{\mathrm{d}t} = i_{\mathrm{dc2}} - i_{\mathrm{dc1}} \tag{14}$$

The average active and reactive power of the grid under the steady-state condition in *d*–*q* frame is defined in (15) and (16)

$$P_{\mathrm{g}} = 1.5\left(v_{\mathrm{gq}} i_{\mathrm{gq}} + v_{\mathrm{gd}} i_{\mathrm{gd}}\right) \tag{15}$$

$$Q_{\mathrm{g}} = 1.5\left(v_{\mathrm{gq}} i_{\mathrm{gd}} - v_{\mathrm{gd}} i_{\mathrm{gq}}\right) \tag{16}$$

Since grid voltage is aligned with *q*-axis v_{gd} becomes zero, and hence grid real and reactive power is controlled by controlling i_{gq} and i_{gd}, respectively.

2.4 *Characterization of Grid Disturbances*

For achieving FRT requirements and to improve the controller performance of WPS, insight knowledge of various grid disturbances that occur at PCC should be known properly. When a fault occurs at any point in the power network, the grid voltage will drop to the lower levels (voltage sag). The type of voltage sag that can be expected at PCC is A, C, D, and F [23]. The main causes of voltage sag are short-circuited and earth faults in the grid. Since the voltage in the 50 Hz system is a phasor quantity with magnitude and phase angle, a short-circuit in the system not only causes a drop in voltage magnitude but also a change in the phase angle of the voltage. The degree of phase-angle jump (φ) for the respective voltage sag is determined by (17) [25]

$$\arg\left(V_{\text{sag}}\right) = \arg(Z_F) - \arg(Z_S + Z_F) \tag{17}$$

where the voltage sag magnitude V_{sag} at PCC during fault is defined with source impedance Z_S and fault feeder impedance Z_F and it is given as

$$V_{\text{sag}} = V_s \frac{Z_F}{Z_S + Z_F} \tag{18}$$

with ($Z_F/Z_S = \tau e^{j\varphi\gamma}$), (18) can be rewritten as

$$V_{\text{sag}} = V_s \frac{\tau e^{j\gamma}}{1 + \tau e^{j\gamma}} \tag{19}$$

where τ is a measure of the electrical distance to the fault and γ the impedance angle, is the measure of angle between Z_S and Z_F, which is constant for any feeder/source combination. As a result, a unique relation between voltage sag magnitude and phase-angle jump exists for a given impedance angle. A typical impedance angle value for transmission, distribution, and offshore wind farms with submarine/underground AC cables is defined to be 0°, −20°, and −60°, respectively [25]. Figure 4 shows the relation between the sag magnitude and the corresponding phase-angle jump for these impedance angles. Hence, it is inferred that for any voltage sag conditions specified in IEGC, the corresponding phase-angle jump can be predicted from Fig. 4. e.g., for 90, 70, and 50% of voltage sag magnitudes, the phase-angle jump is predicted as −5.7°, °9.5°, and −12.5°, respectively, as per tabulated in [24], for the distributed system. Utility harmonics is another important issue to be addressed under grid disturbance, which is the result of modern developments in electricity utility with the use of electronic power conditioning modules.

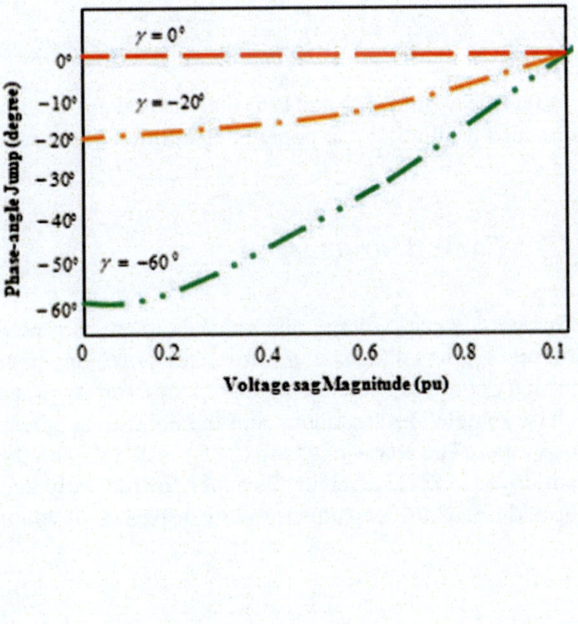

Fig. 4 Voltage sag magnitude versus phase-angle jump for various impedance angles

3 Proposed Control Scheme

3.1 Impact of Grid Voltage Disturbances

The voltage sag of $x\%$ in a—phase of the grid voltage with its associated phase-angle jump creates a double times grid frequency oscillation in the d–q frame voltages. Similarly, the grid voltage in d–q frame with the presence of predominant harmonics of 5th and 7th in the grid voltage has six times grid frequency oscillation [27]. Hence, in general, the grid voltage in PSF under various grid disturbances can be expressed in terms of dc component and oscillating component. The same will hold for the grid currents also. The grid voltage and current in PSF is expressed as

$$v_{\mathrm{gd}}^{+} = v_{\mathrm{gddc}}^{+} + v_{\mathrm{gdosc}}^{+} \tag{20}$$

$$v_{\mathrm{gq}}^{+} = v_{\mathrm{gqdc}}^{+} + v_{\mathrm{gqosc}}^{+} \tag{21}$$

$$i_{\mathrm{gd}}^{+} = i_{\mathrm{gddc}}^{+} + i_{\mathrm{gdosc}}^{+} \tag{22}$$

$$i_{\mathrm{gq}}^{+} = i_{\mathrm{gqdc}}^{+} + i_{\mathrm{gqosc}}^{+} \tag{23}$$

Correspondingly, the real and reactive power of the grid in PSF is given by

$$P_{\mathrm{g}} = 1.5\begin{pmatrix} v_{\mathrm{gqdc}}^{+} i_{\mathrm{gqdc}}^{+} + v_{\mathrm{gddc}}^{+} i_{\mathrm{gddc}}^{+} + v_{\mathrm{gqosc}}^{+} i_{\mathrm{gq}\cos c}^{+} + v_{\mathrm{gdosc}}^{+} i_{\mathrm{gqosc}}^{+} \\ +v_{\mathrm{gqosc}}^{+} i_{gddc}^{+} + v_{\mathrm{gdosc}}^{+} i_{\mathrm{gqdc}}^{+} + v_{gqdc}^{+} i_{\mathrm{gdosc}}^{+} + v_{\mathrm{gddc}}^{+} i_{\mathrm{gqosc}}^{+} \end{pmatrix} \tag{24}$$

$$Q_{\mathrm{g}} = 1.5\begin{pmatrix} v_{\mathrm{gddc}}^{+} i_{\mathrm{gqdc}}^{+} - v_{\mathrm{gqdc}}^{+} i_{\mathrm{gddc}}^{+} + v_{\mathrm{gdosc}}^{+} i_{\mathrm{gdosc}}^{+} - v_{\mathrm{gqosc}}^{+} i_{\mathrm{gqosc}}^{+} \\ +v_{\mathrm{gdosc}}^{+} i_{\mathrm{gddc}}^{+} - v_{\mathrm{gqosc}}^{+} i_{\mathrm{gqdc}}^{+} + v_{\mathrm{gddc}}^{+} i_{\mathrm{gdosc}}^{+} - v_{\mathrm{gqdc}}^{+} i_{\mathrm{gqosc}}^{+} \end{pmatrix} \tag{25}$$

It is clear from (24) and (25) that the real and reactive power injected into the grid contains oscillating components in addition to the required power.

3.2 Control Structure

For various grid voltage disturbances, there is a noticeable oscillation in the d–q frame of grid voltages and currents. Hence, under grid voltage disturbances, two primary controls need to be adopted, one for the proper tracking of the grid voltage phase - angle and frequency and the other to suppress the oscillation in the system quantities. The phase lock loop (PLL) structure with the moving average filter (MAF) employed in [22] is used in this paper, to efficiently track the grid voltage phase angle θg and frequency ωeg under grid disturbances of voltage sag, phase-angle jump, and harmonics. The MAF in discrete-domain is realized in (26) [21]

$$X(Z) = \frac{1}{N}\left(\frac{1 - Z^{-N}}{1 - Z^{-1}}\right)X(Z) \tag{26}$$

Since most practical grid disturbances result in the integer multiples of grid frequency oscillations, the order of the filter N is chosen to filter oscillation with the frequency of 100, 200, 300, etc.

The controller structure for GSC starts with the analytical modelling from (12)–(14). When power loss in both MSC and GSC are ignored, the dc-link power on the GSC is equated to grid real power i.e.

$$P_{\text{dc2}} = i_{\text{dc2}} V_{\text{dc2}} = \frac{3}{2}\left(v_{\text{gq}} i_{\text{gq}} + v_{\text{gd}} i_{\text{gd}}\right) \tag{27}$$

By substituting (27) in (14), i_{sq} is found to be

$$i_{\text{gq}} = \frac{\frac{2V_{\text{dc}}}{3v_{\text{gq}}}\left(V\frac{\text{d}V_{\text{dc}}}{\text{d}t} + i_{\text{dc1}}\right) - v_{\text{gd}} i_{\text{gd}}}{V_{\text{dc}}} \tag{28}$$

Figure 5 gives the block diagram representation of the system along with its controller structure for the dc-link voltage control. It is clear from the block diagram representation that the term $(2V_{\text{dc}} - v_{\text{gd}} i_{\text{gd}})/3v_{\text{gq}}$ must be included in the controller structure for the cancellation of a similar term that presents in the plant modelling. Exclusion of this block may not affect the system performance under normal grid conditions; however, for grid disturbance the oscillation in the system parameters will not cancel out significantly. Hence, this voltage cancellation block must be included in the controller design.

From (24) and (25) it is evident that grid disturbance introduces a considerable amount of oscillation in real and reactive power of the grid. Because of this oscillating power flow, there will be an oscillation in the dc-link voltage. If these oscillations

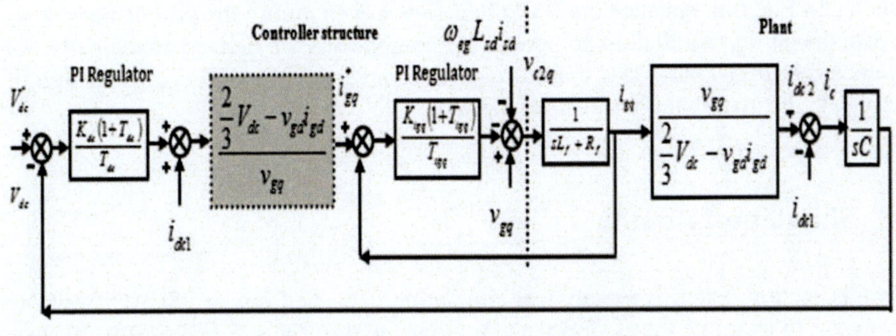

Fig. 5 Block diagram representation of GSC with the dc-link controller

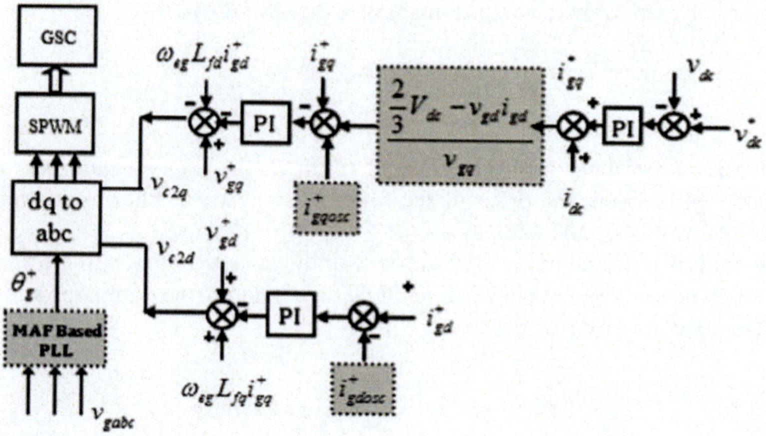

Fig. 6 Proposed controller structure for GSC

are not controlled properly it may lead to a severe problem of degrading the performance of dc-link capacitance and VSCs. Hence, a proper measure should be taken to suppress the oscillation in the dc-link voltage and real and reactive power of the grid. Since the real and reactive power of the grid is controlled by controlling the currents i_{gq} and i_{gd}, respectively, the amount of oscillation in the current components can be determined, and the same can be used to cancel out the oscillation in the grid current. The oscillating current components can be determined by rewriting (24) and (25) in matrix form and the oscillation current components in d–q axis are determined by forcing the oscillation components of real and reactive power of grid (P_{gosc} and $Qgosc$) to be zero [27].

Hence, by knowing the average power P_g from (35) and reactive power Q_g demand by the grid, the oscillating current component in the d–q frame is derived. This current oscillating term is added with the reference current generated in Fig. 5 to cancel out for significant oscillation in the system parameters. Hence, the proposed controller is farmed with two additional oscillation cancellation blocks of voltage and current as shown in Fig. 6 to enhance the PMSG operation even during the grid disturbances. Also, the current oscillation compensation components are derived analytically. As depicted in Fig. 7 the MSC control uses the conventional speed control design to evacuate the maximum power from the wind.

4 Simulation Results

In this section, extensive analytical simulation is carried out in PSCAD/EMTDC software to examine the performance improvement of the 1.5 MW PMSG system under different types of grid disturbance with the proposed controller structure. A comparative analysis is also presented for each case of disturbance with three different

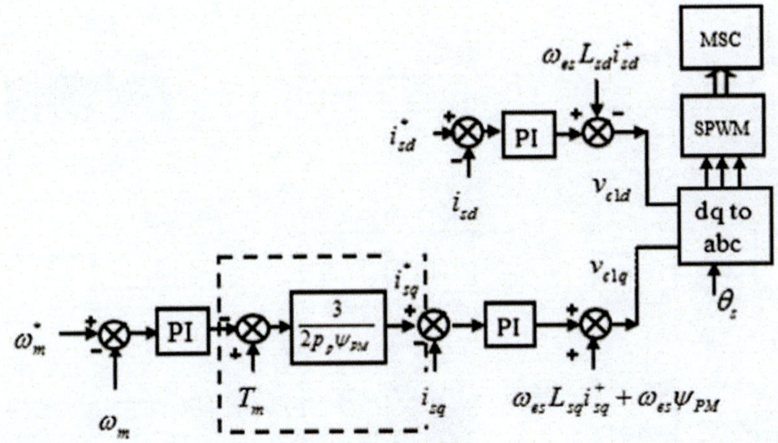

Fig. 7 MSC speed and reactive power controller block

control techniques: (i) conventional vector control; (ii) vector control with current oscillation cancellation method; (iii) proposed control method. The PI controllers for MSC and GSC are designed as per the guidelines presented in [20] with a damping ratio $\varepsilon = 0.707$, and its estimated values are given in the Appendix (Table 2).

Different cases are analysed for the performance evaluation of the system, and the simulated indices are presented below.

Case 1: The transient analysis of PMSG for the change in wind velocity is depicted in Fig. 8. Initially, the system operates with the wind velocity of 12 m/s, and the wind velocity is changed to 8 m/s at t = 10 s. The simulation results are presented only for the proposed controller to show the system performance under normal grid conditions.

Case 2: The fault condition of 10% voltage sag is created in a-phase voltage at $t = 5$ s, and the notable system performance for the rated wind velocity is depicted in Fig. 9. It is clear that with proposed controller the amount of oscillation is considerably suppressed in real and reactive power of the grid.

Case 3: The PMSG operation under symmetrical three-phase fault is tested by imposing a 70% Type-A fault (three-phase fault) with zero phase jump. Figure 10 shows the comparative analysis for three controllers. Controller (iii) maintains the dc-link voltage constant irrespective of the fault condition, whereas controllers (i) and (ii) fails to maintain the dc-link voltage constant. However, the simulation study is also carried out with 90 and 80% of Type-A fault, where the controllers (i) and (ii) maintained the dc-link voltage constant. Hence it is clear that even for higher symmetrical voltage sag the proposed controller works effectively.

Case 4: A system performance index for Type-C voltage sag is tested by creating voltage sag magnitude and phase-angle jump of 80% and −80, respectively. Figure 11

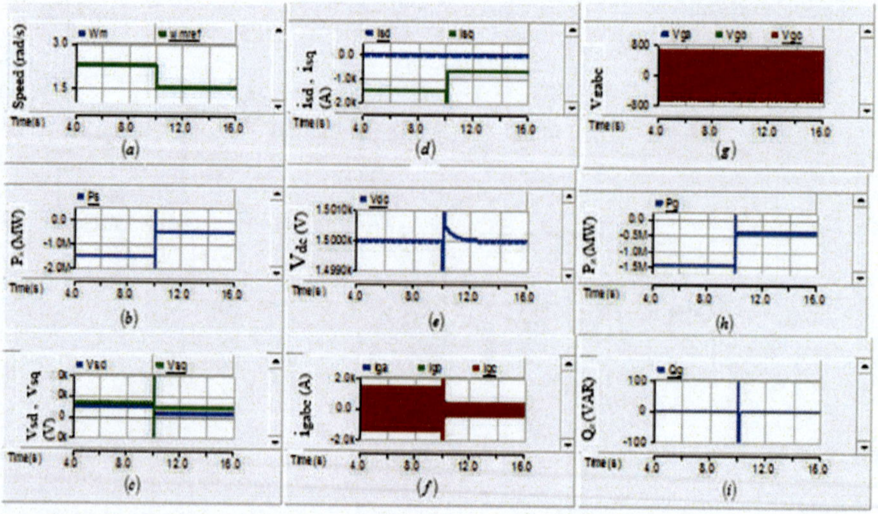

Fig. 8 Performance of PMSG system during wind speed change from 12 to 8 m/s: **a** Actual (ω_m) and reference (ω_{mref}) speed; **b** PMSG Stator Power (MW); **(c)** Stator voltages in *d–q* frame (V); **d** Stator current in *d–q* frame (A); **e** dc-link voltage (V); **f** Grid currents (A); **g** Grid phase voltages (V); **h** Grid real power (kW); **i** Grid reactive power (VAR)

gives the comparative study of three controllers. With controller (iii), the percentage of oscillation in dc-link voltage, real power, and reactive power of gird is reduced considerably compared to other controllers.

Case 5: A 7% of 5th and 7th harmonics are injected into the grid voltage to test the effectiveness of the system performance. Figure 12 shows the dc-link voltage and real power injected into the grid. The amount of oscillation in dc-link voltage and grid real power is measured to be 0.008% and 4.1%, respectively.

Also, the performance analysis is carried out for other different fault conditions of phase to phase (Type C and D) and two-phase to ground (Type F) faults. The numerical value of system parameters comparison for three controllers is presented in Table 1. The simulation study is performed for a voltage sag magnitude of 90%, 70%, and 50% with their corresponding phase-angle jump (φ) of −5.70, −9.70 and −12.50, respectively. For 50% voltage sag magnitude Type F fault with phase – angle jump of 00, controllers (i), and (ii) failed to maintain the dc-link voltage constant. A similar case arises for controller (i) under Type C fault with φ = -12.50.

5 Conclusion

A modified vector controller with voltage and current oscillation cancellation blocks is proposed in this paper to enhance the FRT characteristics of the PMSG-based

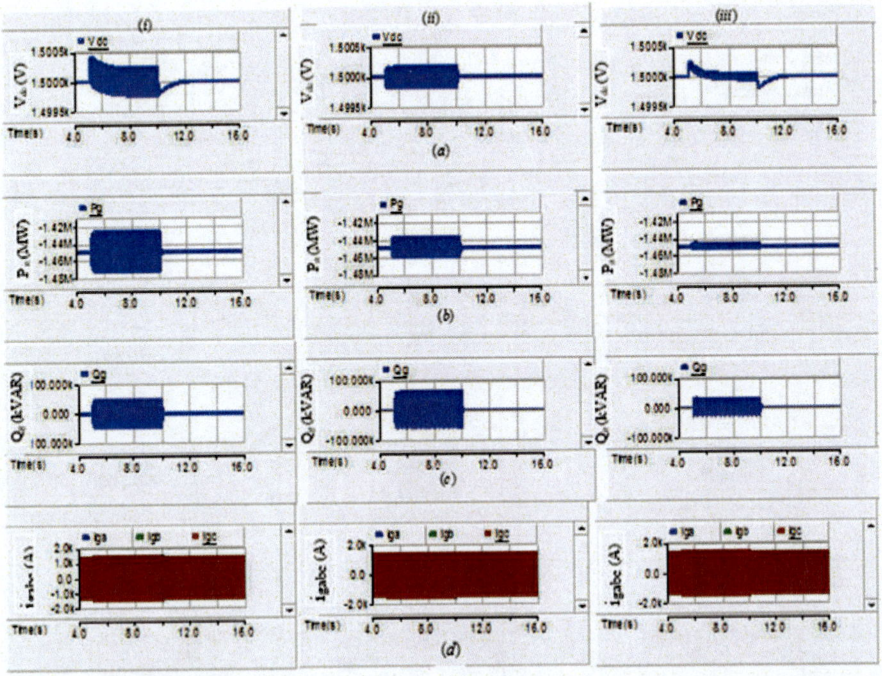

Fig. 9 Performance of controllers (i) Conventional vector control; (ii) current oscillation cancellation method and (iii) Proposed method during a voltage dip of 10% in 'a-phase': **a** dc-link voltage (V); **b** Grid real power (MW); **c** Grid reactive power (VAR); **d** Grid currents (A)

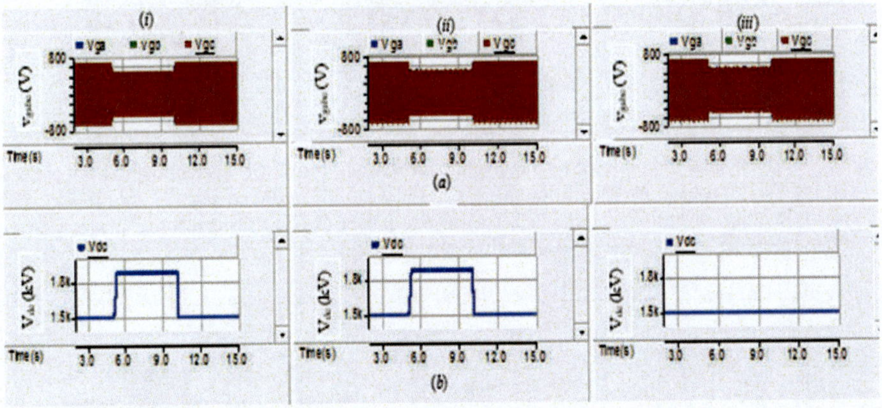

Fig. 10 Performance of controllers (i) Conventional vector control; (ii) current oscillation cancellation method and (iii) Proposed method during Type-A (3ϕ symmetrical) fault with 70% voltage sag magnitude: **a** Grid phase voltages (V); **b** dc-link voltage (V)

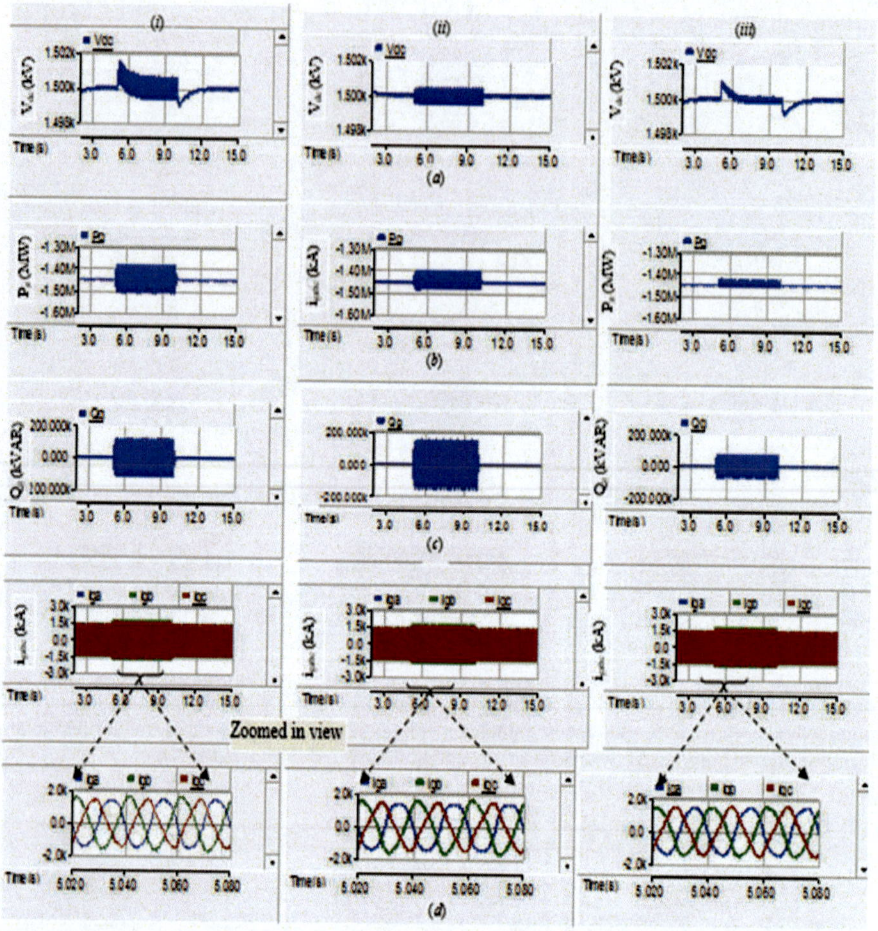

Fig. 11 Performance of controllers (i) conventional vector control; (ii) current oscillation cancellation and (iii) Proposed metho during Type-C fault with 80% voltage sag magnitude and −80 phase-angle jump: **a** dc-link voltage (V); **b** Grid real power (kW); **c** Grid reactive power (VAR); **d** Grid currents (A); **e** Zoomed in view of grid currents (A)

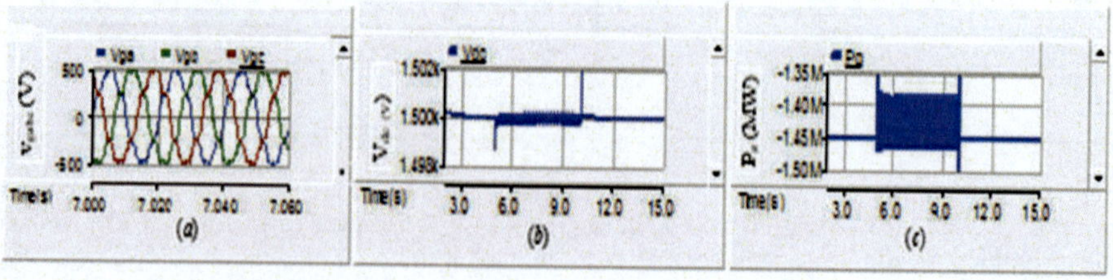

Fig. 12 Simulated results with distorted grid voltage: **a** Grid phase voltages (V); **b** dc-link voltage (V); **c** grid real power (MW)

Table 1 Summarized controller's performance for different types of faults

Type of fault			% of oscillation in V_{dc} and P_g					
			Controller (i)		Controller (ii)		Controller (iii)	
Voltage Sag magnitude of 90%		φ	V_{dc}	P_g	V_{dc}	P_g	V_{dc}	P_g
	Type-C	0°	0.057	6.1	0.023	4.4	0.023	2.6
		−5.7°	0.0307	2.7	0.0307	2	0.0066	1.3
	Type-D	0°	0.024	2.58	0.018	1.42	0.003	0.63
		−5.7°	0.028	2.95	0.039	1.57	0.02	0.6
	Type-F	0°	0.032	3.97	0.025	2.42	0.006	1.24
		−5.7°	0.0169	1.92	0.0124	1.11	0.002	0.054
Voltage Sag magnitude of 70%	Type-C	0°	0.153	10.3	0.063	7.03	0.022	4.2
		−9.5°	0.021	11.3	0.0072	9.7	0.0032	5.7
	Type-D	0°	0.142	9.9	0.06	6.05	0.020	4.1
		−9.5°	0.166	10.52	0.066	6.41	0.023	4.73
	Type-F	0°	0.0413	4.9	0.028	3.17	0.008	1.93
		−9.5°	0.156	8.8	0.05	5.87	0.017	3.9
Voltage Sag magnitude of 50%	Type-C	0°	0.866	32.9	0.233	13.9	0.123	11.4
		−12.5°	–	–	0.124	17.8	0.032	16.6
	Type - D	0°	0.848	33.2	0.2	13.7	0.066	12.7
		−12.5°	0.866	33.5	0.327	14.89	0.066	10.5
	Type-F	0°	–	–	–	–	0.09	10.7

'–' controller failed to maintain dc-link voltage constant

wind power systems. The system rating of 1.5 MW is considered for testing the controller performance. The control scheme is implemented without the need for dual loop vector control, and a simple analytical method is used to determine the current oscillating components. The effectiveness of the control scheme is validated for various types of grid faults with various types of voltage sag and also with the distorted grid voltage. The superior performance of the proposed controller in comparison with the conventional controllers is proved.

Appendix

See Table 2.

Table 2 Types of fault

Types of faults	Voltage expression
A	$V_a = V$ $V_b = -\frac{1}{2}V - j\frac{\sqrt{3}}{2}V$ $V_c = -\frac{1}{2}V + j\frac{\sqrt{3}}{2}V$
C	$V_a = E_1$ $V_b = -\frac{1}{2}E_1 - j\frac{\sqrt{3}}{2}V$ $V_c = -\frac{1}{2}E_1 + j\frac{\sqrt{3}}{2}V$
D	$V_a = V$ $V_b = -\frac{1}{2}V - j\frac{\sqrt{3}}{2}E_1$ $V_c = -\frac{1}{2}V + j\frac{\sqrt{3}}{2}E_1$
F	$V_a = V$ $V_b = -\frac{1}{2}V - j\sqrt{3}(\frac{1}{3}E_1 + \frac{1}{6}V)$ $V_c = -\frac{1}{2}V + j\sqrt{3}(\frac{1}{3}E_1 + \frac{1}{6}V)$

References

1. Central Electricity Regulatory Commission, Ministry of Power Government of India (Indian Electricity Grid Code) New Delhi, Apr 2010
2. Polinder H, Van der Pijl FFA, De Vilder GJ et al (2006) Comparison of direct-drive and geared generator concepts for wind turbines. IEEE Trans Energy Convers 21(3):725–733
3. Chinchilla M, Arnaltes S, Burgos J (2006) Control of permanent-magnet generators applied to variable-speed wind-energy systems connected to the grid. IEEE Trans Energy Convers 21(1):130–135
4. Morneau J, Abbey C, Joos G (2007) Effect of low voltage ride through technologies on wind farm. In: Proceedings of IEEE electrical power conference, Canada, pp 56–61
5. Conroy JF, Watson R (2007) Low-voltage ride-through of a full converter wind turbine with permanent magnet generator. IET Renew Power Gener 1(3):182–189
6. Hansen AD, Michalke G (2009) Multi-pole permanent magnet synchronous generator wind turbines grid support capability in uninterrupted operation during grid faults. IET Renew Power Gener 3(3):333–348
7. Nguyen TH, Lee D-C (2013) Advanced fault ride-through technique for PMSG wind turbine systems using line-side converter as STATCOM. IEEE Trans Ind Electron 60(7):2842–2850
8. Muyeen SM, Takahashi R, Murata T, et.al (2010) A variable speed wind turbine control strategy to meet wind farm grid code requirements. IEEE Trans Power Syst 25(1):331–340)
9. Baroudia JA, Dinavahi V, Knighta AM (2007) A review of power converter topologies for wind generators. Renewable Energy 32(14):2369–2385
10. Yaramasu V, Wu B, Alepuz S et al (2014) Predictive control for low voltage ride-through enhancement of three-level boost and NPC converter based PMSG wind turbine. IEEE Trans Ind Electron 61(12):6832–6843
11. Freire NMA, Marques Cardoso AJ (2014) Fault-tolerant PMSG drive with reduced DC-link ratings for wind turbine applications. IEEE Trans Power Electron 2(1):26–34
12. Fatu Marius, Blaabjerg Frede, Boldea Ion (2014) Grid to standalone transition motion-sensorless dual-inverter control of PMSG With asymmetrical grid voltage sags and harmonics filtering. IEEE Trans Power Electron 29(7):3463–3472

13. Moawwad A, El Moursi MS, Xiao W (2014) A novel transient control strategy for VSC-HVDC connecting offshore wind power plant. IEEE Trans Sustain Energy 5(4):1056–1069
14. Erramia Y, Ouassaid M, Maaroufia M (2013) Control of a PMSG based wind energy generation system for power maximization and grid fault conditions. Energy Procedia 42:220–229
15. Li S, Timothy A, Haskew et al (2012) Optimal and direct-current vector control of direct-driven PMSG wind turbines. IEEE Trans Power Electron 27(5):2325–2337
16. Yuan X, Wang F, Boroyevich D et al (2009) DC-link voltage control of a full power converter for wind generator operating in weak-grid systems. IEEE Trans Power Electron 24(9):2178–2192
17. Geng H, Yang G, Xu D et al (2011) Unified power control for PMSG-based WECS operating under different grid conditions. IEEE Trans Energy Convers 26(3):822–830
18. Kim K-H, Jeung Y-C, Lee D-C (2012) LVRT scheme of PMSG wind power systems based on feedback linearization. IEEE Trans Power Electron 27(5):2376–2384
19. Alizadeh O, Yazdani A (2013) A strategy for real power control in a direct-drive PMSG-based wind energy conversion system. IEEE Trans Power Deliv. 28(3):1297–1305
20. Blasko V, Kaura V (1997) A new mathematical model and control of a three-phase AC–DC voltage source converter. IEEE Trans Power Electron 12(1):116–123
21. Golestan S, Ramezani M, Guerrero JM (2014) Moving average filter based phase-locked loops: performance analysis and design guidelines. IEEE Trans Power Electron 29(6):2750–2763, (2014)
22. Robles E, Ceballos S, Pou J et al (2010) Variable-frequency grid-sequence detector based on a quasi-ideal low-pass filter stage and a phase-locked loop. IEEE Trans Power Electron 25(10):2552–2563
23. Bollen MHJ, Olguin G, Martins M (2005) Voltage dips at the terminals of wind power installations. Wind Energy 8(3):307–318
24. Bollen MHJ, De Graaff RAA (1999) Behavior of AC and DC drives during voltage sags with phase-angle jump and three-phase unbalance. In: Proceedings of IEEE Power Engineering Society Winter Meeting Conference, pp 1225–1230
25. Bollen MHJ (2000) Understanding power quality problems voltage sags and interruptions. IEEE Press, New York
26. Boldea I (2006) Variable speed generators. Taylor & Francis, Boca Raton
27. Vijayapriya R, Raja P, Selvan MP (2018) A modified active power control scheme for enhanced operation of PMSG based WGs. IEEE Trans Sustain Energy 9(2):630–638

Phase Balancing of DG-Integrated Smart Secondary Distribution Network

Swapna Mansani, R. Y. Udaykumar, Santoshkumar, M. A. Asha Rani, and S. Sreejith

Abstract Distribution system is a complicated and significant contributor to the overall losses of power system, wherein the loss due to unbalanced loading at each bus is often overlooked and is vital to consider the same. Unbalance in the system can be mitigated by using one or more of the following techniques: network reconfiguration, Phase Balancing (PB), reactive compensation, distributed generations (DG) integration, etc. With the evolution of smart meters, PB of whole feeder by rephasing Consumer Service Mains (CSM) is achievable. As renewable generation at consumer level is making headway, balancing a distribution system by considering consumer load pattern in DG-integrated system is essential. With this motivation, the effect of PB at each bus in the presence of DG on system losses is examined and investigated in this paper. It is achieved through the development of phase balancing algorithm (PBA) in combination with backward sweep technique requiring an overall lesser computational effort. The developed algorithm is tested in MATLAB considering the IEEE 13 bus feeder along with optimally placed DGs. The performance of the developed algorithm is evaluated with an objective to reduce line losses, the number of rephasings, branch current and neutral current, thereby to enhance the bus voltage profile.

Keywords Distributed generation · Phase Balancing · Secondary distribution network

S. Mansani (✉) · M. A. Asha Rani · S. Sreejith
Department of Electrical Engineering, National Insitute of Technology Silchar, Silchar, Assam, India
e-mail: swapnamansani@gmail.com

R. Y. Udaykumar
Malaviya National Institute of Technology Jaipur, Jaipur, Rajasthan, India

Santoshkumar
School of Electrical and Electronics Engineering, REVA University, Bengaluru, Karnataka, India

N. Zhou and S. Hemamalini (eds.), *Advances in Smart Grid Technology*, Lecture Notes in Electrical Engineering 688, https://doi.org/10.1007/978-981-15-7241-8_23

1 Introduction

The increase in population, urbanization, depleting fossil fuels, and increased environmental concerns has led to the concept of energy saving and DG utilizing green resources. The most effective approach for meeting energy demands (when power demand is more than the generation) is through efficient use of energy and its conservation. The typical power system with different levels of voltage at each step comprising generation, transmission, and distribution (T&D) is shown in Fig. 1.

In the process of T&D of power, we come across a lot of losses due to transformers, lines, protection equipment, power factor, etc. The losses in the distribution system are relatively high as compared to other loss contributors in power system and are to be attended to make the system more efficient by reducing them.

These losses are categorized into: conductor impedance (unbalanced loading, network designing, I^2R loss), improper earthing, low power factor (inductive loads, converters), equipment losses (distribution transformer core and winding losses, energy metering losses), protection and monitoring device losses. The loss due to unbalanced loading is one of the significant contributors to distribution loss, and the methods ameliorating the same need more emphasis.

Distribution networks are often unbalanced due to feeder laterals design/structure (single- and two-phase feeder laterals), and unequal loading by CSM (single, two and three-phase CSM) as shown in Fig. 2. The effects of unbalance on system performance are (i) reduced system peak load supplying capability, (ii) deteriorated transformer's, and other equipment's lifespan, (iii) increased neutral current, (iv) poor voltage regulation, and (v) mal-operation of the protection system.

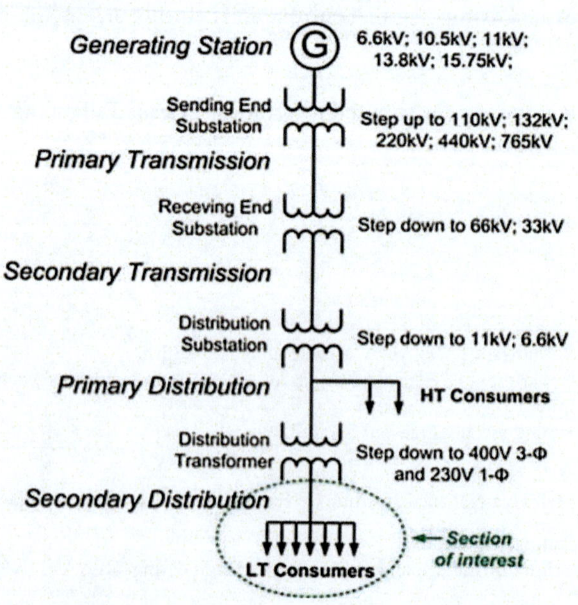

Fig. 1 Single line diagram of the power system

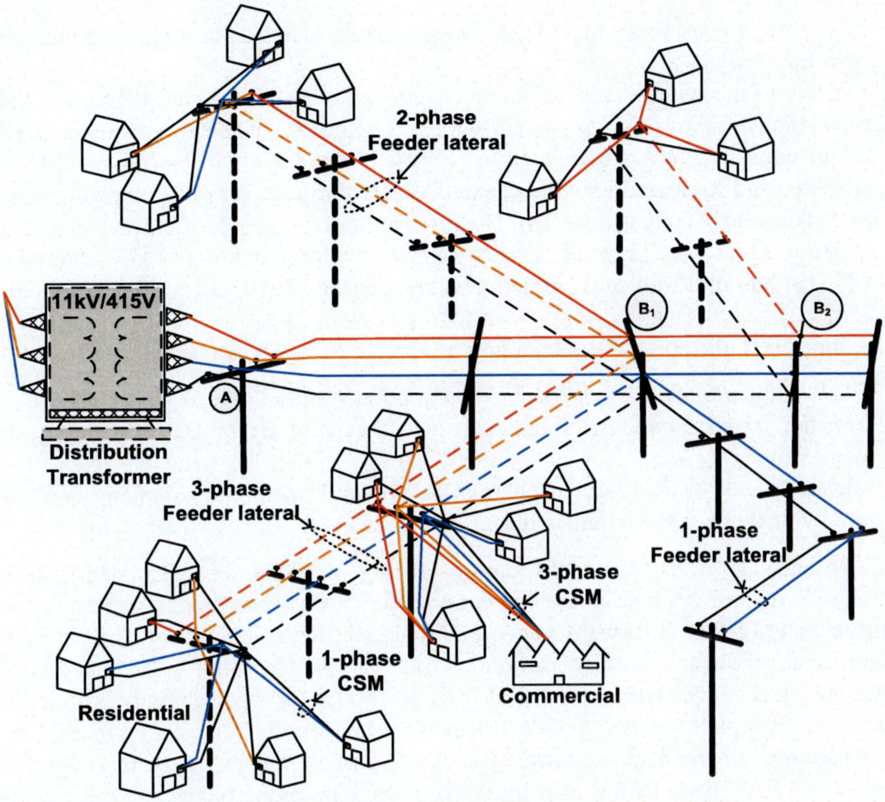

Fig. 2 Schematic diagram of secondary LV distribution network

The load unbalances can be addressed using two approaches: (i) Network reconfiguration: It is carried out at the system level by proper switching of sectionalizing and tie switches. The load from the heavily loaded area is transferred to the lightly loaded area. (ii) PB: It is carried out at feeder level by rephasing the CSM loads among the phases with a motto to have all three phases equally loaded. Among the approaches mentioned above, the former leads to heavy current flow and voltage unbalance in some cases and also may not result in effective balancing among the phases as it addresses unbalance due to feeder lateral. Whereas the latter approach reduces the unbalance problem at the location of unbalance instigation (unbalanced loading by CSMs).

In general, the implementation of such rephasings is achieved through algorithms referred to as PBAs. The various programming technique-based PBAs till date are using mixed-integer programming technique [1], simulated annealing [2], genetic algorithm [3], [4], heuristic backtracking search [5], [6] immune algorithm [7], bacterial foraging oriented by particle swarm optimization [8], fuzzy logic [9], [10, 11], automated mapping, facilities management, geographic information system

[12], hybrid greedy fuzzy logic [13], and self-adaptive hybrid differential evolution algorithms [14].

A few of the widely used objective functions for the above said PBAs are: the critical branch PB of single-phase CSM loads to minimize the maximum unbalanced flow of branch current is computed in [1, 15]. The average unbalance per phase is calculated and compared with the threshold value, further a fuzzy logic technique is applied for load balancing in [9, 16]. The PB to minimize power loss and phase current deviation is presented in [3, 14, 17]. The phase unbalance index (PUI) is mentioned in [5, 18] and multi-objective function is reported in [8, 19, 20], utilization factor [21].

In nutshell, the above PB techniques as applied to Fig. 2 can be categorized as:

Substation alone balancing: Balancing at point A only.
Partial feeder balancing: Balancing at feeder point B1 or B2 individually or collectively.
Whole feeder balancing: In contrast to previous methods, balancing at each individual bus refers to entire feeder balancing.

Recently, PBAs have demonstrated balancing of whole secondary distribution feeder by rephasing single-phase and three-phase CSMs, smart meters, and capacitor placement [22–28]. Backward sweep PBA balances the whole feeder using simple mathematical formulations by considering time-varying individual CSM load profile and movable feeder laterals [23, 24]. The Backward sweep technique is used to achieve a complete balancing of the selected bus accounting downstream loads. The effectiveness of the backward sweep PBA (i.e., whole feeder balancing) over the existing PBAs given in the literature (like substation alone balancing and partial feeder balancing) is demonstrated for the secondary distribution system without DG integration.

Owing to the current integration of many renewable energy sources like a roof-top photovoltaic, electric vehicle, biogas, etc., into the distribution utility, it is imperative to include their presence in PB, which is worth investigating.

This paper aims at developing a backward sweep PBA for DG-integrated smart secondary distribution system. One of the ways to achieve this is by considering such DGs as the negative lumped load in the process of balancing the bus while keeping the remaining aspects intact. The proposed algorithm is applied to the IEEE standard unbalanced 13 bus test feeder with optimally placed DG [29] to demonstrate its performance with the motive of line loss reduction, enhancement of the voltage profile, diminution of neutral current, and a minimum number of phase moves.

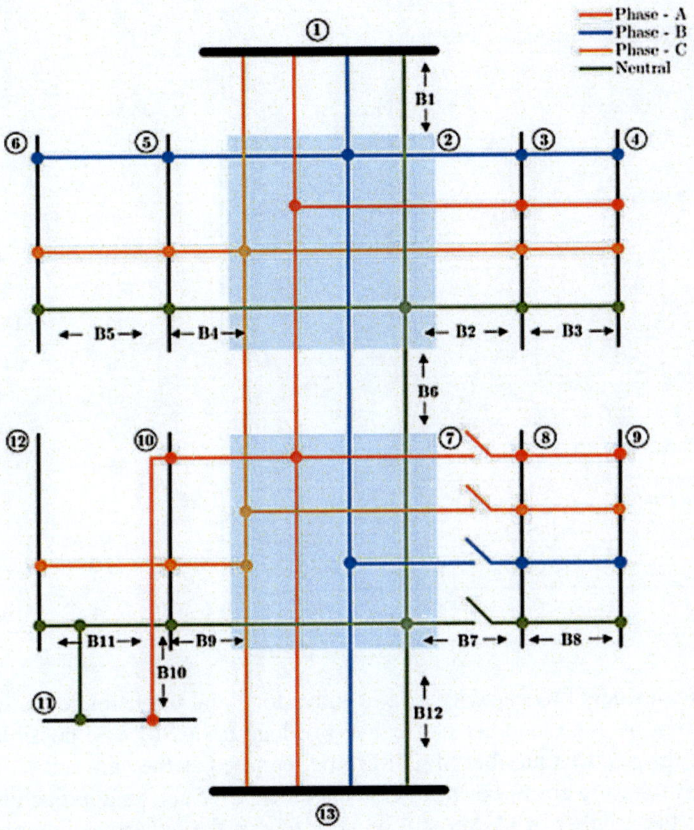

Fig. 3 Renumbered IEEE 13 bus test feeder

2 System Considered for the Study: IEEE 13 Bus Test Feeder

The IEEE 13 bus system [29] shown in Fig. 3 is an unbalanced distribution system with a base voltage of 4.16 kV and load data is enlisted in Table 1. The load data considered for the study is generated randomly (aggregate load at each bus is equal to the load data of IEEE 13 bus test feeder) as PB needs individual CSMs load data at each bus.

3 Proposed PBA Using Backward Sweep Technique

The proposed algorithm aims to obtain the rephasing arrangement of single-phase CSMs and feeder laterals in a DG-integrated secondary distribution system. Here,

Table 1 IEEE 13 bus test feeder load data

Spot load

Bus	Phase A		Phase B		Phase C	
	kW	kVAr	kW	kVAr	kW	kVAr
4	160	110	120	90	120	90
5	0	0	170	125	0	0
6	0	0	230	132	0	0
7	385	220	385	220	385	220
8	0	0	0	0	170	151
9	485	190	68	60	290	212
11	128	86	0	0	0	0
12	0	0	0	0	170	80
Total	1158	606	973	627	1135	753

Distributed load

Bus 1	Bus 2	Phase A		Phase B		Phase C	
		kW	kVAr	kW	kVAr	kW	kVAr
2	7	17	10	66	38	117	68

algorithm considers DG capacity as a negative load and uses simple mathematical terms such as average load per phase, average load per CSM, and possible combinations of the required number of CSM to be rephased as they are self-sufficient to obtain the rephasing arrangement of CSM at each bus. Such an approach is considered since the number of CSMs will be very few as the balancing is performed at each individual bus.

As a first step, the average load per phase and average load per single-phase CSM are computed. The amount of load to be moved in/out of the phase is the difference between the existing load on phase and average load per phase. The approximate number of CSMs to be moved in/out is estimated based on the computed values. Secondly, all the possible combinations of CSMs are computed for the obtained number of CSMs to be moved. Finally, that combination of CSMs resulting in the overall loading equal to the computed amount of load to be moved is selected. As per the selected combination, the respective CSMs are rephased. The same is shown in Figs. 4 and 5, respectively. Case A of Fig. 4 is presented in Fig. 5 and is similar to other cases from B to F.

However, balancing at each bus is considered to treat the instigation of the unbalance problem at its root level. Further to achieve whole feeder balance to reduce all branch currents and neutral current, single-phase CSMs are balanced at each bus using a backward sweep technique by considering single-phase and two-phase feeder laterals as movable from tail end to the feeder origination. In the backward sweep, the load on the rephased bus is considered as a lumped/immovable load at the next monitoring bus. Thereby this combination results in network balance with respect to any bus as swept from the forward end.

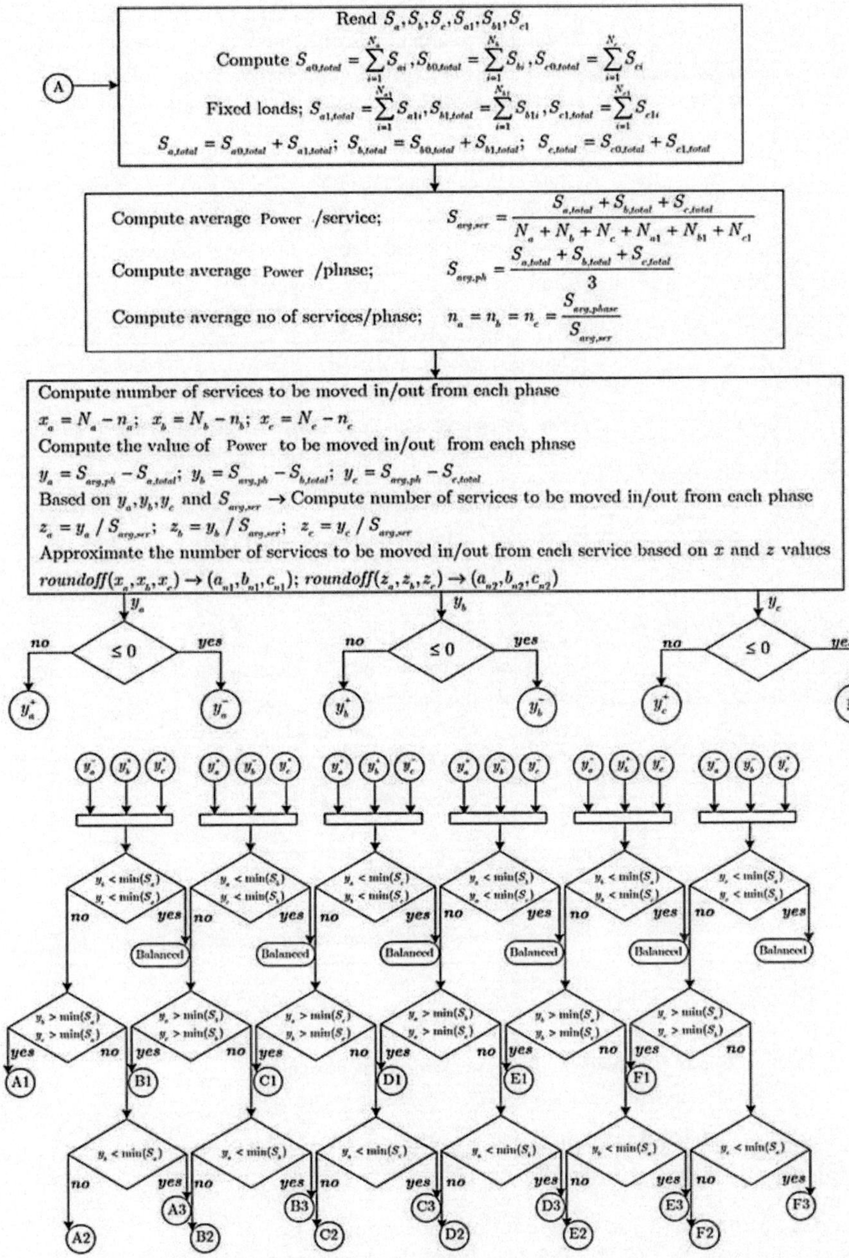

Fig. 4 Proposed single-phase CSMs PBA flowchart

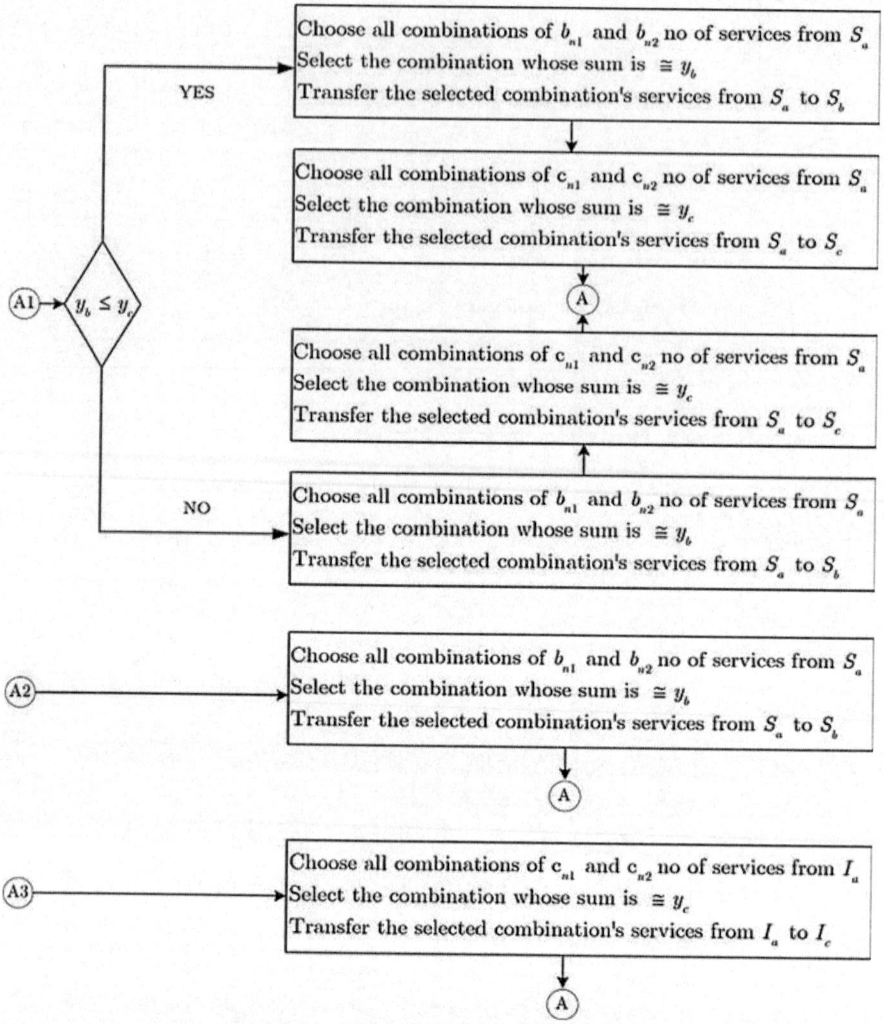

Fig. 5 Continuation of proposed single-phase CSMs PBA flowchart

The process of PB from tail end to the origin of the feeder is explained below by considering IEEE 13-bus test feeder.

- As the load at bus 13 is absent, it does not require PB.
- Since the branches 12–10 and 11–10 are single-phase feeder laterals, bus 11 and 12 does not require PB.
- Bus 9 is of three-phase nature having both three-phase and single-phase CSMs. Balancing of single-phase CSMs is carried out by assuming three-phase consumers load as lumped load irrespective of its balance status.

- Bus 8 is of three-phase natures consisting of only single-phase CSMs on phase C. Thus, a few CSMs on phase C are transferred to phase A and B by accounting load at bus 9 as lumped.
- Bus 7 is of three-phase nature having both three-phase and single-phase CSMs. Balancing of single-phase CSMs is carried out by considering the load at bus 10 as movable and assuming three-phase consumers load and loads at bus 8 as lumped.
- Bus 6 is of two-phase nature consisting of CSMs on phase B alone. While transferring a few loads on phase B to phase C, bus 6 is balanced.
- PB at bus 5 is accomplished in the same manner as that of bus 6 by considering the load at bus 6 as lumped.
- Balancing at bus 4 is accomplished in the similar way as that of bus 9. The load on bus 4 is reflected as lumped load on bus 3 and thus bus 3 does not require balancing since it has no CSMs.
- Bus 2 is of three-phase nature having both three-phase and single-phase CSMs. Balancing of single-phase CSMs is carried out by considering the load at bus 5 as movable and assuming three-phase consumers load and loads at bus 7 and 3 as lumped.

4 Results and Discussion

PB of single-phase CSM using a backward sweep technique algorithm is developed in MATLAB. The performance of the algorithm is evaluated by considering the following figures of merits like line loss, individual bus loading, branch current, neutral current, bus voltage, and average unbalance index (AUI)

$$\left(\frac{|(S_a - S_b)| + |(S_b - S_c)| + |(S_c - S_a)|}{3}\right).$$

The IEEE 13 bus test system with DG is considered and the backward forward sweep load flow method is employed for computing line loss, branch current, neutral current, and bus voltage. In [30], the IEEE 13-node system was tested to identify the optimal locations and capacities for DGs installation, with an objective function including static voltage stability index, three-phase unbalance index, system reliability index, and DG investment cost; the study showed that bus 11 phase a and bus 12 phase c are the best DG location with DG capacities of 200 kW and 150 kW, respectively [30]. The same results are used here to study the effect of PBA in DG integrated smart secondary distribution network.

Here, four different cases are considered to evaluate the performance of the proposed algorithm. They are (i) System without DG and PB, (ii) System with DG and without PB, (iii) System without DG and with PB, (iv) System with DG and PB.

With the application of backward sweep technique for balancing single-phase CSMs alone, the line losses before and after balancing are 622.04 kW and 516.83 kW,

Table 2 Load profile, AUI, and power loss for PB of single-phase CSMs using backward sweep technique

	Pa	Pb	Pc	Qa	Qb	Qc	AUI	Loss
!DG !PB	1175	1039	1252	616	665	821	175.73	622.04
DG !PB	975	1039	1102	616	665	821	147.28	558.22
!DG PB	1223.47	1204.60	1037.93	705.91	765.20	630.88	141.64	563.14
DG PB	1120.31	1107.76	887.93	761.29	709.83	630.88	176.84	516.83

!DG !PB: without DG and PB; DG !PB: with DG and without PB; !DG PB: without DG and with PB; DG PB: with DG and PB; P (in kW); Q (in kVAR); Loss(in kW)

respectively, with a loss reduction of 16.91%. The feeder load profiles, their AUI, and losses are listed in Table 2 and it can be inferred that there is a decrease in loss and the system has achieved an enhanced balance, which is also evident from the plots in Fig. 6. It is evident from the Fig. 6, neutral current has reduced, voltage profile has improved, and unbalance among phase branch currents is reduced by reducing their line losses. Further results demonstrated that backward sweep PBA is applicable for DG-integrated secondary distribution systems also.

5 Conclusion

PB is one of the preferred methods in secondary distribution energy management. In this work, the development of an algorithm for balancing the single-phase CSMs along with feeder laterals to achieve the whole feeder balancing of DG-integrated secondary distribution system is the prime focus.

On this line, an algorithm was developed for balancing single-phase CSMs using a backward sweep technique by considering feeder laterals as movable at each bus, respectively, with an objective of line loss reduction, minimum number of phase moves, branch current and neutral current reduction, and enhanced voltage profile. Performances of the developed algorithms were evaluated by considering the IEEE 13 bus test feeder system with DG.

Following the successful evaluation of algorithm and the corresponding results presented, it is worth mentioning that for the standard IEEE 13 bus test feeder, PB of single-phase CSM using backward sweep technique by considering movable feeder laterals is found to have an appreciable line loss reduction of 16.91%.

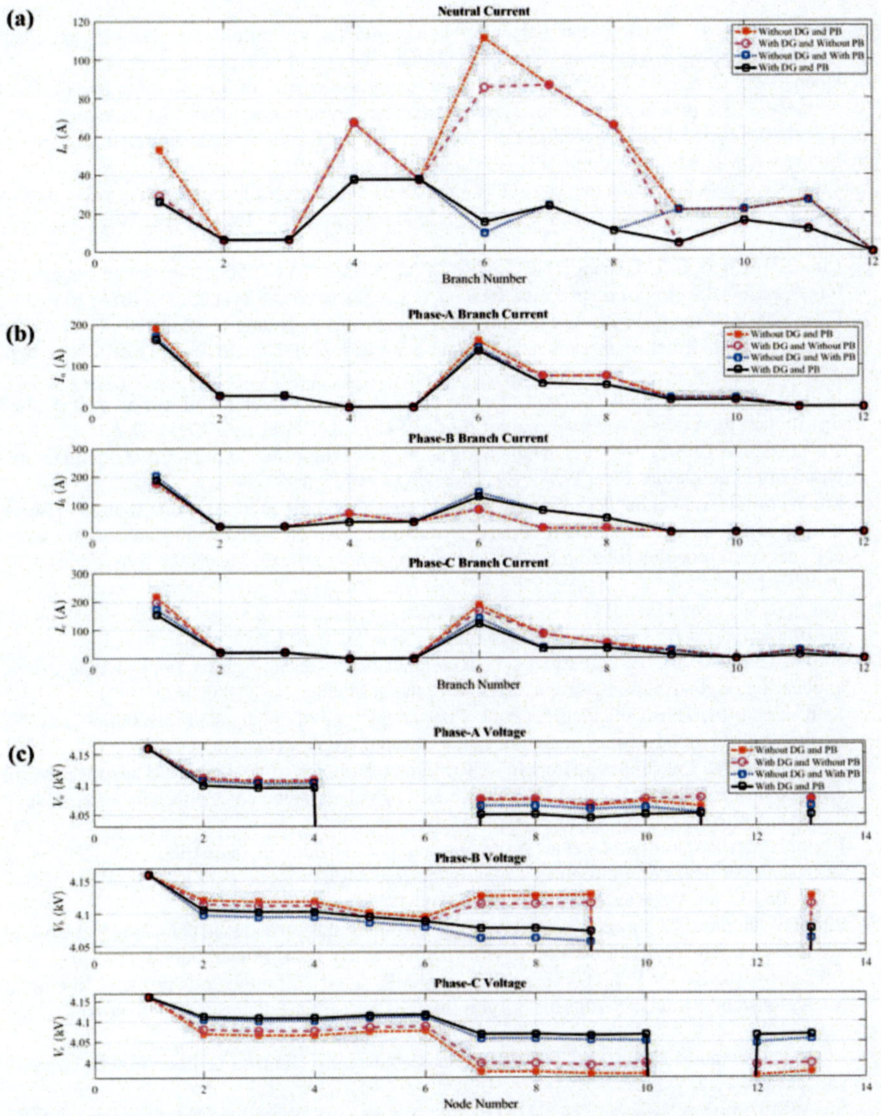

Fig. 6 Results for PB of single-phase CSMs using backward sweep technique : **a** neutral current, **b** branch current, **c** node voltage

References

1. Zhu J, Chow M-Y, Zhang F (1998) Phase balancing using mixed-integer programming distribution feeders. IEEE Trans Power Syst 13(4):1487–1492
2. Zhu J, Bilbro G, Chow M-Y (1999) Phase balancing using simulated annealing. IEEE Trans Power Syst 14(4):1508–1513

3. Gandomkar M (2004) Phase balancing using genetic algorithm. In: 39th International universities power engineering conference (UPEC 2004), vol 1. IEEE, pp 377–379
4. Chen T-H, Cherng J-T (1999) Optimal phase arrangement of distribution transformers connected to a primary feeder for system unbalance improvement and loss reduction using a genetic algorithm. In: Proceedings of the 21st 1999 IEEE international conference on power industry computer applications (PICA'99). IEEE, pp 145–151
5. Lin C-H, Chen C-S, Chuang H-J, Ho C-Y (2005) Heuristic rule-based phase balancing of distribution systems by considering customer load patterns. IEEE Trans Power Syst 20(2):709–716
6. Lin C-H, Chen C-S, Chuang H-J, Huang M-Y, Huang C-W (2008) An expert system for three-phase balancing of distribution feeders. IEEE Trans Power Syst 23(3):1488–1496
7. Huang M-Y, Chen C-S, Lin C-H, Kang M-S, Chuang H-J, Huang C-W (2008) Three-phase balancing of distribution feeders using immune algorithm. IET Gene Trans Distrib 2(3):383–392
8. Hooshmand RA, Soltani S (2012) Fuzzy optimal phase balancing of radial and meshed distribution networks using bf-pso algorithm. IEEE Trans Power Syst 27(1):47–57
9. Ukil A, Siti W (2008) Feeder load balancing using fuzzy logic and combinatorial optimization-based implementation. Elect Power Syst Res 78(11):1922–1932
10. Siti W, Jimoh A, Nicolae D (2009) Phase load balancing in the secondary distribution network using a fuzzy logic and a combinatorial optimization based on the Newton Raphson. In: International conference on intelligent data engineering and automated learning. Springer, Berlin, pp 92–100
11. Chitra R, Neelaveni R (2011) A realistic approach for reduction of energy losses in low voltage distribution network. Int J Electr Power Energy Syst 33(3):377–384
12. Kuo C-C, Chao Y-T (2010) Energy management based on am/fm/gis for phase balancing application on distribution systems. Energy Convers Manage 51(3):485–492
13. Sathiskumar M, Thiruvenkadam S (2011) Phase balancing of unbalanced distribution network through hybrid greedy-fuzzy algorithm. Int J Comput Appl 34(6)
14. Sathiskumar M, Lakshminarasimman L, Thiruvenkadam S et al (2012) A self adaptive hybrid differential evolution algorithm for phase balancing of unbalanced distribution system. Int J Electr Power Energy Syst 42(1):91–97
15. Khodr H, Zerpa I, Matos MA et al (2006) Optimal phase balancing in distribution system using mixed-integer linear programming. In: Transmission & distribution conference and exposition (TDC'06), Latin America. IEEE/PES, IEEE, pp 1–5
16. Siti MW, Nicolae DV, Jimoh AA, Ukil A (2007) Reconfiguration and load balancing in the lv and mv distribution networks for optimal performance. IEEE Trans Power Del 22(4):2534–2540
17. Dilek M, Broadwater RP, Thompson JC, Seqiun R (2001) Simultaneous phase balancing at substations and switches with time-varying load patterns. IEEE Trans Power Syst 16(4):922–928
18. Wang K, Skiena S, Robertazzi TG (2013) Phase balancing algorithms. Elect Power Syst Res 96:218–224
19. Siti WM, Jimoh A, Nicolae D (2011) Distribution network phase load balancing as a combinatorial optimization problem using fuzzy logic and Newton–Raphson. Elect Power Syst Res 81(5):1079–1087
20. Schweickardt GA, Miranda GA, Wiman G (2011) A comparison of metaheuristics algorithms for combinatorial optimization problems. Application to phase balancing in electric distribution systems. J Latin Am Appl Res 41(2):113–120
21. Mansouri K, Hamed MB, Sbita L, Dhaoui M (2015) Three-phase balancing in a lv distribution smart-grids using electrical load flow variation sciences and techniques of automatic control and computer engineering (STA). In: 16th international conference on IEEE, pp 427–431
22. Gheorghe G, Bogdan CN, Florina S (2017) Smart metering based approach for phase balancing in low voltage distribution systems. In: 10th international symposium on advanced topics in electrical engineering, Bucharest, Romania, pp 551–554, 23–25 Mar 2017

23. Swapna M, Udaykumar RY (2017) Backward sweep technique based phase balancing algorithm for secondary distribution system. In: IEEE PES 9th Asia-Pacific power and energy engineering conference (APPEEC), Bengaluru, Karnataka, India)
24. Swapna M, Udaykumar RY (2018) An improved phase balancing algorithm for secondary distribution system with consumer load patterns. J Adv Res Dynam Control Syst 10(5):605–612
25. Ding F, Mirrasoul JM (2018) On per-phase topology control and switching in emerging distribution systems. IEEE Trans Power Delivery 33(5):1–10
26. Hao J, Yifan S, Suzhou G, Tianxuan L (2018) Loss reduction strategy of low-voltage distribution network based on phase swapping. In: 2nd international symposium on resource exploration and environmental science, IOP conference series: earth and environmental science, vol 170, pp 1–6
27. Jorge A, Maria C, Daniel T, Javier G, John EC-B (2019) Historical load balance in distribution systems using the branch and bound algorithm. Energies 12(1219):2–14
28. Laconico KCC, Aguirre Jr RA (2019) Optimal load balancing and capacitor sizing and siting of an unbalanced radial distribution network. In: Proceedings of the 2019 IEEE PES GTD Asia, pp 939–944
29. Dugan R (2017) Distribution test feeders-distribution test feeder working group. In: IEEE PES Distribution System Analysis Subcommittee, EWH IEEE org (2017)
30. Yu Z, Zhao YD, Ke M, Hongming Y, Mingyong L, Kit PW (2017) Multi-objective distributed wind generation planning in an unbalanced distribution system. CSEE J. Power Energy Syst 3(2):186–195

Estimation of Payback Period Incorporating SVC and TCSC in SCUC Problem

S. Sreejith, M. A. Asha Rani, and Swapna Mansani

Abstract This paper focuses in estimating the payback period and percentage recovery with static var compensator (SVC) and thyristor controller series compensator (TCSC) in an SCUC problem applying artificial bee colony (ABC) algorithm. ABC algorithm mimics the foraging behavior of employed, unemployed, and scout bee in a bee colony system. SCUC problem aims in reducing the operating cost satisfying all system, unit, and security constraints. The SCUC problem is disintegrated as master problem (unit commitment—UC) and subproblem (security-constrained economic dispatch—SCED). The capability of SVC and TCSC to regulate the power flow in lines and alleviating the overloading is exploited here. The optimal location of the SVC and TCSC in a power system network is clearly illustrated. The profit obtained with SVC, TCSC, and the payback period of these devices are explained with respect to installation cost. The gain obtained with SVC and TCSC including annual maintenance cost for a certain period of time is illustrated. SVC and TCSC are modeled based on variable reactance modeling technique and incorporated in IEEE 118-bus and south Indian 86 bus systems at their optimal locations. The total generation cost obtained by solving SCUC using the heuristic approach is compared with existing methods in literature.

Keywords Security-constrained unit commitment · Equated monthly installment (EMI) · Artificial bee colony · Line outage

Nomenclature

$\mathrm{BF}_{(i,t)}$	Power flow through ith branch time t (MVA)
$\mathrm{BF}_i^{\max}$	Maximum power capacity limit for branch i (MVA)
$F_i(P_{i,t})$	Total generation cost ($)
a, b, c	Generator (ith) cost coefficient

S. Sreejith (✉) · M. A. Asha Rani · S. Mansani
Department of Electrical Engineering, National Insitute of Technology Silchar, Silchar, Assam, India
e-mail: sreejithsme@gmail.com

N. Zhou and S. Hemamalini (eds.), *Advances in Smart Grid Technology*, Lecture Notes in Electrical Engineering 688, https://doi.org/10.1007/978-981-15-7241-8_24

D_t	Total power demand at time t
fit_p	Fitness function of the solution p
G_{ij}, B_{ij}	Y bus values between bus i and bus j
$I_{(i,t)}$	Commitment status of ith generation unit at tth hour
N	Number of generator units
N_{B}	Number of busses
$N_{\text{B}-1}$	Total number of buses without slack bus
N_{PQ}	Number of load buses
$P_{\text{G}(i,t)}$, $Q_{\text{G}(i,t)}$	Power generation at ith bus at time t
$P_{\text{D}(i,t)}$, $Q_{\text{D}(i,t)}$	Power demand at ith bus at time t
$P_{(i,\min)}$, $P_{(i,\max)}$	Power output limits of ith generator (MW)
$Q_{\text{G}i}^{\min}$, $Q_{\text{G}i}^{\max}$	Reactive power generation limits for i^{th} generating unit
$R_{(i,t)}$	Spinning reserve in MW/hour at tth hour
SU_i^t, SD_i^t	Startup and shut down cost of ith generator unit at time t,
$T_{(i)}^{\text{on}}$, $T_{(i)}^{\text{off}}$	Up and down time of ith generator
T	Generation dispatch time (hours)
$\text{UR}_{(i)}$, $\text{DR}_{(i)}$	Ramp up and ramp down rate limits for ith generator
$V_i^{\min}$, $V_i^{\max}$	Limits of voltage magnitude at bus i (pu)
$X^{\text{on}}(i,t)$	"ON" duration till time t for ith generator
$X^{\text{off}}(i,t)$	"OFF" duration till time t for ith generator
i	Generator index
t	Time index
Ne	Number of bee positions
B_{svc}, X_{TCSC}	Reactance of SVC and TCSC
i	Interest rate per annum
L	Period for which EMI has to be paid in months
S	Range of operation of FACTS devices (MVAR)
$Q2$	Reactive power flowing in transmission line with FACTS devices (MVAR)
$Q1$	Reactive power flowing in transmission line without FACTS device (MVAR)

1 Introduction

Unit commitment problem (UCP) is an important decision-making problem for the steady and economic operation of power system. The aim of UC problem is to identify optimum schedule to commit the generators, while minimizing the cost of power dispatched. This is obtained by satisfying the unit and system constraints. When scheduling of the generators are carried out, it is ensured that the line flows and bus voltages are within the specified limits. Such a commitment is called as security-constrained unit commitment (SCUC) [1]. Flexible AC transmission systems (FACTS) devices are mostly employed for increasing the load ability of the existing transmission lines [2]. UCP along with transmission security limits is termed as SCUCP. In [3], Lagrangian relaxation (LR) method is used to solve SCUC problem, which results in unreasonable relaxations when used for discrete variables.

Hybrid subgradient Dantzig--Wolfe method, an efficient decomposition method, is used to solve long-term SCUC in [4]. Coupled flow and system voltage constraints are incorporated with SCUC in [5]. In [6], the SCUC problem is decomposed into master problem (UC) and subproblem (security checking) using Benders decomposition, which are finally coordinated through Benders cut. But, excess decomposition leads to suboptimal solution. In [7], SCUC is solved incorporating AC constrains. Here, the augmented Lagrangian relaxation (LR) method and dynamic programming (DP) are applied to solve UC. The subproblem checks AC network security constraints. Semi-definite programming-based method is used to solve SCUC in [8]. Heuristic method could fail as generation constraints become more complicated. The installation of FACTS devices is still limited because of complexity in control and cost factor. The analysis of the recovery cost and payback period of FACTS devices is essential to explain the importance.

The restructured power system in which security and economic conditions are considered to be essential, this paper analyzes the effectiveness and payback benefits of SVC and TCSC when incorporated with SCUC problem.

1.1 Proposed Work

In this paper, SCUC problem incorporating SVC and TCSC is solved using ABC algorithm [9]. The cost recovery with FACTS devices and their payback period is also discussed. The following analyses are carried out by incorporating SVC and TCSC into the power system network

(a) Optimal location of SVC and TCSC is identified on an IEEE 118-bus system and 86 bus system.
(b) The role of SVC and TCSC for improving and controlling the power flow in transmission lines while reducing the total generation cost.
(c) The enhancement of system security by SVC and TCSC during a contingency situation. Here, the contingency is created by opening a heavily loaded transmission line. The ability of the FACTS devices to control the power flow without violating the security constraints during the line outage is also studied.
(d) The cost recovery by the FACTS devices, and their payback period is approximately analyzed with respect to the operating cost.

Here, ABC algorithm, which efficiently solves the constrained optimization problems, is used. UC, the master problem, is solved using binary coded ABC, and the subproblem SCED is solved using real coded ABC algorithm [10].

2 Problem Formulation

The objective of SCUC problem is to reduce the operating cost while satisfying the system, network, and security constraints. TCSC and SVC are incorporated in the network for improving the power transmission. Considering the installation cost of SVC and TCSC, it is assumed that an equated monthly installment (EMI) is paid for a specific span of time. This EMI calculation term is incorporated into the SCUC problem formulation. This problem can be formulated as

Minimize

$$\sum_{t=1}^{T}\left(\sum_{i=1}^{N}\left\{F_i(P_{i,t})\cdot I_{i,t}+\mathrm{SU}_i^t\cdot I_{i,t}\cdot[1-I_{i,t-1}]+\mathrm{SD}_i^t\cdot[1-I_{i,t}]\right\} + \frac{Q}{720}\left[i*\left(\frac{(1+i)^L}{(1+i)^L-1}\right)+\frac{M}{12}\right]\right) \tag{1}$$

$$F_i(P_{i,t}) = a + b * P_{i,t} + c * P_{i,t}^2 \tag{2}$$

$$Q = 0.0003S^2 - 0.3051S + 127.38 \quad \text{(For SVC)} \tag{3}$$

$$Q = 0.0015S^2 - 0.7130S + 153.75 \quad \text{(For TCSC)} \tag{4}$$

where $S = |Q_2| - |Q_1|$

Subjected to

1. Power balance constraints

$$P_{\mathrm{G}i,t} - P_{\mathrm{D}i,t} = V_{i,t}\sum_{j=1}^{N} V_{j,t}(G_{ij}\cos\theta_{ij} + B_{ij}\sin\theta_{ij}) \quad i \in N_{\mathrm{B}-1} \tag{5}$$

$$Q_{\mathrm{G}i,t} - Q_{\mathrm{D}i,t} = V_{i,t}\sum_{j=1}^{N} V_{j,t}(G_{ij}\sin\theta_{ij} - B_{ij}\cos\theta_{ij}) \quad i \in N_{\mathrm{PQ}} \tag{6}$$

2. ON/OFF constraints

$$\left[X^{\mathrm{on}}(i,t-1) - T^{\mathrm{on}}(i)\right] * \left[I_{(i,t-1)} - I_{(i,t)}\right] \geq 0 \tag{7}$$

$$\left[X^{off}(i,t-1) - T^{off}(i)\right] * \left[I_{(i,t)} - I_{(i,t-1)}\right] \geq 0 \tag{8}$$

3. Generating capacity and ramp up/down constraints

$$P_{i,t}^{r,\min} = \max\left\{P_i^{\min}, P_{i,t-1} - DR_i\right\} \tag{9}$$

$$P_{i,t}^{r,\max} = \max\left\{P_i^{\max}, P_{i,t-1} + UR_i\right\} \tag{10}$$

$$P_{i,t}^{r,\min} I_{i,t} \leq P_{i,t} \leq P_{i,t}^{r,\max} I_{i,t} \tag{11}$$

4. Spinning reserve constraints

$$\sum_{i=1}^{N} (P_{i,\max} * I_{i,t}) \geq \text{Load}_t + PL_{i,t}, \quad k \in [1, T] \tag{12}$$

5. Voltage and power flow constraints (security)

$$V_i^{\min} \leq V_{i,t} \leq V_i^{\max} \quad i \in N_{\mathrm{B}-1} \tag{13}$$

$$\left|\mathrm{BF}_{i,t}\right| \leq \mathrm{BF}_i^{\max} \quad i \in N_{\mathrm{B}} \tag{14}$$

$$V_{i,r}^{\min} \leq V_{i,r,t} \leq V_{i,r}^{\max} \quad i \in N_{\mathrm{B}-1}, \quad r \in \mathrm{Sc} \tag{15}$$

$$\left|\mathrm{BF}_{i,r,t}\right| \leq \mathrm{BF}_{i,r}^{\max} \quad i \in N_{\mathrm{b}}\ r \in \mathrm{Sc} \tag{16}$$

6. Reactive power limits

$$Q_{\mathrm{G}i}^{\min} \leq Q_{\mathrm{G}i,t} \leq Q_{\mathrm{G}i}^{\max} \tag{17}$$

7. FACTS device limits

$$B_{\mathrm{svc}_{\min}} \leq B_{\mathrm{svc}} \leq B_{\mathrm{svc}_{\max}} \tag{18}$$

$$X_{\mathrm{TCSC}_{\min}} \leq X_{\mathrm{TCSC}} \leq X_{\mathrm{TCSC}_{\max}} \tag{19}$$

The EMI is calculated by

$$R = \frac{P * r * (1 + r)^n}{(1 + r)^n - 1} \tag{20}$$

where

Fig. 1 Static var compensator

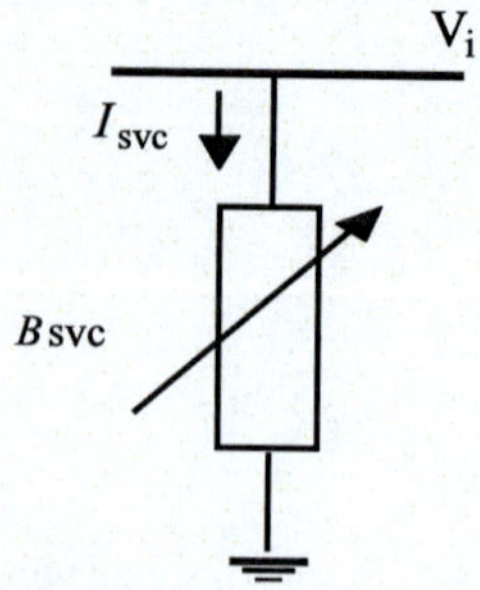

P Cost of installation of FACTS devices, r = Rate of interest per annum,
n Number of EMI payments

3 Modeling of FACTS Devices

3.1 Modeling of SVC

Static var compensator (SVC) can provide variable reactive power to control the variation in system voltages. This improves the stability of the transmission system. If the reactive load in the system is leading, the SVC uses reactors (usually thyristor-controlled reactors) which consumes the reactive power from the network, so that the voltage reduces to the specified value. If the load in the system is lagging, capacitive banks are switched on depending on the requirement, so that the voltage profile is improved. The equivalent circuit of SVC under variable reactance modeling technique [11] is shown in Fig. 1. The power equation [3] is given below.

The reactive power injected by SVC at bus "*i*" is given by

$$Q_{svc} = -V_i^2 B_{svc} \tag{21}$$

where B_{SVC} is the SVC susceptance.

3.2 Modeling of TCSC

TCSC is a series connected device which has the parallel combination of thyristor-controlled reactor (TCR) and capacitor which is shown in Fig. 2. This has two modes of operation namely inductive and capacitive depending on the reactive power requirement.

In inductive mode

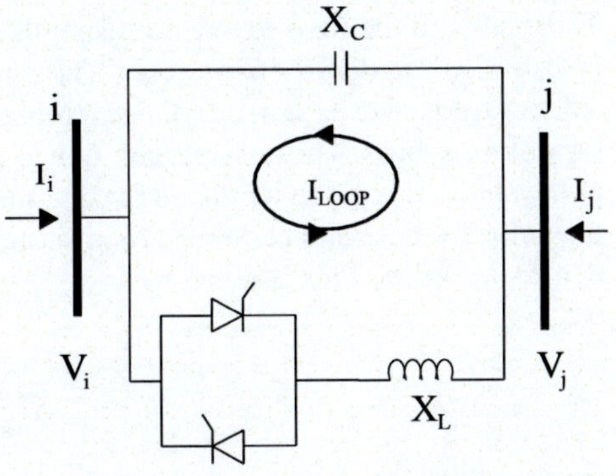

Fig. 2 Thyristor-controlled series compensator

$$B_{ii} = B_{jj} = \frac{-1}{X_{\text{TCSC}}} \tag{22}$$

$$B_{ij} = B_{ji} = \frac{1}{X_{\text{TCSC}}} \tag{23}$$

where X_{TCSC} is the TCSC reactance.

In capacitive mode, the signs are reversed.

At bus *i*, the real and reactive power equations are given by

$$P_i = V_i V_j B_{ij} \sin(\theta_i - \theta_j) \tag{24}$$

$$Q_i = -V_i^2 B_{ii} - V_i V_j B_{ij} \cos(\theta_i - \theta_j) \tag{25}$$

Similarly, at bus *j*,

$$P_j = V_j V_i B_{ji} \sin(\theta_j - \theta_i) \tag{26}$$

$$Q_j = -V_j^2 B_{jj} - V_j V_i B_{ji} \cos(\theta_j - \theta_i) \tag{27}$$

4 ABC Algorithm

ABC algorithm belongs to the category of swarm intelligence in which the intelligent behavior of honey bees for finding sources around their hives is utilized [10]. There are three groups of bees in the colony namely: employed, scout, and onlooker bees. The total number of bees in the colony is divided into employed and onlooker bees equally. Each food position represents the set of parameters of the problem to be optimized. The number of food sources is equal to the number of employed bees.

The quality of the food source corresponds to the fitness value of the function, which is also associated with its position. The employed bees start searching for the food sources by its own experience. Once it finds a source, the information is exchanged to the onlooker bees. The possible solution is given by the food sources, and the quality of the solution is given by the nectar amount. The nectar information is evaluated by the onlooker bees and chooses a food source with respect to the probability value of that food source. This is given by

$$\Pr_{p} = \frac{\text{fit}_p}{\sum_{p=1}^{N_e} \text{fit}_q} \tag{28}$$

where the fitness value of the solution p is given by fit_p. This is proportional to the amount of nectar of the food in that position p. Now, the position of the onlookers is modified using (29), and the nectar amount for the new source is evaluated by

$$v_{pq} = x_{pq} + \varphi_{pq}\left(x_{pq} - x_{kq}\right) \tag{29}$$

where $q \in \{1, 2 \ldots D\}$ and $k \in \{1, 2, \ldots \text{ne}\}$. The production of nearby food sources is controlled by k. If the calculated nectar amount of the new source is better than the previous one, that new position is used. Else, the old position is retained. If the nectar of a food source is not getting improved by a specified number of trials, then that food source is neglected, and that bee is converted to a scout. The new scout bee finds a new food source using

$$x_{pq} = x_{q\min} + \text{rand}(0, 1) \times \left(x_{q\max} - x_{q\min}\right) \tag{30}$$

where $x_{q\min}$ and $x_{q\max}$ are the limits of the parameters.

5 Optimal Location of SVC and TCSC

The optimal location of FACTS device in a power system network is important to reduce the line loss, for controlling the power flow and to reduce the operating cost. Locating the optimal position is not a simple task since the system is complex. The optimal location is identified by considering loss minimization, economic generation cost, and improvement in power flow. The total cost includes the power generation cost and installation cost of SVC and TCSC.

As, the selected line is only compensated, this will reduce the investment on the compensating devices. SVC and TCSC are kept in all buses in the system, and the location corresponding to minimum operating and installation cost, minimum power loss, better power flow control, and voltage profile is considered as the optimal location for installing the device. The total operating cost and the total installation cost for SVC and TCSC are calculated by using (3) and (4). The optimal location,

Table 1 Optimal location, loss, and cost with FACTS devices

Device\parameters	IEEE 118abus system		South Indian 86 bus utility	
	SVC	TCSC	SVC	TCSC
Optimal location	11	65–81	34	66–88
Generation cost	$ 94995.04	$ 94957.96	₹ 715,177.16	₹ 718,880.6
FACTS device cost (*10e5)	$ 0.978	$ 4.63	₹ 1.06	₹ 4.62
Power loss (MW)	142.3	143.46	70.38	66.33

generation cost, device installation cost, and the real power loss for SVC and TCSC are given in Table 1.

6 SCUC Incorporating FACTS Devices

The SCUC is solved without FACTS devices employing ABC algorithm, and the results are discussed and compared with references in literature. SCUC problem is then solved with FACTS devices (SVC and TCSC), and the operation of each device during contingency situation is analyzed. The percentage cost recovered with FACTS devices with certain period of time is illustrated. The payback period of each FACTS devices (SVC and TCSC) and the profit incurred with the installation of them are clearly demonstrated. The parameters of the ABC algorithm are calculated by statistical assessment. The maximum cycle number and colony size are considered as 200 and 300, respectively. The same analysis is also carried out with south Indian 86 bus utility. The system data is available in [16].

6.1 Case 1: SCUC for 118-Bus System Without FACTS Devices

UC is first solved without any security constraints and FACTS devices. Bus voltage and line flow violation are noticed in this case. Line flow violation occurred in line 1–4 during nineth and tenth hour, and voltage violation occurred in bus 2 and 4.

The violation occurred in line 1–4 since the power flow exceeded the maximum capacity of 100 MW. In SCUC solution, economical units such as G_4, G_5, G_{10}, G_{11}, G_{24}, G_{25}, G_{27}, G_{27}, G_{36}, G_{39}, and G_{40} are used as base units. The expensive generating units such as G_6, G_8, G_9, G_{15}, G_{17}, G_{18}, G_{31}, G_{32}, G_{33}, G_{38}, G_{41}, G_{42}, G_{49}, G_{50}, and G_{54} are not committed. According to the demand, the remaining units are committed. The total generation cost per day is $1,713,370.01. The generation cost obtained with the proposed method (Table 2) is less compared to the other methods in literature. Figure 3 shows the convergence characteristics of the proposed algorithm.

Table 2 Total generation cost

S. no.	Solution techniques	Total cost in $ (without FACTS devices)	
1	LR [12]	1,984,081.00	
2	MILP [12]	1,979,625.00	
3	SDP [13]	1,714,255.63	
4	Hybrid ABC	1,713,370.01	
	Incorporating FACTS devices	IEEE 118-bus system		86 bus system
		Without contingency ($)	With contingency ($)	Without contingency (₹)
1	Hybrid ABC (SVC)	171,2667.38	183,8730.21	148,60773.31
2	Hybrid ABC (TCSC)	171,2671.86	184,0171.03	148,61238.2

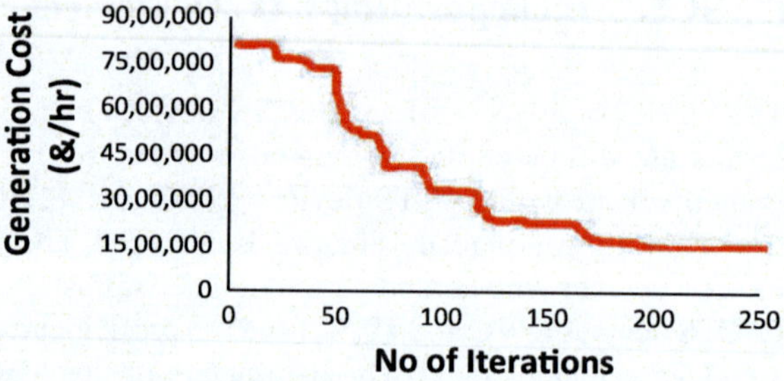

Fig. 3 Convergance characteristics of Hybrid ABC without SVC and TCSC

6.2 *Case 2: SCUC for 118abus System with FACTS Devices*

SVC and TCSC are kept in their optimal locations (Table 1) in both the test systems. The devices are placed individually, SCUC is solved for 10 trails, and the best cost is furnished in Table 2.

The commitment status is found to be same even with SVC and TCSC. The active power dispatch for the generating units with SVC and TCSC during contingency for 14th, 15th, and 16th hour is furnished in Table 3. The difference reflected in the operating cost is because of the reduction in power losses to the incorporation of FACTS devices. As losses are reduced, the active power generation decreases which reduces the generation cost. The cost for 24 h in IEEE 118 bus system is furnished in Table 4. The generation cost incorporating the FACTS devices in 86 bus utility is furnished in Table 2. The generation dispatch for 14th, 15th, and 16th hour in 86 bus system is given in Table 5. However, during the commitment of generators, all the units are committed for dispatch. Since the load demand in 86 bus utility is higher, all generators are turned ON for the demand.

Table 3 Dispatch in MW with SVC and TCSC during contingency

Hour/ Gen no	SVC			TCSC		
	Hour 14	Hour 15	Hour 16	Hour 14	Hour 15	Hour 16
1	030.00	025.28	OFF	030.00	010.33	OFF
2	030.00	022.89	OFF	030.00	015.00	OFF
3	025.00	021.77	OFF	030.00	030.00	OFF
4	300.00	212.00	171.20	300.00	255.34	229.66
5	235.70	229.60	231.45	295.73	232.32	290.11
6	OFF	OFF	OFF	OFF	OFF	OFF
7	100.00	084.38	100.00	100.00	061.61	079.36
8	OFF	OFF	025.58	OFF	OFF	005.00
9	OFF	OFF	OFF	OFF	OFF	OFF
10	300.00	218.00	197.95	151.13	300.00	OFF
11	129.70	350.00	350.00	350.00	116.01	174.30
12	OFF	OFF	OFF	OFF	OFF	OFF
13	100.00	033.80	100.00	100.00	032.09	043.94
14	030.00	030.00	026.49	030.00	020.16	030.00
15	060.49	100.00	051.24	100.00	086.54	085.81
16	OFF	OFF	OFF	OFF	OFF	OFF
17	OFF	OFF	OFF	OFF	OFF	OFF
18	095.05	072.79	030.68	100.00	097.04	079.44
19	117.40	095.05	075.98	250.00	050.00	084.12
20	083.12	250.00	192.13	250.00	247.5	114.91
21	081.11	100.00	026.88	100.00	25.00	066.17
22	036.42	039.53	058.79	100.00	041.81	025.00
23	074.53	103.76	192.03	200.00	200.00	087.13
24	172.50	200.00	200.00	200.00	149.49	107.30
25	085.00	027.00	090.15	100.00	025.00	100.00
26	146.50	118.78	238.05	420.00	265.56	210.81
27	146.30	297.44	143.54	420.00	176.33	213.47
28	168.30	300.00	267.66	241.80	175.79	296.17
29	041.99	052.34	033.79	052.87	059.41	058.52
30	OFF	OFF	OFF	OFF	OFF	OFF
31	OFF	OFF	OFF	OFF	OFF	OFF
32	OFF	OFF	OFF	OFF	OFF	OFF
33	029.48	096.05	034.89	046.56	070.52	067.21
34	028.22	047.03	042.60	056.24	075.69	062.04
35	262.70	170.60	175.34	193.92	215.80	197.81

(continued)

Table 3 (continued)

Hour/ Gen no	SVC			TCSC		
	Hour 14	Hour 15	Hour 16	Hour 14	Hour 15	Hour 16
36	076.50	050.85	066.02	094.12	082.47	035.11
37	OFF	OFF	OFF	OFF	OFF	OFF
38	189.30	155.00	285.87	282.79	174.66	264.20
39	153.80	141.10	095.95	078.42	077.93	083.32
40	012.20	019.90	015.73	016.53	015.61	012.80
41	OFF	OFF	OFF	OFF	OFF	OFF
42	268.80	171.00	141.54	125.55	187.63	104.57
43	163.20	164.80	138.95	217.16	243.29	282.82
44	280.90	103.80	162.20	143.17	192.33	218.71
45	012.90	019.88	019.75	010.19	013.78	019.09
46	041.63	097.02	069.61	091.57	063.07	031.36
47	063.08	080.04	033.80	063.54	039.68	081.56
48	OFF	OFF	OFF	OFF	OFF	OFF
49	OFF	OFF	OFF	OFF	OFF	OFF
50	061.66	066.79	047.44	097.20	031.73	040.95
51	099.09	052.05	071.80	083.48	093.92	059.84
52	032.40	034.74	027.83	087.48	032.96	037.07
53	043.03	033.90	031.54	029.09	032.14	033.56
54	OFF	OFF	OFF	OFF	OFF	OFF

6.3 *Case 3. Performance of SVC and TCSC During Contingency*

In this case, SCUC is solved with single line outage and FACTS devices. From line 17–30 in IEEE 118 bus system, one of the heavily loaded line is opened for contingency. Here, SVC and TCSC are placed in their optimal locations, and their performance in regulating the power flow and reducing the overloading of lines are investigated. SVC and TCSC control the power flow and inject power whenever required, and thus, the stability of the system is maintained. The generation cost for SVC and TCSC during contingency is furnished in Table 4. The operating cost during contingency is more than the cost without contingency since some costly generators are committed to supply power during the contingency stage. The convergence characteristics of the algorithm with TCSC in 86 bus system during contingency devices is shown in Fig. 4. During contingency condition, the number of iterations required to obtain the optimal value is more than the normal case

Table 4 Generation cost with FACTS devices

Device/Hour	Without contingency		With contingency	
	SVC	TCSC	SVC	TCSC
1	77,198.85	70689.94	84,638.58	70,692.00
2	68,077.34	67,887.5	74,638.02	67,889.48
3	67,804.6	77,541.57	74,339.00	77,543.83
4	68,177.83	85,632.5	74,748.20	85,635.00
5	76,851.9	61,337.33	84,258.20	61,339.12
6	72,618.67	94,860.23	79,617.01	94,863.00
7	71,222.06	60,684.81	78,085.80	60,686.58
8	65,735.96	63,148.79	72,071.00	63150.63
9	82,086.56	70,431.94	89,997.33	70,434.00
10	72,168.88	69859.96	79,123.87	69,862.00
11	61,647.01	70,927.8	67,588.00	70,929.87
12	78,518.28	60,711.35	86,085.17	60,713.12
13	66,046.89	70,112.39	72411.90	70,114.44
14	77,563.2	66,242.11	85,038.04	66,244.04
15	73,801.66	76,211.87	80,914.00	76214.10
16	80,246.66	68,042.01	87,980.11	68,044.00
17	72,082.35	69,742.96	79,029.00	69,745.00
18	63,633.92	79,004.69	69,766.38	79,007.00
19	58,864.84	66,298.17	64,537.70	66,300.11
20	85,467.14	77,668.1	93,703.70	77,670.37
21	630,99.08	69,925.54	69,180.00	69,927.58
22	71,013.32	68,681.99	77,856.95	68,684.00
23	67,249.77	80,927.45	73,730.70	80,929.81
24	72,015.77	66,582.06	78,956.00	66,584.00
Total	17,13,235.10	1,713,252.90	1,878,294.66	1,845,334.27

7 Estimation of Cost Recovery and Payback Period

The cost recovered with FACTS devices and the payback period of them are illustrated using two approaches. The first one is the EMI scheme in which a specified EMI is paid for a month. This EMI is calculated using (20). The next approach is the non-EMI scheme in which the installation cost of the FACTS devices along with the rate of interest is used while calculating the profits.

Table 5 Dispatch in MW with FACTS devices (86 bus utility)

Hour/G. no	SVC			TCSC		
	Hour 14	Hour 15	Hour 16	Hour 14	Hour 15	Hour 16
1	075.02	088.53	070.97	075.17	087.37	068.13
2	075.02	088.53	070.97	075.17	087.37	068.13
3	075.02	088.53	070.97	075.17	087.37	068.13
4	075.02	088.53	070.97	075.17	087.37	068.13
5	075.02	088.53	070.97	075.17	087.37	068.13
6	093.77	110.70	088.72	092.55	105.02	108.23
7	093.77	110.70	088.72	092.55	105.02	108.23
8	093.77	110.70	088.72	092.55	105.02	108.23
9	093.77	110.70	088.72	092.55	105.02	108.23
10	093.89	117.40	142.40	096.70	092.01	135.28
11	093.89	117.40	142.40	096.707	092.01	135.28
12	093.89	117.40	142.40	096.70	092.01	109.25
13	077.58	051.93	115.00	076.80	076.64	120.75
14	077.58	051.93	115.00	076.80	076.64	120.75
15	077.58	051.93	115.00	076.80	076.64	120.75
16	077.58	051.93	115.00	076.80	076.64	120.75
17	077.58	051.93	115.00	076.80	076.64	120.75

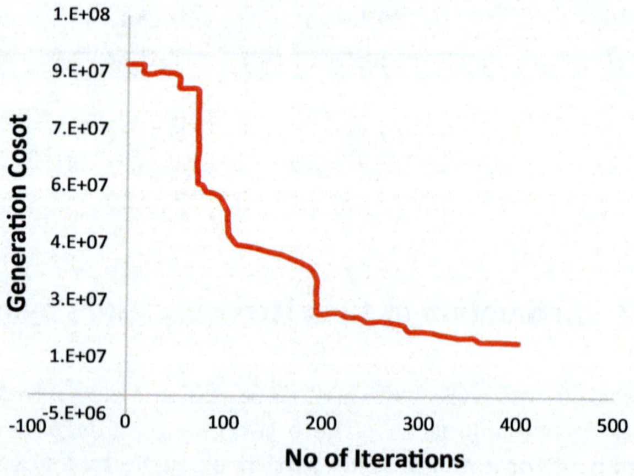

Fig. 4 Convergance characteristics of Hybrid ABC with TCSC in 86 bus system

7.1 *EMI Scheme for SVC*

Assume that the initial cost of SVC is taken as loan with an annual interest of 8%. For this amount, EMI should be paid for a certain period of time. Considering the payback period for the loan amount as 15 years, the calculated EMI is 9368.9 \$/month.

With SVC installed, the total gain obtained in operating cost annually is 2.5646e5 \$. If this is deposited with an annual interest of 8%, then the amount obtained is 3.846900e6 \$ in 15 years. Form the total gain, the total EMI paid is reduced to obtain the profit obtained with SVC. The total gain obtained with the incorporation of SVC after 15 years is 2.1605e6 \$. This is approximately 220.91% of the cost invested for installing SVC and 128.11% of the total EMI paid for the loan amount. The installation cost of SVC can be taken back within a period of 6–7 years, and the cost recovered after the pay back time will be of significant gain. This gain corresponds only to the gain in operating cost without considering the cost due to congestion management, cost of land for construction purpose. For the 86 bus system, the EMI calculated is 1.0442e4 ₹/month. Installing SVC in its optimal location yields a gain of ₹ 51,7205 annually. Assuming that this amount is deposited for 15 years, with a rate of interest of 8%, its total amount obtained will be ₹ 3.846900e6. The total profit obtained is ₹ 2.1605e6 which is 345.78% of the EMI paid for SVC and 133.2% of the initial installation cost of SVC.

7.2 *Non-EMI Scheme for SVC*

In this case, the gain obtained with non-EMI scheme is analyzed. The initial installation cost of SVC for IEEE 118 bus test system is 0.978000e6\$. Assume that this amount is obtained as loan against the rate of interest of 8% per annum. For a span of 15 years, the amount needed to be paid will be 1.056240e6\$. With SVC in its optimal location, the total gain obtained in operating cost annually is 2.5646e5 \$ (Table 6).

Table 6 Installation and recovery cost for SVC and TCSC (IEEE 118 bus system)

Particulars	SVC	TCSC
Device installation cost (M\$)	0.978	4.63
Annual rate of interest	8%	8%
Total span	180 months	180 months
Monthly EMI (\$)	9368.9	44354.7
Annual gain with FACTS devices	2.5646*10e5	2.5482*10e5
Gain for a period of 15 years with 8% interest	3.846900*10e6	3.8220*10e6
Total EMI for 15 years	1.684*10e6	7.98372*10e6
Total profit in 15 years	2.1605e6	–
Profit obtained in %of EMI	128.11	47.88
Profit obtained in % of installation cost	220.91	51.7023
Recovery cost considering only initial installation cost (after payback)	786.69	165.11
Recovery cost considering installation cost and 3% maintenance cost (after payback)	414.05	86.90

Depositing this with an annual interest of 8% yields 3.846900e6$ for a span of 15 years. The gain achieved by SVC over its installation cost in 15 years is 2.1606e6$, which is 204.5% of the installation cost against an interest of 8% per annum. After installation, for every component, a percentage of amount will be allocated as a maintenance cost. Considering an annual maintenance cost of 3% for SVC, the total installation cost will be 1.44810e6$ for 15 years. The gain obtained with SVC considering the maintenance cost will be approximately 2.428800e6 $, and a gain of 167.2% of the installation cost can be obtained. For the 86 bus system, with a time period of 15 years, the amount that is needed to be paid after 15 years with 8% rate of interest will be ₹ 2.332e6. This profit is 249.74% of the device installation cost. The gain from SVC with 8% rate of interest for 15 years will be ₹ 3.846900e6, and the resulted profit is ₹ 1,514,900. Considering an annual maintenance cost of 3% for SVC, the total profit obtained will be 190.07% of the installation cost. This profit is obtained after paying back the installation cost of SVC. Profit cannot be obtained immediately after the installation of FACTS device. The advantages in terms of security and gain will be remarkable and can be obtained only based on a long time basis.

7.3 EMI and Non-EMI Scheme for TCSC

Assume that initial installation cost of TCSC is obtained as a loan having an annual interest of 8%. The total EMI that has to be paid will be 4.4354e4 $. For a period of 15 years, the total EMI paid will be 7.9837e6 $. The gain with TCSC annually will be 2.548e5 $, and for a period of 15 years having an interest of 8% annum, the total amount will be 4.127760e6 $. 47.88% of the TCSC installation cost can be recovered with EMI scheme. With the non-EMI scheme approximately 51.7% of the device installation cost can be recovered. For south Indian system, the initial installation cost is 2.94e6. The EMI paid with 8% interest for 15 years will be 559,656. Approximately, 11.04% of the EMI can be obtained as a profit apart from the payback amount. Considering the installation cost with 8% interest, the profit obtained will be 87.03% of the installation cost, and the profit obtained will be 72.25% if 3% of maintenance cost is included with the installation cost (Table 7).

8 Conclusion

This paper investigates the payback period and percentage cost recovery on the investment cost made on SVC and TCSC in a power system network. The master problem (UC) and the subproblem (SCED) are solved simultaneously incorporating FACTS devices (SVC and TCSC) employing ABC algorithm to obtain the economic operation schedule of the generating units while satisfying unit, system, and security constraints. The cost benefits with SVC and TCSC in operating cost is converted into

Table 7 Installation and recovery cost for SVC and TCSC (south Indian 86 bus)

Particulars	SVC	TCSC
Installation cost of FACTS devices (M ₹)	1.06	2.94
Annual rate of interest	8%	8%
Total span	180 months	180 months
EMI per month EMI (₹)	1.0442e+004	2.8164e+004
Annual gain with FACTS devices	517,205	347,480
Gain for a period of 15 years with 8% interest	8,378,721	5,629,176
Total EMI for 15 years	1,879,560	5,069,520
Total profit in 15 years	6,499,161	559,656
Profit in percentage of EMI after payback	345.78	11.04
Profit in percentage of installation cost	133.28	38.92
Recovery cost considering only device installation cost maintenance cost(after payback)	249.74	87.03
Recovery cost considering device installation and 3% maintenance cost(after payback)	190.07	72.25

monetary values for recovering the investment made on them. The investigations are validated on an IEEE 118abus system and 86 bus utility. The effect of SVC and TCSC in improving the power flow, reducing the losses, and alleviating the overloading of lines during contingency is illustrated. The cost recovery with FACTS devices and their payback period analysis clearly illustrate the significance of installing SVC and TCSC in the network.

References

1. Padhy NP (2004) Unit commitment—a bibliographical survey. IEEE Trans Power Syst 19:1196–1205
2. Acha E (2004) Facts: modelling and simulation in power networks. Wiley, London
3. Zhuang F, Galiana FD (1988) Toward a more rigorous and practical unit commitment by Lagrangian relaxation. IEEE Trans Power Syst 3:763–773
4. Fu Y, Shahidehpour M, Li Z (2005) Long-term security constrained unit commitment: Hybrid Dantzig–Wolfe decomposition and subgradient approach. IEEE Trans Power Syst 20:2093–2106
5. Ma H, Shahidehpour SM (1999) Unit commitment with transmission security and voltage constraints. IEEE Trans Power Syst 14:757–764
6. Tseng CL, Oren SS (1999) A transmission-constrained unit commitment method in power system scheduling. Decision Support Syst 24(3)
7. Fu Y, Shahidehpour M, Li Z (2005) Security-constrained unit commitment with AC constraints. IEEE Trans Power Syst 20:1538–1550
8. Bai X, Wei H (2007) Semi-definite programming-based method for security constrained unit commitment with operational and optimal power flow constraints. IET Gener Trans Distr 3(2):182–197

9. Karaboga D, Basturk B (2007) Artificial bee colony (ABC) optimization algorithm for solving constrained optimization problems. Springer, Berlin, pp 789–798
10. Chandrasekaran K, Hemamalini S, Simon SP, Padhy NP (2012) Thermal unit commitment using binary/real coded artificial bee colony algorithm. Electr Power Syst Res 84:109–119
11. Sreejith S, Simon SP, Selvan MP (2012) Performance comparison of FACTS devices for steady state power flow control. Int Rev Model Simul 5(2):576–588
12. Wu H, Zhai Q, Guan X, Gao F, Ye H (2012) Security constrained unit commitment based on a realizable energy delivery formulation. Math Prob Eng 2012:1–22
13. Bai X, Wei H (2007) Semi-definite programming-based method for securityconstrained unit commitment with operational and optimal power flow constraints. IET Gener Trans Distrib 3(2):182–197

Investigations on Salp Swarm Algorithm to Solve Combined Heat and Power Economic Dispatch

Arunachalam Sundaram, A. Peer Fathima, Morris Stella, and F. Ruby Vincy Roy

Abstract Combined heat and power economic dispatch is an important optimization problem in a power system integrated with cogeneration units. In addition to the optimal solution satisfying the power balance equality constraint and the bounds of the thermal units, it must also lie within feasible operating region of the cogeneration units. This increases the complexity of the problem, and a potent meta-heuristic algorithm is required to solve the problem. This paper investigates the optimal solutions of the combined heat and power economic dispatch problem obtained by a recent meta-heuristic salp swarm algorithm. Transmission losses of the power system and valve point loading have been taken into consideration in this work. The algorithm is tested on standard test system available in literature. The results indicate there is scope for improvement of the salp swarm algorithm to solve combined heat and power economic dispatch problem.

Keywords Economic dispatch · Meta-heuristic algorithm · Salp swarm algorithm

1 Introduction

In many countries around the world, to reduce the pollution caused by the thermal power plants, regulations are imposed on power system companies [1]. To improve

A. Sundaram
Jubail Industrial College, Al Jubail 31961, Kingdom of Saudi Arabia
e-mail: sundaram_a@jic.edu.sa

A. Peer Fathima (✉) · F. Ruby Vincy Roy
School of Electrical Engineering, Vellore Institute of Technology, Chennai 600127, India
e-mail: peerfathima.a@vit.ac.in

F. Ruby Vincy Roy
e-mail: rubyvincy.royf2019@vitstudent.ac.in

M. Stella
Department of Electrical & Electronic Engineering, Univeristi Tunku Abdul Rahman, Sungai Long, Malaysia
e-mail: stellam@utar.edu.my

N. Zhou and S. Hemamalini (eds.), *Advances in Smart Grid Technology*, Lecture Notes in Electrical Engineering 688, https://doi.org/10.1007/978-981-15-7241-8_25

the efficiency of the heat conversion process, combined heat and power (CHP) units also called as cogeneration units are integrated to the existing power system [2]. The increase in efficiency of the heat conversion process is significant when CHP units are used to supply power and heat simultaneously [3]. Incorporation of CHP units to existing power system can reduce air pollutants by 18% and cost savings up to 40% [4, 5].

Many meta-heuristic algorithms have been employed to solve combined heat and power economic dispatch (CHPED) in the literature. A genetic algorithm (GA) with penalty functions to handle the heat power dependency of CHP units is proposed in [6], improved GA in [7], and self-adaptive GA to solve this problem is proposed in [8]. The greedy heuristics of the ant colony algorithm is used in [9] to find the optimal solutions of the CHPED problem. A explicit formula is developed in [10] to find the direct solution of CHPED problem. Bee colony optimization algorithm in [11], artificial immune system in [12], particle swarm optimization (PSO) with penalty function in [13], PSO with time varying acceleration in [14], two-level flexible algorithm in [15], improved group search optimization in [16], opposition-based group search optimization in [17], cuckoo optimization in [18, 19], gravitational search algorithm in [20], gray wolf optimization in [21], teaching learning algorithm in [22], harmony search algorithm in [5], hybrid harmony search with Nelder-Mead algorithm in [23], differential evolution in [24], integrated civilized swarm optimization in [25], whale optimization in [3], and hybrid bat-artificial bee colony in [26] have all been successfully employed to solve this highly important and complicated single-objective problem. Moreover, the literature survey indicates salp swarm algorithm is not implemented to solve this CHPED problem.

No free lunch theorem [27] states that there is no specific algorithm which can produce best results for all optimization problems. There is always a scope for new algorithm to solve a specific problem better than an existing algorithm is the main motivation for applying a recent salp swarm algorithm proposed in [28] to solve this highly nonlinear CHPED problem. The rest of the paper is organized as follows: The next section briefs the problem formulation of CHPED problem. Section three briefs the stochastic salp search algorithm. Section four investigates the results obtained by the algorithm for test systems, and the last section concludes the paper.

2 Formulation of the CHPED Problem

This CHPED formulation considers Np coal-fired power plants, Nc CHP units, and Nh heat only units. The decision variable vector x in (1) contains the real power production of power only units given by P, the power and heat produced by CHP units given by O and H respectively, and heat produced by heat only units given by T.

$$x = [P_1, P_2, P_3, \ldots, P_{NP}, O_1, O_2, \ldots O_{NC}, H_1, H_2, \ldots H_{Nc}, T_1, T_2, \ldots T_{Nh}]^t \tag{1}$$

The fuel cost objective function of the ith coal-fired power unit with the valve point effect is given by $C_{pi}(P_i)$, the fuel cost objective function of ith cogeneration unit is given by $C_{ci}(O_i, H_i)$ and the fuel cost objective of ith heat only unit is given by $C_{hi}(T_i)$ [29]. The total fuel cost is represented by $f_1(x)$ in (2). The optimal solution obtained by solving the CHPED problem will minimize the total fuel production cost given by $f_1(x)$ and will also satisfy the power balance equality constraint $h_1(x)$ in (3), heat balance operating constraint in (4), feasible operating region (FOR) inequality constraints given by (5)–(8), and bounds of the thermal units given by (9), and heat only units given by (10). The total power demand in the system is Pd and total heat demand in the system is Hd. The transmission loss Pl in the system is calculated using B-loss coefficients as in (11).

Objective Function

To minimize $f_1(x)$

$$\begin{aligned} f_1(x) &= \sum_{i=1}^{Np} C_{\mathrm{P}i}(P_i) + \sum_{i=1}^{Nc} C_{\mathrm{c}i}(O_i, H_i) + \sum_{i=1}^{Nh} C_{\mathrm{h}i}(T_i) \\ &= \sum_{i=1}^{Np} \left[a_i P_i^2 + b_i P_i + c_i + \left| d_i \sin\left\{ e_i (P_i^{\min} - P_i) \right\} \right| \right] \\ &\quad + \sum_{i=1}^{Nc} [\alpha_i O_i^2 + \beta_i O_i + \gamma_i + \delta_i H_i^2 + \varepsilon_i H_i + \zeta_i O_i H_i] + \sum_{i=1}^{Nh} [\eta_i T_i^2 + \theta_i T_i + \lambda_i] \end{aligned} \tag{2}$$

Subject to constraints

$$h_1(x) = \sum_{i=1}^{Np} P_i + \sum_{i=1}^{Nc} O_i - P_{\mathrm{d}} - Pl = 0 \tag{3}$$

$$h_2(x) = \sum_{i=1}^{Nc} H_i + \sum_{i=1}^{Nh} T_i - Hd = 0 \tag{4}$$

$$g_1(x) = P_i - P_i^{\max}(H_i) \le 0; \quad i \in 1, 2, \ldots, Nc \tag{5}$$

$$g_2(x) = P_i^{\min}(H_i) - P_i \le 0; \quad i \in 1, 2, \ldots, Nc \tag{6}$$

$$g_3(x) = H_i - H_i^{\max}(P_i) \le 0; \quad i \in 1, 2, \ldots, Nc \tag{7}$$

$$g_4(x) = H_i^{\min}(P_i) - H_i \le 0; \quad i \in 1, 2, \ldots, Nc \tag{8}$$

$$P_i^{\min} \le P_i \le P_i^{\max}; \quad i \in 1, 2, \ldots, Np \tag{9}$$

$$H_i^{\min} \le H_i \le H_i^{\max}; \quad i \in 1, 2, \ldots, Nh \tag{10}$$

Transmission loss calculation

$$Pl = \sum_{i=1}^{Np}\sum_{j=1}^{Np} P_i B_{ij} P_j + \sum_{i=1}^{Np}\sum_{j=1}^{Nc} P_i B_{ij} O_j + \sum_{i=1}^{Nc}\sum_{j=1}^{Nc} O_i B_{ij} O_j + \sum_{i=1}^{Np} B_{0i} P_i + \sum_{i=1}^{Nc} B_{0i} O_i + B_{00} \tag{11}$$

List of Symbols

Np	The number of power only units.
Nc	The number of CHP (cogeneration) units.
Nh	The number of heat only units.
P_i	The power generation of *i*th power only unit.
O_i	The real power produced by the *i*th cogeneration unit.
H_i	The heat generated by the *i*th cogeneration unit.
T_i	The heat generated by the *i*th heat only unit
$h_i(x)$	The *i*th equality constraint
$g_i(x)$	The ith inequality constraint
Pl	The transmission loss for the power network
C_{pi}	The fuel cost of the *i*th power only unit
C_{ci}	The fuel cost of the *i*th cogeneration unit
C_{hi}	The fuel cost of the *i*th heat only generator
a_i, b_i, c_i	The fuel cost coefficient of the *i*th generator
d_i, e_i	The valve point coefficients of the *i*th generator
$\alpha_i, \beta_i, \gamma_i, \delta_i, \varepsilon_i, \zeta_i$	The fuel cost coefficient of the *i*th CHP unit.
$\eta_i, \theta_i, \lambda_i$	The fuel cost coefficient of the *i*th heat only unit.
$P_i^{\min}$	The lower bound for power generation of unit *i*
$P_i^{\max}$	The upper bound for power generation of unit *i*
$H_i^{\min}$	The lower bound for heat produced by *i*th heat only unit
$H_i^{\max}$	The upper bound for heat produced by *i*th heat only unit
B_{ij}	The transmission loss coefficients of the line connecting the buses *i* and *j*.

3 Salp Swarm Algorithm

To model the swarming behavior of the salps chains, the entire population of the salps are divided into two groups: leaders and followers [28]. The foraging behavior of the salps in an *n*-dimensional search space to find the food source F is stored in a two-dimensional matrix X. The leader's position of the first slap in the jth dimension is updated using (12). c_1, c_2, and c_3 are random numbers.

$$X_j^1 = \begin{Bmatrix} F_j + c_1((ub_j - lb_j)c_2 + lb_j)c_3 \geq 0 \\ F_j - c_1((ub_j - lb_j)c_2 + lb_j)c_3 < 0 \end{Bmatrix} \tag{12}$$

The global and local search capability of salp search algorithm is balanced by coefficient c_1 and is defined in (13). Maximum number of iteration is indicated by L ,and the current iteration of the algorithm is indicated by l in (13).

$$c_1 = 2\mathrm{e}^{-\left(\frac{4l}{L}\right)^2} \tag{13}$$

The coefficients c_2 and c_3 is in the range of 0–1 define the movement of the salp in the next iteration. The ith follower in the dimension j is updated using (14)

$$X_j^i = \frac{1}{2}at^2 + v_0 t;\ i \geq 2;\quad X_j^i = \frac{1}{2}\left(X_j^i + X_j^{i-1}\right) \tag{14}$$

The initial speed of salp is v_0, and $a = \frac{v_{\text{final}}}{v_0}$, $v = \frac{x - x_0}{t}$. The detail salp swarm algorithm (SSA) is available in [28].

4 Case Study

4.1 Test Case 1

This test case consists of four power units, two CHP units, and one heat only unit. The data of this widely used seven units test case including transmission loss coefficients is available in [11] and is used in [14, 22, 24, 26]. The FOR of the two CHP units is also available in [22]. The system heat and power demands are 150 MWth and 600 MW, respectively. The best solution obtained from 20 runs of salp algorithm is presented in Table 1, and the solution is compared with other algorithm available in the literature. The best solution provided by salp algorithm is 10153.037$ which is better than most of the algorithms, but TVAC-PSO in [14] and TLBO, OTLBO in [22] outperform the salp swarm algorithm (SSA). The convergence curve of the salp swarm algorithm (SSA) is shown in Fig. 1.

4.2 Test Case 2

This test system consists of thirteen thermal units, six cogeneration units, and five heat only units. The data for this test system and the FOR of the CHP units are

Table 1 Optimal solutions obtained by different algorithms for test case 1

Variable	PSO [11]	EP [11]	DE [24]	RCGA [11]	TLBO [22]	TVAC-PSO [14]	SSA
P_1 (MW)	18.4626	61.361	44.2118	74.6834	45.2660	47.3383	10.0006
P_2 (MW)	124.2602	95.1205	98.5383	97.9578	98.5479	98.5398	98.7516
P_3 (MW)	112.7794	99.9427	112.691	167.231	112.679	112.673	147.813
P_4 (MW)	209.8158	208.732	209.774	124.907	209.828	209.815	209.631
O_1 (MW)	98.814	98.8000	98.8217	98.8008	94.4121	92.3718	94.5416
O_2 (MW)	44.0107	44.0000	44.0000	44.0001	40.0062	40.000	40.000
H_1 (MWth)	57.9236	18.0713	12.5379	58.0965	25.8365	37.8467	25.3967
H_2 (MWth)	32.7603	77.5548	78.3481	32.4116	74.9970	74.9999	75.000
T_1 (MWth)	59.3161	54.3739	59.1139	59.4919	49.1666	37.1532	49.6032
Total Power (MW)	608.142	607.956	608.037	607.5801	600.7392	600.737	600.737
Total Heat (MWth)	150	150	150	150	150	150	150
Fuel Cost ($)	10,613	10,390	10,317	10,667	**10,094.83**	**10,100**	10,153

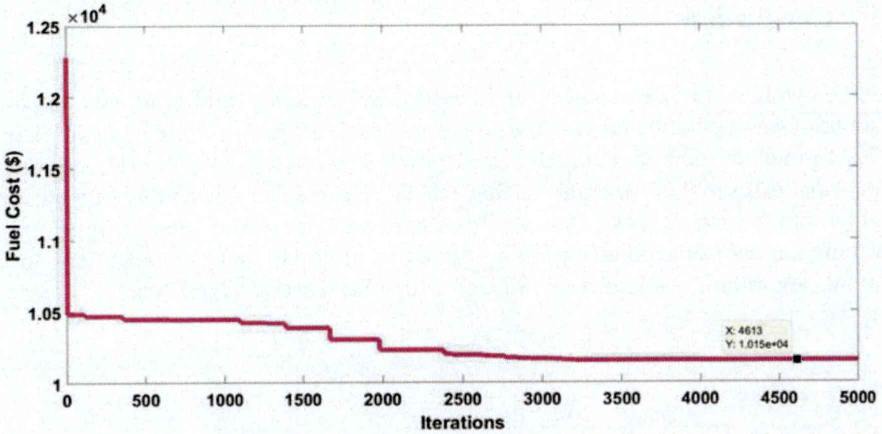

Fig. 1 The convergence curve for SSA algorithm for test case 1

available in [22]. In this test case, transmission losses are not considered. The total heat demand in this test system is 1250 MWth and the power demand is 2350 MW. The best solution obtained by SSA in 20 runs is given in Table 2. The fuel cost obtained by SSA for the schedule shown in Table 2 is 58,807.74 $. When compared to the results obtained by the existing methods, this fuel cost is better than CPSO (59,736.26 $) in [14], but inferior to TVAC-PSO (58,122.74 $) in [14], and TLBO (58,006.99$) in [22].

Table 2 Schedules obtained for test system 2 using SSA

Variable	SSA	Variable	SSA	Variable	SSA
P_1 (MW)	631.1234	P_{11} (MW)	40.0000	H_2 (MWth)	74.8999
P_2 (MW)	360.0000	P_{12} (MW)	55.0000	H_3 (MWth)	104.8000
P_3 (MW)	360.0000	P_{13} (MW)	55.0000	H_4 (MWth)	75.0000
P_4 (MW)	60.0187	O_1 (MW)	81.0000	H_5 (MWth)	40.0000
P_5 (MW)	101.8503	O_2 (MW)	40.0077	H_6 (MWth)	19.1504
P_6 (MW)	60.0000	O_3 (MW)	81.0000	T_1 (MWth)	490.1657
P_7 (MW)	60.0000	O_4 (MW)	40.0000	T_2 (MWth)	57.6411
P_8 (MW)	60.0000	O_5 (MW)	10.0000	T_3 (MWth)	59.9312
P_9 (MW)	180.0000	O_6 (MW)	35.0000	T_4 (MWth)	116.4431
P_{10} (MW)	40.0000	H_1 (MWth)	104.8000	T_5 (MWth)	107.1686

5 Conclusion

In this work, salp swarm algorithm is used to solve highly nonlinear and complex single-objective combined heat and power economic dispatch optimization problem. The formulation of the combined heat and power economic dispatch considered transmission losses and valve point loading effects. The results obtained by salp swarm algorithm for two test cases indicate this algorithm outperforms some existing algorithms, but inferior to other algorithms. In future, to further improve the performance of this algorithm, it would be hybridized with a local search algorithm.

References

1. Tillman DA (2018) Coal-fired electricity and emissions control: efficiency and effectiveness. Butterworth-Heinemann, Oxford
2. Keirstead J, Samsatli N, Shah N, Weber C (2012) The impact of CHP (combined heat and power) planning restrictions on the efficiency of urban energy systems. Energy. 41:93–103. https://doi.org/10.1016/j.energy.2011.06.011
3. Nazari-Heris M, Mehdinejad M, Mohammadi-Ivatloo B, Babamalek-Gharehpetian G (2019) Combined heat and power economic dispatch problem solution by implementation of whale optimization method. Neural Comput Appl 31:421–436. https://doi.org/10.1007/s00521-017-3074-9
4. Nazari-Heris M, Mohammadi-Ivatloo B, Gharehpetian GB (2018) A comprehensive review of heuristic optimization algorithms for optimal combined heat and power dispatch from economic and environmental perspectives. https://www.sciencedirect.com/science/article/pii/S1364032117309668. https://doi.org/10.1016/j.rser.2017.06.024
5. Nazari-Heris M, Mohammadi-Ivatloo B, Asadi S, Geem ZW (2019) Large-scale combined heat and power economic dispatch using a novel multi-player harmony search method. Appl Therm Eng 154:493–504. https://doi.org/10.1016/J.APPLTHERMALENG.2019.03.095
6. Song YH, Xuan QY (1998) Combined heat and power economic dispatch using genetic algorithm based penalty function method. Electr Mach Power Syst 26:363–372. https://doi.org/10.1080/07313569808955828
7. Zou D, Li S, Kong X, Ouyang H, Li Z (2019) Solving the combined heat and power economic dispatch problems by an improved genetic algorithm and a new constraint handling strategy. Appl Energy 237:646–670. https://doi.org/10.1016/J.APENERGY.2019.01.056
8. Subbaraj P, Rengaraj R, Salivahanan S (2009) Enhancement of combined heat and power economic dispatch using self adaptive real-coded genetic algorithm. Appl Energy 86:915–921. https://doi.org/10.1016/j.apenergy.2008.10.002
9. Song YH, Chou CS, Stonham TJ (1999) Combined heat and power economic dispatch by improved ant colony search algorithm. Electr Power Syst Res 52:115–121. https://doi.org/10.1016/S0378-7796(99)00011-5
10. Rao PSN (2006) Combined heat and power economic dispatch: a direct solution. Electr Power Components Syst 34:1043–1056. https://doi.org/10.1080/15325000600596775
11. Basu M (2011) Bee colony optimization for combined heat and power economic dispatch. Expert Syst Appl 38:13527–13531. https://doi.org/10.1016/J.ESWA.2011.03.067
12. Basu M (2012) Artificial immune system for combined heat and power economic dispatch. Int J Electr Power Energy Syst 43:1–5. https://doi.org/10.1016/j.ijepes.2012.05.016
13. Murugan R, Mohan MR, Arunachalam S (2013) A PSO based hybrid algorithm with adaptive penalty function approach for the combined heat and power economic dispatch. Arch Des Sci 66:3

14. Mohammadi-Ivatloo B, Moradi-Dalvand M, Rabiee A (2013) Combined heat and power economic dispatch problem solution using particle swarm optimization with time varying acceleration coefficients. Electr Power Syst Res 95:9–18. https://doi.org/10.1016/j.epsr.2012.08.005
15. Sashirekha A, Pasupuleti J, Moin NH, Tan CS (2013) Combined heat and power (CHP) economic dispatch solved using Lagrangian relaxation with surrogate subgradient multiplier updates. Int J Electr Power Energy Syst 44:421–430. https://doi.org/10.1016/J.IJEPES.2012.07.038
16. Hagh MT, Teimourzadeh S, Alipour M, Aliasghary P (2014) Improved group search optimization method for solving CHPED in large scale power systems. Energy Convers Manag 80:446–456. https://doi.org/10.1016/J.ENCONMAN.2014.01.051
17. Basu M (2015) Combined heat and power economic dispatch using opposition-based group search optimization. Int J Electr Power Energy Syst 73:819–829. https://doi.org/10.1016/J.IJEPES.2015.06.023
18. Mellal MA, Williams EJ (2015) Cuckoo optimization algorithm with penalty function for combined heat and power economic dispatch problem. Energy 93:1711–1718. https://doi.org/10.1016/J.ENERGY.2015.10.006
19. Nguyen TT, Nguyen TT, Vo DN (2018) An effective cuckoo search algorithm for large-scale combined heat and power economic dispatch problem. Neural Comput Appl 30:3545–3564. https://doi.org/10.1007/s00521-017-2941-8
20. Beigvand SD, Abdi H, La Scala M (2016) Combined heat and power economic dispatch problem using gravitational search algorithm. Electr Power Syst Res 133:160–172. https://doi.org/10.1016/j.epsr.2015.10.007
21. Jayakumar N, Subramanian S, Ganesan S, Elanchezhian EB (2016) Grey wolf optimization for combined heat and power dispatch with cogeneration systems. Int J Electr Power Energy Syst 74:252–264. https://doi.org/10.1016/J.IJEPES.2015.07.031
22. Roy PK, Paul C, Sultana S (2014) Oppositional teaching learning based optimization approach for combined heat and power dispatch. Int J Electr Power Energy Syst 57:392–403. https://doi.org/10.1016/J.IJEPES.2013.12.006
23. Feng Z-Y, Guo H, Liu Z-T, Xu L, She J (2017) Hybridization of harmony search with Nelder-Mead algorithm for combined heat and power economic dispatch problem. In: 2017 36th Chinese Control Conference (CCC), pp 2790–2795. IEEE. https://doi.org/10.23919/ChiCC.2017.8027787
24. Basu M (2010) Combined heat and power economic dispatch by using differential evolution. Electr Power Components Syst 38:996–1004. https://doi.org/10.1080/15325000903571574
25. Narang N, Sharma E, Dhillon JS (2017) Combined heat and power economic dispatch using integrated civilized swarm optimization and Powell's pattern search method. Appl Soft Comput 52:190–202. https://doi.org/10.1016/J.ASOC.2016.12.046
26. Murugan R, Mohan MR, Asir Rajan CC, Sundari PD, Arunachalam S (2018) Hybridizing bat algorithm with artificial bee colony for combined heat and power economic dispatch. Appl Soft Comput 72:189–217. https://doi.org/10.1016/J.ASOC.2018.06.034
27. Wolpert DH, Macready WG (1997) No free lunch theorems for optimization. IEEE Trans Evol Comput 1:67
28. Mirjalili S, Gandomi AH, Mirjalili SZ, Saremi S, Faris H, Mirjalili SM (2017) Salp swarm algorithm: a bio-inspired optimizer for engineering design problems. Adv Eng Softw 114:163–191. https://doi.org/10.1016/j.advengsoft.2017.07.002
29. Chiang CL (2005) Improved genetic algorithm for power economic dispatch of units with valve-point effects and multiple fuels. IEEE Trans Power Syst 20:1690–1699. https://doi.org/10.1109/TPWRS.2005.857924

A Review on Topological Aspects of Transformerless Dynamic Voltage Compensators

Mohanasundaram Ravi and R. Chendur Kumaran

Abstract The necessity of Power Quality (PQ) is always in demand by the power consumers. This paper presents a review on Transformerless Dynamic Voltage Compensators (TDVC) topologies for voltage related PQ enhancement. Transformerless topology ensures that the device is less cost, smaller in size compared to ordinary compensators. This paper gives the knowledge of all past and recent TDVCs which helps engineers and researchers to find out which transformerless topology suites better for their compensation.

Keywords Power quality · Dynamic voltage restorer (DVR) · Transformerless DVR · AC-to-AC converters

1 Introduction

Due to the rapid growth of modern electrical and electronic loads, power suppliers are unable to deliver the quality of power to their customers. Engineers and researchers are continuously searching solutions for the Power Quality (PQ) improvement along with cost and size effectiveness [1, 2]. Other than these loads, integration of Distributed Generation (DG) systems and Renewable Energy Source (RES) with power grid leads to PQ problems to the customers; which affect the sensitive loads to malfunction. Some IEEE standards are followed to avoid malfunctioning and increase in the life span of the loads [3, 4]. Some of the common PQ problems faced by customers are voltage sag, swell, unbalance, flickering and harmonics. To compensate these aforementioned PQ problems Custom Power Device (CPD)s are introduced [5]. CPDs are made up of power electronics components basically derived from Flexible Alternating Current Transmission System (FACTS) devices [6]. CPDs are of three types; series compensation device called Dynamic Voltage Restorer (DVR) [7, 8], shunt compensation device called D-STATCOM and Unified PQ Conditioner (UPQC) which consist of a combination of DVR and D-STATCOM

M. Ravi (✉) · R. Chendur Kumaran
School of Electrical Engineering, Vellore Institute of Technology, Chennai, Tamil Nadu, India
e-mail: mohana.sundaramr2014@vit.ac.in

N. Zhou and S. Hemamalini (eds.), *Advances in Smart Grid Technology*, Lecture Notes in Electrical Engineering 688, https://doi.org/10.1007/978-981-15-7241-8_26

with common dc link. DVR is dominant in solving voltage quality related problems whereas D-STATCOM is dominant in solving current quality related problems and UPQC does both. A regular DVR consists of a current transformer for the purpose of grid synchronization through which compensation voltage is injected to the system. This requires DVR to take more space, costly and more weight [9]. Hence, in recent years transformerless dynamic voltage compensator (TDVC) was introduced to overcome the aforementioned transformer issues. Hence knowledge of TDVC topologies should be known by engineers and researchers before the implementation.

2 DVR Basic Working

Voltage sag, swell, unbalance, and flickering cause damages to the sensitive loads. Voltage interruption causes production loss for the customers. Since DVR is connected in series, voltage quality related problems can be mitigated effectively [10, 11]. Configuration of conventional DVR with power system is shown (see Fig. 1). Power grid supplies two loads in which load 2 is considered as sensitive load. The source voltage is denoted as V_{source} followed by line impedance R_s and L_s are connected respectively. A DVR consists of a DC-link capacitor fed by DC energy storage device or rectifier support from an auxiliary AC supply [12, 13], a converter which converts DC to AC supply, a filter circuit, a series-connected injection transformer and a controller. The controller generates PWM for converter switches such that injection voltage V_{inj} should synchronous with PCC voltage V_{pcc} to make constant load voltage V_{load} magnitude as per Eq. (1). The controller also gets feedback signal of DC-link capacitor voltage V_{dc} to make constant during converter operation.

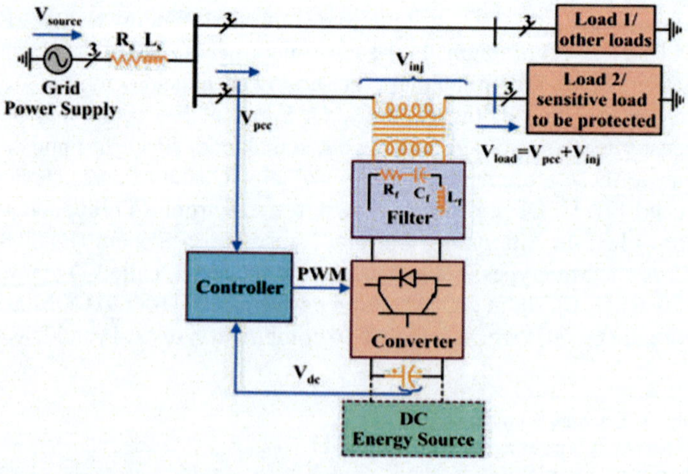

Fig. 1 Conventional dynamic voltage compensator

$$V_{\text{load}} = V_{\text{pcc}} + V_{\text{inj}} \quad (1)$$

3 Topologies of TDVC

Categories of TDVC topologies are shown (see Fig. 2). It is broadly classified into two main categories: TDVC based on DC-to-AC converter and TDVC based on AC-to-AC converter. DC/AC converter classified into H-Bridge inverter and multilevel inverter (MLI)-based TDVC. H-Bridge inverter-based TDVCs are of two types: full-bridge-based TDVC and half-bridge-based TDVC. MLI-based TDVC are of classified into two groups. They are conventional MLI-based TDVC and new MLI topology-based TDVC. Cascaded MLI and flying capacitor MLI are the conventional MLI-based TDVC. Reduced switch count MLI-based TDVC is the new MLI topology-based TDVC. On the other side, parasitic boost converter-based TDVC and AC-to-AC buck boost converter-based TDVC are the AC/AC converter-based TDVC.

3.1 TDVC Based on H-Bridge Converter

Configuration of TDVC achieved by series-connected voltage source synchronized with the grid power supply by means of capacitor or bypass switch for the power of voltage quality enhancement.

Half Bridge. Transformerless Dynamic Voltage compensator (TDVC) based on half bridge inverter integrated with the power system is shown (see Fig. 3a). Grid power supplied to a sensitive load in which source voltage is denoted as v_s followed by

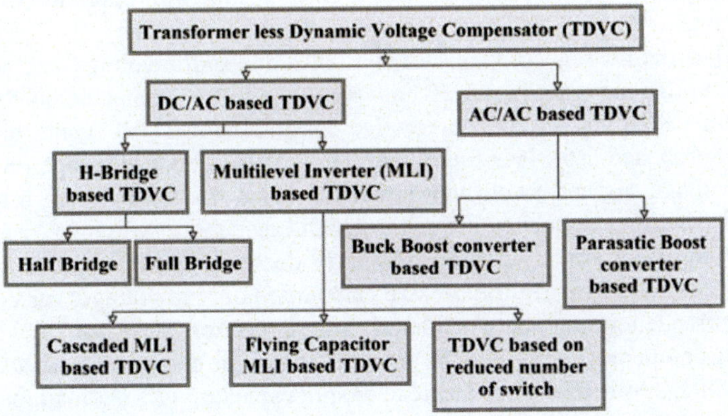

Fig. 2 Transformerless dynamic voltage compensator topology classification

Fig. 3 **a** Half bridge converter-based TDVC. **b** Full-bridge converter-based TDVC

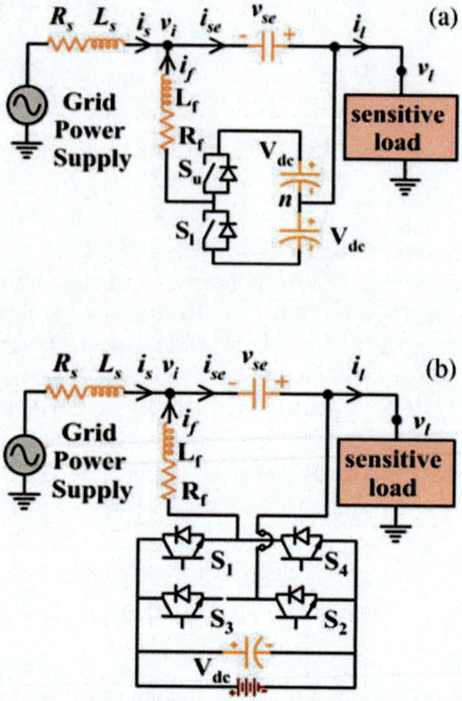

line impedance R_s and L_s are connected respectively. v_{se} and v_l are the voltage injected through a series capacitor and load voltage respectively. The half bridge inverter consists of series-connected capacitor, filter inductor L_f and resistor R_f respectively. Half bridge inverter supplied by series-connected split capacitor V_{dc}. Upper switch S_u and lower switch S_l operates such that half bridge inverter is in phase during sag operation and 180° phase shift during swell operation to maintain V_{dc} constant.

In [14] dynamic sag compensator was introduced for voltage sag mitigation with half bridge inverter in series and parallel diodes connected to the supply for capacitor charging. In 2015 predictive controller for TDVC using half bridge inverter is proposed [15]. It is found that TDVC based on half bridge inverter can mitigate voltage sag and swell with small filter size. However, when filter size is decreased switching frequency increases which cause more switching loss. Considering medium filter size, switching frequency decrease and decrease in switching loss. Hence for choosing right filter size calculation of load voltage, harmonics and switching frequency must be considered. Split capacitor made the DC voltage by half during compensation. Hence the power rating of the compensator is also less. In [16] two voltage sources are used instead of split capacitor with sliding mode control made the system more reliable.

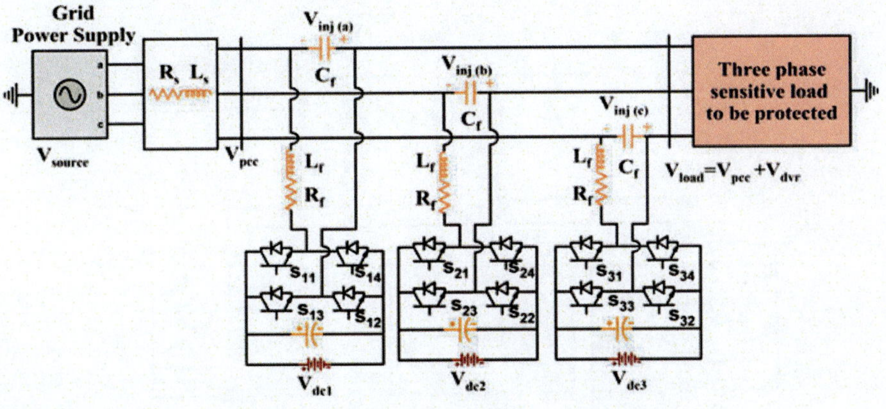

Fig. 4 Three-phase full-bridge transformerless dynamic voltage compensator

Full bridge. The system configuration of TDVC based on full-bridge inverter is same as half bridge except the inverter topology is shown (see Fig. 3b). Full-bridge inverter consists of four switches S_1, S_2, S_3 and S_4. In which S_1 is equal to S_2 and S_3 is equal to S_4. Where S_3 is compliment of S_1. DC capacitor ensures ripple free input DC voltage source for the compensator during compensation. Power rating and compensation level of the compensator is high. In [17] V_{dc} is controlled by another converter associated with the energy storage system and made the system as a self-charging system. TDVC based on full-bridge inverter for three-phase power is proposed is shown (see Fig. 4). In [18, 19] proposed three individual single phase inverter is connected in series with grid power supply makes three-phase TDVC. Three individual DC voltages for each phase is denoted as V_{dc1}, V_{dc2} and V_{dc3}.

3.2 *TDVC Based on MLI*

Transformerless dynamic voltage compensator (TDVC) based on multilevel inverter (MLI) are of classified into two main categories.

1. TDVC based on conventional MLI topology
2. TDVC based on new MLI topology.

Under conventional topology cascaded multilevel inverters are mostly used for TDVC. Other than cascade MLI, Flying capacitor multilevel inverters are also used for TDVC.

TDVC based on Cascaded MLI. Power distribution system with integrated TDVC based on Cascaded MLI is shown (see Fig. 5). Where grid power supply is denoted as V_{source} followed by line impedance R_s and L_s are connected respectively. Cascaded MLI topology consists of series-connected H-Bridge inverter with independent DC voltage source denoted as V_{dc}. Number of H-Bridge connected in series varies

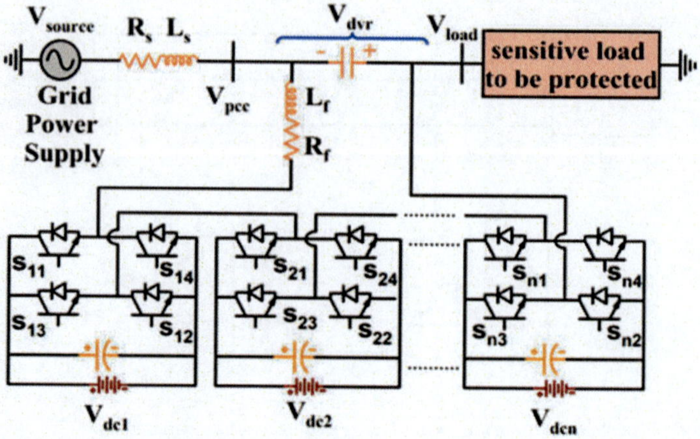

Fig. 5 Cascaded MLI-based TDVC

depending on desired output voltage levels. Hence the number of DC sources increases with the number of inverters increases from V_{dc1} to V_{dcn}. Each inverter consist of four switches, inverter 1 consist of switches namely S_{11}, S_{12}, S_{13} and S_{14}. In which S_{11} is equal to S_{12} and S_{13} is equal to S_{14}. Where S_{13} is compliment of S_{11}. The number of switches of MLI increases when the number of inverters connected increases and it is denoted as S_{n1}, S_{n2},S_{n3} and S_{n4}. Output MLI voltage is filtered using the passive filter R_f and L_f. Cascaded MLI connected to the system through a series-connected capacitor. The output voltage of MLI is denoted as V_{dvr}. MLI injects V_{dvr} such that it synchronous with grid supply and compensate voltage quality issues at PCC V_{pcc} to maintain smooth sine waveform with maintained load voltage V_{load} [20–23, 12]. Error voltage signal v_{error} for controlling MLI is given by Eq. (2). Where v_{ref} and v_{pcc} are reference voltage and PCC voltage respectively.

$$V_{error} = V_{ref} - V_{pcc} \tag{2}$$

In [24, 19] cascaded multilevel inverter-based TDVC was introduced. For Investigation of cost-effective TDVC based on cascaded MLI by eliminating the large injection transformer for compensating 250 kVA load proposed in [21]. Results proved that control is easy and compensated for different voltage sag levels. Followed by in 2014 HVDVR (High Voltage Dynamic Voltage Restorer) using P+ resonant and Posicast compensators as a controller for tracking reference voltage effectively and compensated transient during voltage sag without injecting transformer is proposed [22]. During 2012 a scheme for TDVC based on cascaded MLI for compensating unbalanced load and voltage swell is proposed [23]. In [25] an adjustable DC-link DVR based on cascaded MLI in which common DC supply is adjusted by a boost converter is proposed.

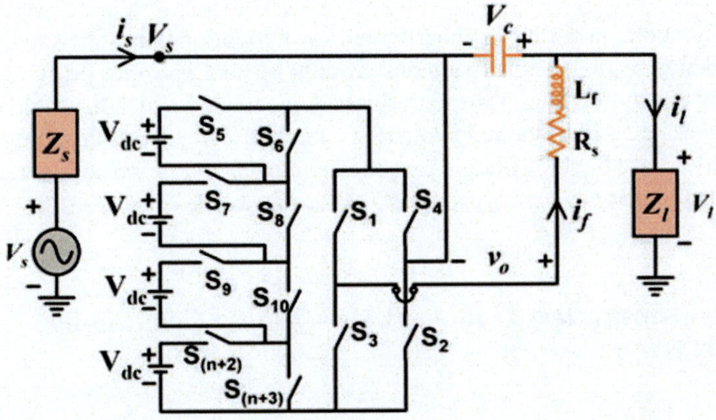

Fig. 6 TDVC based on MLI with reduced number of switches

TDVC based on flying capacitor MLI. In 2014 a Stacked Multicell converter (SM) Converter based on Multicell flying capacitor converter for the purpose of TDVC [26]. Voltage sag and voltage unbalance is compensated using this proposed converter. Energy stored in the capacitor is compared between SM and FCM for the same compensation level and found out SMC requires less energy than FCM.

TDVC-based MLI with less switch count. In 2010 Transformerless dynamic voltage restorer with a reduced number of topology for compensating voltage sag and swell is proposed [27] (see Fig. 6). TDVR connected with the power system through a series-connected capacitor. The grid voltage is denoted as V_s followed by line impedance Z_s and grid current is denoted as i_s. The voltage across a series capacitor is denoted as V_c connected to load impedance Z_l. Reduced number of switch topology consists of Multilevel DC circuit (MLDC) and Voltage source inverter (VSI). MLDC consist of several cells, each cell consists of a switch in the positive side of input DC source and a parallel switch to the DC source. S_1, S_2, S_3 and S_4 are the VSI switches. S_5, S_7, S_9 up to $S_{(n+2)}$ are the MLDC positive switches and S_5, S_7, S_9 up to $S_{(n+3)}$ are the MLDC negative switches.

3.3 *Transformerless Series Active Power Filter (TSeAPF)*

In [28] three-phase TseAPF is proposed for protecting the sensitive load from voltage harmonics. TseAPF consists of three isolated DC supply for three single phase VSI connected in series with the grid. In [29, 30] TseAPF with shunt connected passive filter is proposed for mitigating PQ problems related to electric transportation and energy management. A controller is designed to enhance PQ to protect the sensitive load and also for improving grid power factor. Results proved that proposed hybrid TseAPF able to mitigate sag, swell, voltage and current harmonics. In [31] same aforementioned hybrid filter is investigated for mitigating harmonics and PQ problems generated by house hold machines. Sliding Mode Controller (SMC) is used for controlling this hybrid filter. Harmonic analysis are made for all house hold machines and compensated within the range of IEEE standards. In [32] diode clamped multilevel inverter-based hybrid filter is proposed controlled by propositional resonant (P + R) controller for enhancing power PQ of a smart home.

3.4 *Transformerless Unified Power Quality Conditioner (TUPQC)*

The system configuration of TUPQC with power system is shown in (see Fig. 7). In [33] Configuration TUPQC consists of TDVC and D-STATCOM linked with a common DC supply. Series connection of TDVC achieved through a series capacitor. Filter inductors $\frac{L_e}{2}$ are connected in parallel to a series capacitor. Converter 2 is parallel connected with filter inductor denoted as $\frac{L_h}{2}$ and a filter capacitor. Converter 1 current, converter 2 current and load current are denoted as i_e, i_h and i_l respectively. Figure 9 shows system configuration of three phases TUPQC with isolated DC link. In [34] [35] configuration of TUPQC in which converter 1 act as TDVC based on half bridge inverter and converter 2 act as transformerless shunt active power filter (see Fig. 8). There is no common DC link for this type of compensator. Harmonics created by three-phase nonlinear load is mitigated by shunt connected active filter by injecting negative harmonics to the system. Voltage quality problems are compensated by injecting voltage through series-connected converter 1. S_1, S_4 and S_5 are the upper side switches of converter 1 and S_2, S_5 and S_6 are the lower side switches of the same converter. S_a, S_b and S_c are the upper end switches of converter 2 with each 120° phase shifted. $S_{a'}$, $S_{b'}$ and $S_{c'}$ are the lower end switches of converter 2. Lower end switches are the compliment of upper end switches in both the converter 1 and converter 2.

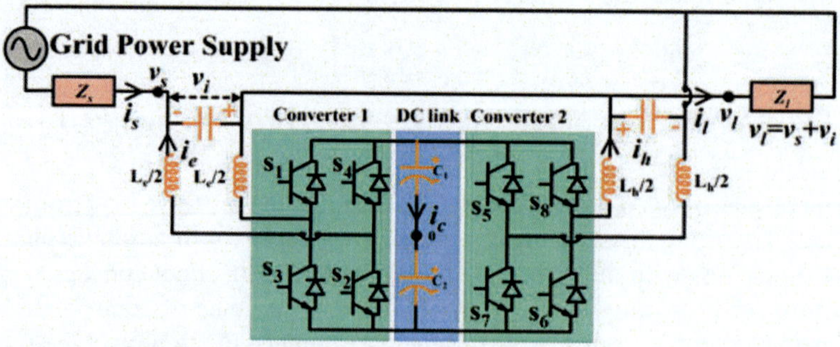

Fig. 7 TUPQC with common DC link

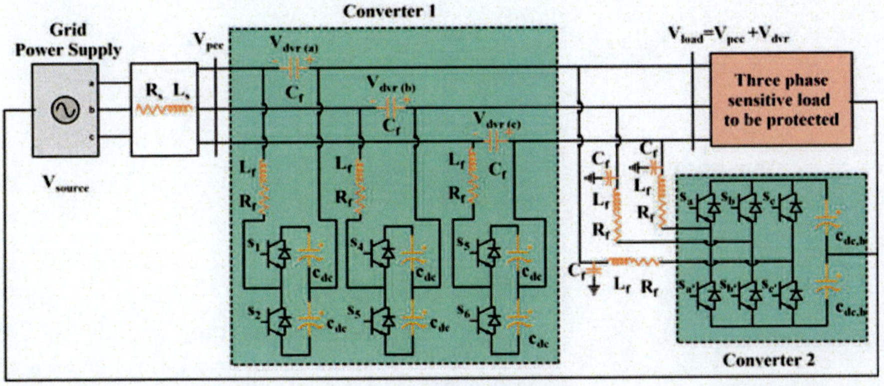

Fig. 8 TUPQC with isolated DC link

In [36], three-phase multiconverter UPQC is proposed with a common DC bus of 800 V_{dc}. A parallel converter connected to DC bus is used for renewable energy interface. In [37] flexible PQ conditioner is proposed for voltage sag, swell and harmonics mitigation. A model predictive controller is used for controlling and pulse generation. [38] presents common dc link with six switch converter-based TUPQC.

3.5 *Transformerless Interline Dynamic Voltage Compensator (TIDVC)*

Interline dynamic voltage restorer with one end transformerless connection is shown (see Fig. 9). In [39] TDVC connected in series through a series-connected capacitor with grid power supply 1. A conventional dynamic voltage compensator is connected through a series injection transformer with grid power supply 2. Converter 1 and converter 2 are made up of cascaded MLI topology coupled with a common DC capacitor V_{dc} Voltage injected by TDVC and conventional DVR for compensation is denoted as V_{inj1} and V_{inj2} respectively. R_{s1} and L_{s1} are the line impedance of supply 1; R_{s2} and L_{s2} are the line impedance of supply 2 respectively. V_{inj1} and V_{inj2} are the injected voltages by compensators such that sensitive load 1 and sensitive load 2 should maintain desired load voltage magnitude V_{load1} and V_{load2} constant by synchronizing with V_{pcc1} and V_{pcc2} respectively as per Eqs. (3, 4).

$$V_{load1} = V_{pcc1} + V_{inj1} \tag{3}$$

$$V_{load2} = V_{pcc2} + V_{inj2} \tag{4}$$

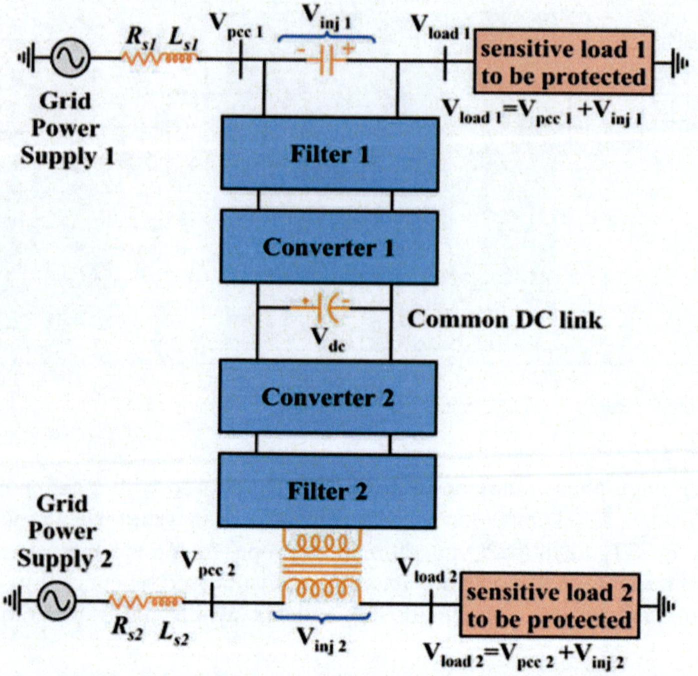

Fig. 9 IDVC with one end transformerless

4 Topologies of TDVC Based on AC-to-AC Converter

Basics of AC-to-AC voltage regulators are used in the application of TDVC for voltage quality improvement. Some of these types of TDVC are as follow.

4.1 Parasitic Boost Converter-Based TDVC

In [40] Simplified Parasitic Boost Active Voltage Quality Regulator (SPB-AVQR) and PB-AVQR and are proposed on the basis of AC-to-AC voltage regulator. Configuration of PB-AVQR shown (see Fig. 10a). Deep voltage sag is mitigated through improved shunt connected boost converter. It is long duration voltage sag compensator under 50% and it is transformerless. T_1 and T_2 are the bypass switch operates during normal condition when there is no sag or swell event occurring. T_3 and T_4 are the shunt connected converter switch. Voltage injection is done through a series capacitor and a filter inductor L_f.

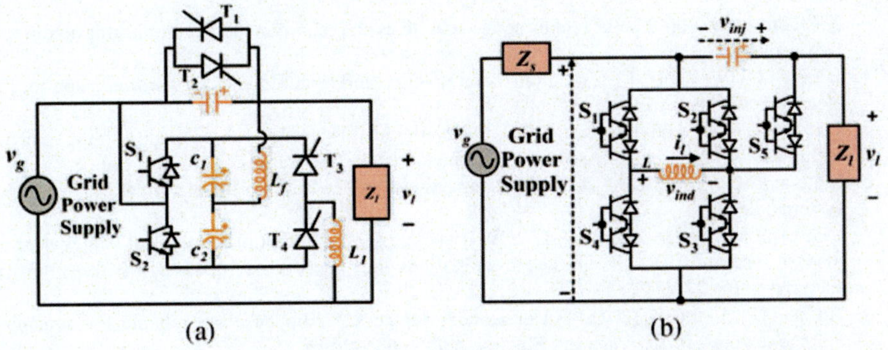

Fig. 10 **a** TDVC based on PB-AVQR, **b** TDVC based on AC-to-AC buck boost converter

4.2 Buck Boost Converter-Based TDVC

In [41] AC-to-AC buck boost converter supported TDVC is proposed for compensation of different voltage quality problems such as voltage sag, swell, flickering without any bounce in phase angle. The configuration of this compensator with power system is shown (see Fig. 10b). Since the boost or buck voltage during compensation is directly delivered from the grid hence no necessary of DC-link capacitor. S_1, S_2, S_3, S_4 and S_5 are the bi-directional IGBT switches operates such that compensating voltage v_{inj} injected through a series capacitor. The voltage across buck boost inductor is denoted as v_{ind} and the current flowing through the inductor is denoted as i_L.

5 Conclusion

A review on various topologies of TDVC has been presented. Topologies are classified in a clear manner such that beginners, researchers and engineers get knowledge before the implementation of any TDVC. Performance of different topologies with respect to cost, size, and compensation capabilities are analyzed. Each TDVC topology consists of merits and demerits from which best topology should be chosen to get the desired output.

References

1. Arindam G, Gerard L (2002) Power quality enhancement using custom power devices. Kluwer Academic Publishers
2. Dugan RC, McGranaghan MF, Santoso S (2004) Electrical power systems quality, 2nd edn. McGraw-Hill

3. IEEE Std 1159 (2009) IEEE recommended practices and requirements for monitoring electrical power quality
4. IEEE Std 519 (2014) IEEE recommended practice and requirements for harmonic control in electric power systems
5. Om Prakash M, Abdul Gafoor S (2016) Topological aspects of power quality improvement techniques: a comprehensive overview. Renew Sustain Energy Rev 58:1129–1142
6. Mathur RM, Varma RK (2002) Thyristor-based facts controllers for electrical transmission systems. Wiley
7. Christoph M, De Doncker RW, Li YW, Blaabjerg F (2008) Optimized control strategy for a medium-voltage DVR—theoretical investigations and experimental results. IEEE Trans Power Electron 23(6):2746–2754
8. Abdul MR, Vinod K (2015) An enhanced voltage sag compensation scheme for dynamic voltage restorer. IEEE Trans Ind Electron 62(5):2683–2692
9. Wuthikrai C, Krischonme B (2011) The effect of series-connected transformer in DVR applications. In: 9th eco-energy and materials science and engineering symposium 2011, energy procedia, vol 9. Elsevier, pp 306–315
10. Samet B, Hasan K (2016) Optimized sliding mode control to maximize existence region for single-phase dynamic voltage restorers. IEEE Trans Ind Inf 12(4):1486–1497
11. Mohammad F-K, Ebrahim B, Frede B (2017) A comprehensive review of dynamic voltage restorers. Int J Electr Power Energy Syst 92:136–155
12. Guodong C, Zhu M, Cai X (2014) Medium-voltage level dynamic voltage restorer compensation strategy by positive and negative sequence extractions in multiple reference frames. IET Power Electron 7(7):1747–1758
13. Lam C-S, Wong M-C, Han Y-D (2008) Voltage swell and overvoltage compensation with unidirectional power flow controlled dynamic voltage restorer. IEEE Trans Power Deliv 23(4):2513–2521
14. Sidelmo MS, Sandro ÉS, de Souza Reis A, Cardoso Filho BJ (2005) Analysis of a dynamic voltage compensator with reduced switch-count and absence of energy storage system. IEEE Trans Ind Appl 41(5):1255–1262
15. Chandan K, Mahesh KM (2015) Predictive voltage control of transformerless dynamic voltage restorer. IEEE Trans Ind Electron 62(5):2693–2697
16. Hasan K, Samet B (2017) Time-varying and constant switching frequency based sliding-mode control methods for transformerless DVR employing half-bridge VSI. IEEE Trans Ind Electron 64(4):2570–2579
17. Kian E, Kenneth S, Choi SS, Mahinda Vilathgamuwa D (2004) Analysis of series compensation and dc-link voltage controls of a transformerless self-charging dynamic voltage restorer. IEEE Trans Power Deliv 19(3):1511–1518
18. Mohanasundaram R, Chendur Kumaran R (2017) Modelling and simulation of transformer less dynamic voltage restorer for power quality improvement using combined clark's and park's transformation technique. Adv Intell Syst Comput 467:513–521
19. Li BH, Choi SS, Vilathgamuwa DM (2002) Transformerless dynamic voltage restorer. IEE Proc Gener Transm Distrib 149(3):263–273
20. de Almeida Carlos GA, dos Santos EC Jr, Jacobina CB, Mello JPRA (2016) Dynamic voltage restorer based on three phase inverters cascaded through an open-end winding transformer. IEEE Trans Power Electron 31(1):188–199
21. Abraham JV, Enslin JHR, Mouton HT (2002) Transformerless series sag compensation with a cascaded multilevel inverter. IEEE Trans Ind Electron 49(4):824–831
22. Poh CL, Vilathgamuwa DM, Tang SK, Long HL (2004) Multilevel dynamic voltage restorer. IEEE Power Electron Lett 2(4):125–130
23. Mostafa I, Mareia, A, Eltantawyb B, Ahmed AE-S (2012) An energy optimized control scheme for a transformerless DVR. Electric Power Syst Res 83:110–118
24. Herbert S, Andreas B, Hideo O (1998) Transformerless reactive series compensators with voltage source inverters. IEEE Japan Trans Ind Appl 118(10):1–7

25. Ebrahim B, Mohammad FK, Mehran S (2014) Dynamic voltage restorer based on multilevel inverter with adjustable dc-link voltage. IET Power Electron 7(3):576–590
26. Vahid D, Arash KS, Mohammad RAP, Shoulaie A (2012) DC (direct current) voltage source reduction in stacked multicell converter based energy systems. Energy 46:649–663
27. Farhadi M, Kangarlu SH, Hosseini EB, Khoshkbar Sadigh A (2010) Transformerless DVR topology based on multilevel inverter with reduced number of switches. In: 1st Power electronic & drive systems & technologies conference 2010. Tehran, pp 271–375
28. Pinto JG, Helder C, Bruno E, Carlos C, João LA (2011) Transformerless series active power filter to compensate voltage. In: Proceedings of the 14th European conference on power electronics and applications 2011. Birmingham, pp 1–6
29. Alireza J, Handy FB, Kamal A-H (2012) A novel transformer less hybrid series active filter. In: IECON 2012—38th annual conference on IEEE industrial electronics society. IEEE, Montreal, QC, pp 5312–5317
30. Alireza J, Kamal A-H (2015) A single-phase active device for power quality improvement of electrified transportation. IEEE Trans Ind Electron 62(5):3033–3041
31. Alireza J, Abdelhamid H, Woodward L, Al-Haddad K (2016) Experimental investigation on a hybrid series active power compensator to improve power quality of typical households. IEEE Trans Ind Electron 63(8):4849–4859
32. Alireza J, Abdelhamid H, Auguste N, Kamal A-H (2017) Power quality enhancement of smart households using a multilevel-THSeAF with a PR controller. IEEE Trans Ind Smart Grid 8(1):465–473
33. Ricardo Nogueira Santos W, Roberto Cabral da Silva E, Jacobina CB (2014) The transformer-less single-phase universal active power filter for harmonic and reactive power compensation. IET Power Electron 29(7):3563–3572
34. Venkatraman K, Moorthi S, Selvan MP (2017) Modelling and control of transformer-less universal power quality conditioner (TUnPQC): an effective solution for power quality enhancement in distribution system. J Control Autom Electr Syst 28(1):123–134
35. Venkatraman K, Kiran Kumar n, Moorthi S, Selvan MP (2019) Universal power quality conditioner (UnPQC) for enhancing the power quality in distribution system: an effective solution for power quality enhancement in distribution system. J Power Technol 99(3):195–203
36. Bin Y, Kangli L, Sen Z, Jianfeng Z (2018) Design and implementation of novel multi-converter-based unified power quality conditioner for low-voltage high-current distribution system. Energies 11(11):3150
37. Jian Y, Gooi HB, Benfei W, Xinan Z (2019) A new flexible power quality conditioner with model predictive control. IEEE Trans Ind Inf 5(5):2569–2579
38. Genu LGB, Limongi LR, Cavalcanti MC, Bradaschia F, Azevedo GMS (2020) Single-phase transformerless power conditioner based on a two-leg of a nine-switch converter. Int J Electr Power Energy Syst 117:1–11
39. Masoud S, Hossein I-E (2016) Improving the performance of a cascaded H-Bridge-based interline dynamic voltage restorer. IEEE Trans Power Deliv 31(3):1160–1167
40. Yong L, Guochun X, Lei B, Xuanlv W, Sihan Z (2014) A transformerless active voltage quality regulator with the parasitic boost circuit. IET Power Electron 29(4):1746–1756
41. Dariush N, Mohammadreza F, Hafez N (2017) Transformer-less dynamic voltage restorer based on buck-boost converter. IET Power Electron 10(13):1767–1777

A Novel Index Method for Distributed Generator Placement in a Radial Distribution System Using Pandapower Python Module

C. M. Thasnimol and R. Rajathy

Abstract A novel index method to find out the optimal location and optimal size of DGs in a distribution system is presented with the objective of loss minimization and reduction in voltage deviation. To highlight the importance of nodes in terms of loss reduction and improvement in voltage profile, a new metric called Aggregate Loss and Voltage Sensitivity Index (ALVSI) is proposed. The optimal location and size are found out to maximize the Aggregate Loss and Voltage Sensitivity Index (ALVSI). A detailed analysis was presented for the CIGRE medium voltage distribution system benchmark. The simulation platform used is the Pandapower Python module.

Keywords DG placement · Loss sensitivity index · Voltage Sensitivity Index

Abbreviations

$K_{BC}P_L$	Base case power loss at kth line
$BC\Delta V_i$	Base case voltage deviation at ith bus
i	Bus number iterator
j	Bus at which solar generator is integrating
k	Bus number iterator
$\Delta V_i G_j$	Voltage deviation at ith bus with DG on jth bus
P_{LKGJ}	Power loss on kth line with DG on jth bus
N_c	Candidate bus for DG placement
N_b	Number of buses
N_l	Number of lines
N_{DG}	Number of DGs

C. M. Thasnimol (✉) · R. Rajathy
Pondicherry Engineering College (PEC), Pillaichavadi, Puducherry 605014, India
e-mail: cmthasni@pec.edu

N. Zhou and S. Hemamalini (eds.), *Advances in Smart Grid Technology*, Lecture Notes in Electrical Engineering 688, https://doi.org/10.1007/978-981-15-7241-8_27

1 Introduction

The traditional power grid is characterized by centrally located power plants and one-way power flow. Due to the increase in power demand, it became necessary to go for additional energy resources. Building supplemental large power plants is not a practical option due to high cost involved. The large transmission power loss, need for additional transmission lines and increasing awareness about global warming and carbon emission forced us to rethink the conventional energy sources. The remedy is to generate power locally near the load center.

An important factor that decides the power transfer capability is the line losses. For maintaining voltage stability, it is important to reduce the line losses. By providing active and reactive power support locally, we can improve voltage regulation and minimize power transmission losses. Distributed generators (DG) like wind, solar CHP and micro-generators can do this job at a lower cost. It is not a practical option to give support to every bus of the distribution grid. Therefore, it is necessary to select an appropriate site and sizing for DGs in such a way to get maximum benefits. The optimal DG placement problem is one of the key steps in the distribution planning process. If this is not done effectively, it will reduce DG owner's profit and also will cause severe problems to the grid.

The objectives for DG sizing and placement can be loss minimization [1–4], congestion management [5], voltage profile improvement [6], voltage stability enhancement [7–9], increasing system loadability [10], harmonic minimization [11, 12], etc. The level of DG penetration is defined as the ratio of total active power generated by DG to the active power demand in the distribution network (DN). As per Government of India regulations, the maximum DG penetration level in the distribution system should be between 40 and 50% [13]. Another factor that limits the DG penetration level is the transient stability of the system [14]. Under the low level of penetration, it will not affect the transient stability of the system but as the DG penetration level increases, the impact on transient stability also increases [15]. Another important factor is the steady-state voltage rise of the network [16].

To model, the effect of solar generation an Aggregate Loss and Voltage Sensitivity Index (ALVSI) is formulated which can picture the ability of each bus generation to improve the system parameters. The differential evolution (DE) optimization algorithm is developed to find out the optimal location of DGs based on ALVSI. The proposed method was tested using the CIGRE 14 bus radial distribution system using the Pandapower Python module. This paper is organized as follows. The first section gives an introduction and some of the prominent researches related to the DG placement problems. Second section discusses the performance indexes employed in this study. A load flow-based method which employs the DE optimization technique is developed to find out the optimal location as well as the optimal size simultaneously using the proposed performance index. Fifth section presents the results of the DG placement problem. The conclusion of the study is presented in sixth section.

2 Problem Formulation

The distribution system considered in the study is radial, and for the optimal location of DG, the objective is taken as simultaneous minimization of the loss and voltage deviation. In this section, a novel performance index called Aggregate Loss and Voltage Sensitivity Index (ALVSI) is formulated to indicate the importance of each bus in terms of its capability in minimization of loss and reduction in voltage deviation. Metrics called loss sensitivity index (LSI) and voltage sensitivity index (VSI) are also explained briefly.

2.1 *Loss Sensitivity Index (LSI)*

Loss sensitivity index for line *k*, with respect to bus *j* (LSI_{kj}), is defined as the reduction in *k*th line loss as a fraction of base case loss for the integration of solar power in bus *j*. It characterizes the influence of *j*th bus generation on *k*th line's loss.

LSI for *k*th line, with respect to *j*th bus can be represented as given in Eq. 1:

$$LSI_{kj} = \frac{P_{LkBC} - P_{LkGj}}{P_{LkBC}} \tag{1}$$

$$LSI_{jAg} = \sum_{j,k=1}^{N_l} LSI_{kj} \tag{2}$$

LSI_{jAg} (Eq. 2) is the aggregate value of LSI for a bus and it shows the effect of solar integration in *j*th bus in reducing the line losses in the entire distribution system.

2.2 *Voltage Sensitivity Index (VSI)*

VSI is defined as the change in voltage deviation with and without DG. VSI for *i*th bus with DG on *j*th bus can be represented as given in Eq. 3:

$$VSI_{ij} = \frac{\Delta ViBC - \Delta ViGj}{\Delta Vi} \tag{3}$$

$$VSI_{jAg} = \sum_{j,i=1}^{N_b} \frac{VSI_{ij}}{N_b} \tag{4}$$

VSI_{jAg} as given in Eq. 4 indicates the power of the respective bus generation in improving the whole system voltage profile.

2.3 Aggregate Loss and Voltage Sensitivity Index (ALVSI)

In this section, a new metric Aggregate Loss and Voltage Sensitivity Index (ALVSI) is proposed to find out the optimal location of DG in the given DN. To consider the effectiveness of DG in loss reduction and voltage profile improvement, a new metric named Aggregate Loss and Voltage Sensitivity Index (ALVSI) is formulated. ALVSI, which is the weighted sum of VSI_{jAg} and LSI_{jAg}, gives a good indication of the capability of each bus generation for loss reduction and voltage profile improvement. ALVSI of jth bus at a penetration level of G can be written as:

$$ALVSI_jG = w_1 * LSI_{jAg} + w_2 * VSI_{jAg} \tag{5}$$

As both LSI and VSI are dimensionless quantities, they can be added together without a further scaling operation. As in this study, both loss reduction and voltage enhancement are given with equal priority, the constants w_1 and w_2 are taken as 0.5. The obtained ALVSI values are in the range between 0 and 1. Table 1 depicts the ranges and inferences of ALVSI values obtained through power flow solutions.

3 Algorithm for Selecting the Optimal Location and Size for Installing Solar PV

The optimal DG location and sizing problem are explained briefly in this section using a novel performance index called as Aggregate Loss and Voltage Sensitivity Index (ALVSI). The optimal DG placement problem is a multi-objective problem with objective of loss minimization and reduction in voltage deviation. In this paper, these two objectives are mapped together into a single performance index known as Aggregate Loss and Voltage Sensitivity Index (ALVSI). The problem of finding out the optimum size and location of DG is modeled as an optimization problem to maximize the performance index ALVSI and is solved by using differential evolution (DE) algorithm.

The objective function for DG allocation problem is given by:

$$\text{Maximize } ALVSI_jG \tag{6}$$

Table 1 Inference from ALVSI values

ALV SIjG	Inference
0	No further decrease in losses and voltage deviation from the base case
1	losses and voltage deviation became zero
$0 \leq ALVSIjG \leq 1$	Losses and voltage deviation are less than that of the base case
$ALV\,SIjG \leq 0$	Losses and voltage deviation are more than that of the base case

Subject to the constraints:

$$V_{\min} < V_k < V\max V_{\min} = 0.95\,\text{pu}\; V\max = 1.05\,\text{pu}$$

The detailed procedure for the optimal DG allocation problem is given in Fig. 1. The first step in the DE algorithm is initialization. A population of DG allocation vectors is generated randomly during this stage. The trial vectors formed by mutation and crossover operation are compared with the target vectors during the selection stage. The vectors which give the highest ALVSI are selected for the next generation. The process stops when the maximum number of generations is reached.

4 Results and Discussion

A case study on CIGRE 14 bus distribution system [17] is given in this section. The bus connected to the external grid and buses without load is not considered in this study for DG placement. Bus 2 does not have any load connected to it, and therefore, it is not considered for DG placement.

4.1 *Loss Sensitivity Index*

The present analysis has been carried out in CIGRE Medium Voltage 14 bus test system. LSI for one unit of solar power penetration is shown in Table 2. One unit of solar generation is integrated with the first bus, and the line losses are calculated in every line. The change in line loss from the base case value (line loss before incrementing solar generation) is calculated. The same process will be repeated in each bus in the system, and the corresponding LSI is calculated. An LSI greater than 0.3 represents a strong interdependency between a line and a bus. It is clear from Table 2 that lines 0, 1, 6, 7, 8 and 9 strongly depend on bus 10. It may be highlighted that a change in the solar generation on bus 10 will highly influence the line losses in these lines. For a unit increment in solar generation, there will be a loss reduction of 39.99% in line 0, 39.62% in line 1, 79% in line 6, 92.1% in line 7 and 65.89% in line 9. Further, the lines 2, 3 and 4 are strongly depended on bus 6, where a unit increment in solar generation resulted in line losses decline of 79.48% in line 2, 91.06% in line 3 and 32.9% in line 4.

The study further tried to calculate the aggregate loss sensitivity index (LSI_{Ag}) of each bus. LSI_{Ag} quantifies the influence of a particular bus in reducing the losses of the entire distribution system. It is found from the LSI_{Ag} that bus 10 is the most influencing bus in the system with a LSI_{Ag} of 3.254. Therefore, bus 10 can be selected as an optimal site for installing solar generation for this particular solar energy penetration level. The next most participating bus is bus 6 with LSI_{Ag} Index of 2.9016. Therefore, bus 6 is the most suitable bus to install the next DG.

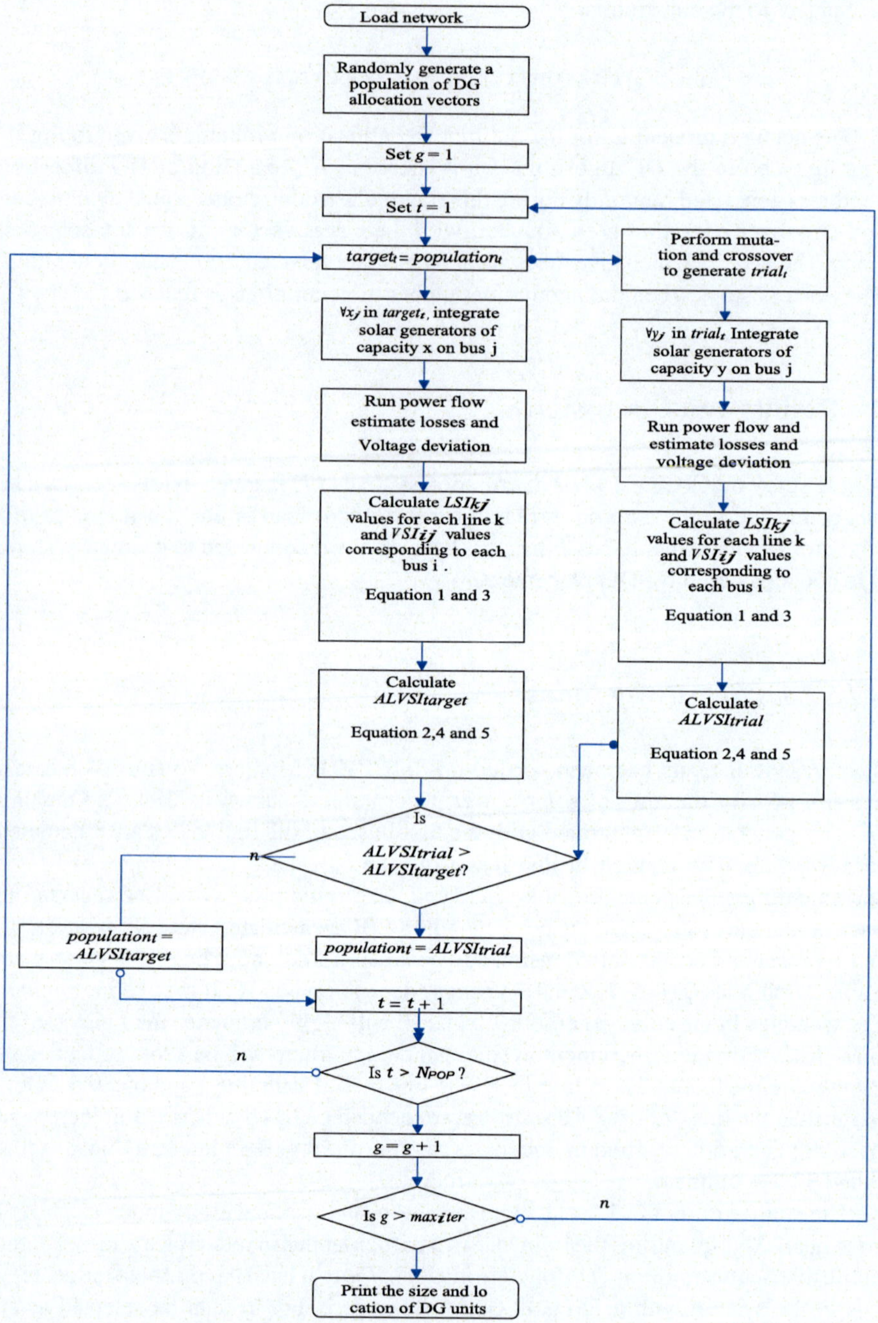

Fig. 1 DE algorithm for optimal DG allocation

Table 2 LSI for CIGRE 14 bus DS

Lines bus no	1	2	3	4	5	6	7	8	9	10	11	12	13	14
0	0.001	0.37	0.3948	0.3960	0.3967	0.3968	0.3970	0.3981	0.3986	0.3990	0.3990	0	0	0
1	0.001	0.01	0.3920	0.3931	0.3938	0.3939	0.3942	0.3953	0.3958	0.3962	0.3961	0	0	0
2	0.001	0.01	0.0266	0.7936	0.7947	0.7948	0.0269	0.0271	0.0272	0.0272	0.0272	0	0	0
3	0.001	0.01	0.0266	0.0285	0.9104	0.9106	0.0270	0.0271	0.0272	0.0273	0.0273	0	0	0
4	0.001	0.01	0..0267	0.0286	0.0303	0.3293	0.0271	0.0272	0.0273	0.0274	0.0274	0	0	0
5	0.001	0.01	0..0299	0.0301	0.0302	0.0302	−116	0.0346	0.0347	0.0348	0.0347	0	0	0
6	0.001	0.01	0..0269	0.0271	0.0272	0.0272	0.0309	0.0311	0.7894	0.7902	0.7901	0	0	0
7	0.001	0.01	0.0268	0.0270	0.0271	0.0271	0.0308	0.0310	0.0320	0.9213	0.9213	0	0	0
8	0.001	0.01	0.0269	0.0271	0.0272	0.0272	0.0309	0.0311	0.0321	0.0343	−2.8479	0	0	0
9	0.001	0.01	0.0270	0.0272	0.0273	0.0273	0.6557	0.6574	0.6582	0.6589	0.6587	0	0	0
10	0	0	0	0	0	0	0	0	0	0	0	0.001	0.38	0.38
11	0	0	0	0	0	0	0	0	0	0	0	0.01	0.01	0.2415
LSIAg	0.0123	0.45	0.95	1.7218	2.6066	2.9016	−115.21	1.6008	2.3623	3.254	0.3703	0.0011	0.3817	0.6015

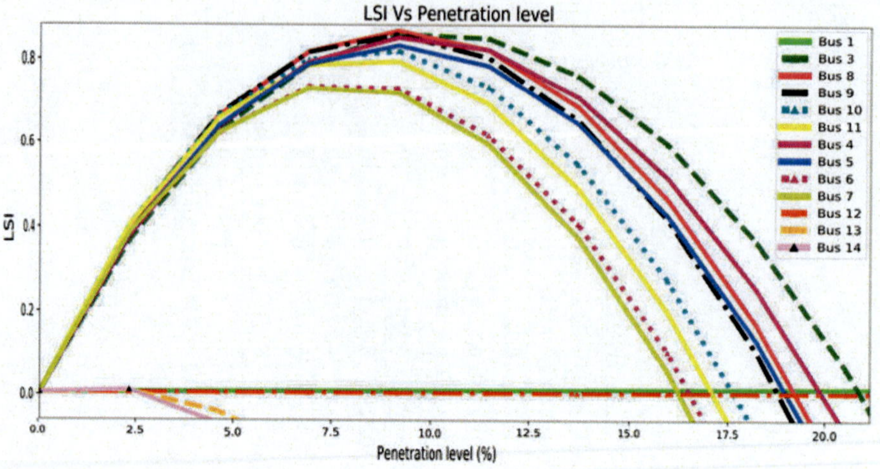

Fig. 2 LSI versus penetration

Figure 2 shows the variation of LSI with penetration level. Up to a penetration level of 2.5% bus, 11 is the most influencing bus in terms of loss reduction. After that bus 10 predominates up to around 5%. The optimal location is bus 9 from about 5 to 7.5% of penetration level, and after that, bus 8 has a maximum value of LSI and hence it predominates. From Fig. 2, it is clear that the optimum penetration level is 10%, and the optimal location is bus 8. For a penetration level greater than 10%, bus 3 predominates up to 25% and after that bus 1 is the only bus capable of making loss reduction. LSI value of all the other buses is negative, which means that installing PV on these buses results in increasing losses beyond the base case value.

4.2 *Voltage Sensitivity Index (VSI)*

Table 3 shows the Voltage Sensitivity Index of each bus in the distribution system. VSI depicts the interdependency between system voltage profile and generation profile. From the table, we can see that the voltage profile of buses 1, 2, 3, 8 and 9 strongly depends on solar generation on bus 10. The voltage profile of buses 4 and 5 greatly depends on the bus 6 generation. Similarly, buses 12 and 13 depend on generation on bus 14. The Aggregate Voltage Sensitivity Index (VSI_{Ag}) of each bus is calculated further. VSI_{Ag} indicates the power of the respective bus generation in improving the whole system voltage profile. Penetrating solar panel in the bus having high VSI_{Ag} will improve the system voltage profile. Further, it improves the voltage stability of the whole system. It is clear from Table 3 that for a penetration level of one unit of solar power, VSI_{Ag} of bus 11 is the maximum, and hence, it is the ideal bus for PV placement. The next optimal position is bus 10 with VSI_{Ag} of 2.0789. Variation of VSI with penetration level is shown in Fig. 3.

Table 3 VSI for CIGRE 14 bus DS

	Bus nos.													
Bus nos.	1	2	3	4	5	6	7	8	9	10	11	12	13	14
1	0.0799	0.1123	0.1650	0.1673	0.1687	0.1690	0.1696	0.1717	0.1727	0.1736	0.1735	0.0000	0.0000	0.0000
2	0.0208	0.1494	0.1812	0.1826	0.1834	0.1836	0.1839	0.1852	0.1858	0.1864	0.1863	0.0000	0.0000	0.0000
3	0.0100	0.0721	0.1795	0.1807	0.1815	0.1816	0.1819	0.1831	0.1836	0.1841	0.1840	0.0000	0.0000	0.0000
4	0.0098	0.0704	0.1752	0.1878	0.1886	0.1888	0.1775	0.1787	0.1792	0.1796	0.1796	0.0000	0.0000	0.0000
5	0.0096	0.0692	0.1723	0.1848	0.1959	0.1960	0.1746	0.1757	0.1762	0.1767	0.1766	0.0000	0.0000	0.0000
6	0.0095	0.0680	0.1691	0.1813	0.1922	0.2202	0.1714	0.1725	0.1729	0.1734	0.1733	0.0000	0.0000	0.0000
7	0.0093	0.0670	0.1666	0.1678	0.1685	0.1686	0.2217	0.1933	0.1939	0.1944	0.1943	0.0000	0.0000	0.0000
8	0.0094	0.0672	0.1672	0.1683	0.1690	0.1692	0.1928	0.1940	0.1945	0.1951	0.1950	0.0000	0.0000	0.0000
9	0.0092	0.0664	0.1652	0.1663	0.1670	0.1672	0.1905	0.1917	0.1979	0.1984	0.1983	0.0000	0.0000	0.0000
10	0.0091	0.0654	0.1627	0.1639	0.1645	0.1647	0.1876	0.1888	0.1949	0.2088	0.2087	0.0000	0.0000	0.0000
11	0.0091	0.0653	0.1624	0.1635	0.1642	0.1643	0.1872	0.1884	0.1945	0.2084	0.2140	0.0000	0.0000	0.0000
12	0.0000	0.0000	0.0000	0.0000	0.0000	0.0000	0.0000	0.0000	0.0000	0.0000	0.0000	−3.6350	−3.6764	−3.6665
13	0.0000	0.0000	0.0000	0.0000	0.0000	0.0000	0.0000	0.0000	0.0000	0.0000	0.0000	0.1143	0.5484	0.5475
14	0.0000	0.0000	0.0000	0.0000	0.0000	0.0000	0.0000	0.0000	0.0000	0.0000	0.0000	0.0719	0.9137	0.5743
$VSIAg$	0.18569	0.8727	1.8664	1.9143	1.9435	1.9732	2.0387	2.0231	2.0461	2.0789	2.0836	−3.4488	−2.2044	−2.5546

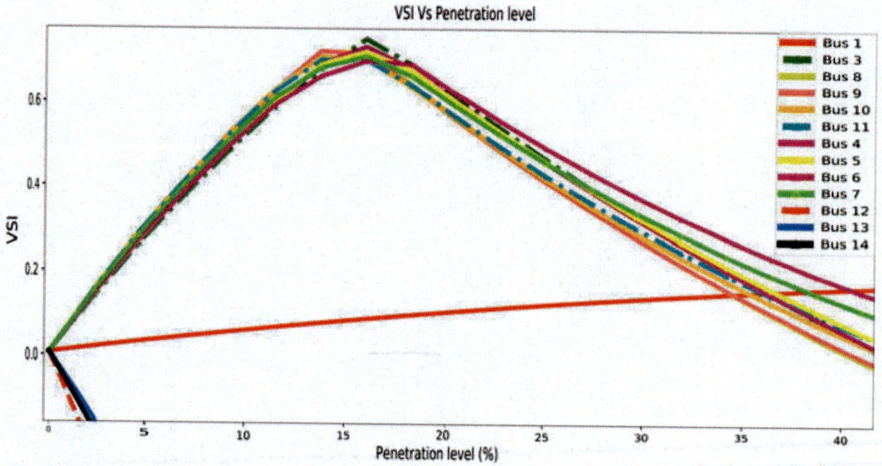

Fig. 3 VSI versus penetration

4.3 *Aggregate Loss and Voltage Sensitivity Index (ALVSI)*

A graph of ALVSI versus penetration level is shown in Fig. 4. Up to a penetration level of 4% bus, 11 has a maximum value of ALVSI and is the optimum position of DG. Bus 10 is the most influencing bus for a penetration level up to 9.27%. After that bus 9 predominates up to 13.6%. From 13.6 to 14.6%, bus 8 is the most influencing bus. After that, bus 3 is dominating. After 31% of penetration level, bus 1 is the only bus which has a positive value of ALVSI. This value of all the other buses is negative

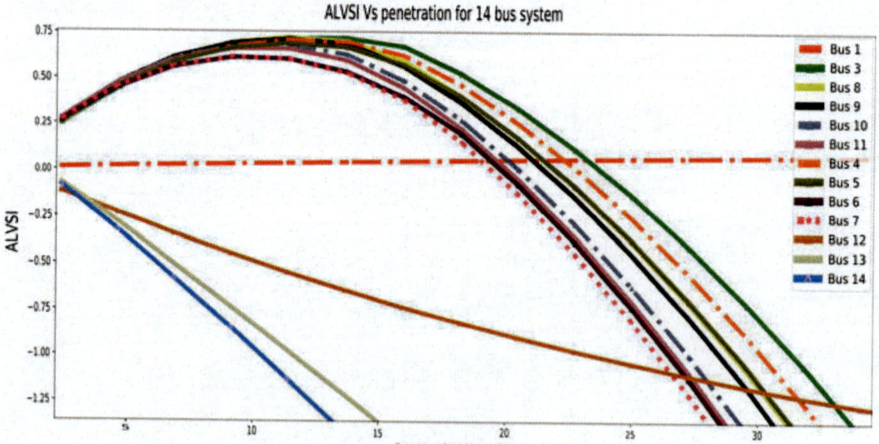

Fig. 4 ALVSI versus penetration

which indicates that increasing the DG penetration beyond this value in those buses will increase the line losses and deteriorate the voltage profile compared to the base case without DG.

The ranking of buses for each penetration level together with corresponding ALVSI values is given in Table 4. From the table, it is understood that the most influencing bus will be the one which is at the edge of a feeder at low penetration levels (bus 11), and as the penetration level increases, the interior buses (bus 9, bus 10) will become more superior in terms of loss reduction and voltage improvement. At high penetration levels, the bus at the beginning of a feeder will be the dominant one.

After a particular penetration level (34.3%), only the starting bus which is fed directly from the external grid (bus 1) is capable of doing loss reduction and voltage improvement that too at a very little extent (ALVSI value is only 0.1456). All the

Table 4 Bus ranking and ALVSI values for different penetration level in CIGRE 14 bus distribution system

Penetration (%)	Size (MW)	Rank 1	Rank 2	Rank 3	Rank 4	Rank 5
2.2	1	(11, 0.1753)	(10, 0.175)	(9, 0.1724)	(7, 0.1709)	(8, 0.1705)
4.4	2	(10, 0.3272)	(11, 0.327)	(9, 0.3238)	(8, 0.3208)	(7, 0.3182)
6.8	3	(10, 0.4591)	(11, 0.4578)	(9, 0.4565)	(8, 0.4531)	(7, 0.4445)
9	4	(10, 0.5726)	(9, 0.5722)	(11, 0.5698)	(8, 0.5689)	(7, 0.5517)
11.3	5	(9, 0.6722)	(8, 0.6697)	(10, 0.6694)	(11,0.6645)	(5, 0.6427)
13.3	6	(9, 0.7566)	(8, 0.7564)	(10, 0.7353)	(4, 0.7274)	(5, 0.7247)
15.9	7	(3, 0.7815)	(4, 0.758)	(8, 0.742)	(5,0.7398)	(9, 0.7325)
18.18	8	(3, 0.7056)	(4,0.6877)	(5,0.6769)	(6,0.6631)	(8, 0.645)
20.45	9	(3, 0.6091)	(4, 0.589)	(5, 0.577)	(6,0.5618)	(8, 0.5415)
22.7	10	(3, 0.511)	(4,0.4888)	(5,0.4756)	(6,0.4589)	(8, 0.4366)
25	11	(3, 0.4116)	(4,0.3873)	(5,0.3728)	(6,0.3545)	(8, 0.3302)
31.81	14	(1, 0.1395)	(3,0.1043)	(4,0.0736)	(5,0.0554)	(6, 0.0323)
34.09	15	(1, 0.1456)	(4, −0.034)	(5, −0.0535)	(6, −0.0783)	(3, −0.0012)

other buses will aggravate the situation than the base case (ALVSI values for all the other buses are negative). A negative value for ALVSI indicates that increasing the DG penetration beyond this value will intensify the line losses and deteriorate the voltage profile than the base case without DG.

4.4 Optimal DG Placement Based on ALVSI

Comparison of DG placement based on LSI, VSI and ALVSI are given in Table 5. In the above base case, the losses and voltage deviation are, respectively, 0.233 MW and 0.7468 without DG. The impact of DG placement on voltage stability is analyzed by drawing the PV curve as shown in Fig. 5a, b.

From Fig. 5a, it is clear that the maximum loadability limit of the base case without DG is only 20 and the corresponding voltage stability margin is 19. When single DG placement is done through LSI, the loadability margin is increased to 28, the voltage

Table 5 DG allocation and comparison of system parameters. number of DG performance index bus no: size loss (MW) voltage deviation voltage stability margin

Without DG	–	–	–	0.233	0.7468	19
1 DG	LSI VSI ALVSI	8 3 3	4.0905 7.0724 6.1	0.0329 0.098 0.06068	0.2684 0.0625 0.1219	27 29 29
2 DG	LSI VSI ALVSI	8 5 8 5 9 5	2.07 2.30 3.58 2.94 3.59 2.6	0.0228 0.0710 0.059	0.2468 0.053 0.0690	25.66 28.0033 29.0033

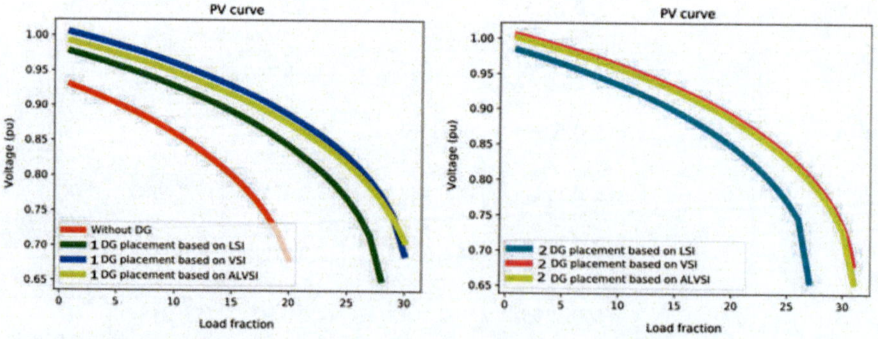

(a) Single DG placement (b) 2DG placement

Fig. 5 Impact of DG placement using different indices on maximum loadability and voltage stability margin at bus 6 stability margin is increased to 27, the total system losses are decreased to 0.0329 and the voltage deviation has decreased to 0.2684 from the base case. In the case of VSI-based single DG placement, the losses are more (0.098 MW) and voltage deviation (0.0625) is less than that of LSI based one.

Further, the loadability limit is 30 and the stability margin is 29 for VSI-based optimum DG placement. ALVSI-based DG placement results in reduced losses compared to VSI-based DG placement. It also achieves the same system loadability that results in the same voltage stability margin as that of the VSI-based DG placement.

LSI-based 2DG placement results in a total system loss of 0.0228 and a voltage deviation of 0.2468. The voltage stability margin corresponding to LSI based 2DG placement is 25.66. VSI-based 2DG placement improves the system voltage profile than that of 1DG case but the voltage stability margin (28) is less than that of VSI and ALVSI-based 1DG placement. ALVSI-based 2DG placement gives a high voltage stability margin of 29.003 with reduced total system losses of 0.059 than that of VSI based counterpart.

5 Conclusion

A novel performance index for the optimal placement of distributed generators in a radial distribution system is proposed with the objective of loss reduction and reduction in voltage deviation. The capability of the DGs that optimally placed based on the three indices LSI, VSI and ALVSI for improving the various system parameters is analyzed. Results are presented for CIGRE 14 bus distribution system.

References

1. Sharma R, Gupta AR (2016) Seasonal loading impact on performance of solar dg for loss reduction based on capacity factor. Proc Technol 25:751–758
2. Asmi RY, Salama M, Ahmad M et al (2018) Distributed generation's integration planning involving growth load models by means of genetic algorithm. Arch Electr Eng
3. Remha S, Chettih S, Arif S (2018) A novel multi-objective bat algorithm for optimal placement and sizing of distributed generation in radial distributed systems. Adv Electr Electron Eng 15(5):736–746
4. Arulraj R, Kumarappan N (2018)Optimal single and multiple dg installations in radial distribution network using SLPSO algorithm. In: Intelligent and efficient electrical systems. Springer, pp 89–96
5. Kashyap M, Kansal S (2018) Hybrid approach for congestion management using optimal placement of distributed generator. Int J Ambient Energy 39(2):132–142
6. Sujatha M, Roja V, Prasad TN (2019) Multiple DG placement and sizing in radial distribution system using genetic algorithm and particle swarm optimization. In: Computational intelligence and big data analytics. Springer, pp 21–36

7. Antony NR, Baby S (2013) Optimal dg placement considering voltage stability enhancement using PSO. In: 2013 international conference on control communication and computing (ICCC). IEEE, pp 394–399
8. Aravinth A, Vatul VA, Narayanan K, Muthukumar K, Senjyu T (2019) A multi objective frame work to improve voltage stability in a distribution network. Int J Emerg Electric Power Syst 20(1)
9. Ettehadi M, Ghasemi H, Vaez-Zadeh S (2012) Voltage stability-based dg placement in distribution networks. IEEE Trans Power Deliv 28(1):171–178
10. Nuri M, Miveh MR, Mirsaeidi S, Gharibdoost MR (2012) Distributed generation placement to maximize the loadability of distribution system using genetic algorithm. In: IEEE proceedings of 17th conference on electrical power distribution, pp 1–5
11. Din FU, Ahmad A, Ullah H, Khan A, Umer T, Wan S (2019) Efficient sizing and placement of distributed generators in cyber-physical power systems. J Syst Arch
12. HassanzadehFard H, Jalilian A (2018) Optimal sizing and location of renewable energy based dg units in distribution systems considering load growth. Int J Electr Power Energy Syst 101:356–370
13. Miller N, Ye Z (2003) Report on distributed generation penetration study. Tech Rep. National Renewable Energy Lab (NREL), Golden, CO
14. Slootweg J, Kling W (2002) Impacts of distributed generation on power system transient stability. IEEE Power Eng Society Summer Meet 2:862–867
15. Reza M, Schavemaker P, Slootweg J, Kling W, Van Der Sluis L (2004) Impacts of distributed generation penetration levels on power systems transient stability. In: IEEE power engineering society general meeting, pp 2150–2155
16. Shayani RA, de Oliveira MAG (2010) Photovoltaic generation penetration limits in radial distribution systems. IEEE Trans Power Syst 26(3):1625–1631
17. Bhutto GM, Bak-Jensen B, Mahat P (2013) Modeling of the CIGRE low voltage test distribution network and the development of appropriate controllers. Trans Smart Grid Clean Energy 2:184–191

Smart Eco Structures and Systems

Haar FCM with DEMATEL Technique to Analyze the Solid Waste Management

A. Felix and Saroj Kumar Dash

Abstract An interesting attempt is taken to form a hybrid model by integrating Fuzzy Cognitive Map (FCM), the Delphi Method, and the DEMATEL method through the Haar ranking of Hexagonal Fuzzy Number. To examine this model, the problem of solid waste management is taken. Solid waste arises from human and animal activities can generate major health problems and horrible living environment when they are not dumped of safely and appropriately. The proper disposal of solid waste helps to reduce the terrible impacts on both human health and environment to sustain economic growth and better quality of life. Therefore, this present study analyzes the solid waste management through the Haar FCM with DEMATEL technique.

Keywords Fuzzy logic · Hexagonal fuzzy number · Haar ranking · FCM · Delphi method · DEMATEL

1 Introduction

Multiple attribute decision making (MADM) method is the well know branch of decision making. DEMATEL, AHP, ANP, VIKOR, TOPSIS, etc., are powerful tools for solving multi attribute decision making problems in fuzzy environment. DEMATEL is one of the effective techniques, designed by the Battelle Memorial Institute through its Geneva Research Centre [8, 9]. It is a practical tool which is used for visualizing the complex structure into a simple structure. It can also be defined by the matrices or directed graphs. Recently, DEMATEL has been attracted by the many researchers and applied successfully in software system design [11], hospital service quality [20], and social related problems [2]. When designing this model, the relationships among the factors are usually assigned by the crisp values. Nevertheless, in many situations, human decisions with preferences are often complex in nature so crisp values are

A. Felix · S. K. Dash (✉)
Mathematics Division, School of Advanced Sciences, VIT Chennai, Chennai, Tamil Nadu, India
e-mail: sarojkumar.dash@vit.ac.in

N. Zhou and S. Hemamalini (eds.), *Advances in Smart Grid Technology*, Lecture Notes in Electrical Engineering 688, https://doi.org/10.1007/978-981-15-7241-8_28

inadequate due to inherent uncertainties in the data. Hence, linguistic terms facilitate the decision makers to describe the correlation between criteria [23, 24].

Fuzzy Cognitive Map (FCM) was designed by Kosco [13] from the Cognitive maps. It is one of the simple and powerful techniques in decision making, which motivated by the cognition of human brain. FCM is a directed graph relating the concepts. The relationship between concepts is determined by the weights of the link joining the concepts. In FCM, fuzzy weights are used rather than crisp values to describe the strength of the connection between concepts. It is one of the prominent techniques to acquire the experts' opinion in a natural manner. Classical cognitive maps take the connection weights only from $\{-1, 0, 1\}$ whereas FCM takes intervals, linguistic terms or fuzzy numbers as the connection weights.

The current research has paid more attention on uncertain linguistic terms to solve decision making problems where linguistic terms are employed as the input variables in decision making problems and fuzzy numbers are used to assign the values for the linguistic terms. Fuzzy DEMATEL method was extended using different fuzzy numbers triangular [2, 15], trapezoidal [21]. Currently, few of the researchers are interested in designing the hybrid model to have more flexibility and to reduce the time complexity. The hybrid models such as DEMATEL-CETD [19], FCM is combined with VIKOR [3] and with DEMATEL [7] TOPSIS-FCM [6], TOPSIS-DEMATEL [1], etc. were designed in the recent years. Hence, a new attempt is made to design a hybrid system called Fuzzy Haar FCM with DEMATEL using Hexagonal fuzzy number [18] and Haar ranking [4]. To examine this method, the problem of solid waste management is chosen.

It is stated by Britannica that "Solid-waste management, the collecting, treating, and disposing of solid material that is discarded because it has served its purpose or is no longer useful. Improper disposal of municipal solid waste can create unsanitary conditions, and these conditions in turn can lead to pollution of the environment and to outbreaks of vector-borne disease that is, diseases spread by rodents and insects." In the past two decades, many researchers have designed multi criteria decision making models on solid waste management which is briefed as follows. Multi criteria decision analysis for the proper land fill capacity [10], land planning of regional solid waste management systems [12], anticipation of solid waste generation in a fast-growing metropolitan cities [5], A sustainable municipal solid waste (MSW) management [16], selecting appropriate municipal solid waste disposal methods [14], causal structure of urban solid waste management [22], waste management system in metropolitan city councils in Melbourne [17]. From the review it is observed that all the researchers have focused either triangular or trapezoidal fuzzy number to represent the value for linguistic term. Hence, this paper designs the hybrid system Haar FCM with DEMATEL using hexagonal fuzzy number to analyze the solid waste management in Chennai.

2 Theoretical Background

Definition 2.1 A fuzzy set $\tilde{A}$ is a subset of a universe of discourse X, which is characterized by a membership function $\mu_{\tilde{A}}(x)$ representing a mapping $\mu_{\tilde{A}}:X \rightarrow [0, 1]$. The function value of $\mu_{\tilde{A}}(x)$ is called the membership value, which represents the degree of truth that x is an element of fuzzy set $\tilde{A}$.

Definition 2.2 A fuzzy set $\tilde{A}$ defined on the set of real numbers R is said to be a fuzzy number and its membership function $\tilde{A}$ is mapping from the real line R to the closed interval [0, 1], holds $\tilde{A}$ is convex, normal and piecewise continuous.

Definition 2.3 The α-cut of the fuzzy set $\tilde{A}$ of the universe of discourse X is defined as $\tilde{A}_\alpha = \{x \in X/\mu_{\tilde{A}}(x) \geq \alpha\}$, where $\alpha \in [0, 1]$.

Definition 2.4 A triangular fuzzy number $\tilde{N}$ can be defined as a triplet (l, m, r) and the membership function $\mu_{\tilde{N}}(x)$ is defined as:

$$\mu_{\tilde{N}}(x) = \begin{cases} 0 & x < l \\ \left(\frac{x-l}{m-l}\right) & l \leq x \leq m \\ \left(\frac{r-x}{r-m}\right) & m \leq x \leq r \\ 0 & x > r \end{cases}$$

where l, m, r are real numbers and $l \leq m \leq r$.

Definition 2.5 A trapezoidal fuzzy number $\tilde{A}$ can be defined as $(a_1, a_{2,}, a_3, a_4)$ and the membership function is defined as

$$\mu_{\tilde{A}}(x) = \begin{cases} \left(\frac{x-a_1}{a_2-a_1}\right) & a_1 \leq x \leq a_2 \\ 1 & a_2 \leq x \leq a_3 \\ \left(\frac{a_4-x}{a_4-a_3}\right) & a_3 \leq x \leq a_4 \\ 0 & a_1 \leq 0 \text{ and } a_4 \geq 0 \end{cases}$$

where a_1, a_2, a_3, a_4 are real numbers $a_1 \leq a_2 \leq a_3 \leq a_4$.

Definition 2.6 Linear hexagonal fuzzy number with symmetry (LHFNS): A linear hexagonal fuzzy number is written as $\tilde{A}_{LS} = (a_1, a_2, a_3, a_4, a_5, a_6; h)$ whose membership function is written as

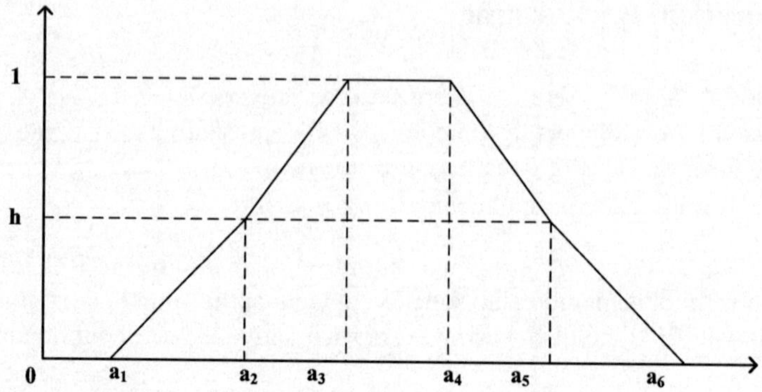

Fig. 1 Hexagonal fuzzy number

$$\mu_{\tilde{A}_{LS}}(x) = \begin{cases} h\left(\frac{x-a_1}{a_2-a_1}\right), & a_1 \le x \le a_2 \\ 1-(1-h)\left(\frac{x-a_2}{a_3-a_2}\right), & a_2 \le x \le a_3 \\ 1, & a_3 \le x \le a_4 \\ 1-(1-h)\left(\frac{a_5-x}{a_5-a_4}\right), & a_4 \le x \le a_5 \\ h\left(\frac{a_6-x}{a_6-a_5}\right), & a_5 \le x \le a_6 \\ 0, & x \le a_1 \text{ and } x \ge a_6 \end{cases}$$

Definition 2.7 Alpha cut-symmetry linear hexagonal fuzzy number (SLHFN): Given a fuzzy set A defined on X and any number $\alpha \in [0, 1]$, the α-cut, then we define α_A the crisp sets as $\alpha_A = \{x/A(x) \ge \alpha\}$

$$A_\alpha = \begin{cases} A_{1L}(\alpha) = a_1 + \frac{\alpha}{h}(a_2 - a_1)\ for\ \alpha \in [0, h] \\ A_{2L}(\alpha) = a_2 + \left(\frac{1-\alpha}{1-n}\right)(a_3 - a_2)\ for\ \alpha \in [h, 1] \\ A_{2R}(\alpha) = a_5 - \left(\frac{1-\alpha}{1-h}\right)(a_5 - a_4)\ for\ \alpha \in [h, 1] \\ A_{1R}(\alpha) = a_6 - \frac{\alpha}{h}(a_6 - a_5)\ for\ \alpha \in [0, h] \end{cases}$$

where $A_{1L}(\alpha), A_{2L}(\alpha)$ are increasing functions with respect to α and $A_{3R}(\alpha), A_{2R}(\alpha)$ decreasing functions with respect to α (Fig. 1).

3 Proposed Methodology

This present study designs the Fuzzy Haar DEMATEL-FCM system by integrating the significant features of FCM and DEMATEL methods through Haar ranking of Hexagonal Fuzzy number. This system is executed via three stages.

Stage-1 FCM Development:

(i) construct the initial linguistic uncertain direct relation matrix.

Let $C = \{C_i, i = 1, 2, \ldots, n\}$ be the attributes and $E = \{E_i, i = 1, 2, \ldots, k\}$ be the experts. Then, the linguistic uncertain direct relation matrices are framed with the aid of experts from the linguistic set $L = \{$No Influence (NI), Very Low (VL), Low (L), Medium (M), High (H), Very High (VH)$\}$.

(ii) Convert the initial linguistic uncertain relation matrix $\hat{Z}_k = [\hat{z}_{kij}]_{n\times n}$ into hexagonal fuzzy matrix $\tilde{Z}_k = [\tilde{z}_{kij}]_{n\times n}$ using the given table

(iii) Form a group uncertain direct relation matrix $\tilde{Z}_k = [\tilde{z}_{kij}]_{n\times n}$.

All the uncertain direct relation matrices $\tilde{Z}_1, \tilde{Z}_2, \ldots, \tilde{Z}_m$ are aggregated to frame a total uncertain direct relation matrix $\tilde{Z}_k = [\tilde{z}_{kij}]_{n\times n}$.

Stage-2 Converting Hexagonal Fuzzy Number into Haar Hexagonal Fuzzy Number

Let $\tilde{A} = (a_1, a_2, a_3, a_4, a_5, a_6)$ be considered as the Hexagonal fuzzy number. For using Haar wavelet ranking, the Hexagonal fuzzy number is rewritten as $\tilde{\tilde{A}} = (a_1, a_2, a_3, a_3, a_4, a_4, a_5, a_6)$. The average and detailed coefficients namely the scaling and wavelet coefficients of Hexagonal fuzzy number can be calculated as follows.

(i) Group the Hexagonal fuzzy numbers in pairs. $[a_1, a_2]$, $[a_3, a_3]$, $[a_4, a_4]$, $[a_5, a_6]$

(ii) The first four elements of $\tilde{\tilde{A}}$ with the average of these pairs (approximation coefficients) and replace the last 4 four elements of $\tilde{\tilde{A}}$ with half of the difference of these pairs (detailed coefficients).

$$\alpha_1 = \left(\frac{a_1 + a_2}{2}\right),\ \alpha_2 = \left(\frac{a_3 + a_3}{2}\right),\ \alpha_3 = \left(\frac{a_4 + a_4}{2}\right),\ \alpha_4 = \left(\frac{a_5 + a_6}{2}\right)$$

$$\beta_1 = \left(\frac{a_1 - a_2}{2}\right),\ \beta_2 = \left(\frac{a_3 - a_3}{2}\right),\ \beta_3 = \left(\frac{a_4 - a_4}{2}\right),\ \beta_4 = \left(\frac{a_5 - a_6}{2}\right)$$

The $\tilde{\tilde{A}}_1$ changed into $\tilde{\tilde{A}}_1 = (\alpha_1, \alpha_2, \alpha_3, \alpha_4, \beta_1, \beta_2, \beta_3, \beta_4)$

(iii) Group the pair of approximation coefficients of $\tilde{\tilde{A}}_1$. Then, find the new approximation coefficients and the detailed coefficients for the pair of approximation coefficient of $\tilde{\tilde{A}}_1[\alpha_1, \alpha_2]$, $[\alpha_2, \alpha_3]$

$$\gamma_1 = \left(\frac{\alpha_1 + \alpha_2}{2}\right), \gamma_2 = \left(\frac{\alpha_2 + \alpha_3}{2}\right);$$

$$\eta_1 = \left(\frac{\alpha_1 - \alpha_2}{2}\right), \eta_2 = \left(\frac{\alpha_2 - \alpha_3}{2}\right)$$

The $\tilde{\tilde{A}}_1$ changed into $\tilde{\tilde{A}}_2 = (\gamma_1, \gamma_2, \eta_1, \eta_2, \beta_1, \beta_2, \beta_3, \beta_4)$

(iv) Find the pair of approximation coefficient in $\tilde{\tilde{A}}_2$. Then, find the new approximation and detailed coefficients for the pair of approximation coefficient of $\tilde{\tilde{A}}_2$. $[\gamma_1, \gamma_2]$ $\delta_1 = \left(\frac{\gamma_1+\gamma_2}{2}\right)$, $\delta_2 = \left(\frac{\gamma_1-\gamma_2}{2}\right)$

The $\tilde{\tilde{A}}_2$ changed into $H(\tilde{\tilde{A}}) = (\delta_1, \delta_2, \eta_1, \eta_2, \beta_1, \beta_2, \beta_3, \beta_4)$

(v) Determine the sum of $H(\tilde{\tilde{A}})$. (i.e.)

$$D = \text{Sum}(H(\tilde{\tilde{A}})) = (\delta_1 + \delta_2 + \eta_1 + \eta_2 + \beta_1 + \beta_2 + \beta_3 + \beta_4)$$

Stage-3 DEMATEL Process

(i) Create the uncertain overall relation matrix $\tilde{T} = [t_{ij}]_{n\times n}$ using $[t_{ij}]_{n\times n} = D(I - D)^{-1},\ i, j = 1, 2, \ldots, n$

(ii) Determine the overall intensities of influencing and influenced criteria C_i

Let c_i be the correlation between C_i and C_j and it can be obtained by $c_i = \sum_{j=1}^{n} t_{ij},\quad i = 1, 2 \ldots n$

Let h_i be the overall strength the that criteria C_i is induced by other criteria, and it is got by $h_i = \sum_{j=1}^{n} t_{ji},\quad i = 1, 2 \ldots n$

(iii) Determine the prominence and relation of each criteria p_i and c_i

The prominence p_i of the criteria is evaluated by $p_i = c_i + h_i,\quad i = 1, 2, \ldots n$, which provides the importance of the criteria. The relation r_i is obtained by $r_i = c_i - h_i,\quad i = 1, 2, \ldots n$, which separates the cause and effect group of the criteria. If r_i is +ve, then C_i is a cause criteria. If r_i is −ve, then C_i is an effect group.

(iii) Construct the causal diagram

Causal diagram is plotted based on prominence p_i and relation r_i.

4 Adaptation of the Problem to the Proposed Model

To illustrate this model, the problem of solid waste management is taken. Solid waste refers to the variety of garbage generating from human and animal activities

Table 1 Hexagonal linguistic scale

Alternative assessment	Fuzzy number
No influence (NI)	(0, 0, 0, 0.06, 0.12, 0.18)
Very low (VL)	(0.06, 0.12, 0.18, 0.24, 0.3, 0.36)
Low (L)	(0.24, 0.3, 0.36, 0.42, 0.48, 0.54)
Medium (M)	(0.42, 0.48, 0.54, 0.6, 0.66, 0.72)
High (H)	(0.6, 0.66, 0.72, 0.78, 0.84, 0.9)
Very high (VH)	(0.78, 0.84, 0.9, 1, 1, 1)

such as industrial, residential and commercial activities, etc., that are categorized into renewable and non-renewable. The most important aim of solid waste management is to diminish the terrible impacts on both human health and environment to sustain economic growth and better quality of life. To identify the most important factor of solid waste management, the following factors are identified. C_1—Local resources consumption and reproducibility, C_2—Compatibility with environmental and geological characteristics, C_3—Atmospheric emissions, C_4—Water pollution and wastewater, C_5—Waste production, C_6—Land use and occupation, C_7—Fuel or non-renewable energy consumption, C_8—Water consumption, C_9—Non-renewable raw materials use, C_{10}—Respect for local culture, C_{11}—Percentage of collection and population served, C_{12}—Separated management of organic, hazardous or recyclable waste, C_{13}—Living conditions of local community would be a useful advance in the practice of solid waste management.

FCM Process: At this process, the initial uncertain direct relation matrices are constructed with the aid of two experts—educationist, scavenger from the municipal. They are called to give their judgments on the interrelation among the factors from S = {No Influence, Very Low, Medium Low, Low, High, Medium High, Very High}. The Initial uncertain direct relation matrices are then transformed into hexagonal fuzzy numbers using Table 1. Then, the direct relation matrix $\hat{Z}_k = [\hat{z}_{kij}]_{n\times n}$ changed into hexagonal fuzzy matrix $\tilde{Z}_k = [\tilde{z}_{kij}]_{n\times n}$. Next, the group uncertain direct relation matrix is constructed by aggregating all two matrices.

Converting Hexagonal Fuzzy Number into Haar Hexagonal Fuzzy Number: At this stage, Hexagonal Fuzzy Number is converted into Haar Hexagonal Fuzzy Number. Subsequently, using the Haar ranking algorithm for Hexagonal fuzzy Number, the group uncertain direct relation matrix $\tilde{Z}_k = [\tilde{z}_{kij}]_{n\times n}$ is changed into Haar Hexagonal direct relation matrix $H(\tilde{\tilde{A}}) = [\tilde{\tilde{a}}_{ij}]_{n\times n}$. Next, Haar Hexagonal direct relation matrix $H(\tilde{\tilde{A}}) = [\tilde{\tilde{a}}_{ij}]_{n\times n}$ is summed to get the Sum of Haar Hexagonal Fuzzy Number matrix $D = [d_{ij}]_{n\times n}$.

DEMATEL Process: At the final stage, the uncertain overall relation matrix $\tilde{T} = [\tilde{t}_{ij}]_{n\times n}$ is obtained. Further, the overall intensities of influencing and influenced criteria are determined. Also, the prominence and relation ate calculated, which are given in Table 2.

The causal diagram (Fig. 2) is depicted based on the prominence p_i and relation r_i. The prominence p_i elucidates the importance of the criteria. From the result, the

Table 2 Computational results

	c	h	$p = c + h$	$r = c - h$
C_1	−1.894	−1.103	−2.997	−0.791
C_2	−2.291	−1.521	−3.812	−0.77
C_3	−3.089	−2.026	−5.115	−1.063
C_4	−2.493	−1.691	−4.184	−0.802
C_5	−2.081	−1.25	−3.331	−0.831
C_6	−0.892	−1.708	−2.6	0.816
C_7	−1.226	−1.975	−3.201	0.749
C_8	−1.147	−2.54	−3.687	1.393
C_9	−0.718	−1.83	−2.548	1.112
C_{10}	−0.169	−1.117	−1.286	0.948
C_{11}	−1.147	−1.355	−2.502	0.208
C_{12}	−1.178	−1.23	−2.408	0.052
C_{13}	−2.608	−1.587	−4.195	−1.021

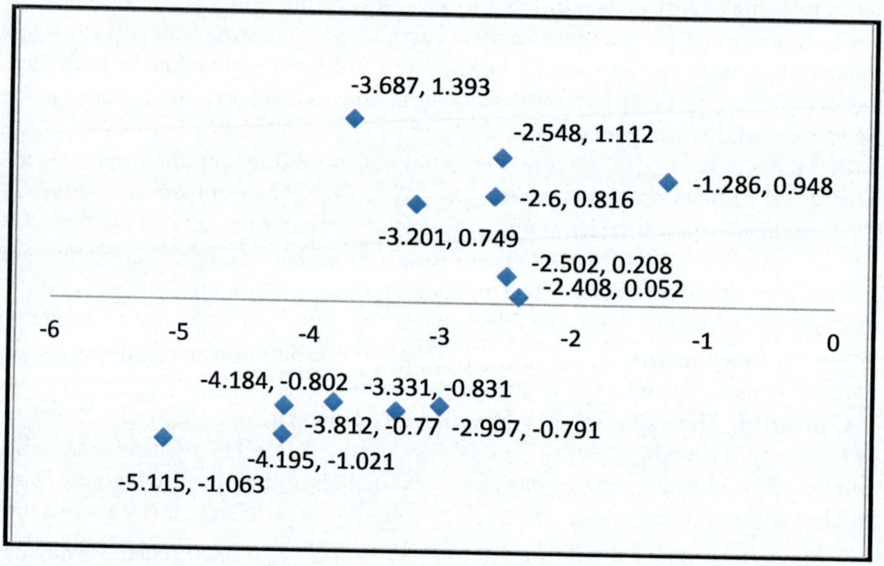

Fig. 2 Causal diagram

following ranking is obtained $C_8 > C_4 > C_1 > C_3 > C_6 > C_9 > C_7 > C_5 > C_{10} > C_{13} > C_{11} > C_{12} > C_2$. Here, C_8—Water consumption and C_4—Water pollution and wastewater are the most influencing criteria for the cause of solid waste. Moreover, the relation r_i divides the cause and effect group. if r_i is +ve, then C_i is a cause criteria. If r_i is −ve, then C_i is an effect group. It is observed that the cause group of solid waste includes

C_6—Land use and occupation, C_7—Fuel or non-renewable energy consumption, C_8—Water consumption, C_9—Non-renewable raw materials use, C_{10}—Respect for local culture, C_{11}—Percentage of collection and population served, C_{12}—Separated management of organic, hazardous or recyclable waste, and the effect group includes C_1—Local resources consumption and reproducibility, C_2—Compatibility with environmental and geological characteristics, C_3—Atmospheric emissions, C_4—Water pollution and wastewater, C_5—Waste production C_{13}—Living conditions of local community.

5 Conclusion

In this present investigation, a novel Haar FCM with DEMATEL technique is proposed to determine the major influenced factor and to obtain the cause and effect criteria of solid waste management. By adapting this system, the complicated structure can be visualized into a simple structure to take a appropriate decision by the decision makers. Furthermore, the proposed technique can be useful to obtain the most feasible solution in all MCDM problem with full of uncertainty.

References

1. Aseervatham S, Devadoss AV (2015) Analysis on criteria based emotive music composition selection using a new trapezoidal fuzzy DEMATEL—TOPSIS hybrid technique. J Fuzzy Set Valued Anal 2(2015):122–133
2. Devadoss AV, Felix A (2013) A fuzzy DEMATEL approach to study cause and effect relationship of youth violence. Int J Comput Alg 02:363–372
3. Devadoss AV, Prabhakaran R, Felix A (2018) A hybrid scenario FCM with VIKOR technique for ranking the factors. Int J Pure Appl Math 119(9):233–244
4. Dhanasekar S, Hariharan S (2017) Haar hungarian algorithm to solve fuzzy assignment problem. Int J Pure Appl Math 113(7):58–66
5. Dyson B, Chang NB (2005) Forecasting municipal solid waste generation in a fast-growing urban region with system dynamics modeling. Waste Manag 25(7):669–679
6. Felix A, Kumar SR, Parthiban A (2019) Soft computing decision making system to analyze the risk factors of T2DM, AIP conference proceedings, 2112, 020086-1–020086-12
7. Felix A, Christopher S, Mani R (2017) Fuzzy cognitive DEMATEL technique for modeling cause and effect of climate change. Int J Pure Appl Math 117(4):655–661
8. Fontela E, Gabus A (1976) The DEMATEL observer, DEMATEL 1976 report. Battelle Geneva Research Centre, Geneva
9. Gabus A, Fontela E (1972) World problems an invitation to further thought within the framework of DEMATEL. Battelle Geneva Research Centre, Geneva
10. Hokkanen J, Salminen P (1997) Choosing a solid waste management system using multicriteria decision analysis 98:19–36
11. Hori S, Shimizu Y (1999) Designing methods of human interface for supervisory control systems. Control Eng Pract 7(11):1413–1419
12. Jarrah OA, Qdais HA (2006) Municipal solid waste landfill siting using intelligent system 26:299–306

13. Kosko B (1986) Fuzzy cognitive maps. Int J Man Mach Stud 24(1):65–75
14. Khan S, Faisal MN (2008) An analytic network process model for municipal solid waste disposal options 28:1500–1550
15. Lin W (2008) A causal analytical method for decision making under fuzzy environment. Exp Syst Appl 205–213
16. Morrissey JA, Browne J (2004) Waste management models and their application to sustainable waste management 24:297–308
17. Rahman S, Yuan Q (2013) A fuzzy DEMATEL approach to assess determinants of efficient kerbside waste management in an urban context. In: Proceedings of the people and the planet conference, pp 1–14
18. Rajarajeshwari P, Sudha SA, Karthika R (2013) A new operation on hexagonal fuzzy number. Int J Fuzzy Logic Syst 3(3):15–26
19. Sankar RJ, Felix A (2016) An improved fuzzy DEMATEL technique and its applications. Int J Pharm Technol 8(3):19122–19134
20. Shieh IJ, Wu HH, Huang KK (2010) A DEMATEL method in identifying key success factors of hospital service quality. Knowl Syst 23(3):277–282
21. Suo LW, Feng B, Ping Z (2012) Fan, extension of the DEMATEL method in an uncertain linguistic Environment. Soft Comput 16:471–483
22. Tseng LM, Lin HY (2009) Application of fuzzy DEMATEL to develop a cause and effect model of municipal solid waste management in Metro Manila 158:519–533
23. Zadeh AL (1965) Fuzzy sets. Inform Control 8(3):338–353
24. Zadeh AL (1975) The concept of a linguistic variable and its application to approximate reasoning, (Part II). Inform Sci 8:301–357

NARX-EMFO Based Controller Optimization for pH Neutralization in Wastewater Treatment

Akshaykumar Naregalkar and D. Subbulekshmi

Abstract In order to meet the growing water demand, two sustainable options are available—desalination and wastewater recycle and reuse. The wastewater recycling and reuse concept, which is already prevalent in most of Europe and America, is currently in high demand in India with Swachh Bharat Mission and smart cities program of Government of India. The necessity to treat wastewater for recycling and reuse is slowly gaining grounds in industries, municipalities, and residential segments. The wastewater can be treated with different means—physical process like separation of suspended solids, chemical process like pH neutralization, and biological process to decompose organic substances. In pH neutralization of wastewater, the incoming untreated wastewater has a wide range of pH, and it is difficult to treat wastewater with such a high variability of pH. Neutralization is the process used for adjusting pH to optimize treatment efficiency. Neutralization may take place in a holding, rapid mix, or an equalization tank. In this work, a novel approach using NARX (Nonlinear Auto-Regressive eXogenous) model and enhanced moth flame optimization (EMFO) is proposed for pH neutralization of the wastewater and its performance is compared with ARX (Auto-Regressive eXogenous), ARMAX (Autoregressive Moving Average eXogenous), and BJ's (Box-Jenkins) model performance in terms of ISE, IAE, MSE, settling time, and peak overshoot for evaluating controller performance. The outcomes are then evaluated for both proposed and existing approaches to prove the effectiveness of the proposed scheme for the case under study.

Keywords Wastewater pH neutralization · Model · NARX · Controller optimization · EMFO

A. Naregalkar · D. Subbulekshmi (✉)
School of Electrical Engineering, Vellore Institute of Technology, Chennai, India
e-mail: subbulekshmi.d@vit.ac.in

N. Zhou and S. Hemamalini (eds.), *Advances in Smart Grid Technology*, Lecture Notes in Electrical Engineering 688, https://doi.org/10.1007/978-981-15-7241-8_29

1 Introduction

The necessity to treat wastewater for recycling and reuse is slowly gaining grounds in industries, municipalities, and residential segments. The wastewater can be treated with chemical process like pH neutralization and biological process to decompose organic substances. In pH neutralization of wastewater, the incoming untreated wastewater has a wide range of pH, and it is difficult to treat wastewater with such a high variability of pH. Neutralization is the process used for adjusting pH to optimize treatment efficiency. Acids such as sulfuric or hydrochloric may be added to reduce pH or alkalis such as dehydrated lime or sodium hydroxide may be added to raise pH values. Neutralization may take place in a holding, rapid mix, or an equalization tank. It can be carried out at the end of the treatment also to control the pH of discharge in order to meet the standards.

The pH level control is a trivial process in the treatment of wastewater, pharmaceuticals, chemical industries, and biotechnology [1]. Once the wastewater is discharged in the environment, the value of pH should be sustained at range of 6–8 so as to safeguard the life of aquatic and human. The pH neutralization shows a vibrant part in medicinal industry to make sure the product's value. Fine-tuning the pH value is challenging because of the corresponding time-varying nature along with nonlinear features [2]. The pH neutralization method generally comprises of the measurement of pH value of an acid base biochemical response at which the hydroxide ions and hydrogen ions are neutralized by everyone to produce water, whereas the other ions that are convoluted keep on unaffected [3].

The pH control approaches are usually allocated into three foremost kinds. The first type is an open-loop controller in which the rate of control valve opening is preserved at some assured location for a particular time period. The other type established on response control opinion in which there is a straight correlation among the progressive pH value and the opening of control valve. The 3rd control technique is extensively utilized for the pH mechanism is feed forward mechanism, where the regulator will recompensate for any of the dignified disruption before the development got disturbed. The pH control is required in various manufacturing methods. The simple approach is for neutralizing the pH fluid, i.e., achieve a pH value near to 7 [4]. The Continuously Stirred Tank Reactor—CSTR is utilized for pH neutralization process. As per the value of pH the resolution identified by pH sensor, base or acid has been included to attain the solution of neutralized solution [5]. The Wastewater pH which is below 4.5 and beyond 9 might significantly moderate the movement of the microbes which treat the water and might not maintain their lifetime at all. In this broadside hydrogen chloric acid (HCL = 3.01) utilized in wastewater management abilities to regulate alkalinity (NaOH = 10.98). Sodium Hydro Oxide base exploited to keep pH of method in base section. In pH neutralization method two cases are deliberated Servo and Monitoring difficulties [6].

The later proficiencies both fast and enormous changes inside the controller channel which result in significant variations in process improvement over an insignificant kind of pH. Furthermore, spontaneous pH controls have to compete with the certain development instabilities, improper valve sizing and in certain circumstances poorly intended methods [7].

So as to recognize the modeling structure of ARMAX and ARX, the black box mathematical modeling structures are unconstrained. In common, the system is acknowledged by means of black box. Here, the input has been signified as $u(t)$ given to the system model and both input $u(t)$ and output $y(t)$ are recorded for modeling the system. Conversely, the model system is concerned by disturbance. Usually, the outputs and inputs are totally or partly assessable whereas the system disturbance hasn't been categorized openly. This has some effect on output. The modeling error $e(t)$ has been unconstrained as the outcome of the variance among the model and system output. The over-all configuration for black box models in an input/output simple correlation as follows:

$$A\left(q^{-1}\right) \cdot y(t) = \frac{B\left(q^{-1}\right)}{F\left(q^{-1}\right)} \cdot u(t) + \frac{C\left(q^{-1}\right)}{D\left(q^{-1}\right)} e(t) \quad (1)$$

where $y(t)$, $u(t)$, and $e(t)$ are represented as the output, input, and noise. Altogether A, B, C, D and F are considered as the polynomials. For black box models, the algorithm generic structures are well-defined below.

ARX model is one of the simplest parametric structures. The structure of ARX can be signified as follows:

$$A\left(q^{-1}\right)y(k-n) = q^{-d}B\left(q^{-1}\right)u(k-n) + e(k) \quad (2)$$

In this, the polynomial $C\left(q^{-1}\right) = 1$.

ARMAX Model is similar to ARX model, however the additional terms signifies the moving error in an average. This model was useful on disturbance domination which are entering in the initial stage of the process. The generic structure of this model is depicted as follows:

$$A\left(q^{-1}\right)y(k-n) = q^{-d}B\left(q^{-1}\right)u(k-n) + C\left(q^{-1}\right)e(k) \quad (3)$$

where

$A\left(q^{-1}\right)$, $B\left(q^{-1}\right)$ and $C\left(q^{-1}\right)$ are the polynomial parameters that were to be estimated.

However, due to time-varying and nonlinear features of pH neutralization process, it is hard to control the pH concentrations effectively with ARX, ARMAX, and BJ's model. So an efficient system is required to maintain the pH range. At last, an efficient NARX model has been discovered applying best optimization techniques to maintain the pH range of CSTR. So, in order to overcome these limitations the NARX has been implemented where the ANN (Artificial neural network) controller is integrated for maintaining the pH of the wastewater. The structure of NARX can be represented as

$$y(t) = f\left[u(t-1), \ldots u(u-n_u), y(t-1), \ldots u(u-n_y))\right] \quad (4)$$

The predefined NARX model is taken into account with the integration of ANN controller and EMFO (Enhanced Moth Flame Optimization) for optimizing this controller so as to maintain the desired pH value. Here, ANN is one of the significant controllers preferred to reduce the fluctuations in pH of wastewater thereby maintaining the range of pH. In this work instead of moth flame optimization (MFO), EMFO is preferred as MFO gives the default lower bound and upper bound values for evaluating fitness function but suffers from poor exploration and slow convergence problem. In few complex optimization problems like multimodal problems in nonlinear systems, MFO doesn't perform well in terms of convergence. An EMFO is a nature-inspired algorithm with a simple structure. In this case with EMFO, the best fitness function can be evaluated via setting the lower and upper bound ranges as per required pH range and to maintain it. The integration of ANN and EMFO in NARX model attains an effective pH range that to be maintained.

1.1 Related Work Review

This section provides the detailed description of various existing process on the pH neutralization approach.

A model location adaptive controller system is intended for the mechanism of the level variables and pH in a method called as neutralization in CSTR. In such adaptive pattern, the parameter and the device evaluation laws which were extracted depending on the reference type approach for the nonlinearly parameterized systems. On choosing the constraint, more appropriate controller laws have been updated, the processes are reduced for this problem of pH. There was insufficient equipment in order to implement level control remains as the major downside of this technique [8]. Formation of the pH neutralization method was essential for proper control which in turn offers the usage of Hammerstein–Wiener model for identifying the process of pH neutralization. There was a need for enhancement in the fitness function values by means of employing the extreme machine learning dependent Hammerstein technique [9]. CDM-PI controller for the neutralization process as (FOPTD) first order plus time delay model was implemented for calculating the control factors. But this requires more complex calculations [10].

A nonlinear model analytical controller structure depending on the (SQP) successive quadratic programming process was implemented for neutralization of pH. But it requires to solve complex quadratic equations for solving optimization problems [11]. The mission toward the pH value tuning, predominantly concerning a progression by a constant supplies, is relatively hard because of essential nonlinearity and high understanding to improbability disruption. The process of detection in the taken up study had been related with the means of the state-space method, nonlinear methodology of ARX, Hammerstein–Wiener methodology, and ARX methodology. For the pH system this technique may lack the design of controller which was the downside of this technique [12]. The pH should be monitored and controlled in the neutral region by applying some tuning methodological factors to design the PID controller. By applying the operations such as smith operation, marlin and also the Branica defined operation, there could be a large variation in outcome on altering the small deviations of income. The major applications of the pH neutralization were of more important in the wastewater retreatment process, medical applications, and food processing methods and so on. Initially, material balance with the hydroxyl ion and hydrogen compositions became challenging to be noted. This issue was in accordance with the water splitting and recording the variations in the composition of the water concentration. Next, the material balance could be carried out on every atomic class and all additional equilibrium associations. The electro neutrality principle was utilized for simplifying the equations [13]. The deviations in pH were compensated by applying a feedback control loop. The manipulation of the alkaline flow rate helped to maintain the pH level of the process at a preferred level during the operations of transient. This was the main motive of this pH control but the application of advanced control schemes required numerous parameters of the process which were optimized [14].

The brain spiritual learning dependent on intelligent controller (BELBIC) was demonstrated and introduced to implement the artificial intelligent controller and utilized it in jacketed CSTR in both presence and absence of noise [15]. The enhanced SOFLC for pH controller by means of performance alteration table was used with (FLC) fuzzy logic controller in addition to the essentials of the fuzzy performance alteration table. The subsequent enhanced SOFLC was related with improved FLC for servo and monitoring controller. The advantages of this approach were as follows: reduced overshoot, settling time reduction, and less rate of rejection [16]. Model predictive controller based on piecewise model was employed in pH neutralization. This employment needs the information, over extent or by assessment, of the situations of the scheme of neutralization, together producing numerous difficulties. At this time the problem was addressed by comprising essential accomplishment in the controller to recompense for the unmeasured situations. However, this approach faces the issue of time delay factor [17]. A fuzzy logic dependent PID controller for the process of neutralization of the pH values was implemented for considered hydro management unit. The robustness of the arrangement controller was reliant on the flow rate control tap, pH value meter, concentration monitoring sensors and flow transmitter as these instruments are the influencing variables all through the present controller scheme and application. This was the downside of this approach [18]. A predictive control for the pH neutralization technique depending on the fuzzy inverse approach was employed in a comprehensive analytical control. One of the benefits of this controller organization was the deliberation of categorical constriction in system control which also comprised the ambiguous projecting mechanism. The presence of disruption and improbability was the limitation of this system [19]. The adaptive pH neutralization for the strong acid and base system was depicted for the maintenance of pH was with much abundant significance in the progressions of chemical, biotechnological productions, and several other regions. High presentation along with the strong maintenance of pH counteraction was hard to attain because of the time-varying and nonlinear method features. The method expansion differs at greater demand of amount over a small series of pH [20]. The performance Study of Neutralization pH operation for Conservative PI Regulator and IMC Established PI maintainer. The wide-ranging intention was to preserve the pH rate in a liquid at a particular level. The main aim of this method was to defuse the liquid in assessment by fluctuating the substance solution stream level till the mix alleviates at fixed point [21].

A dynamic forming of neutralization of pH method with fuzzy concept centered pH maintenance system for the operation of neutralization was established but lacks optimization techniques [22]. Enhanced moth flame optimization (EMFO) was proposed to overcome the problem of poor global searching in moth flame optimization algorithm. EMFO also having ability of global search [23]. It was shown that MFO algorithm gives optimum function value using standard deviation compared to other constraint optimization algorithms like Genetic Algorithms and artificial bee colony [24].

Several existing techniques related to the process of pH neutralization were examined and from this analysis it was evident that there some drawbacks in the existing techniques. Hence, it could be overcome by the implementation of some effective methodologies. Due to features of time-varying and the nonlinear methods, it was hard to control the pH concentrations. So an efficient system is required to maintain the pH range. For at last, an efficient NARX model has been discovered applying best optimization techniques to maintain the pH range of CSTR or pH neutralization in the process tank.

To reveal these drawbacks, the paper also implements the existing designs like, and NARX. The novelty of the proposed system is to combine ANN with NARX. The best results are then gained by using optimization methods (EMFO). The outcomes are then evaluated for both proposed and existing approaches like ARX, BJ and ARMAX to prove the effectiveness of the proposed scheme. Objective is to design a mathematical model for the case under study. To determine the best outcomes of controller parameters, to achieve the system requirements and to focuses on the simplification of design and implementation. To design new algorithm for parameter estimation and model identification and accordingly implementing controller algorithm to adapt for nonlinear changes in nonlinear/MIMO systems

2 Proposed NARX Based Approach for pH Neutralization Process Under Study

To apply the proposed algorithm, a pH neutralization process is considered as shown in Fig. 1. In this process of neutralization, the wastewater is introduced into the process tank. Then the pH of this wastewater is measured using pH sensor, if pH value is below 7, i.e., acidic then alkaline flow (Q_b) with solution NaOH is introduced into the process tank using control valve. If the pH value is greater than 7, i.e., alkaline in nature, then Acid flow (Q_a) with HCl can be introduced into the process tank. Thus based on pH value of the wastewater, chemical treatment of neutralization can be applied to treat the water and the treated water can be recycled for further usage.

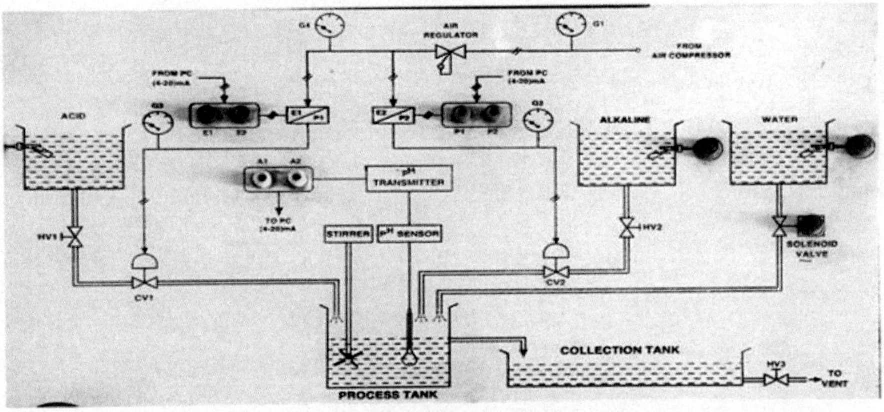

Fig. 1 pH neutralization process for chemical treatment of wastewater

Table 1 Parameters for pH process

Parameter	Description	Value
V	Volume of CSTR	4 Liters
F_a	Flow rate of base stream (NaoH)	13.88 e^{-6} (m^3/s)
F_b	Flow rate of acid stream (acetic acid)	4.204 e^{-6} (m^3/s)
C_a	Concentration of acetic acid	0.0185 g mol l^{-1}
C_b	Concentration of NaoH	0.0056 g mol l^{-1}

In this neutralization process, the existing control algorithm is made effective with proposed technique with NARX and EMFO optimization for the improvement of the controller performance.

The pH neutralization plant under study consists of parameters as mentioned in Table 1.

3 Parameter Estimation of pH Neutralization Process

The neutralization of pH is usually challenging because of its nonlinearity in titration curve, essentially when strong bases and acids are convoluted. The pH is correlated to the $[H^+]$ ions utilizations over the subsequent logarithmic relation:

$$\mathrm{pH} = -\log_{10}\left[\mathrm{H}^+\right] \tag{5}$$

This investigated operation deals with the good acid effluent neutralization with HCL in a CSTR with the help of a good base like NaOH.

$$\begin{cases} \text{Hcl} + \text{NaOH} \rightarrow \text{H}_2\text{O} + \text{Nacl} \\ \text{H}^+ + \text{OH}^- \rightarrow \text{H}_2\text{O} \end{cases} \tag{6}$$

If the combination is faultless and rapid, the ionic considerations $[\text{cl}^-]$ and $[\text{Na}^+]$ at CSTR are interrelated with the streams of base Q_b and acid Q_a and also towards the input reflections $[\text{cl}^-_{\text{in}}]$ along with $[\text{Na}^+_{\text{in}}]$ in relation to the subsequent calculations:

$$V\frac{d}{dt}[\text{cl}^-_{\text{in}}] \cdot Q_a - [\text{cl}^-] \cdot Q_{\text{out}} \tag{7}$$

$$V\frac{d}{dt}[\text{Na}^+_{\text{in}}] \cdot Q_b - [\text{Na}^+] \cdot Q_{\text{out}} \tag{8}$$

where V is the corresponding values of the fluid in the CSTR.

This concentration should fulfill the equivalence of lector-neutrality.

$$[\text{Na}^+] + [\text{H}^+] = [\text{cl}^-] + [\text{OH}^-] \tag{9}$$

However, the dissociation equation of the water is as follows:

$$[\text{H}^+] \cdot [\text{OH}^-] = k_w = 10^{-14} \tag{10}$$

NaOH neutralizes the acid solution which emanates into CSTR/process tank by the flow using pH controller and actuators with control valves.

Differential equations below describes pH process mathematically,

$$V \cdot \frac{d}{dt} b = V_F \cdot b_F - (V_F + V_A) \cdot b + V_A + b_A \tag{11}$$

$$x(t) = \frac{V_F \cdot b_F}{V} - \frac{V_F}{V} \cdot x(t) - \frac{V_F}{V} \cdot x(t) \cdot u(t) + \frac{b_A}{V} \cdot u(t) \tag{12}$$

The outcome of the taken up plant is actually not the variable 'b', but the outcome is value of pH in the water. Thus, the utilization of quadratic relation helps to get the desired output equation in terms of pH as shown in equations below.

$$C[\text{H}^+] = \frac{1}{2}\left(b + \sqrt{b^2 + 4.10^{-14}}\right) \tag{13}$$

After applying the quadratic relation as desired, the above relation becomes,

$$\text{pH} = -\log_{10}\frac{1}{2}\left(b + \sqrt{b^2 + 4.10^{-14}}\right) \tag{14}$$

Then, we consider the below fundamental relation, wherein, b is nothing but the phase acid or base flow '$x(t)$' and pH is the desired output '$y(t)$',

$$y(t) = -\log_{10}\frac{1}{2}\left(b + \sqrt{x^2(t) + 4.10^{-14}}\right) \tag{15}$$

In general, at the point of occurrence of equilibrium, uniform phase point of the considered plant is selected (i.e.) the reacting element would have alike constitution similar to the source path. Since, the system was majorly nonlinear, the chosen option might not go well because the desire was to maintain the value of pH (about 10th level of pH). Hence, it's well and good if 10th level of pH is chosen as the point of linearization and further, there is a need to explore for the exact solution. '*b*' is just determined from the pH as follows:

$$b(t) = \frac{c^2[H^+] - 10^{-14}}{C[H^+]} \tag{16}$$

On transforming it, we can get

$$y(t) = \frac{10^{-2pH} - 10^{-14}}{10^{-pH}} \tag{17}$$

Thus, after the estimation of parameter the identification of the system is made.

3.1 *System Identification of pH Neutralization Process with NARX Model*

NARX is the nonlinear controller in which the input data is selected randomly. The discrete time for the polynomial model is estimated and the integration of ANN based NARX controller with the estimator is done in this system. By this integration, the feed forward approach is made in which the output is given back to the input side which helps in the appropriate estimation of the controller.

In this work as shown in Fig. 2, the two layer neural network structure is used. This two layer neural network structure consists of a hidden layer and an output layer. The hidden layer consists of neurons and each neuron contains defined threshold function and the neuron gets activated as soon as the weighted sum of inputs reaches

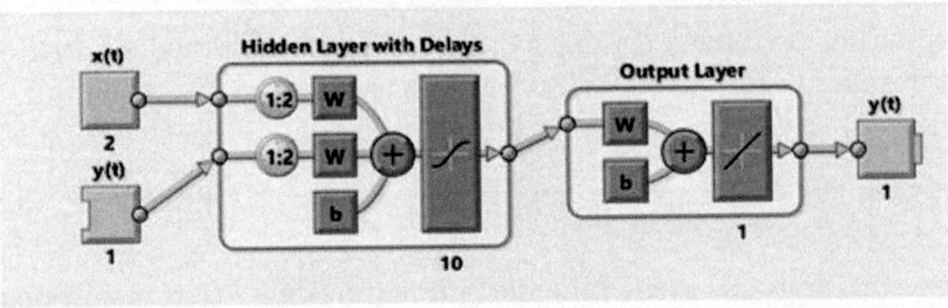

Fig. 2 NARX-neural network structure

a particular threshold value. The optimal number of neurons required in the ANN structure can be determined by comparing validation results with different number of hidden neurons. The two input channels are formed by regressors in the input layers. These regressors consist of linear and nonlinear terms. The output signal $y(t)$ is the one-step ahead output $y(t-1)$ that passes through a delay marked with 1:2. The output layer is used to filter the weighted sum of output from the hidden layer.

The mathematical model of the NARX is shown below:

$$y(t) = f[u(t-1), \ldots u(u-n_u), y(t-1), \ldots u(u-n_y))] \tag{18}$$

where y is the output variable and u is the externally determined variable.

$$\begin{bmatrix} x(t) = Ax(t) + bu(t) \\ y(t) = cx(t) + du(t) \end{bmatrix} \tag{19}$$

On applying the Laplace transformation on the above relation, the basic model becomes which is indicated in the below relation,

$$\begin{bmatrix} x(s) \cdot s = Ax(s) + bu(s) \\ y(s) = cx(s) + du(s) \end{bmatrix} \tag{20}$$

A, b, c and d are the simple constants.

After applying the algebraic operation, the Laplace transformed relation becomes,

$$x(s) = \frac{b \cdot u(s)}{s - A} \tag{21}$$

Then,

$$Y(s) = \left[\frac{c \cdot b}{s - A} + d\right] u(s) \tag{22}$$

The transfer function can be indicated by the following relation,

$$G(s) = \frac{y(s)}{u(s)} = \frac{c \cdot b}{S - A} + d \tag{23}$$

On substituting the intended values on the earlier transformed relation, we get the below output relation for the combined system as,

$$Y(t) = 1(t) + 7(t) - \frac{1}{10s + 0.8} + 7 - \frac{1}{0.5s + 1} + 20 - \frac{1}{0.2s + 1} + \frac{1}{s} \tag{24}$$

The above relations are used while designing the elements of the proposed ANN-NARX model.

Table 2 Polynomial model structure

Model	n_a	n_b	n_c	n_d	Input delay n_k
ARX	2	2	–	–	1
ARMAX	2	2	2	–	1
BJ	–	2	2	1	1

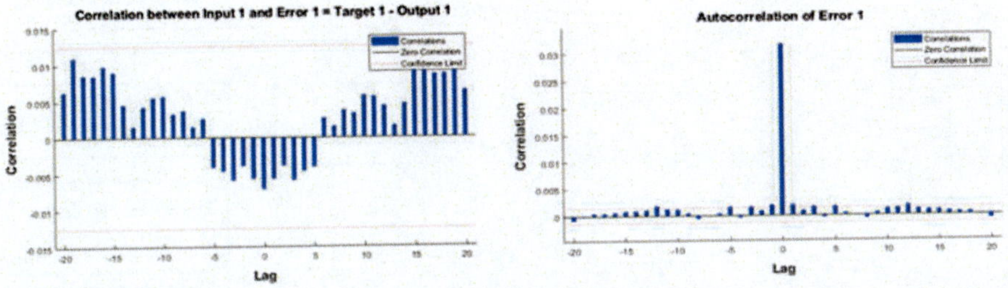

Fig. 3 Correlation and autocorrelation analysis of NARX model

For comparing performance of proposed NARX model, other models like ARX, ARMAX and BJ are used with following parameter configuration. In Table 2, n_a, n_b, n_c, n_d polynomial order of the polynomial $A(q-1)$, $B(q-1)$, $C(q-1)$ and $D(q-1)$ respectively. The model orders can be set to orders n_a, n_b, n_c, n_k. For example the parameters n_a, n_b, and n_c are the orders of the ARMAX model and n_k is the delay.

In NARX design with Neural Network (ANN), the no. of hidden neurons are as selected 10 and the no. of delays are selected as 2 due to better performance with these values. The neural network for NARX design is created with 70% trained data, 15% validation data and 15% testing data. The training algorithm to train this NARX with ANN is selected here is Levenberg–Marquardt as this algorithm takes less time and performance calculates in terms of Mean Square Error. The training stops as soon as the MSE value decreases to the lowest one. The lowest MSE value and Regression value nearest to 1 like 0.99 or 0.98 is used during training the NARX based on neural networks. The designed model correctness can be checked from correlation and autocorrelation analysis as shown in Fig. 3. It can be seen that the most of the residues fall in the desired range to correlate input with output.

3.2 *Controller Optimization with EMFO*

A process of filtering is carried out by the use of Kalman filter in order to filter out the parameters that were attained after the process of estimation. The filtered output is then optimized by means of an Enhanced moth flame optimization algorithm (EMFO). The filtered output is optimized in order to provide some better output level of pH present in the water. Thus, the pH of the water should be reduced and are

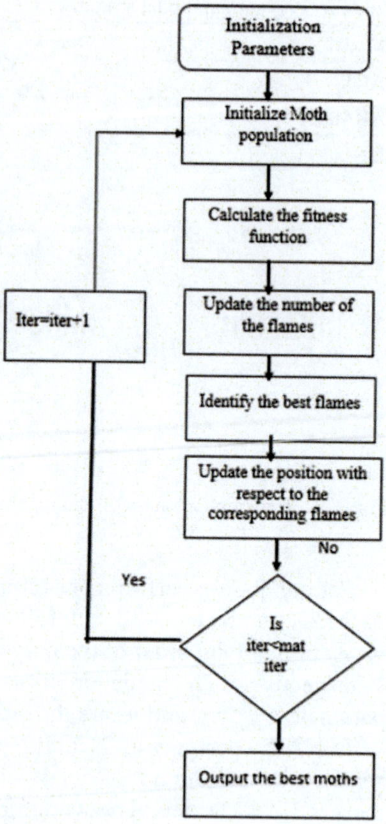

Fig. 4 Flow chart for EMFO implementation

maintained at an optimum level for the healthy usage. For optimizing the output of NARX, Enhanced MFO (EMFO) is employed. Following this, the fitness function is attained for pH and is given as input size. The default MFO count/limit is tuned by varying the number of iterations, lower bound and upper bound constraints. This level of pH should be maintained at a constant rate.

The flow chart of applying enhanced mouth flame optimization algorithm is shown in Fig. 4. The fitness function is estimated by repeating the number of iterations. Once, the fitness function that is estimated is updated then the iteration process is repeated thereby applying the sorting functions. Therefore, the optimized result is obtained. The flow chart of MFO is shown below:

4 Result Analysis

This result analysis shows the extensive performance analysis of the existing system along with the proposed system. Figure 5 shows the performance of four models

Fig. 5 pH estimation with ARX, ARMAX, BJ and proposed NARX models

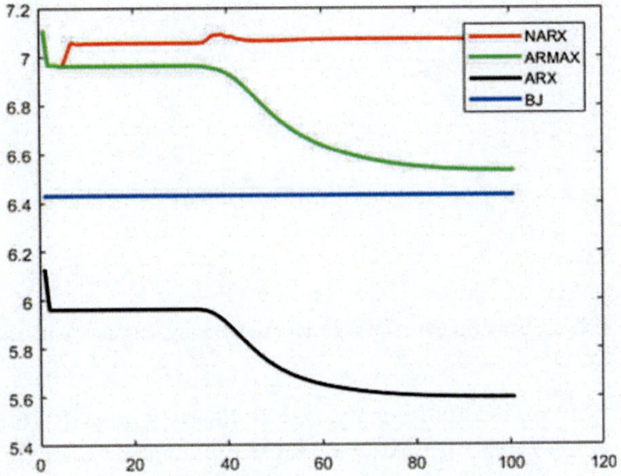

ARX, ARMAX, BJ and proposed NARX in estimation of pH coefficients. It can be seen from Fig. 6 that the proposed NARX model gives better estimation of pH for the wastewater in process tank.

The comparative analysis as shown in Table 3 and Fig. 6 is done for maintaining the desired value of pH with the both proposed NARX model and existing techniques ARX, ARMAX and BJ is done in terms of ISE, IAE, MSE, overshoot and settling time.

Figure 6 shows the comparative analysis of the proposed NARX and the existing techniques in terms of maintaining desired pH level in the process tank. The pH value of proposed technique is attained as 6.8 which was the suitable range for the reuse of the wastewater. The pH controller based on this proposed NARX–MFO algorithm is applied for the pH neutralization of the wastewater and as shown in Fig. 7, it is evident that the pH of wastewater is controlled around the set point, i.e., pH = 7.

Fig. 6 Comparative analysis of the proposed NARX and the existing techniques

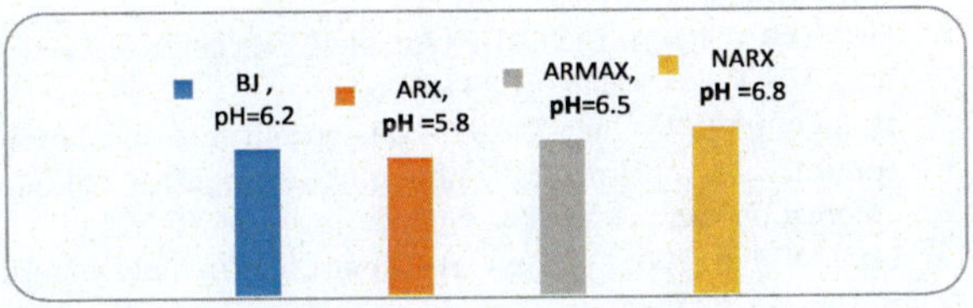

Table 3 Comparison of error rate of pH

Model	ISE	IAE	MSE	Overshoot (%)	Settling time (s)
ARX	1.4801	2.4108	1.4	35.43	1.114
ARMAX	0.4777	0.5398	0.437	14.02	1.694
BJ	1.2974	1.9326	0.57	19.007	0.694
NARX	0.2654	0.3536	0.06	0.694	0.685

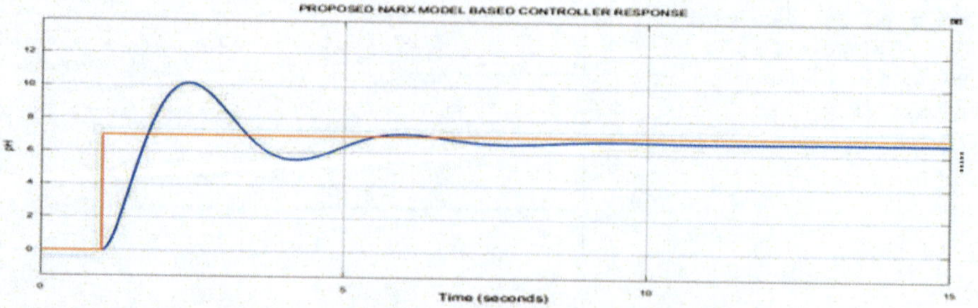

Fig. 7 The pH controller response in tracking set point with the proposed method

From the above analysis, it is clearly shown that the proposed NARX model that optimized with EMFO achieves efficient pH maintenance as per optimized values assigned using the Enhanced MFO where the fitness function can be fixed with the required range of pH.

References

1. Arun UP et al (2015) Control of pH neutralization process using fuzzy logic: a review. Int Res J Eng Technol 2(3):1081–1084
2. Anbu S, Jaya N (2014) Design of adaptive controller based on Lyapunov stability for a CSTR. Int J Electron Commun Eng 8(1):183–186
3. Agarwal V et al (2014) A model predictive controller using multiple linear models for continuous stirred tank reactor (CSTR) and ITS implementation issue. In: 2014 fourth international conference on communication systems and network technologies (CSNT), pp 1001–1005 (2014)
4. Hermansson A, Syafiie S (2015) Model predictive control of pH neutralization processes: a review. Contr. Eng. Pract. 45:98–109
5. Sharmila B, Vidhyanandhan L (2016) Modeling and designing of controllers for pH process. J Adv Chem 12(15):4872–4883
6. Shekhar C, Sharma MS (2017) An intelligent controller designing for pH neutralization process. Int J Adv Res Sci Eng 6(6):446–454
7. Nsengiyumva W, Chen X (2018) Design and implementation of a novel self-adaptive fuzzy logic controller for a pH neutralization process. In: 2018 3rd international conference on electrical, automation and mechanical engineering (EAME 2018)
8. Loh AP et al (2001) pH and level controller for a pH neutralization process. Industr Eng Chem Res 40:3579–3584
9. Abinayadhevi P, Prasad S (2015) Identification of pH process using Hammerstein-Wiener model. In 2015 IEEE 9th international conference on intelligent systems and control (ISCO), pp 1–5
10. Meenakshipriya B, Manikandan S (2015) Wiener model-based CDM-PI controller for pH neutralisation process. Int J Model Identif Contr 24(2):127–137
11. Mahmoodi S et al (2009) Nonlinear model predictive control of a pH neutralization process based on Wiener-Laguerre model. Chem Eng J 146:328–337
12. Rattanawaorahirunkul R et al (2016) Nonlinear system identification of pH process using Hammerstein-Wiener model. In: 2016 international conference on electronics, information, and communications (ICEIC), pp 1–4

13. Ram SS et al (2016) Designing of PID controllers for pH Neutralization process. Indian J Sci Technol 9(12):1–5
14. Jose T et al (2013) pH Neutralization in CSTR using model reference neural network and fuzzy logic adaptive controlling schemes. Int J Adv Res Technol 2(1):1–6
15. Rajagopalen A (2013) Identification of an effective controller for a stirred tank heater process. Int J Eng Adv Technol 3(1):271–279
16. Singh PK et al (2013) Optimized and self-organized fuzzy logic controller for pH Neutralization process. Int J Intell Syst Appl 5:99–112
17. Hermansson AW, Syafiie S (2014) Control of pH neutralization system using nonlinear model predictive control with I-controller. In: 2014 IEEE international conference on industrial engineering and engineering management (IEEM), pp 853–857
18. Rammohan T (2014) Fuzzy logic based PID controller for pH neutralization process. Int J Comput Appl 95(6):1–5
19. Shaghaghi D et al (2013) Generalized predictive control of pH neutralization process based on fuzzy inverse model. In: 2013 13th Iranian conference on fuzzy systems (IFSC), pp 1–6
20. Jaleel JA et al (2014) Adaptive and neural pH neutralization for strong acid-strong base system. In: Global trends in intelligent computing research and development, pp 496–515
21. Prasad P, Mathew J (2016) Performance analysis of pH neutralization process for conventional PI controller and IMC based PI controller. Int J Innov Res Sci Technol 3:262–267
22. Singh PK et al (2013) Neural control of neutralization process using fuzzy inference system based lookup table. Int J Comput Appl 61(9):16–22
23. Kaur K, Singh U, Salgotra (2018) An enhanced moth flame optimization. Neural Comput Appl 1–35
24. Narottam J et al (2016) Moth-Flame optimization Algorithm for solving real challenging constrained engineering optimization problems. In: 2016 IEEE students' conference on electrical, electronics and computer science (SCEECS), pp 1–5

Review of Particulate Matter Filters

Nerella Venkata Sai Charan, S. Krithiga, and Partha Sarathi Subudhi

Abstract Increase in particulate matter (PM) in air causes a biggest threat in the twentieth century. In order to filtrate PM from air, filters are employed. These filters are called as particulate matter filter (PMF). These filters separate the PM from the polluted air. The processes followed by PMF to separate the PM from the polluted air are filtration and adsorption. Different filters are developed for this purpose. This chapter presents a critical review of different types of particulate filters. Porous- and fibrous-type filters are discussed in this chapter. This chapter explains the fabrication process of different PMF along with its advantages and limitations. This chapter helps researchers in choosing PMF for their specific application.

Keywords Particulate matter · Particulate matter filter · Adsorption technique · Filtration process

1 Introduction

Increase in air pollution has brought revolution in industrial and transportation sectors. Industries and transportation sectors produce particulate matter (PM) in large scale. PMs are generated not only from industries and transportation sectors but also by natural calamities like volcanoes and dust storms. PMs are very small in size and are able to penetrate into living beings and plants. The presence of PM in human body causes deadly diseases like heart stroke and asthma. In plants, PM blocks the stomatal openings which restricts the exchange of gases and leads to the death of the plant. PM can mix with blood of human beings due to its smaller size. Thus, it is very harmful for both living beings and plants. Every year, a large percentage of total deaths is caused because of the PM present in air. Particulate matter filters (PMF) help in restricting the presence of PM in surrounding air. In order to filtrate PM from air, dust collectors were manufactured during earlier stage. In 1921, scientist Wilhelm Beth invented the dust collector. PMF efficiency gets affected by the

N. V. Sai Charan · S. Krithiga (✉) · P. S. Subudhi
Vellore Institute of Technology, Chennai, Tamil Nadu, India
e-mail: s_krithiga@yahoo.com

N. Zhou and S. Hemamalini (eds.), *Advances in Smart Grid Technology*, Lecture Notes in Electrical Engineering 688, https://doi.org/10.1007/978-981-15-7241-8_30

presence of moisture since it covers a large area in the PMF surface which limits the capability of the filter to absorb PM. Moisture should be collected separately from the surface of the PMF for its better performance.

Initially, porous filter was developed which has the capability of filtering PM of size up to 2.5 μm [1]. These filters are able to filter PM as per the number of pores available in them which is the biggest limitation of porous-type PMF. This motivated researchers to invent fibrous-type filters which use improved technology in order to filter PM of size <2.5 μm in the surrounding air. The paper is structured into four sections. Sections 1 and 4 explain the introduction and conclusion of the paper. Section 2 explains about different types of PMF and explanation of various types of fibrous-type PMF is given in Sect. 3.

The contribution of this review article is as follows: (i) Review of different PMFs is carried out and (ii) Advantages and limitation of different PMF are explained in order to determine a proper filter for a specific application.

2 Review on Different Types of Particulate Matter Filters

In order to filter the PM present in surrounding, PMFs are necessary. This section describes two types of PMF viz. porous and fibrous type.

2.1 Porous Type

The filtering capability of porous filters is mainly depending on the pores and the adsorption strengths of the walls of the pores. Pores are in circular shape and its size is very small in order to filter the PM as shown in Fig. 1. PMs which are of larger than the pore size are filtered through the porous-type PMF whereas the PM which is of smaller in size compared to the size of the pores, has a chance to pass through the PMF. The adsorption capability of pores adsorbs some of the small size PM on the surface of the walls of the pores. This adsorption capability helps to filter most of the PM in air. The pore size is determined by using the bubble point test. This test is performed to check the filter quality. This test checks the filter capability of withstanding the air pressure. The pore size of a porous filter is designed as follows [2]:

$$D = \frac{4 * \gamma * \cos(x)}{P} \tag{1}$$

Where D is pore diameter (μm), γ is surface tension of the liquid (N/m), P is bubble point pressure (Pascal), X is angle between liquid and filter

The adsorption of PM through the pore wall follows the process like impaction, interception, diffusion, sedimentation, and electrostatic attraction [2]. During filter

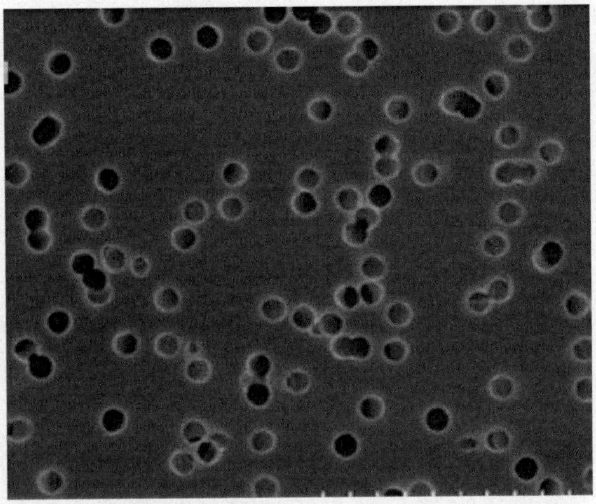

Fig. 1 Porous-type filter with 1 μm diameter pores by William G. Lindsley, Ph.D., NIOSH [2]

process, the size of pore reduces gradually after adsorbing the PM. Porous-type filter requires more energy to filter the polluted air which causes reduction in the filter efficiency. As pore diameter is limited, PM of certain diameter size can only be filtered through the porous-type filter. Also, the adsorption capability of the pore walls decreases when more layer of PM is deposited on the walls.

Limitations of the porous-type filter due to the size of the pores and the adsorption capability of the pore wall restrict its application to certain area only. These limitations of the porous-type filter lead to the invention of fibrous-type filter.

2.2 Fibrous Type

Fibrous-type filters work on the basic principle called adsorption. These filters use fibres to filter the PM present in air. The fibre provides larger surface area for adsorption of particles. Fibrous-type filters are also hydrophobic in nature which allows easy separation of moisture from the surface of PMF. Details regarding fibrous-type filter are explained in the following section.

3 Fibrous-Type Particulate Matter Filter

Detailed study of three types of fibrous-type filters, namely activated carbon fibre (ACF) filter, carbon nanotube (CNT) filters, and transparent antibacterial (TAB) filter is explained in this section. Fabrication process, experimental setup, and PM filtering capability of these fibrous-type filters are also explained in this section.

Fig. 2 Photograph of the pristine ACF filter [9]

3.1 *Activated Carbon Fibre Filter*

These filters are fabricated by using pristine ACF. These filters have a capability of filtering PM of diameter 0.3 μm [3]. Photograph of ACF is shown in Fig. 2. These filters are mainly used in the residential environment. These filters can also be used to filter ozone. Ozone on reacting with oxide groups present on the surface of activated carbon (AC) filter gets separated from the polluted air [4, 5].

Due to humidity on the surface of the PMF, pressure resistance increases which in turn require more energy to pass air through these filters. This reduces the efficiency of the filter. Also, secondary pollutants are generated from the reaction of the ozone with the oxides present on the surfaces of AC filter which limits the usage of ACF.

These limitations of the ACF lead researchers to invent an advanced filter. This improved filter is termed as CNT filter. Details regarding CNT filter are explained in the following section.

3.2 *Carbon Nanotube (CNT) Filter*

The limitations of the ACF are overcome by fabricating carbon nanotubes on the surface of ACF. This newly invented filter is termed as CNT/ACF fibrous filter. Structural designs of CNT/ACF PMF are: CNT coat, CNT sheet, hierarchical CNT structure, and agglomerated CNT fluidized bed [6–8]. Photograph of CNT/ACF is shown in Fig. 3. Fabrication of CNT/ACF is explained in the following subsection.

Fig. 3 Photograph under scanning electron microscope of CNT/ACF particulate matter filter [9]

3.2.1 Fabrication of the CNT/ACF PMF

Floating catalyst chemical vapor deposite method is used to fabricate the CNT. The substrate for the process is prestine ACF filter of 50 mm diameter and 2.94 mm thickness. It is placed in the middle of the quartz tube of 38 mm inner diameter and 180 cm length. Ferrocene (purity >99.99%) is used as catalyst and is introduced at entrance side of the quartz tube inside of the heater. Protection parameters are used as argon gas at 600 ml/min and hydrogen at 100 ml/min. Temperature inside the tube is gradually increased to 760 °C and then it is maintained. As temperature raises, ferrocene gets evapourated and adsorbed on the surface of ACF medium. Ethylene is used (200 ml/min for 40 min) as the carbon source for CNT growth on catalyzed ACF medium and then it is cooled to room temperature in argon and hydrogen atmosphere and CNTs/ACF was fabricated. The morphology and structure were observed under scanning electron microscope and Raman spectrometer and is shown in Figs. 3 and 4, respectively.

3.2.2 Experimental Setup of the CNT/ACF PMF

In order to measure the efficiency of the fabricated filter, it is fitted to a fixture with 32 mm inner diameter and the particle velocity is 6.21 cm/s. Carrier PM particles are generated by atomizing 0.01 g/ml NaCl solution using atomizer (TSI 3076) with nitrogen. Concentration and size of PM at fixture are determined by scanning mobility particle sizer, consisting of electrostatic classifier, a long differential mobility, and an ultra condensational particle counter. The atomizer can only produce more number of particles of size ranging 20 to 300 nm. For varying velocities of flow linearly from

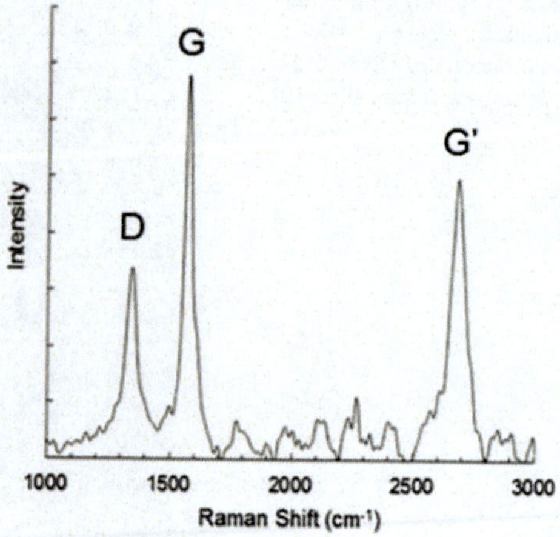

Fig. 4 Analysis of CNT/ACF PMF with Raman spectrometer [9]

0.5 to 5.0 L/min, the efficiency is calculated as follows [9]:

$$\eta = \left\{1 - \left(\frac{C_{\text{out}}}{C_{\text{in}}}\right)\right\} * 100\% \tag{2}$$

Where C_{in} is inlet concentration of PM in air, C_{out} is oultet concentartion of PM in air. Quality factor, Q_f of the PMF with different materials is determined by [9]:

$$Q_f = -\frac{\ln(1 - \eta)}{\Delta P} \tag{3}$$

Where, ΔP is pressure resistance. From Eq. 3, it is observed that, the quality factor of the PMF increases with the decrease in the pressure resistance of the PMF.

Comparison between the pristine ACF and CNT/ACF PMF is provided in the following section.

3.3 *Comparison of ACF and CNT/ACF*

This section explains the structural comparison of ACF with CNT/ACF. Raman spectrography comparison of both PMF is also provided in this section. This section also bestow the comparison of PM removal efficiency of ACF and CNT/ACF PMF. Structures of the pristine and manufactured CNTs/ACFs are not viewable in naked eyes as the color of both is black which is shown in Fig. 5a, b, respectively. Scanning electron microscope (SEM) image of the two PMF is depicted in Fig. 6a, b, respectively. From Figs. 5 and 6, it is viewable that the carbon nanotubes are placed on the substrate-activated carbon. The Raman spectrum analysis of the two PMF is shown in Fig. 7a, b, respectively. It is observed from the spectrum that, in ACF filter, the

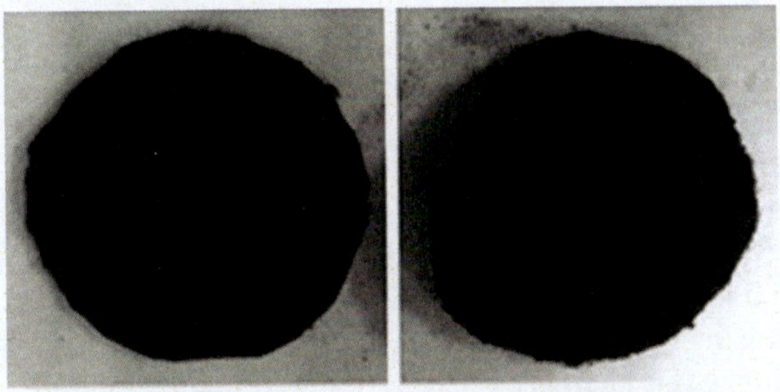

(a) (b)

Fig. 5 **a** Images of pristine ACF and **b** images of CNTs/ACF filter [9]

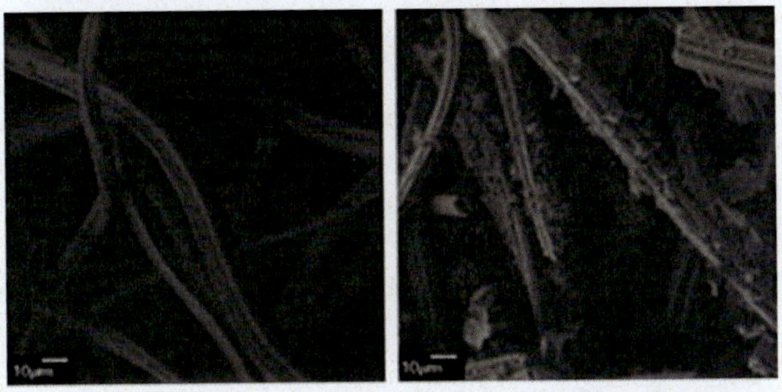

(a) (b)

Fig. 6 **a** Scanning electron microscope (SEM) images of pristine ACF and **b** CNTs/ACF filter [9]

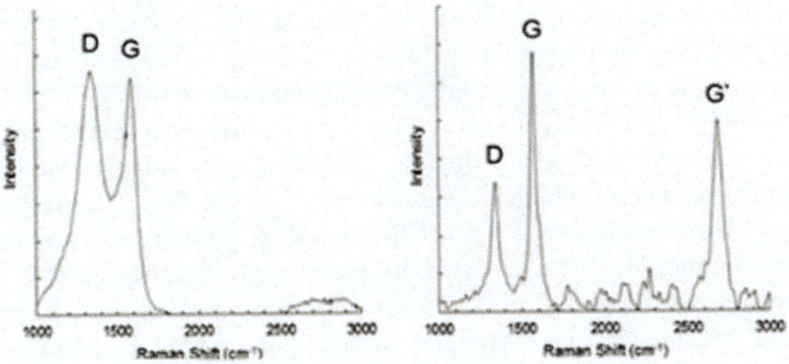

(a) (b)

Fig. 7 **a** Raman spectrum for pristine ACF and **b** Raman spectrum for CNTs/ACF filter [9]

peak intensity of the reflected waveform is limited whereas in the CNT/ACF filter, the particle intensity is spreaded over a wide band. This shows that, the fabrication of the carbon nanotubes on the activated carbon filter increase its capability to filter PM of very less size.

In order to do a effciency comparison analysis of these two filters, particles of range 10 to 350 nm are used. Generally, penetration particle size is common for the two filters in this range. It is observed that, pristine ACF filter shows very less efficiency with particles of larger size and in worst case, efficiency is of 75.66% (for 260.9 nm). Whereas, for ACF/CNT PMF, shows better filteration efficiency at any size of PM. The efficiency is observed as 95.97% for Most Penetrating Particle Size (MPPS) of 145.9 nm which is efficient compared to ACF PMF.

The revolution in PMF motivated researchers to develop an improved PMF. Details regarding this improved PMF is explained in the following section.

3.4 *Transparent Antibacterial Nanofiber Air Filter*

The polymide nanofibers do not have the capability to avoid adsorption of the water. Therefore, researchers developed hydrophobic filter-based PMF. Hydrophobic filters have hydrophobicity property which helps in avoiding the water adsorption on the surface of the PMF. Improvement in the quality of the material used for fabricating the PMF has led researchers to invent Poly Methyl MethAcrylate (PMMA) and Poly Di Methyl Siloxane (PDMS) nanofibers. This PMF has high hydrophobic capability which does not allow water adsorption on the surface of the fiber. Later stage 'chitosan' a poly saccharide from chitin was used to fabricate the PMF [11]. The material has advantages such as: biodegradability, nontoxic, and antibacterial capability. This fiber also has strong polarity which enables it to polarize easily for good filtration. Then, transparent antibacterial nanofiber air filter is developed which is a bi-layered fibrous filter which contains electrospun super hydrophobic fibers which prevents water to adsorb on the fiber and chitosan. These filters mainly work on adsorption principle. In order to make a better and highly efficient and transparent filter, electro spinning method is used. PM removal is done using different types of fiber materials like Poly AcryloNitrile (PAN), poly vinyl pyrrolidone, poly styrene, and poly vinyl alcohol and poly propylene [10]. This filter is fabricated using polyimide nanofibers substrate which has high thermal stability. This PMF has a capability of filtering PM of size less than 2.5 μm. Efficiency of the filter remains unaltered upto a temperature range of 370° C. As it is discussed that efficiency of PMF decreases with the presence of humidity, it is necessary to avoid the adsorption of water on the filter surface. Mechanism and fabrication of this filter are explained in the following section.

3.4.1 Mechanism and Fabrication of Transparent Antibacterial Filter

Mesh and nanofibers are used while fabricating these filters. Design process of this filter is depicted in Fig. 8a. It is observed from Fig. 8a that, while fabricating, a wire mesh is used as the base of the PMF. A small amount of the voltage is applied to the solution followed by electro spinning process in order to fabricate the filter [12].

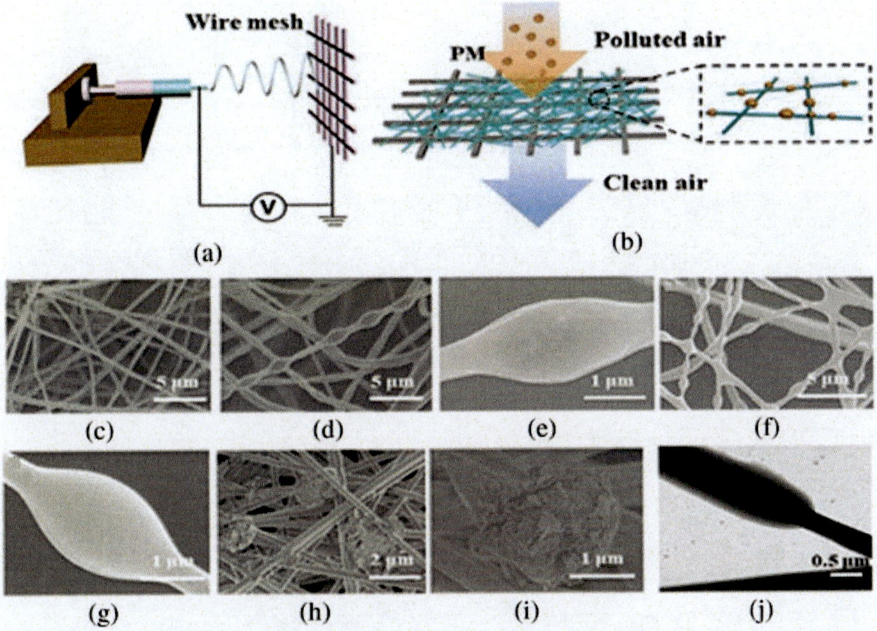

Fig. 8 **a** Fabrication process of the PMF, **b** filtration mechanism, **c** SEM images filter before filtration, **d** and **e** PDMS/PMMA fiber after filtration from burning incense at different magnifications, **f** and **g** chitosan PMMA fiber after filtration from burning incense at different magnifications, **h** and **i** SEM images of filter with captured dust particles and **j** morphologies of PM captured on fibers [1]

Figure 8b shows the animated diagram of the filtering process. Figure 8c to j shows the SEM photograph of the filter at different instants with different magnifications. It is observed from the figures that, gradually PMs are adsorbed on the surface of the fiber and the layer formed on the surface due to PM again supports another layer of PM to be adsorbed on it.

It is observed from the surface morphology which is depicted in Fig. 8j that, the filter contains organic particles on its outer surface which resembles the amorphous carbon. Further analogy Fourier transform infrared spectography is conducted and it displays enhanced peaks of C = O, C–O and C–N at 1.757, 1.238 and 7.386 cm^{-1}, respectively. C, N, and O were detected when energy dispersive x-ray spectography is conducted. It is observed that, the smoke consists of organic compounds. High polar functional groups were on the outer surface. The dipole-dipole and induced dipole interactions between chitosan fiber and nitrogen group is high. The filtration can be performed by using columbic forces also.

The particulate matter capturing process by using this filter is shown in Fig. 9. It is observed from the Fig. 9 that, as time increases, a greater number of PM particles get adsorbed on the nanofiber, altogether increasing the diameter of the nanofiber. This provides more surfaces to the new particles to get adsorb, that is the new particles need not to be attached directly to the fiber.

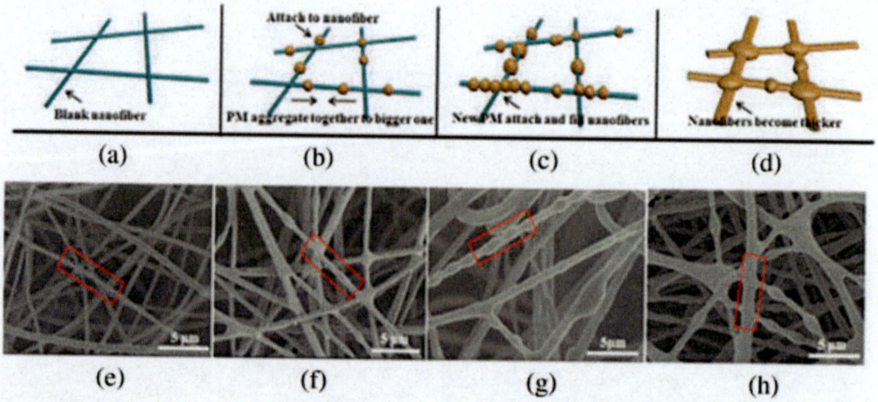

Fig. 9 PM capturing process of PDMS/PMMA composite fiber filter [1]

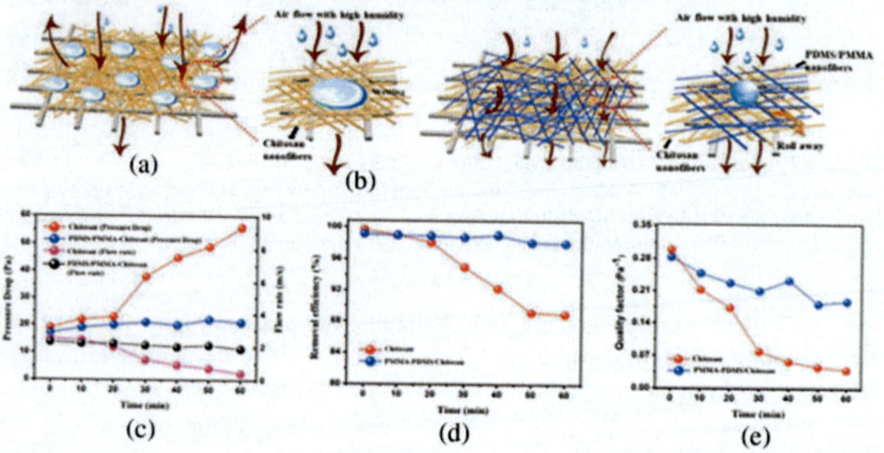

Fig. 10 **a** Mechanism of filtration with chitosan fiber, **b** mechanism of filtration with composite of chitosan and PDMS/PMMA fiber, **c** pressure drop and flow rate of chitosan as well as PDMS/PMMA with time, **d** removal efficiencies of both chitosan and composite fiber with time, **e** quality factor of both chitosan and composite fiber with time [1]

In order to validate the properties of the fabricated filter, its behavior was studied in different conditions. These behavioral analyses are mentioned in the following sections.

Investigation Under High Humid Conditions

Figure 10a to d shows the systematic air flow with chitosan fiber with PDMS/PMMA-chitosan composite. Due to the high humid conditions, there were a greater number of chances of water condensing on chitosan fiber due its capillarity and the hydrophilicity of these nanofibers. This affects the air flow through the filter. PDMS/PMMA fibers are added on chitosan fibers and properties are observed. It is observed that, all tiny water droplets collage together to form a bigger droplet and it is rolled out of the filter which does not affect the air flow and filtration process. It is also observed that, because of super hydrophobicity of the PDMS/PMMA, the water was easily be removed from the filter surface. While comparing the two fibers, i.e., chitosan fiber and composite of chitosan fiber and PDMS/PMMA fiber, it is observed that, Chitosan fiber filled completely with water but composite fiber does not have any change in its surface. It is also observed that, there is a sharp increment in pressure of a chitosan fiber during the filtration process which is about 19 to 56 pa. Due to this increment in pressure, there is fall in its flow rate which is about 2.5 to 0.4 ms^{-1}. The removal efficiency is also decreased from 99.66 to 88.92%. Very less pressure drop is observed with composite PDMS/PMMA fiber which is of 17 to 21 pa and the flow rate fall to 1.9 ms^{-1}.

PM Adsorption and Disorption Capability of PDMS/PMMA Composite Fiber

Composite of chitosan and PDMS/PMMA filter is easy to clean. This process is carried out by using water with detergent at 45 °C for 10 min under 16% transmittance level with incense smoke. While cleaning, the force of attractions between PM particles and dipole of the detergent is high which helps the PM particle to come out of filter. The filter is white and smooth when it is fabricated and after the adsorption of the particle, it turns into light yellow and after the cleaning processes, it will again turn to white and also it is observed that there is damage to the fibers during cleaning process. As per the graphs, the efficiency is deviated in milli order which is negligible. Photograph of the filter during different stages is shown in Fig. 11.

Mechanical Durability and Long-Term Performance Test of PDMS/PMMA Composite Filter

The quality checking process is done in two steps and they are sand abrasion and waterfall test. In sand abrasion test, 50 gm of fine sand is impinged from a height of 50 cms for 30 s. This process is again done in two stages: with and without presence of chitosan fiber. From the test, it is observed that, huge damage to the fibers in the absence of the chitosan happens which increases the durability of the filter. Water fall test is carried out after releasing 5000 droplets from the height of 50 cm on the surface of the filter for two hours. It is observed from the water fall test that, there was no damage to the filter surface after the process. The filter is also tested for long-term performance test. In order to perform long-term performance test, the filter is kept at 54% admittance level and the filter tested for 100 h in heavy pollution condition with PM 2.5 > 1500 μgm^{-3} and PM 10 > 2000 μgm^{-3}. After 100 h test, the collection bottle looks very clear which says that the filter is allowing very less

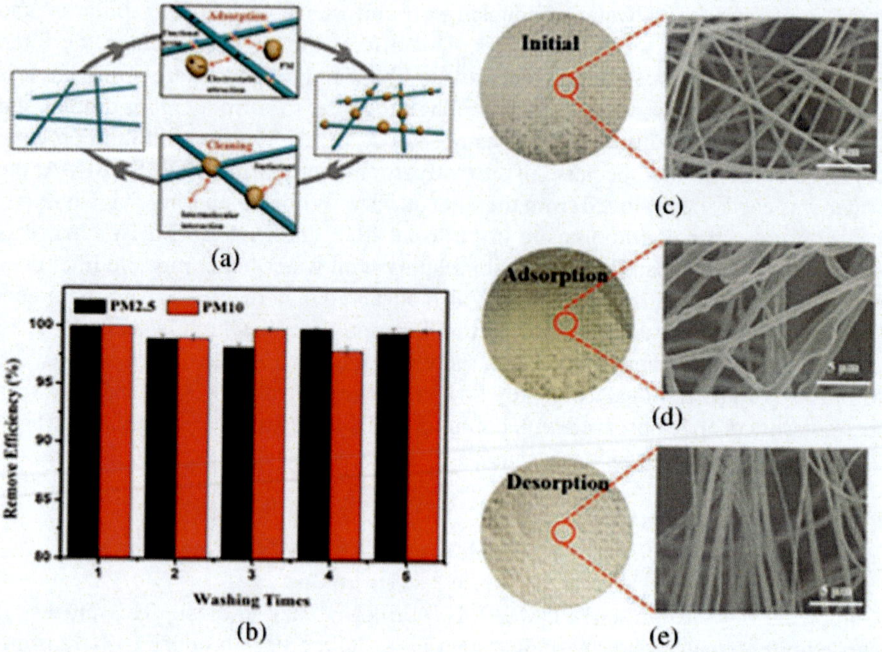

Fig. 11 **a** Animation of cyclic process of adsorption and disorption of PDMS/PMMA composite filter. **b** Efficiency of the filter before and after washing time. **c** SEM of initial stage of the filter. **d** SEM of the filter after adsorption and **e** SEM of filter after desorption [1]

particles and it is maintaining very high efficiency of 91% for PM 2.5 and 99% for PM 10. Photograph of the different tests is depicted in Fig. 12.

Among all the filters discussed in this paper, transparent antibacterial filter has more advantages over other filters. PDMS/PMMA composite filter has high durability, offers low air resistance, antibacterial, and transparent in nature compared to all the filters reviewed in this paper.

4 Conclusion

This paper explains different types of PMF. The advantages and disadvantages of porous-type and fibrous-type filters were discussed. Fibrous-type filter shows better performance than the porous-type filters. In fibrous-type filter, ACF, CNT, CNT/ACF, PMMA, PDMS, and transparent antibacterial fibre filters were discussed. The design and fabrication of the filters were also discussed. It is concluded from this review work that, transparent antibacterial fibre filters show better performance among all the filters. This review article will help researchers to design and fabricate PMF for their particular application.

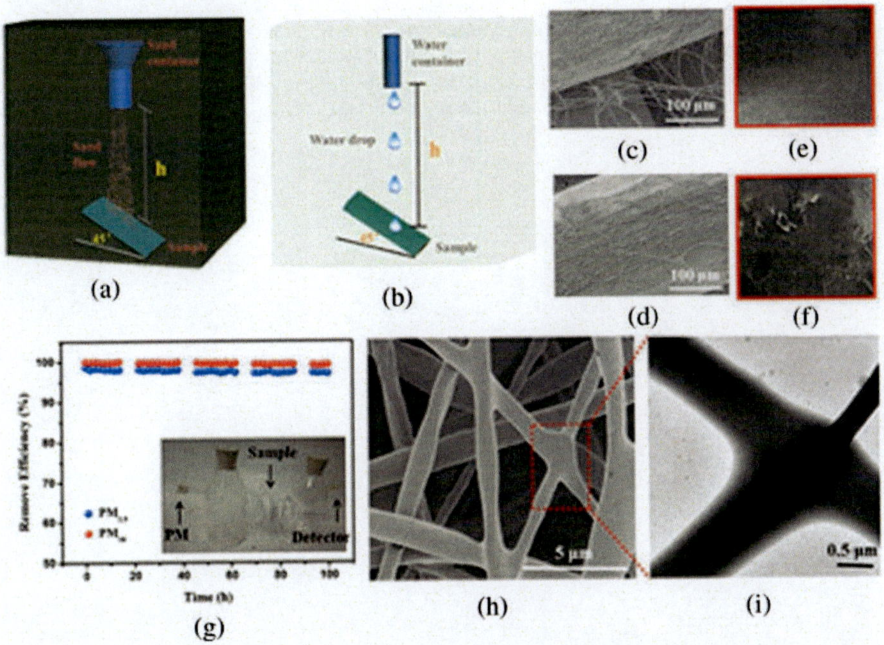

Fig. 12 **a** The sand test, **b** the water drop test, **c–f** the images of the filter before and after the test, **g** time versus removal efficiency graph for PM2.5 and PM10 and **h–i** magnified images of the damage [1]

References

1. Liu H, Huang J, Mao J, Chen Z, Chen G, Lai Y (2019) Transparent antibacterial nano fibre air filters with highly efficient moisture resistance for sustainable particulate matter capture. Science 19:214–223
2. Lindsley WG Pore size and aerosol sample collection. NIOSH manual of analytical methods (NMAM), 5th edn
3. Bennett A (2012) Standards and testing: meeting standards in filter media. Filtr.+ 49:22–25
4. Darling E, Morrison GC, Corsi RL (2016) Passive removal materials for indoor ozone control. Build Environ 106:33–44
5. Deitz VR, Bitner JL (1973) Interaction of ozone with adsorbent charcoals. Carbon NY 11:393–401
6. Wang C, Li P, Zong Y, Zhang Y, Li S, Wei F (2014) A high efficiency particulate air filter based on agglomerated carbon nanotube fluidized bed. Carbon NY 79:424–431
7. Yildiz O, Bradford PD (2013) Aligned carbon nanotube sheet high efficiency particulate air filters. Carbon NY 64:295–304
8. Viswanathan G, Kane DB, Lipowicz PJ (2004) High efficiency fine particulate filtration using carbon nanotube coatings. Adv Mater 16:2045–2049
9. Yang S, Zhu Z, Wei F, Yang X (2017) Carbon nanotubes/activated carbon fiber based air filter media for simultaneous removal of particulate matter and ozone. Build Environ 125:60–66
10. Zhang S, Tang N, Cao L, Yin X, Yu J, Ding B (2016) Highly integrated polysulfone/polyacrylonitrile/polyamide-6 air filter for multilevel physical sieving airborne particles. ACS Appl Mater Interf 8:29062–29072

11. Bhattarai N, Edmondson D, Veiseh O, Matsen FA, Zhang M (2005) Electrospun chitosan-based nanofibers and their cellular compatibility. Biomaterials 26:6176–6184
12. Zhang B, Zhang Z-G, Yan X, Wang X-X, Zhao H, Guo J, Feng J-Y, Long Y-Z (2017) Chitosan nanostructures by in situ electrospinning for high-efficiency PM2.5 capture. Nanoscale 9:4154–4161

Data Science and Data Analytics

Survey on Crime Analysis and Prediction Using Data Mining and Machine Learning Techniques

P. Saravanan, J. Selvaprabu, L. Arun Raj, A. Abdul Azeez Khan, and K. Javubar Sathick

Abstract Crime is an unlawful event which affects the harmony of humanity. Whoever got victimized in a crime, it affects them both physically and mentally. Hence, they are haunted by the memories throughout their life. Due to the limitations, traditional data collection and analysis methods are not very effective now. Yesteryears, the researches were concentrating on demographic features of the population. Nowadays, the dynamic characteristics of individual or specific group could easily be extracted from the search engines, social media, e-commerce platforms, mobile applications, IOT devices, surveillance cameras, sensors and geographical information systems. The recent technological advancements are helpful in integration of data from various sources, classification of information into granular level, identification of crime sequences and designing a framework. Particularly, the artificial intelligence methodology called deep learning imitates the functions of human brain and able to acquire knowledge from unstructured data. It makes revolutionary changes in crime forecasting, predictive policing and legal strategy formulations. The following survey explores the possibilities of scrutinizing the data from huge repositories, analyzing various socioeconomic factors associated to the crime incidents, identifying the outliers, categorizing the patterns and designing effective computational models to predict crimes by using data mining and machine learning techniques.

P. Saravanan (✉) · J. Selvaprabu · L. Arun Raj · A. Abdul Azeez Khan · K. Javubar Sathick
B.S. Abdur Rahman Crescent Institute of Science and Technology, Chennai, India
e-mail: saravanan_cse_2019@crescent.education

J. Selvaprabu
e-mail: selva_cse_phd_17@crescent.education

L. Arun Raj
e-mail: arunraj@crescent.education

A. Abdul Azeez Khan
e-mail: abdulazeezkhan@crescent.education

K. Javubar Sathick
e-mail: javubar@crescent.education

N. Zhou and S. Hemamalini (eds.), *Advances in Smart Grid Technology*, Lecture Notes in Electrical Engineering 688, https://doi.org/10.1007/978-981-15-7241-8_31

Keywords Data mining · Mining algorithms · Machine learning algorithms · Classification · Spatial–temporal factors · Predictive modeling

1 Introduction

Crimes are real trammels to the societies irrespective of their cultural, social, economic background. Since the crime incidents are considered as negative indicators of the wellness of any civilization, they played a key role in the life of an individual, family and state. Punishment is one of the ways to condense the crime rates. The general public in the society always look for preventive measures which would stopover the fortuity of crime. Crime prevention could only be done by implementing effective strategies. Those strategies may perhaps be derived from the analysis of existing data. The law enforcement agencies practiced these organized processes for the past few centuries. It was called by the name of hot spot analysis, which required substantial number of human resources and abundant amount of time.

People always dreamed about a predictive mechanism, which correctly identifies and forecasts the crime events that may occur in the future. The concept of crime prediction is a cradle for many science fiction books. The Minority Report, a novel by Philip K Dick, was published in 1950s, and then, the same was adapted as a movie in 2002. That story accomplished a massive success globally in both the forms in spite of cultural, linguistic and geographic differences of readers and viewers. Since safety is a universal factor, everyone has an expectation about predicting and stopping crime. It was a fantasy or fictional until the last decade, however not now [1] (Table 1).

There are prolific quantities of crime-related data available in law enforcement agencies like police department, judicial organizations and NGOs. They have information about the type, spatial–temporal properties of crimes along with socioeconomic background. The above-listed data was gleaned from NCRB-India [2]; it shows 85% of augmentation in cognizable crime incidents in the past 35 years. The trend clearly indicates the possibilities of increment in crime occurrences in the future. The crime controlling strategies could only be designed by cramming the data. Hence, we have to use data mining and machine learning algorithms in the process of exploring historical information and predicting crimes. Nevertheless, there are few challenges

Table 1 IPC and SLL crimes in India

Year	Total number of cognizable crimes covered under either Indian penal code (IPC) or special and local laws (SLL)
1980	3952229
1990	4898012
2000	5167750
2010	6750748
2015	7326099

in the accuracy of prediction and decision making. Those obstacles would have been resolved by the forthcoming methodologies.

2 Survey Classification

Crime data mining is an interdisciplinary field; it comprises criminology, data mining algorithms and statistics. Presently, many researchers are working on designing an effective crime prediction models. Comprehensive observation about variegated extensive researches in crime data mining is presented here. This review involves the identification of techniques used, methodologies implemented, conclusions obtained and forthcoming routings directed in the existing works. The specified articles are classified based on the following factors:

2.1 Predicting crimes based on socioeconomic, spatial–temporal, demographic factors.
2.2 Predicting crimes based on their nature (i.e., grading by severity)
2.3 Predicting crimes by the characteristics of victims (i.e., women, children, elderly people)
2.4 Predicting crimes by using different methodologies (i.e., machine learning, deep learning, transfer learning).

3 Literature Survey

Cyber forensic is a process of finding the evidences linked to a technical crime. There are numerous possibilities of e-crimes which include hacking user accounts, financial frauds, data theft, phishing and blackmailing. These unlawful activities are severely affecting the people, organizations and society. In most cases, the detection of cybercrime and locating the criminals are considered as intricate tasks. Whenever the technological advancements got rolled out, it elevates the risk of crime occurrence. To tackle this issue, the technical forensic experts are studying the past data, identifying the risks associated and trying to fix them by new approaches. Prevention of cybercrime would save millions of dollar every year. Once any e-crime happened, it would make irreversible damage to the system. Even though there are limitations, data mining and machine learning models are considered as the most successful methodologies in e-crime prediction. The emerging methodologies like deep learning are giving hope that these challenges could be fixed through effective forecasting. Since the deep learning is a methodology of artificial intelligence, it tries to imitate like a human brain during the execution. It could also be effective with processing unstructured data.

Karie et al. [3] are proposing a framework to aid the cyber forensic process by implementing deep learning algorithm—shortly called as DLCF framework. It contains various layers, including initialization, data sources identification, deep learning-enabled investigation, reporting, decision making and closure. After the evidences are identified, the relationship between entities would be established by DL. The authors hope that the algorithm could study various incidents, derive the pattern and deliver superior predictions.

Comparatively, urban crime data mining researches are outnumbered the rural ones. Frequency of crime occurrences, availability of datasets and density of population are the few factors attracting researchers to concentrate on issues related to urban community. There are very few researches on small cities, town and rural areas. The characteristics of urban and rural societies are contrasting with each other, hence the crimes. The approaches toward urban crime prediction may not be suitable for the rural and vice versa. Bolger and Bolger [4] performed a survey on fear of crime in a small town. Fear of crime had been scrutinized by dissecting individual demographic factors or community-level effects or combination of both. They have taken the vulnerability model and the incivilities model for their analysis.

The vulnerability model focuses on the susceptible demographic factors and their influence on fear of crime, i.e., females are more likely to be victimized in sexual crimes like rape. The incivilities model tries to explain the effects of neighborhood disorder on fear of crimes. The social disorder and dissatisfaction with law enforcement agencies are increasing the fear. The authors examined the trend of fear by analyzing neighborhood and demographic variables. They predict that social disorder is a strongest factor which induces the fear of crime. Based on the findings, it confirmed that the presence of police or other legal authorities reduce the fear of crime in town and rural population.

Crime is a spasmodic event, associated with many external factors including economy of the country. The relationship between crime and economy was examined by various scholars in the past. The USA had gone through recession in 1930s, and the crime rate was dipped. Later, in 1950s, the economy got boomed, and the crime rate too upraised and rumbled. During 1980s and 1990s, there were ups and downs in crime rates. Hence, the researchers superannuated the connection between economy and crime rate. Few others who still believe in the correlation of these factors are working on the same.

Mittal et al. [5] deliberated the economic crisis in India and try to prove the influence of economy on crime rates. The crime data and the economic indicators like Gross State Domestic Product (GSDP), Net State Domestic Product (NSDP), Per capita Net State Domestic Product, unemployment rate and Consumer Price Index (CPI) of the years 2004–2013 were elicited from various government portals. Independent variables (Gross District Domestic Product (GDDP) and the unemployment rate) and dependent variables (theft, burglary and robbery) were evaluated by the decision trees, random forest, linear regression and neural networks algorithms. Out of four, linear regression had generated results with high accuracy. It corroborated the correlation between unemployment and robbery. This paper started the debate about the influence of economic factors on crime incidents again.

Population of the area is an important factor in crime prediction. Mostly, predictions are derived by referring the specific residential population. However, the influence of ambient population and other related factors on crime rate is not negligible.

Kadar and Pletikosa [6] collected the data about population from various resources like crime data from police department, census information, location-based social networks, subway rides and taxi rides. There are two types of models practiced in crime prediction. Long-term crime prediction models are aiming at the accumulated crime rates over 1–5 years. Instead the other one, short-term crime prediction models are pointing at the short period which may vary from one day to one month.

The location-based social networks like Foursquare are providing information about people on move and their purpose of transportation. Data about public had been correlated with the crime data elicited from police department. Census, spatial and temporal features were evaluated by three different tree-based machine learning models. The residential population along with spatial–temporal feature would delineate the characteristics of ambient population and that leads to effective crime prediction by the machine learning models. The outcome of this human mobility data-based model is superior than census-based models.

Cyber-attack is an action executed against a computer or entire network. The invaders primarily try to access the data available in the computer or knocked out the system or partially disabled the utilities. These types of activities could cause a catastrophic effect in national security. Hence, it is called as 'cyberterrorism.' The experts estimated that cybercrimes are going to cost six trillion dollar annually by 2021 [7]. There are many types of cyber-attacks including phishing attacks, drive-by-download attack, denial-of-service attack, man-in-the-middle attack, password attack, snooping attack, cryptographic attack, malware attack, zero-day exploits, cryptojacking, SQL injection attack, cross-site scripting attack and spoofing attack.

A cyber-attack could possibly damage the reputation of a person, collapse a company or even create a socioeconomic warfare. Guns are muted in modern society, instead cybercrimes firing the shots. Okutan et al. [8] analyzed the unconventional signals which were perceived from Global Database of Events, Language, and Tone (GDELT), Open Threat Exchange (OTX) and Twitter data. Then, they appraised the predictive signal imputation (PSI), aggregating signals with significant lags (ASL) and SMOTE++ for imbalanced data techniques against the incidents. The missing data in some attributes was filled with possible values to increase the quantity of training data and reduce the insignificance. Their model had addressed the problem of predicting the type of cyber-attack.

CCTV cameras are installed in public places to ensure the safety of individuals, organizations and institutions. The technological advancements lead to mass production of budget-friendly CCTV cameras. Hence, people engrossed to install these devices in and around their property. The government is also concerned about public safety; the respective public departments are allotting money for installing surveillance cameras. There are millions of cameras installed throughout the globe. They are producing enormous amount of data every second. Usually, these data were referred while the investigation process to collect the proof. There are few technologies which are automatically identifying the suspicious people by studying their actions. Nevertheless, the effectiveness of those technologies is still questionable.

Collecting data from video surveillance system, background subtraction, identification of objects and detecting the moving ones are the few steps followed in video data mining process. Video surveillance systems concentrate on region of interest (ROI); their focus is on the particular area or object including human. The decisions would be made based on the suspicious action or movement of object. In real scenario, these systems mostly failed to identify the exact object. Kim et al. [9] propounded a new approach with the help of Gaussian mixture model (GMM) and convolutional neural networks (CNNs). GMM was used for extraction of moving objects with background subtraction, and CNN algorithm classified the objects inside the ROI. The above said model even identified small objects which are moving in distance, and the accuracy is comparatively better than the existing approaches.

Social disorganization theory suggests that the possibilities of a person to be a criminal lie on the location where he lives. It argues that ecological or neighborhood factors are contributing more on designing a person's character than any other factors. Vomfell et al. [10] had taken social disorganization theory as a tool for analyzing crime data. They have considered the population of locality, point of interest (POI), taxi flow and the tweets of people who resides in the specified location. Crime analysis has been done with social and structural point of view. The collected data was analyzed by the spatial linear regression, Poisson generalized linear model (GLM) and machine learning methods. They have deliberated the property and violent crimes and then confirmed the correlation between crime and spatial factors. Hence, it would be effective in crime prediction process as well as designing policing strategies.

Organized crimes like corruption, drug distribution, arms trafficking, smuggling, blackmailing, human trafficking, money laundering, robbery, gambling and murder are mostly committed by the gangs. Since these crime organizations are running in a tight-knit environment, their activities are not perceptible. Most of the time, the legal agencies received partial information about these gang crimes. It is really a byzantine task to predict a crime with the availability of partial data. Specially trained police officers are deployed in the criminal identification process. They analyze the circumstances, suspect, victim, type of crime and come to conclusion about particular incident. However, it is neither a time competent approach nor an effective way to resolve cases. Instead the machine learning algorithms are identifying patterns by analyzing the relationship between attributes. If there is any missing data, it would affect the learning process and accuracy of prediction. Seo et al. [11] proffered a solution to resolve this issue with the help of partially generative neural network (PGNN) architecture. It is effective in generating missing values and improves the performance of classification process and prediction.

In case, regions with same demographic trait were analyzed by the algorithm, then it would obviously yield the similar results. To overcome this challenge, inter- and intra-temporal–spatial patterns should be considered as primary factors. Zhao and Tang [12] recommended a new approach that deals with two regions with common characteristics. The model would be trained in one region or administrative block. Then, the learning would be transferred and tested with another region.

Crime partition could be done in many ways; one of such ways is segmenting the crimes by their severity. Mohd et al. [13] studied the filtering methods for crime data features selection. The factors like race, income class, age, family structure, education, population, locality and unemployment rate were considered as important entities in crime subdivision. The influences of these features in crime would differ case by case; hence, finding out the relevant features of crime is considered as a complex process. Appropriately identified features would persuade the outcome of prediction algorithm. In this research, hybrid features selection method was also implemented. Later, it was identified as a best performer compared to other features selection methods.

Hardyns et al. [14] used the term intelligence-led policing (ILP) in the sense of integrative approach. It includes explication of tactical strategies and intelligent pre-emptive policing. The crime hot spots are identified by the knowledge-based smart model. Amsterdam Police Department-based Crime Anticipation System (CAS) is able to process 200 attributes and results with 3% of high risk cells. The ensemble model is a combination of logistic regression and neural networks. In this study, the ensemble model could manage to achieve a better balance between hit rate and precision. The specific predictive policing mechanisms assist the city law enforcement agencies by providing early forecasting.

Classification is a supervised learning approach, which helps computers to learn from the input data. There are many types of machine learning classification algorithms including linear classifiers, kernel estimation, support vector machines, decision trees, quadratic classifiers, neural networks and learning vector quantization.

Vural and Gok [15] had chosen the decision tree, Naive Bayes classifier and assessed their capability. The decision tree algorithm denotes possible decisions in a tree-like model. The tree comprises roots, leaf nodes and branching; it also encompasses the sets and subsets. They are representing the attributes of data. Naive Bayes classifier is a linear classification algorithm, and it is working based on the Bayes' theorem. It reconnoiters the probability of an event happening while the evidence occurred. It assumes that every feature is independent and equal. Mehmet Sait Vural et al. found a correlation between the size of data and the accuracy of predictions. If there is an elevation in data size, the accuracy would also get improved. The Naive Bayes classifier outperformed the decision tree algorithm in the precision. It also has more sensitivity while processing more amounts of data. Effectiveness of Naive Bayes classifier has already proven in other domains like medical, genetics and phishing. Through the outcome of this research, it confirmed that the same could be used in crime prediction. Even though it surpassed the other machine learning classifier by its high accuracy rate, still there is a necessity for improvement.

Whenever crime hot spot identification process got initiated, it would deal with the spatial–temporal attributes. None of the crime could be defined without the place and time of the specific event. The combination of spatial–temporal data and socioeconomic factor would be the main attributes in crime hot spot identification. The socioeconomic factors would have hardly changed in the specific period of time. In case, the user wants to predict the crime hot spot weekly by giving the above said two attributes, and then, the system would not indicate any variation in its prediction. Hence, Ding et al. [16] were not only concentrated on hot spot identification; they have also focused the borders on hot spots. They have tried three different options of RNN architectures and propose spatio–temporal neural network (STNN)-based model. During the evaluation, authors used deep recurrent neural network algorithm and it outperformed the traditional machine learning algorithms.

Location and environment are the prime factors of any crime. The characteristic of urban crimes is varying from rural crimes. These days, the crime prediction and prevention methods are included in urban planning processes. Urban areas are equipped with public transportation facilities like buses, taxies, trains. At the same time, the transportation facilities are concomitant with crime incidents. These risks affect the day-to-day activities of public whoever using that provision. Geographic information systems (GIS), spatial clustering and artificial neural networks are collectively utilized by the transportation risk identification tools. The data mining methods, specifically the neural network-based algorithms, are used to assess the risks associated to the transportation and forecast the crimes. As Kouziokas [17] emphasized, scaled conjugate gradient algorithm is considered as a quick learning and time effective compared with others. Optimum neural network model categorized the crime hot spot areas by examining spatial data.

Predictive analytical process consists of statistical and analytical tools which study the historical and current data. These tools are designed based on the algorithms; they would carefully anatomize the relationship between different attributes and impart predictions about future. Russell [18] suggested that the effectiveness of predictive modeling could be measured by validity, equity, reliability and usefulness. Validity is determined by the successful outcome derived from the tests executed. The receiver operating characteristic (ROC) curve contains true positive and false positive values of the tests. True positive rates designate how many positive predictions have been envisioned by the model. False positive rates represent how many incorrect positive predictions have been speculated by the model. True negative and false negative rates also played a vital role in validity. The performance of model in sub-population is measured by equity. The model should result equivalent outcome while analyzing the same information by different users. It is called as reliability. The usefulness of model is very important in the end user point of view irrespective of its effectiveness (Table 2).

4 Research Gap

Criminology is one of the oldest intellectual departments in the world. Crimes are evolving with the human civilization. They challenge the society with their intensified catastrophic effects. As a response, the society tries to build systems that would eradicate the reoccurrence of crime. The systems like policing, law and firewall could only be designed by analyzing the existing crime incidents. The role of identified crime features in classification and prediction process is inimitable. The crime features are depending on socioeconomic-culture-demographic factors. Particularly, the influence of socio-economic factors on crimes is inevitable. These factors are varying based on the social structure and hierarchy.

Table 2 Key findings and interpretation

S. No.	Author(s)	Methodology used	Pros	Cons
1	Karie et al. [3]	Neural networks, deep learning cyber-forensics (DLCF) framework	Inclusive approach of DLCF framework	Conceptual methodology
2	Bolger and Bolger [4]	Ordinary least squares (OLS) regression	Confirmed the impact of demographic factors on fear of crime with limited resources	Sample population under-represent the linguistic and racial minorities and younger residents
3	Mittal et al. [5]	Decision trees, random forest, linear regression, neural networks	Analyzed various crime variables in detail with the help of different algorithms. Achieving the better outcome using linear regression	Influence of demographic features on economic factors was not considered in the study
4	Kadar and Pletikosa [6]	Random forest regressor, extremely randomized tree regressor, gradient-boosting regressor	Proved the influence of ambient population on crime occurrence of particular area. 65% on a geographical out-of-sample test set and up to 89% on a temporal out-of-sample test set	Lack on long-term crime prediction
5	Okutan et al. [8]	k-nearest neighbor algorithm, k-NN algorithm, predictive signal imputation technique, signal aggregation approach, SMOTE++ algorithm	Predicted the cyber-attacks (>85%) effectively through integrated approaches	The complexity in the integration of algorithms
6	Kim et al. [9]	Background subtraction algorithm, gaussian mixture model (GMM), convolutional neural networks (CNNs)	Detecting very small objects persuasively	Issues in prediction performance, the outcome of background subtraction process is yet to be improved

(continued)

Table 2 (continued)

S. No.	Author(s)	Methodology used	Pros	Cons
7	Vomfella et al. [10]	Spatial linear regression, poisson generalized linear mixed models (GLMM), simultaneous autoregressive (CAR) model	Behavioral-based crime prediction technique	Urbanized prediction model
8	Seo et al. [11]	Logistic regression, support vector machine (SVM), decision tree (DT), generative variational module (GVM), neural networks (NN), partially generative neural network (PGNN)	Better in missing feature values generation.	The performance of PGNN was not measured with other crime types.
9	Zhao and Tang (2017) [12]	Transfer learning	The crime could be predicted through analyzing the data of particular administrative block.	Changes in layers drastically affect the results
10	Jalil et al. [13]	The filter method, the human expert selection method, Naive Bayes, multi-layer perceptron, super vector machine, logistic, multi-class classifier, random forest, decision stump, random tree	The proposed feature selection method gives the highest accuracy with a mean of 96.94%	Specific feature-based methodology
11	Rummens et al. [14]	Logistic regression model, neural network, ensemble model	Grid-level crime prediction has been successfully achieved	Applied only for limited crime types
12	Vural and Gok [15]	Naive Bayes theorem	Crime prediction has been forecasted at the incident-level, and the success rate of the model is 78.05%	Feature selection process is yet to be implemented.

(continued)

Table 2 (continued)

S. No.	Author(s)	Methodology used	Pros	Cons
13	Zhuang et al. [16]	Recurrent neural network (RNN), Gated recurrent unit (GRU), long short-term memory (LSTM), spatiotemporal neural network (STNN)	The spatial–temporal factors-based neural network algorithm outperformed the other machine learning approaches	Non-demographic
14	Kouziokas [17]	Spatial clustering and analysis, artificial neural network, the scaled conjugate gradient algorithm	Faster learning algorithm	Domain-specific approach
15	Russell [18]	Predictive analytics	Importance and possibilities of predictive analytics	Lack in embodiment

Most of the western world is constructed by class hierarchy. Hence, the researchers from developed countries concentrated on the association of crimes and economic factors. However, Indian society still got influenced by the castes and religions. There are many subcultural groups inside these social structures. Few of these groups are insisting others to follow some impractical customs and values. When there is a resistance against this compulsion, it leads to social unrest and crimes. The subcultural theory of criminology describes that these crimes are committed through learned behavior. The youth population of the problematic subgroups has to be treated with strict legal and social policies to resolve these issues.

In India, the hate crimes are upraising in the past few years. Amnesty International report [19] declares 721 such cases from 2015 to 2018. They have been committed against religious minority, dalits, sexual minority, women and children. The association of subcultural groups with these crimes could no longer be ignored.

During the feature identification process in data mining, the dominance of subgroup on crimes should be considered as an element along with other social factors. The hate crimes should be classified under distinct category (Ex: Murder and honor killing should not be classified under a same group). The distinct categorization of hate crimes and spatial–temporal information along with other data would be helpful to improve the accuracy of predictions (Fig. 1; Table 3).

5 Conclusion and Future Work

The features of unlawful act play a significant role in crime occurrences. Since the features are determined by sociocultural factors, the feature identification process

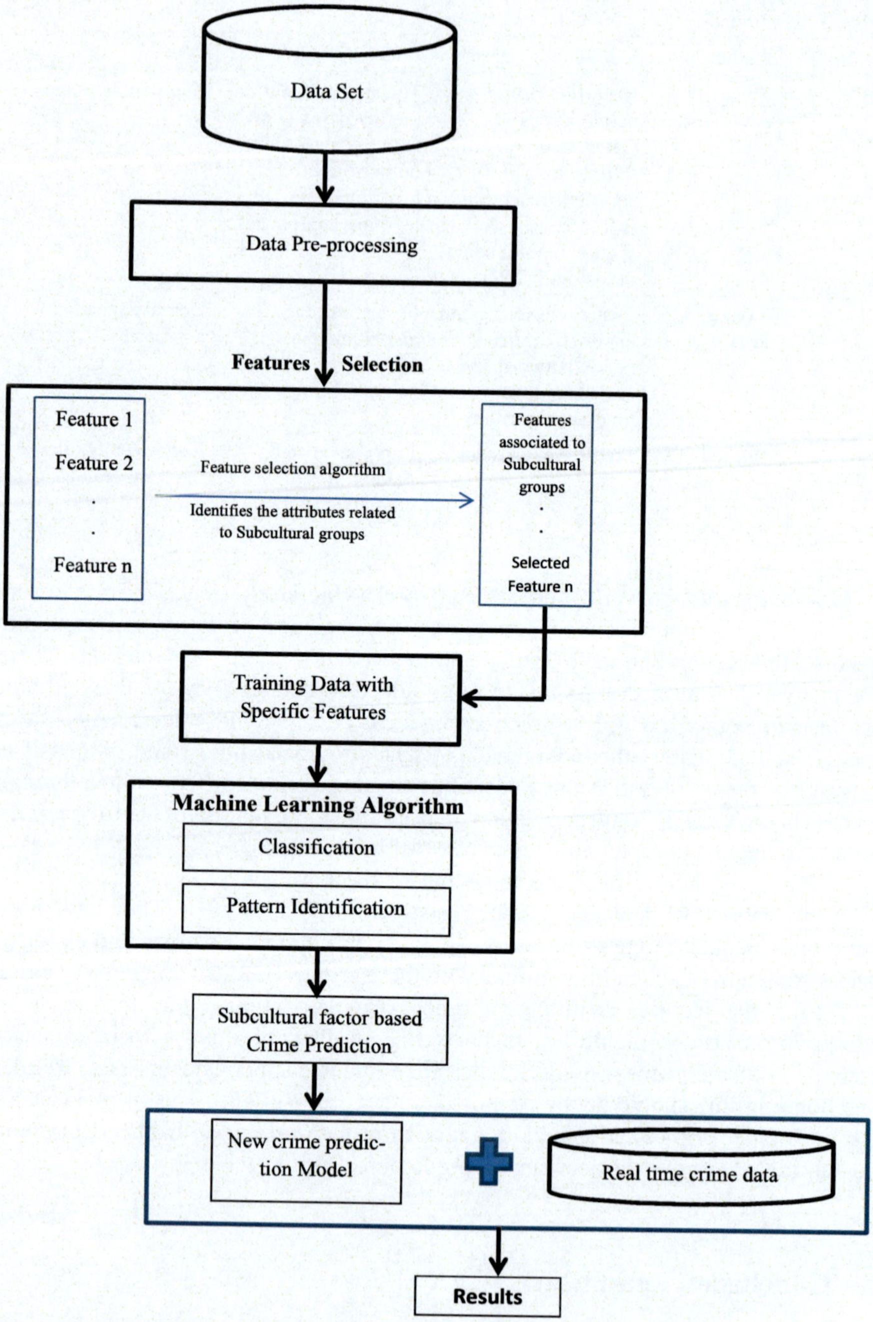

Fig. 1 Inclusion of attributes related to subcultural groups in mining process

Table 3 Pseudocode of the workflow

Data: crime data from National Crime Records Bureau (NCRB)	
Result: build a crime prediction model using subcultural factors	
1.	Set C as the source data: *C → Comprises of crime data
2.	Set D as the feature selection model
3.	Let E be the entity
4.	Let A be the attributes in the source data C
5.	Let FD be the featured dataset
6.	Define E based on the source data: C
7.	Establish the relationship between entities → E by mapping the selected Subcultural group based attributes → A.
8.	Build FD based on the data model D
9.	Pass FD to the mining engine
10.	Apply the selected classification technique on FD
11.	Identify the crime pattern
12.	Develop the crime prediction model and train the same with FD data
13.	Assess the trained model using validation dataset

should be localized. The influence of economic factors on crimes was investigated by many Indian researchers. As the cultural factors also make considerable effect, the impact of those factors is yet to be intensely navigated. Hence, this study proposed a new conceptual data mining approach toward crime feature identification.

In future, the impact of cultural factors on different crimes should be deliberated by machine learning algorithms. The crime data has to be classified based on the impact of features. The deep learning algorithms are going to be used for knowledge extraction, pattern identification and prediction. Based on the accuracy of prediction, a new crime prediction model would be designed and implemented. This feature-based crime prediction model would be helpful in predictive policing and legal policy making.

References

1. HARVARD University—"Minority Report" a Reality? The NYPD's big data approach to predicting crime. Available at https://digital.hbs.edu/platform-rctom/submission/minority-report-a-reality-the-nypds-big-data-approach-to-predicting-crime/. Accessed 08 Nov 2019
2. NCRB—Crime in India—All Previous Publications. Available at http://ncrb.gov.in/StatPublications/CII/PrevPublications.htm. Accessed 08 Nov 2019
3. Karie NM, Kebande VR, Venter HS (2019) Diverging deep learning cognitive computing techniques into cyber forensics. Forens Sci Int Syn 1:61–67
4. Bolger MA, Colin Bolger P (2019) Predicting fear of crime: results from a community survey of a small city. Am J Crim Just. 44(2):334–351

5. Mittal M, Goyal LM, Sethi JK, Hemanth DJ (2019) Monitoring the impact of economic crisis on crime in India using machine learning. Comput Econ 53(4):1467–1485
6. Kadar C, Pletikosa I (2018) Mining large-scale human mobility data for long-term crime prediction. EPJ Data Sci 7(1):26
7. Cyber security Ventures: Official Annual Cybercrime Report. Available at https://cybersecurityventures.com/2015-wp/wp-content/uploads/2017/10/2017-Cybercrime-Report.pdf. Accessed 08 Nov 2019
8. Okutan A, Werner G, Yang SJ, McConky K (2018) Forecasting cyber-attacks with incomplete, imbalanced, and insignificant data. Cyber Secur 1(1):15
9. Kim C, Lee J, Han T, Kim YM (2018) A hybrid framework combining background subtraction and deep neural networks for rapid person detection. J Big Data 5:22
10. Vomfella L, Hardle WK, Lessmann S (2018) Improving crime count forecasts using twitter and taxi data. Dec Supp Syst 113:73–85
11. Seo S, Chan H, Brantingham PJ, Leap J, Vayanos P, Tambe M, Liu Y (2018) Partially generative neural networks for gang crime classification with partial information. In: Proceedings of the 2018 AAAI/ACM conference on AI, ethics, and society, pp 257–263
12. Zhao X, Tang J (2017) Exploring transfer learning for crime prediction. In: IEEE international conference on data mining workshops, pp 1158–1159
13. Jalil MA, Mohd F, Noor NMMZ (2017) A comparative study to evaluate filtering methods for crime data feature selection. Proced Comput Sci 116:113–120
14. Rummens A, Hardyns W, Pauwels L (2017) The use of predictive analysis in spatiotemporal crime forecasting. Appl Geogr 86:255–261
15. Vural MS, Gok M (2017) Criminal prediction using Naive Bayes theory. Neur Comput Appl 28(9):2581–2592
16. Zhuang Y, Almeida M, Morabito M, Ding W (2017) Crime hot spot forecasting: a recurrent model with spatial and temporal information. In: IEEE, International Conference on Big Knowledge, pp 143–150
17. Kouziokas GN (2016) The application of artificial intelligence in public administration for forecasting high crime risk transportation areas in urban environment. Transp Res Proced 24:467–473
18. Russell J (2015) Predictive analytics and child protection: constraints and opportunities. Child Abuse Negl 46:182–189
19. Amnesty International India—Over 200 Alleged Hate Crimes In 2018, Reveals 'Halt The Hate' Website. Available at https://amnesty.org.in/news-update/over-200-alleged-hate-crimes-in-2018-reveals-halt-the-hate-website/. Accessed 08 Nov 2019

Crowd Management Using Ambient Intelligence

V. Jacob Jebaraj, S. Surya, and Velmathi Guruviah

Abstract Ambient Intelligence (AmI) is the Artificial Intelligence which is completely human-centric. It deals with a new world where computing devices are spread everywhere, allowing the human being to interact in physical world environments in an intelligent and unobtrusive way. These environments should be aware of the needs of people, customizing requirements and forecasting behaviors. AmI involves integrating the technology invisible in day-to-day life. According to Information Society Technologies Advisory Group (ISTAG) 1999, AmI1 should be the result of the convergence of three key technologies: 'Ubiquitous Computing,' 'Ubiquitous Communication' and 'Intelligent User-Friendly Interfaces.' AmI is unobtrusive and often invisible, being embedded in everyday objects such as furniture, clothes, vehicles, roads and smart materials. People will be surrounded by intelligent and intuitive interfaces recognizing and responding to the presence of individuals. Intelligence embedded in everyday objects and the surrounding environment such that the use of these smart objects is intuitive to the inhabitants of the environment. Combining both mobile and sensing technologies for providing a pervasive and unobtrusive intelligence and environment in supporting the main activities and interactions of the users. This forms the base of Ambient Intelligence. Technologies like face-based interfaces and affective computing are inherent ambient intelligence technologies. Many features being the crucial element in setting up an effective and efficient Ambient Intelligence environment, intelligent user interface is the key component. A vision of society of the future, where the people will find themselves in an environment of intelligent and intuitively usable interfaces, ergonomic space in a broad sense, encompassing better, secure and active living environment around them, capable of aiding them with daily chores and professional duties by recognizing the presence of individuals, reacting to it in a non-disturbing, invisible way, fully integrated into the particular situation.

Keywords Ambient intelligence · Crowd management

V. Jacob Jebaraj · S. Surya · V. Guruviah (✉)
School of Electronics Engineering, Vellore Institute of Technology, Vandalur-Kelambakkam Road, Chennai 600127, India
e-mail: velmathi.g@vit.ac.in

N. Zhou and S. Hemamalini (eds.), *Advances in Smart Grid Technology*, Lecture Notes in Electrical Engineering 688, https://doi.org/10.1007/978-981-15-7241-8_32

1 Ambient Intelligence (AmI) in Crowd Management

Crowd management in India is very poor compared to many foreign countries. This is mainly due to the lack of discipline in people. Many incidents prove this fact. In fact, it is merely impossible to manage the crowd with human force alone in India with such a huge population. Therefore, many technological developments have been introduced for the better management of crowds in many festivals and public gatherings. But not all of them were effective in achieving the desired task. Hence, ambient intelligence came into effect. As discussed earlier, Ambient Intelligence (AmI) is the Artificial Intelligence which is completely human centric. Most of the AIs are used for automation and smart thinking in computer and software areas. But, AmI involves the smart thinking capability of AI in places where human life and activities are closely related. Thus, in crowd management AmI can be a very good solution. AmI will provide video analytics service using Artificial Intelligence (AI) for crowd control, several powerful CCTVs will be set up to monitor the situation where crowd has to be controlled. Advanced control panels are also set up for effective use of AmI in crowd management. Different other techniques like Global Positioning System (GPS), Infrared cameras (IR), Image Processing techniques can be used for crowd management in festivals and programs in India. RFID is one of the most efficient crowd numbering and sourcing techniques for managing huge crowds in a closed architecture. It always gives the most accurate information about crowd density in a particular sector with RFID scanner in the entrance and exit (Fig. 1).

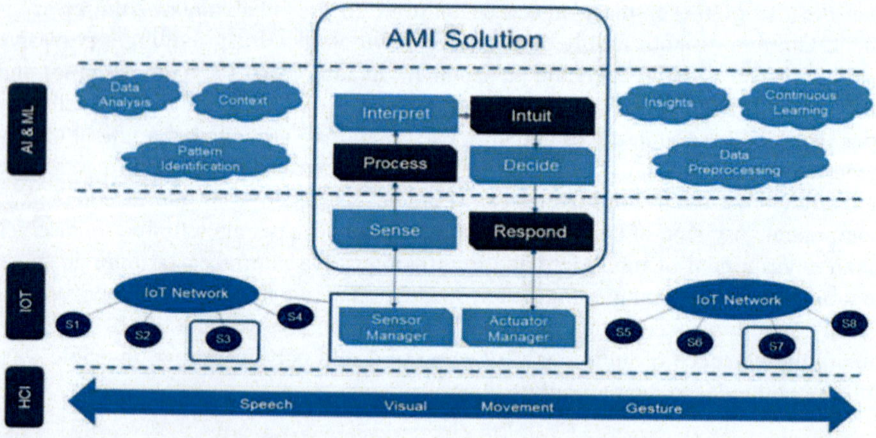

Fig. 1 Mind map of AMI solutions

2 History of Stampede in India

Stampede in India have proved to be very deadly in the past few decades and has claimed more than thousand lives in the last decade.

Let us take a look at some of the deadliest stampedes in the recent years in India:

1. Rajahmundry Stampede, 2015

On the opening day of the pushkaralu festival there was a huge crowd which rushed into the pushkar ghat to take a holy dip. It killed nearly 27 pilgrims out of which at least 13 of them were women.

2. Datia Stampede, 2013

In the month of October 2013, during the Navratri, a stampede broke out on the bridge near the Ratangarh Mata temple in Madhya Pradesh killing around 115 people.

3. Sabarimala Stampede, 2011

It is one of the worst stampedes in India killing almost 106 devotees and more than 100 were injured.

4. Jodhpur Stampede, 2008

During the Navratri, about 25,000 pilgrims were visiting the Chamundi Devi temple in Jodhpur, Rajasthan. At the time of opening of the temple doors, a tragic stampede caused the death of more than 224 people and left many more injured (Figs. 2 and 3).

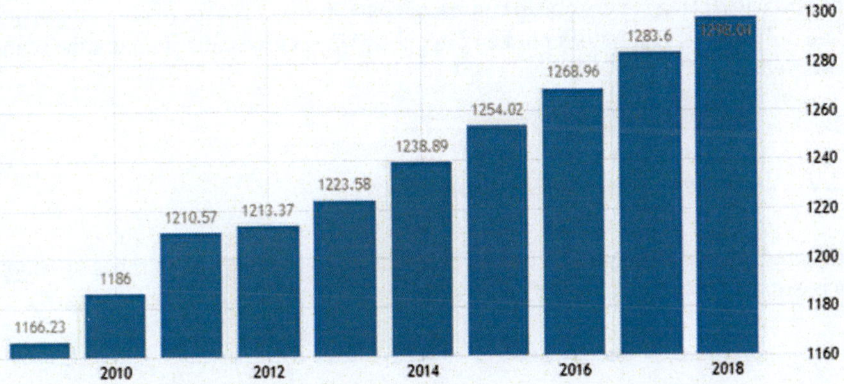

Fig. 2 Exponential increase in rate of deaths

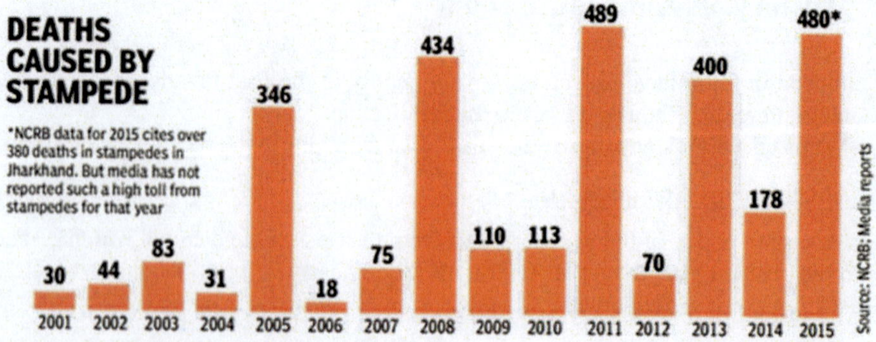

Fig. 3 Deaths caused by stampede

3 Ways to Manage the Crowd

As we have seen, the enormous crowd in religious gatherings like Sabarimala, Kumbh Mela, Navratri, etc. is increasing year by year. People nowadays are more religious than before and like to attend most of the festival gatherings.

3.1 RFID TAGS

This crowd can be effectively controlled using techniques like RFID protocols. The cost of a RFID tag is very cheap compared to the entry ticket for many of the religious and holy places. But it does not work in open places as we cannot mount the scanners everywhere.

3.2 Infrared Thermal Imaging System

Infrared Thermal Image Processing can be used to identify the number of people in a place. But this cannot work under sunlight as it affects the imaging (Fig. 4).

3.3 Global Positioning System (GPS)

The GPS module in the smart phones is used to predict traffic in applications such as Google Maps. The same method is used here to predict the crowd density in a speculated area. But it would not work in places where there is no network coverage (Fig. 5).

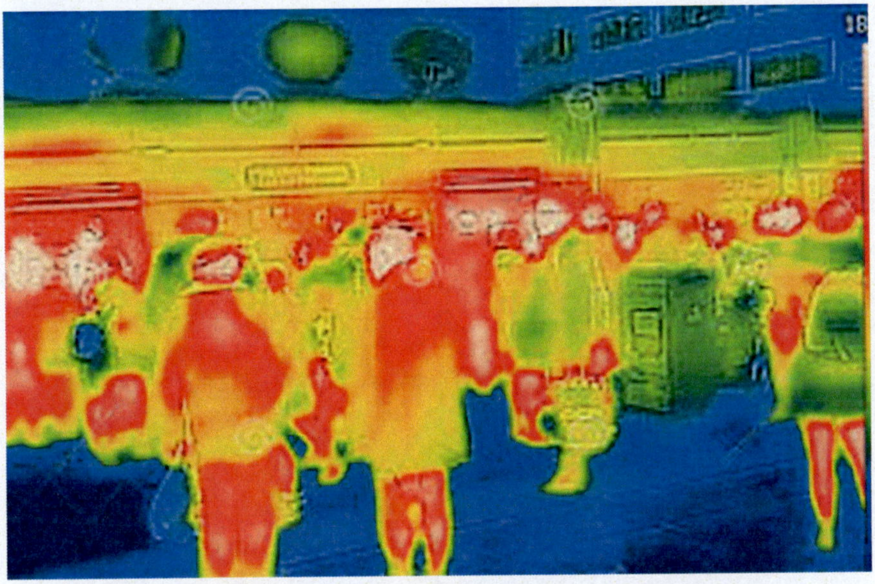

Fig. 4 Image from an IR camera of a crowded area

3.4 CCTV

Using image recognition technique we can find the total head count in a particular area and can also detect if any person have fell down. But it too cannot be that accurate and sometimes it leads to misjudgments.

The cost of a RFID tag is very cheap compared to the entry ticket for many of the religious and holy places. In case of a closed and surrounded place, RFID tagging is effective in detecting the crowd density. For example: In a closed structure like Tirumala Tirupati Devasthanam, crowd is scattered and separated initially based on the cost of the entry ticket whereas on approaching further, they converge into a single crowd. This can be controlled by knowing the exact crowd density in a particular region and regulating the inflow of the pilgrims and rescheduling their visiting time in case of overpopulation. RFID readers can be placed at the entrance of each section and when a pilgrim crosses by, it detects the entry of a devotee in that section.

This is one example adopted for crowd management in closed sectioned structures. But this is not the case in a wide-open land. RFID cannot be used in such places. Infrared thermal image processing can be used to identify the number of people in a place. Scanning of already known stampede prone hotspots in closed loop fashion using Thermal Cameras can give the risk of a stampede in that region. But this also is not very effective since the Thermal Image cannot give a clear picture of the number of people under the influence of sunlight since we are considering an open place.

Global Positioning System (GPS) is one of the best ways in open places for management and calculation of crowd density. Nowadays, almost everyone has a

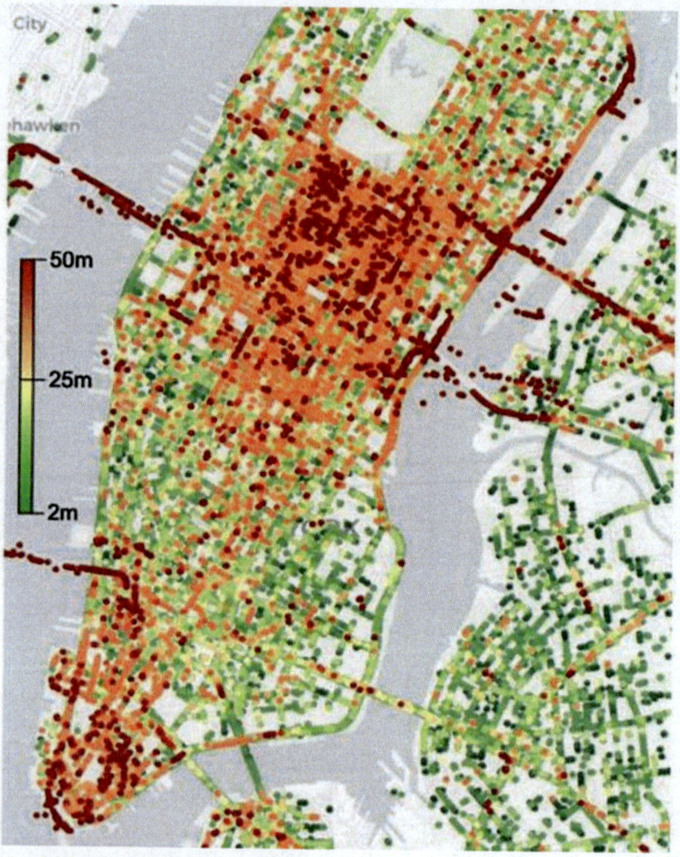

Fig. 5 Image from a satellite of a crowded area

smartphone. The GPS module in the smart phones is used to predict traffic in applications such as Google Maps. The same method is used here to predict the crowd density in a speculated area. If the crowd density exceeds the set threshold, a warning message is sent to the police officers for controlling or it can also be displayed in a Big heads-up display warning the people automatically without any human intervention. Through the above-mentioned steps, stampede can be effectively controlled in Indian festivals thereby preventing deaths and injury of thousands of people every year.

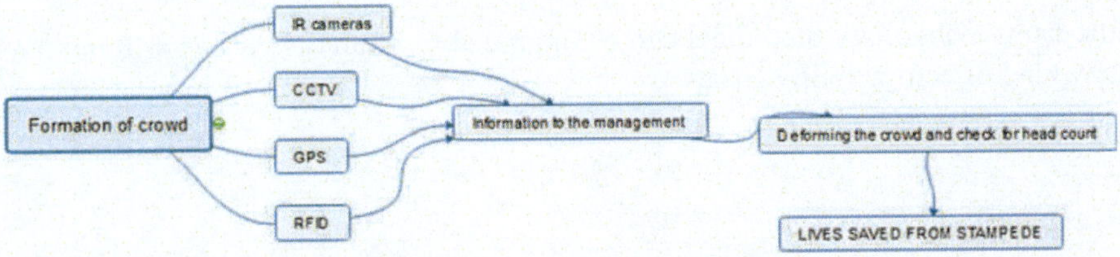

Fig. 6 Block diagram of the system

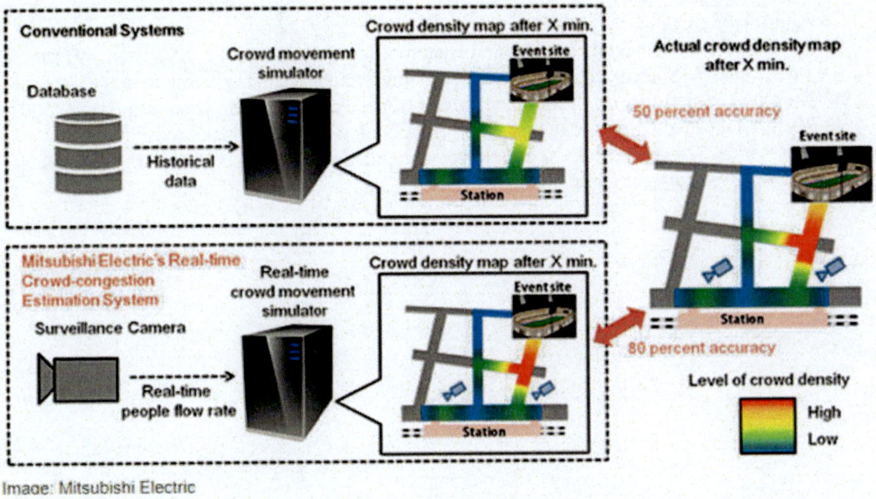

Fig. 7 Working of the system

4 Block Diagram

In the block diagram mentioned below gives a clear explanation of how the system works. First whenever there is formation of crowd then that particular area gets monitored by four of the subsystems, i.e., IR cameras, CCTV, GPS and RFID scanners. Whenever one of the system gets triggered it sends an alert information to the nearby management to clear or manage the crowd and they have to stop more people moving into the crowd (Figs. 6 and 7).

5 Conclusion

Hence Ami in crowd management in really important as they help in saving lives from unnecessary accidents caused by man himself. This helps in improving and organizing crowd management in a better way and helps people live better. As it uses the latest technology the model can be upgraded to future needs and will not have any kind of compatibility issues.

References

1. Sudhi KS (2015) Crowd management crisis looms large over Sabarimala. The Hindu, KOCHI, 25 Dec 2015
2. Bradshaw L (2011) Crowd management plan for an existing outdoor event research paper. Custom Writing.Com 3 Sept 2011
3. Doukas SP Crowd management: past and contemporary issues. Sport J ISSN: 1543-9518
4. Ali MF, Bashar A, Shah A. (2015) SmartCrowd: novel approach to big crowd management using mobile cloud computing. In: Conference: international cloud computing conference ICCC 2015, At Riyadh Saudi Arabia https://doi.org/10.1109/cloudcomp.2015.7149656
5. Marine Profile Sweden AB, Crowd management. Academia.edu [6]. globalco_glob, Research paper on crowd management plan, 31 Mar 2015

Analysis of Wind Speed Data in Tadipatri Region in Andhra Pradesh

R. Reshma Gopi, A. Chitra, Pujari Harish Kumar, and R. Mageshvaran

Abstract Wind energy is one of the renewable energies, which has been utilized effectively to maintain the ecological balance of the environment. The wind speeds (m/s) at the 10 m level above the earth's surface in Aluru Kona site (latitude of 14.920, longitude of 78.020) and Ellutala site (latitude of 14.559, longitude of 77.779) for a period of 01.01.2019 to 01.02.2019 are used to estimate wind power density. The benchmark data is obtained from MINES ParisTech Web portal. The research work is limited in the regions of Andhra Pradesh: Anantapur district, Tadipatri, for Aluru Kona village and Ellutala village wherein the wind power plants have located. In this work, an empirical method is being utilized for the evaluation of the Weibull probability distribution parameters (shape parameter and scale parameter) and cumulative function which provides an estimate of the wind power density.

Keywords Weibull distribution · Wind power density · Wind power potential · Cumulative function

1 Introduction

Every day the world blazes enormous amounts of fossil fuels to meet the never-ending demand for electrical energy. Coal, petroleum and natural gas are the most important fossil fuels, and their supply is unable to meet the growing demand in developing countries. Conserving non-renewable sources of energy and considering the use of renewable energy sources as an alternative has thus become crucial. Among the natural renewable energy sources, the wind is the most abundant source, which does not cause any pollution. So, this research work has mainly focused on wind energy, which has become essential for social, economic and environmental stature [1]. The work provided an estimation for the wind speed distribution at 10 m above the earth's surface and evaluated wind power density (WPD) using the mean value of wind information, considering the 24 h wind speed variations for the period of

R. R. Gopi · A. Chitra (✉) · P. H. Kumar · R. Mageshvaran
School of Electrical Engineering, Vellore Institute of Technology, Vellore, Tamil Nadu 632014, India
e-mail: chitra.a@vit.ac.in

N. Zhou and S. Hemamalini (eds.), *Advances in Smart Grid Technology*, Lecture Notes in Electrical Engineering 688, https://doi.org/10.1007/978-981-15-7241-8_33

01-01-2019 to 01-02-2019. ReNew power limited has installed 119.7 MW of wind power plant in Ellutala in Tadipatri, Andhra Pradesh, India.

Estimation of WPD of a particular location aids in the choice of an appropriate wind turbine. The higher the WPD of a location, the more electricity is being generated from the wind turbines. The WPD of locations: Aluru Kona and Ellutala village, Tadipatri region of Andhra Pradesh, India are being estimated in this paper using the data taken from the Web portal of MINES ParisTech (France). Weibull probability density function parameters of the two locations are estimated using the empirical method [2, 3] and used for calculating WPD. These calculations are useful to the industries and also policymakers for wind energy enhancement for the coming years across Andhra Pradesh, India.

1.1 Methodology

For each site, the daily mean value for the highest wind speed in a day above 10 m ground level is calculated using data considering 24 h in a day, for two months at two different locations/sites. Correspondingly, standard deviation values have derived from the daily mean value. The following equations are used to compute the arithmetic mean and standard deviation [2, 3] for statistical investigation.

Arithmetic mean and standard deviation is given by Eqs. (1) and (2), respectively, are calculated by considering highest wind speed in a day.

$$\mu_{\text{mean}} = \frac{\sum X_i}{\text{N}} \tag{1}$$

$$\sigma_\mu = \sqrt{\frac{1}{N} * \left\{ \sum_{i=1}^{n} (X_i - \mu_{\text{mean}})^2 \right\}} \tag{2}$$

Mathematical inspection of wind speed data. Depending on the terrain, the attributes (wind speed and energy) vary in different locations. The shape parameter of Weibull distribution can alter the appearance of the curve and its mean value. In [2, 3], the conventional numerical methods for calculating scale and shape parameters have been discussed. In this paper, the empirical method (EM) has been imposed for calculating the shape parameter (**k**) and scale parameter (**c**) by the following equations:

$$k = \left(\frac{\sigma_\mu}{\mu_{\text{mean}}} \right)^{-1.086} \tag{3}$$

$$c = \left[\frac{\mu_{\text{mean}}}{\Gamma(1 + \frac{1}{k})} \right] \tag{4}$$

where Γ is the gamma function and is given by [4]

$$\Gamma = \int_0^{\infty} e^{-t} * t^{x-1} dt \quad (5)$$

The formulae for analyzing Weibull probability distribution function are [5]

$$h(\mu) = \left[\frac{k}{c} * \frac{\mu_{\text{mean}}}{c} * \exp\left\{-\left(\frac{\mu_{\text{mean}}}{c}\right)^k\right\}\right] \quad (6)$$

The Weibull cumulative distribution function has calculated as

$$H(\mu) = \left(1 - \exp\left\{-\left(\frac{\mu_{\text{mean}}}{c}\right)^k\right\}\right) \quad (7)$$

where μ_{mean} defines the highest value of mean wind speed in a month. The measured wind power density (WPD) in W/m^2 is calculated by [6]

$$WPD = 0.5 * \rho * v^3 \quad (8)$$

where v is the average value of wind speed. In [7] estimated WPD of a location, derived from Weibull distribution as

$$\text{WPD} = 0.5 * \rho * c^3 * \Gamma\left(1 + \frac{3}{k}\right) \quad (9)$$

where ρ represents the air density. An error value between the measured and the estimated values of WPD is calculated by [8],

$$\text{Error} = \frac{\text{measured value-estimated value}}{\text{measured value}} \quad (10)$$

The hub height of installed turbines in specified regions is 30, 60 and 90 m. The mean wind velocity at those heights is estimated by the wind power law as shown in Eq. (11).

$$\frac{v_2}{v_1} = \left(\frac{h_2}{h_1}\right)^{\propto} \quad (11)$$

where α is the wind shear coefficient, h_2 is the hub height, h_1 is the anemometer height, v_1 is the mean wind speed velocity at anemometer height, and v_2 is the mean wind speed velocity at rotor-hub height. In this work, h_1 is taken as 10 m as wind speed is assessed at 10 m [9]. The value of α is computed by using Eq. (12):

$$\alpha = \frac{0.37 - 0.0881 \ln(c_1)}{1 - 0.0881 \ln\left(\frac{h_1}{10}\right)} \quad (12)$$

where c_1 is the scale parameter at 10 m height. The value of shape and scale parameter at hub heights is estimated by following equations:

$$c_2 = c_1\left(\frac{h_2}{10}\right)^{\propto} \tag{13}$$

$$k_2 = k_1 * \frac{\left[1 - 0.088\ln\left(\frac{h_1}{10}\right)\right]}{\left[1 - 0.088\ln\left(\frac{h_2}{10}\right)\right]} \tag{14}$$

where k_1, k_2 and c_1, c_2 are shape and scale parameters at 10 m height and hub height, respectively. The wind power density at specified hub height is calculated by Eq. (8), where v is mean wind speed at corresponding rotor-hub height.

The reliability analysis of wind particulars on Weibull distribution is done by plotting the Q–Q plot. The Q–Q plot is a graphical way of estimating the linear relationship between wind speed data and the distribution of sampled data [10]. The Weibull parameters for plotting the graph are calculated from a fitted least square line. The *p*-quantile for sorted data is calculated using the expression:

$$Pi = \frac{i - 0.5}{n} \tag{15}$$

where $i = 1, 2, \ldots n$. The quantile of Weibull distribution (*z*-score) is calculated for each *p*-quantile by using the formula:

$$z\text{-}score_i = k\left[-\log(1 - p_i)\right]^{\frac{1}{c}} \tag{16}$$

The quantile of Weibull distribution plotted against sorted data samples to check the plot follows the 45° reference line.

2 Result and Discussions

The wind speeds at the altitude of 10 m, from Jan 2019 to Feb 2019 in Aluru Kona and Ellutalla are used for calculation. Over the two locations, the highest wind speed obtained is above 6 m/s. Air density ($\rho = 1.29$) is taken at standard temperature and pressure. Table 1 depicts measured monthly average wind speed, standard deviations, shape parameter, scale parameter using Empirical method, WPD at the average wind speed and also highest wind speed in a month.

Figure 1 shows the time series of highest wind speed, WPD respectively in Aluru Kona site for January 2019.

Figure 2 shows the time series of highest wind speed, WPD respectively in Aluru Kona site for February 2019.

Table 1 Comparison of calculated parameters at two locations

Parameters	Month and year	Aluru kona site	Elutalla site
Mean wind speed (m/s)	Jan 2019	4.727	4.959
	Feb 2019	5.008	5.103
Standard deviation	Jan 2019	0.698	0.552
	Feb 2019	0.865	0.862
Shape parameter (k)	Jan 2019	7.983	10.85
	Feb 2019	6.733	6.902
Scale parameter (c)	Jan 2019	5.020	5.195
	Feb 2019	5.364	5.458
Wind power density at mean wind speed in a month (W/m^2)	Jan 2019	67.82	78.69
	Feb 2019	81.02	85.76
Highest wind speed in month (m/s)	Jan 2019	6.200	6.410
	Feb 2019	6.550	6.570

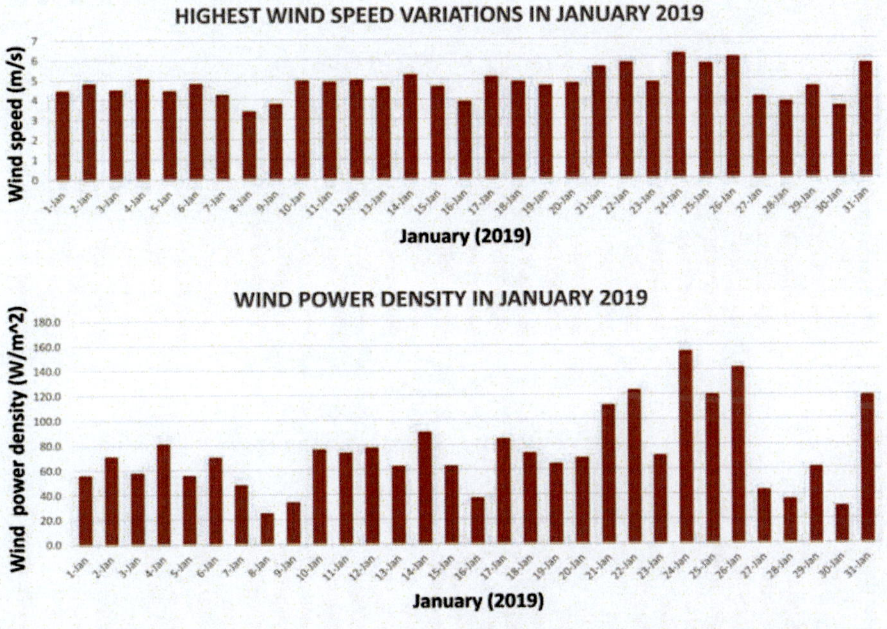

Fig. 1 Highest wind speed variation and wind power density in January 2019 in Aluru Kona site

Figure 3 shows the time series of highest wind speed, WPD respectively in Elutalla site for January 2019.

Figure 4 shows the time series of highest wind speed, WPD respectively in Elutalla site for February 2019.

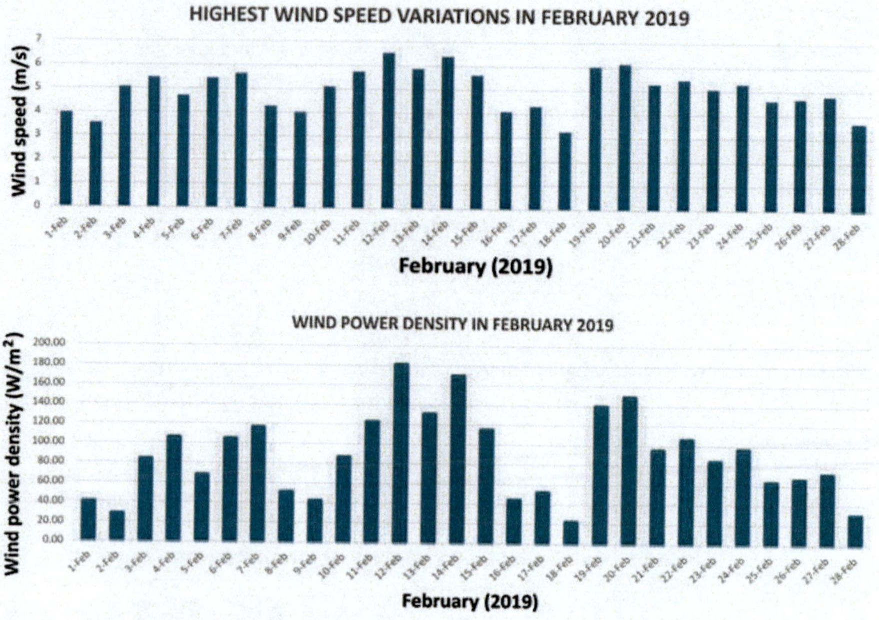

Fig. 2 Highest wind speed variation and wind power density in February 2019 in Aluru Kona site

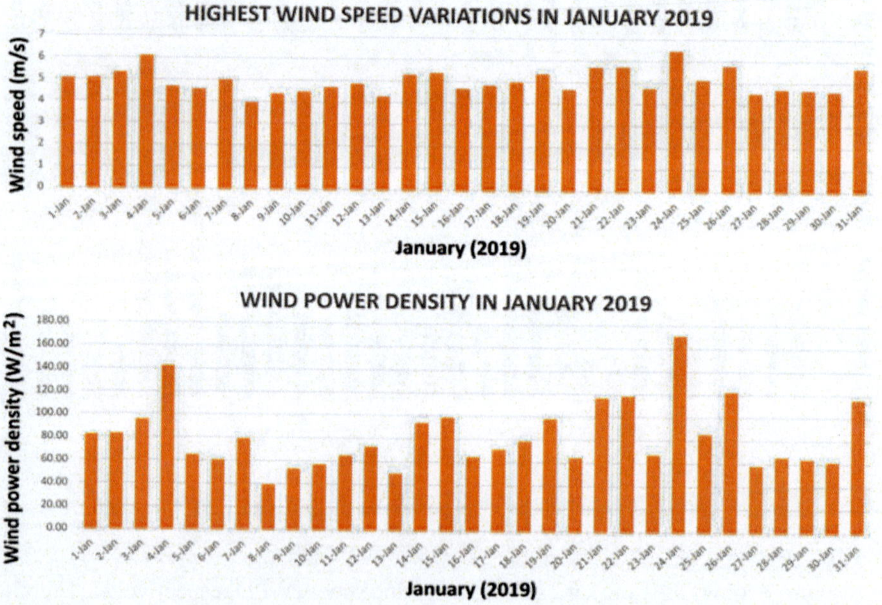

Fig. 3 Highest wind speed variation and wind power density in January 2019 in Elutalla site

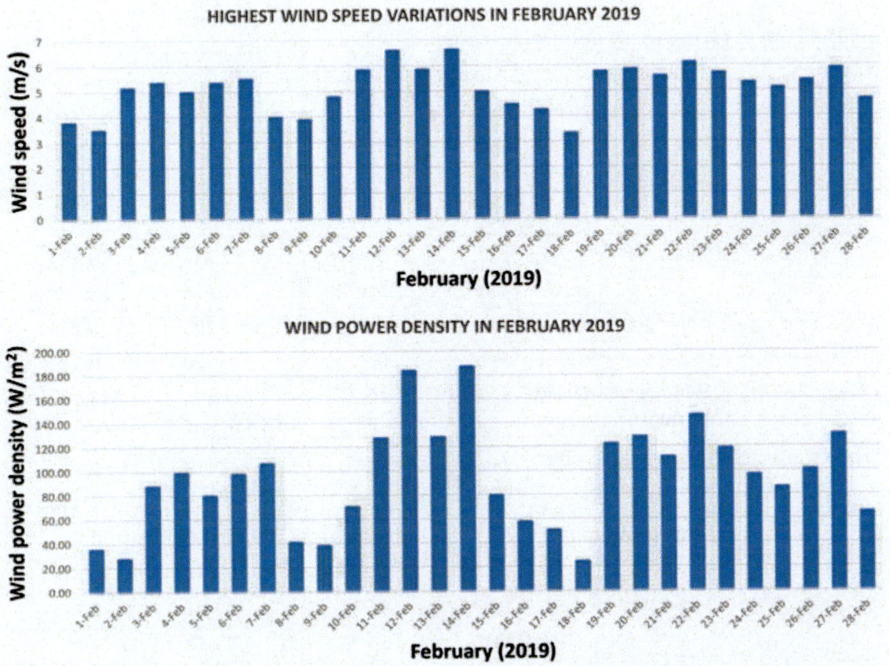

Fig. 4 Highest wind speed variation and wind power density in February 2019 in Elutalla site

Table 2 depicts the measured monthly WPD, estimated WPD at the site by using Eq. (9) as well as the % error Eq. (10). The results show that the Weibull probability distribution is ranging between 0.1 and 0.7 and cumulative function between 0.4 and 0.5, respectively. The percentage error has a negative value. The Elutalla site has higher wind potential compared to the Aluru Kona site.

Table 2 Measured and estimated WPD values

Parameters	Month and year	Aluru kona site	Elutalla site
Measured WPD (W/m^2)	Jan 2019	67.820	78.650
	Feb 2019	81.022	85.750
Estimated WPD (W/m^2)	Jan 2019	72.400	81.520
	Feb 2019	88.170	94.030
Percentage error	Jan 2019	−6.335	−3.649
	Feb 2019	−8.820	−9.650
Weibull probability distribution	Jan 2019	0.562	0.723
	Feb 2019	0.451	0.453
Weibull Cumulative distribution function	Jan 2019	0.462	0.454
	Feb 2019	0.467	0.468

Table 3 List of parameters at different heights

Location	Parameters	Month	30 m	60 m	90 m
Elutalla site	Shape parameter	Jan	12.012	12.870	13.460
		Feb	7.460	8.173	8.565
	Scale parameter	Jan	6.650	7.772	8.509
		Feb	6.943	8.100	8.852
	Mean speed	Jan	6.340	7.414	8.123
		Feb	6.497	7.574	8.278
	Shear coefficient	Jan	0.223	0.223	0.223
		Feb	0.220	0.220	0.220
	Wind power density (W/m^2)	Jan	164.37	262.85	345.70
		Feb	176.31	280.20	365.87
Aluru Kona site	Shape parameter	Jan	8.770	9.478	9.899
		Feb	7.454	7.995	8.348
	Scale parameter	Jan	6.440	7.550	8.266
		Feb	6.845	7.984	8.734
	Mean speed	Jan	6.075	7.091	7.784
		Feb	6.390	7.450	8.150
	Shear coefficient	Jan	0.227	0.227	0.227
		Feb	0.222	0.222	0.222
	Wind power density (W/m^2)	Jan	144.50	229.98	304.21
		Feb	168.40	266.70	349.16

Table 3 shows Weibull parameters and wind power density at 30, 60 and 90 m which are calculated from assessed wind data by using the power law method. The scape parameter and scale parameter exhibit an increase in value as rotor-hub height is varied from 30 to 90 m. The range of scale parameters in the Elutalla site is varied from 6.65 to 8.852. In the Aluru Kona site, it shows a variation from 6.44 to 8.734. During February, the shape parameter in Elutalla is estimated as 7.460, 8.173 and 8.565 at 30 m, 60 m, and 90 m heights accordingly. In February, the mean wind speed for 30 m, 60 m, and 90 m heights is determined as 6.497 m/s, 7.574 m/s and 8.278 m/s, respectively, in Elutalla site. The wind shear coefficient remains constant as it is independent of hub height. The wind power density in Elutella is higher than Aluru Kona and is 365.87 W/m^2 at 90 m height in February. The lowest wind power density is observed in the Aluru Kona site and is 164.37 W/m^2 in January.

The Q-Q plot obtained for both the locations during February is a straight line along the 45° reference line as shown in Fig. 5. A linear relationship between collected wind data and sampled data shows that Weibull distribution is acceptable for analyzing the wind data in current locations.

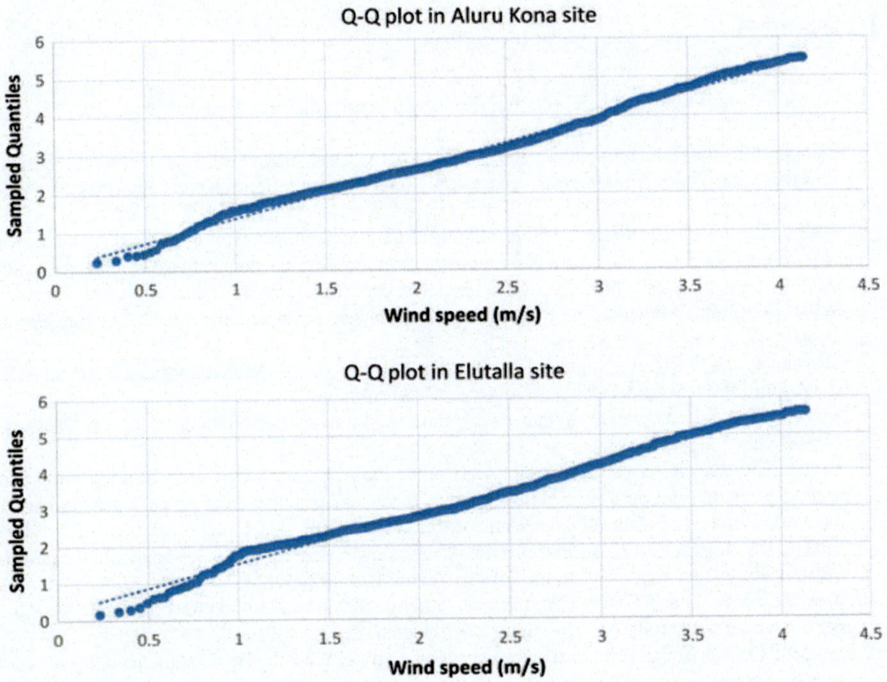

Fig. 5 QQ plot in Elutalla site and Aluru Kona site

3 Conclusion

This work provides a detailed insight into the process of evaluation for the wind power density assessment in Tadipatri region at Andhra Pradesh, India. The empirical method used for evaluating the Weibull probability distribution parameters helped to provide a better estimation in evaluating the WPD. From the results attained, it is evident that the mean wind speed varies in the span of 4.5–5 m/s in the area of Aluru Kona and Ellutala, and WPD measured is changing between the range of 65–80 W/m^2. The goodness of fit test is done by plotting the Q–Q plot. This work may be utilized by the industry, policymakers, and the appropriate agencies for designing the wind control and assessment in the Tadipatri region in Andhra Pradesh, India.

Acknowledgements The authors would like to thank MINES ParisTech, Center for Observation, Impacts, Energy (O.I.E.), France for issuing wind data. The authors are extremely grateful to the Chancellor, Vice-Chancellor and Vice-Presidents, of VIT University, Vellore, for providing the excellent infrastructure facilities.

References

1. Al-Mohamad A, Karmeh H (2003) Wind energy potential in Syria. Renew Energy 28(7):1039–1046
2. Dongbum Kang K, Kyungnam K, Jongchul H (2018) Comparative study of different methods for estimating Weibull parameters: a case study on Jeju Island, South Korea. Energies 11(2):1–19
3. Rocha P, Sousa R, Andrade C, Silva M (2012) Comparison of seven numerical methods for determining Weibull parameters for wind energy generation in the northeast region of Brazil. Appl Energy 89(1):395–400
4. Wilks DS (2006) Statistical methods in the atmospheric sciences, 2nd edn. Academic Press, London
5. Satyanarayana GC, Lucy Supriya RH, Bhaskar Rao DV (2019) Wind energy assessment over the Andhra Pradesh and Telangana regions. Meteorol Appl 26:14–29
6. Mani A, Mooley DA (1983) Wind energy data for India. Allied Publishers Private Limited, New Delhi
7. Saxena BK, Rao KVS (2015) Comparison of Weibull parameters computation methods and analytical estimation of wind turbine capacity factor using polynomial power curve model: a case study of a wind farm. Renew Wind, Water, Solar 2(3):1–11
8. Saxena BK, Rao KVS (2016) Estimation of wind power density at a wind farm site located in Western Rajasthan Region of India. Proced Technol 24:492–498
9. Kang D, Ko K, Huh J (2018) Comparative study of different methods for estimating Weibull parameters: a case study on Jeju Island South Korea. Energies 11:6
10. Scholz F (1999) Weibull reliability analysis. University of Washington, Department of statistics, Washington

A Literature Survey: Semantic Technology Approach in Machine Learning

L. Rachana and S. Shridevi

Abstract Semantic technology approach in machine learning is an emerging technique to solve the problems in the machine learning. Semantic technology has been the improvised from decades according to the human needs and industrial demands. This new era is all about teaching a machine to learn on its own and to make it understand the concept and the purpose for what it is used, using algorithms. This paper, condenses the work of semantic technology approach in machine learning and its idea put forward. The introduction details with brief explanation followed by description of the semantic technology and machine learning, important role. The literature survey contains summarized view of the papers with a graph plotted on the analysis of paper throughout the decade; a table with summary of the related works and concluded with review analysis.

Keywords Semantic technology · Machine learning · Ontology

1 Introduction

The approach of the semantic technology in machine learning is just like icing on a cake, the technology enables to provide them with an interface that allows them to directly model their knowledge in the system, and to review the knowledge in the system. Semantic systems also excel at provability. In a well-structured semantic system, you can get formal logical proofs, expressed in fairly understandable terms, backing up the answers produced by the system. You can have a very high degree of confidence in the answers which the system produces. Of course, sometimes there are flaws or holes in the underlying ontology but when there are, they're generally easier to identify. The inclusion of semantic technology has also given great opportunity to the machine learning to interact with human, has for machine learning is

L. Rachana (✉) · S. Shridevi (✉)
School of Computing Science and Engineering, Vellore Institute of Technology, Chennai, India
e-mail: rachana.l2019@vitstudent.ac.in

S. Shridevi
e-mail: shridevi.s@vit.ac.in

N. Zhou and S. Hemamalini (eds.), *Advances in Smart Grid Technology*, Lecture Notes in Electrical Engineering 688, https://doi.org/10.1007/978-981-15-7241-8_34

quiet complex for human to understand at one go. The datasets are categorized and processed as well as the relationship of the data within the sets is found. In this paper it deals the method used to solve the problems which evolved during the comparison of data for more accurate values.

1.1 Semantic Technology

The term "semantic" refers to the sense of words. Semantic technology uses Artificial intelligence to simulate the understanding and processing of language information by people. Semantic software reads and tries to grasp language and words in its context with respect to the approach. Technically speaking, this method is based on various levels of research: analysis of morphology and grammar. In Cogito's case, the semantics design requirements include various integrated elements. The following are the most important elements

I. Parser performs morphological, grammatical and syntactic study of the expression
II. Lexicon understands terms and all their meanings memory keeps track of analysis outcomes
III. Memory records empirical outcomes information represents real-world knowledge
IV. Content representation is language content in the context of a theoretical and mental graph.

The lexicon includes the so-called "semantic network." We name our collection of semantic networks "Sensigrafo" at Expert System. It is not an ordinary dictionary but tools that have been designed for programmatic use, where word types are knots connected to each other by multiple links denoting semantic or lexical relationships. As human beings, it's easy to understand our everyday language and words' meanings. It's not that easy to pass these same skills to a computer. It takes time to learn, and the same applies to a computer. There are no shortcuts or magic formulas: learning a language is challenging, and it takes time and work for automated processes [18].

1.2 Machine Learning

Machine learning is an artificial intelligence application that provides machines with learning skills and automatically learns from experience without direct programming. Machine learning focuses on building software tool for processing and using data for their own purposes. The learning process starts with observations, evidence including examples, direct experience, and preparation based on our examples, to interpret knowledge patterns and make better future decisions. The main goal is to make intelligent machines without human intervention to understand and change the behavior.

Some other methods in machine learning are often referred to as supervised or non-supervised algorithms for machine learning.

I. Techniques of supervised machine learning can use defined occurrences to predict future events to use what has already been observed from new data. Beginning with the evaluation of the existing learning data collection, the learner algorithm produces a system that is supposed to predict the performance values. The device will set a goal for any new input after proper training. The analysis of algorithm can also adapt its output to the correct output and detect errors so that the design can be improved accordingly.
II. In contrast, when the software used for training is not labeled or numbered, unguarded algorithms are used for machine learning. Unsupervised training investigates how constructs in order to explain a secret framework could derive a feature from unlabeled data. The computer does not find the correct performance, but analyzes the information and can extract details from data sets to clarify the secret properties of unlabeled software.
III. Half-supervised machine learning algorithms fall between supervised and non-supervised learning because they use marked and unlabeled training information, usually with a small proportion of labeled and a large proportion of unlabeled information. The programs that use this approach can greatly enhance reading accuracy. Semi-supervised training is usually handled when it is needed to educate and benefit from the information gained when the programs are educated or relevant. Typically, however, additional resources are not required to collect unlabeled information.
IV. Enhancing machine learning algorithms is a form of learning that interacts with its environment and finds errors and rewards. Evaluation and error detection and delayed compensation are the most important features for optimizing training. This approach allows devices and computer agents to immediately examine the optimal actions to improve their efficiency in a particular context. Clear feedback is needed to let the agent know what behavior is needed, known as the reinforcement signal.

Machine learning involves the storage of massive data. Using the computer, although it usually delivers faster and more accurate results to detect favorable opportunities and harmful threats, additional time and resources may also be required to train properly. The combination of machine learning with AI and cognitive technology can make it even more productive to produce large amounts of information.

1.3 Role of Semantic Technology in Machine Learning

The Semantic Web offers structured templates for both information and ontological context awareness representation. Semantic Web concepts are used to define Meta data, but as a general framework for interaction and information management, they also have great potential. Machine learning will play an extremely important role in a wide range of possible applications: Machine learning alternatives have been evolved to support ontology management, semi-automatic analysis unstructured data, and Web mining integration of semantic data. Increasingly, machine learning will be used to evaluate distributed data sources represented in semantic Web formats and to help theoretical system reasoning and querying.

2 Literature Survey

The survey is all about to understand the approach of semantic technology in machine learning and realize the state of art of the works in this field for finding the research gap, so that it can motivate for better research problems.

Marques et al. [1], "Semi-automatic Semantic Annotation of Images Using Machine Learning Techniques", in order to semi-automatically annotate objects using machine learning methods, we suggested a three-layer structure. The research expands the previous work into a Semantic Web-oriented approach in the related field of content-based picture recovery. The proposed architecture allows for a good degree of independence between efforts to improve its supporting technologies and algorithms, such as: development of better algorithms for visual feature extraction; implementation of better clustering algorithms; improved communication methods inquiry; deployment of intelligent ontology agents; enlargement, preservation, and binding of ontologies with keywords. It is also recognized that the quality of the algorithms—especially those used to obtain visual characteristics, to collect, to detect similarity and to map the bottom-middle layer of our model is an important factor in the success of the methodology proposed.

Taylor et al. [2], "Autonomous Classification of Knowledge into an Ontology", The problem of how to automatically determine where new knowledge can be placed in existing ontologies is seen, though distinct, in this work. Alternatively, it is increasingly important for machine learning technologies to simplify a job rather than relying on human knowledge engineers to carefully identify information. The rates of ontology building through automatic knowledge acquisition techniques are increasing. This paper compares three well-established machine learning approaches and illustrates that they can be used effectively for this function. In the Cyc knowledge base program our processes are fully implemented and evaluated.

Shein [3], "Ontology based combined approach for Sentiment Classification", In order to enhance the sentiment classification we suggested the combination of POS tagging, FCA-based domain ontology, and SVM classification. By using this approach, we can take a closer look at the strength or weakness of the products or artifacts and we hope that it will be useful for further creation and enhancement of the product or object growth. We need to experiment with a large amount of data sets as the future work and need more learning and classification to solve the comparative sentence problem.

Shein et al. [4], "Sentiment Classification based on Ontology and SVM Classifier", the suggested hybrid method of POS labeling, FCA-based domain ontology and SVM classifier to boost the identification of feelings.

By using this method, we may take a closer look at the strength or weakness of the goods or artifacts and we expect that it will be helpful for further creation and enhancement of the product or artifact growth. This approach is still being further developed.

Nezhadi et al. [5], "ontology alignment using machine learning techniques", This paper proposes a method of effective ontological consistency based on the combination of various categories of similarity in one input sample. The proposed model specifies the coordination mechanism without having to have ontology instances beforehand, making it easier to coordinate. Our proposed exemplary does not require user intervention and it has a consistent performance, both for aligned and non-aligning entities, through a detailed operational enchantment process. AdaBoost (DT) model provides the best overall accuracy, particularly when using feature selection scheme. Experimental results indicate that up to 99% of F-measurement parameters were higher than other relevant research trials that used up to 23 similarity tests.

Although this research uses only eight similarities measurements in its optimal model, it has compensated for the potential impacts of the decrease of feature through the use of ontology observations, the increase in sample diversity and the choice of more effective similarity measures. This provides greater precision and lower computing costs, making a model suitable even for a task to align ontology online.

Pinto et al. [6], "A semantic-based approach for Machine Learning data analysis", the wide range of technology and products are primarily centered on the intelligent presentation of data obtained in the atmosphere by heterogeneous sensors. Traditional Machine Learning (ML) approaches often do not go beyond a simple description without a clear description of the events observed. This paper delivers an early draft of a semantic-enhanced machine learning study on sensor flow information, doing much better on ubiquitous smart objects that are resource-constrained. The architecture blends ontology-driven features of statistical data delivery with non-standard matching resources. Encourages the identification of fine-grained events by addressing the standard asset discovery classification issue of ML.

Syamala Devi et al. [7], "machine learning techniques with ontology for subjective answer evaluation", in this study, the methods examined and applied show a high association with human Performance. This is because the duration of the response is largely influenced by human interpretation. Presence of keywords and keyword meaning. Use of Ontology, searches for keywords as well as keywords that occur in the correct Context. This dimension is absent in ontology-free techniques. Use Ontology tests for term existence, synonyms, correct word meaning, and scope of all definitions. It is assumed that, owing to systematic analysis, utilizing ML strategies with Ontology provides satisfactory results.

Imsombut et al. [8], "An Alternative Technique for Populating Thai Tourism Ontology from Texts Based on Machine Learning", This study proposed the use of Conditional Random Fields (CRFs) to identify examples of concepts with dictionary knowledge and a cue word list. In contrast, the post processing of instance abstraction is used as the method to boost the effects of NE boundary recognition, and the lexico-syntactic template is used to classify the relationships between instances. The experiments were carried out in the tourism domain on Thai language Web documents.

The method can derive instances with appropriate results according to these preliminary results.

Xu et al. [9], "An OWL ontology representation for machine-learned functions using linked data", Machine-learned tasks in the world of big information are incredibly useful. It is difficult to create new learned functions, but linguistic awareness can be used to allow anyone to replicate current functions effectively. We demonstrated the use of OWL ontology to canonize variable names and function properties, such as characteristics and performance metrics. Using hierarchical relationships between variables, in the semantic web, we may find missing function values based on other data provided by the client. We also have a database interface that allows users to use these generic objects to submit information.

Kim et al. [10], "Augmented Ontology by Handshaking with Machine Learning", we are exploring two methods and finding a good solution that maximizes and mitigates the advantages of both strategies. This study indicates a new idea of convergence to make up for each innovation with the other: that is semantic filtering. This paper involves a semantic simulation of toys and an adaptation of a machine learning functionality to understand the suggested term, semantic sorting. Recognition of the situation is AI's secret and the essence of AI is a data analysis. This paper suggests smooth and secure implementation in the IoT world for machine learning and semantic software. Overall framework needs to be refined as a further work, and real integrated implementation needs to be followed.

Dorc et al. [11], "An Approach for Automatic and Dynamic Analysis of Learning Objects Repositories through Ontologies and Data Mining Techniques for Supporting Personalized Recommendation of Content in Adaptive and Intelligent Educational Systems", Adaptive and Intelligent Educational Systems are valuable resources to support teaching-learning activities. These systems utilize innovative strategies to tailor educational content to students' real needs. With the growing abundance of educational content, there is a good reason to believe that smart data analytics and machine learning technologies will become essential ingredients of educational advancement. Hence, this research suggests a method to systematic and complex study of databases of learning artifacts in which ontology models the relationship between properties of learning objects and styles of learning. Better outcomes are collected, and in this study they were discussed.

Saranya et al. [12], "Onto-based sentiment classification was using Machine Learning Techniques", In this article, for better outcomes, the use of semantics and ontology to define text is coupled with machine learning techniques. The experiment shows that SVM, NB and kNN classifiers in the existing literature have been shown to be more useful for emotional research. The process began by defining and removing the affective class hierarchy in WordNet. The next move was to attribute emotions to the linguistic positions of the derived emo-words to establish a hierarchy of emotions. With more accurate results, the machine learning algorithm or a mixture of the algorithms (hybrid algorithm) can be implemented. Semi structured text grouping and clustering have some dispute and new opportunities.

Liu et al. [13], "The Extension of Domain Ontology Based on Text Clustering", This work discusses the relevant principles and approaches of constructing ontology and expanding ontology, introducing an automated ontology extension approach focused on supervised training and text clustering. This method uses the K-means clustering algorithm to isolate the information of the domain and to direct the construction of the Naive Bayes classification training collection. Terms are applied to the goal ontology in the nominee array, while noise terms are introduced to the stop-word dictionary at the same time. This method's input system was designed to promote the structure of the consistency of ontology, and eventually it will semi-automatically expand ontology. Ontology editors and other development tools are also necessary throughout the infrastructure construction of automatic cycle of ontology. Future work will also include attempts to improve the reliability of proposed methods and implement the Fuzzy Inference Rules training process.

Hwang et al. [14], "Autonomous Machine Learning Modeling using a Task Ontology", we extracted essential keywords from articles and textbooks on machine learning in this paper to establish ontology. In addition, we have designed a MEX-based ontology task. We have also researched the automated machine learning system workflow. For automated workflows at a specified autonomous stage, the following protocol applies. Therefore, non-experts are able to perform complex tasks by the system and can quickly apply the machine learning framework in a particular application.

Rutaa et al. [15], "Machine Learning in the Internet of Things: a Semantic-enhanced Approach", In this paper we presented a new approach for semantic-enhanced machine training in the Internet of things on heterogeneous data streams. Carrying raw data on ontology-based labels of concept provides a low-level linguistic understanding of the numerical distribution of information, whereas the conjunctive aggregation of concept components enables automated representation of events during model training phases to be rich and substantive. Finally, the use of non-standard match-making inferences allows the discovery of finest occurrences by treating the topic of ML identification as an asset seek.

Fu et al. [16], "Using Behavior Data to Predict User Success in Ontology Class Mapping—An Application of Machine Learning in Interaction Analysis", Representation in ontology usually consists of one-size-all approaches that offer the same representation for a given task/system to each client. Not every visualization technique, however, can result in a successful task outcome for each user, and thus it may be beneficial to develop systems that can provide adaptive assistance during visualization use. We must first determine if a client is likely to succeed in a job to create these scalable systems. For this reason this article provides a possible solution for the avoidance of user failure by exploring how effectively, in terms of accuracy, completeness and overall performance, can be accomplished through a user-interaction information (including eye tracking and click-data). We execute three classification tests to forecast success based on a data set obtained in a client sample utilizing two separate ontology modeling methods and two different types of activities. Results showed that for these predictions communication information could potentially be used and all three classifiers statistically outperform simple classifiers substantially.

Sacha et al. [17], "VIS4ML: An Ontology for Visual Analytics Assisted Machine Learning", we suggest an ontology (VIS4ML) in this paper for a VA sub-area, called "VA-assisted ML." VIS4ML is intended to identify and explain current VA workflows used in ML as well as to recognize deficiencies in ML processes and the ability to apply new VA techniques to such processes. Ontologies are commonly used in biology, medicine, and many other disciplines to map the reach of a subject. We follow the empirical methodologies for VIS4ML development, including ontology definition, conceptualization, formalization, deployment, and validation. In general, to include model-development workflows, they reinterpret the conventional VA pipeline.

In order to formulate VIS4ML, we add the requisite concepts, laws, syntaxes and graphical notes and use semantic Web technology to incorporate it in the Web Ontology Language (OWL). VIS4ML incorporates the information from past workflows at a high level where VA is used to aid in ML. It is aligned with the existing VA principles and will continue to evolve in tandem with potential VA and ML innovations.

While this ontology is an effort to build the theoretical basis for VA, clinicians can use it in real-world applications to refine model-development workflows by routinely analyzing the potential benefits which computer and person ability may bring about.

2.1 *Review Analysis*

After the analysis of the above the we can say that every experiment has its own drawbacks the such that, algorithms requires the storage of massive data sets, which should be diverse/ impartial and of good quality. There may also be occasions when they have to wait for the generation of new information. The next key challenge is the ability to analyze algorithm-generated findings correctly. You also need to pick the algorithms carefully for your intent. Mistakes may set up a chain of errors which can go undetected for prolonged periods of time. And it takes quite a while to identify and even lengthy to correct the source of the problem when it becomes noticed. Figure 1 allows us to track the number of publications during the last decade.

Table 1 summarizes the works of few authors of recent publication to analyze the research gap of semantic technology approach in machine learning.

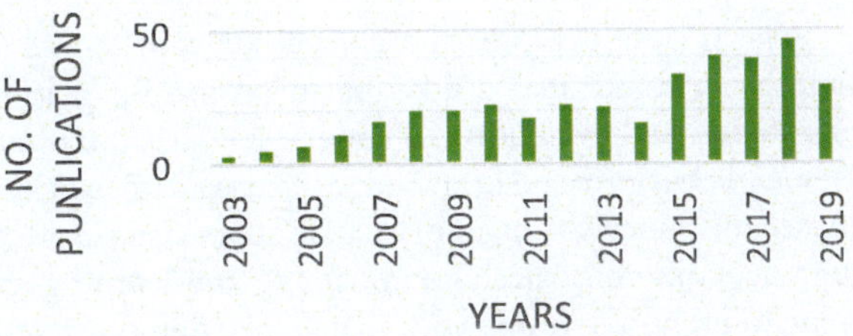

Fig. 1 Number of publications during the last decade

Table 1 Summary of the work related to the survey

S. No.	Paper details	Role of machine learning	Role of semantic technology
1	Shein et al. [4] 2010, sentiment classification based on ontology and SVM classifier	• Linear SVM is used to classify functions • Parameters are classified by a linearly divided hyper plane into the binary category	• Domain ontology to extract the related concepts and attributes
2	Nezhadi et al. [5], ontology alignment using machine learning techniques	• Algorithm of machine learning are used for aligning the ontology fields • Algorithm are used individually and in combined form to find the best out-come	• Training and generating model and; classification process • Alignment is solely base on comparing the entities of two ontologies
3	Syamala Devi et al. [7] 2016, machine learning techniques with ontology for subjective answer evaluation	• Create bag-of-concept to improve the classification of documentation	• Using ontology relevant words can be found • To match against an exhaustive knowledge base
4	Xu et al. [9] An OWL ontology representation for machine-learned functions using linked data	• Techniques of machine learning and data mining is used to improve the information of project	• Idea is that general computer functions are expressed in the semantine Internet, so that they can be used by anybody
5	Dorc et al. [11] an approach for automatic and dynamic analysis of learning objects repositories through ontologies and data mining techniques for supporting personalized recommendation of content in adaptive and intelligent educational systems	• K-means and EM techniques are better suited • K-means algorithm has been overcome with regard to the metric F	• To encapsulate this pedagogical awareness, an ontology was created • The LOM fields and LS interactions are modelled in ontology as SWRL laws

The review explores the current state of the art of semantic machine learning approaches and also guides us to propose a platform for facilitating the implementation and evaluation of supervised structured machine learning methods using semantic background knowledge. The goal of such system is to provide an ontology-based machine learning tool to solve supervised learning tasks and support knowledge engineers in constructing knowledge and learning about the data they created. We plan to propose a framework that can be demonstrated to explore the construction of semantic knowledge from existing ontologies and that knowledge can be used as a semantically annotated training set for learning algorithms to identify patterns. The review gave opportunities for how these artificial intelligence technologies can be combined to develop such intelligent systems.

3 Conclusion

This paper concludes the work of the survey on semantic technology approach on machine learning with ideas of the approach in real time hence with help of this we can summarizes the work stating that with semantic tools data can be handled better. The above survey gives an insight to recommend ontological data streams for better semantic machine learning. Tracking down raw data to ontological framework labels, gives a low-level semantic understanding of numerical classified information, while the combination of concept components enables a robust and solid description of events to be constructed effectively during the testing period. Ultimately, the use of non-standard corresponding inferences allows fine-grained incidents to be detected by approaching the ML classification issue as an object discovery.

References

1. Marques O, Barman N (2003) Semi-automatic semantic annotation of images using machine learning techniques", ISWC
2. Taylor ME, Matuszek C, Klimt B, Witbrock M (2007) Autonomous classification of knowledge into an ontology. Am Assoc Artif Intell
3. Shein KPP (2009) Ontology based combined approach for sentiment classification. In International conference on communications and information technology
4. Shein KPP, Nyunt TTS (2010) Sentiment classification based on ontology and SVM classifier. In: International conference on communication software and networks
5. Nezhadi AH, Shadgar B, Osareh A (2011) Ontology alignment using machine learning techniques. Int J Comput Sci Inf Technol (IJCSIT), 3(2)
6. Pinto A, Scioscia F, Loseto G, Ruta M, Bove E, Di Sciascio Politecnico di Bari 2(2015) A semantic-based approach for Machine Learning data analysis. International conference on semantic computing (IEEE ICSC 2015)
7. Syamala Devi M, Mittal H (2016) Machine learning techniques with ontology for subjective answer evaluation. Int J Natural Lang Comput (IJNLC) 5(2)
8. Imsombut A, Sirikayon C2 (2016) An alternative technique for populating thai tourism ontology from texts based on machine learning. ICIS

9. Xu J, Wang H, Trimbach H (2016) An OWL ontology representation for machine-learned functions using linked data. In: International congress on big data
10. Kim M, Kang H, Kwon S, Lee Y, Kim K, Pyo CS (2017) Augmented ontology by handshaking with machine learning. ICACT
11. Dorc FA, Carvalho VC, Mendes MM, Aráujo RD, Ferreira HN, Catellan RG (2017) An approach for automatic and dynamic analysis of learning objects repositories through ontologies and data mining techniques for supporting personalized recommendation of content in adaptive and intelligent educational systems. In: International conference on advanced learning technologies
12. Saranya K, Jayanthy S (2017) Onto-based sentiment classification using machine learning techniques. In: International conference on innovations in information embedded and communication systems (ICIIECS)
13. Liu F, Li G (2018) The extension of domain ontology based on text clustering. In: International conference on intelligent human-machine systems and cybernetics
14. Hwang KS, Park KS, Lee SH, Kim K, Lee KM (2018) Autonomous machine learning modeling using a task ontology. In: Joint international conference on soft computing and intelligent systems and international symposium on advanced intelligent systems
15. Rutaa M, Sciosciaa F, Losetoa G, Pintoa A, Di Sciascioa E (2018) Machine learning in the internet of things: a semantic-enhanced approach. IOS Press and the authors
16. Fu B, Steichen B (2019) Using behavior data to predict user success in ontology class mapping—an application of machine learning in interaction analysis. In: International conference on semantic computing (ICSC)
17. Sacha D, Kraus M, Keim DA, Chen M (2019) An ontology for visual analytics assisted machine learning. IEEE Trans Visual Comput Graph 25(1)
18. Tresp V, Bundschus M, Rettinger A, Huang Y (2008) Towards machine learning on the semantic web". Springer, Berlin Heidelberg

Deep Learning Approaches for Fall Detection Using Acoustic Information

John Sahaya Rani Alex, M. Abai Kumar, and D. V. Swathy

Abstract Senior citizens are prone to accidents due to their old age. The accidents may cause severe injuries and even to death if it is not identified and treated within a short period of time. Also, it is more risk if they stay alone in their homes. To mitigate the risk, an alert system is to be designed to alert the caretaker about the occurrence of the accident. By mounting three aerial microphones and one-floor acoustic sensor (FAS) in their room and monitoring the acoustic information received from the microphone and FAS, the acoustic information of the fall event is recorded. The acoustic features such as energy, spectral centroid, spectral flux, zero-crossing rate and Mel-frequency cepstral coefficients (MFCC) are extracted from the acoustic signal. Support vector machine (SVM) network and deep learning neural networks (DNN) with more than two hidden layers are trained with a reduced set of features obtained with principal component analysis (PCA) from the acoustic features. DNN classifier is proved to be better than SVM classifier. The obtained accuracy for DNN is 97%, the accuracy of the SVM classifier with MLP kernel and RBF kernel is 50% and 83%, respectively.

Keywords MFCC · Fall event · Acoustic cues · SVM · DNN

1 Introduction

Fall is a serious threat for elderly people which may lead to a cause of death. If there is a delay in the identification of the fall, their condition gets worse and it leads them to a critical condition. To avoid this risk, many researchers are developing automatic fall detection using sensors like accelerometer, gyroscope, and also placing a camera

J. S. R. Alex (✉) · M. Abai Kumar · D. V. Swathy
School of Electronics Engineering, Vellore Institute of Technology, Chennai, India
e-mail: jsranialex@vit.ac.in

M. Abai Kumar
e-mail: abaikumar88@gmail.com

D. V. Swathy
e-mail: d.v.swathy@gmail.com

N. Zhou and S. Hemamalini (eds.), *Advances in Smart Grid Technology*, Lecture Notes in Electrical Engineering 688, https://doi.org/10.1007/978-981-15-7241-8_35

in their room for monitoring their activities. But, there are many constraints and hindrances to the elderly person by placing the sensors and cameras [1, 2]. They are: for a wearable device, the power supply for the device is needed always. The senior person might have a tendency to forget to wear it, with the camera their privacy gets affected. To overcome the privacy and wearable device disadvantage, a microphone is used in this paper. Three microphones are separated by 4 cm and positioned on a table 80 cm high. In this also, there are constraints and they are the noise created by fan, television, different persons will fall in a different direction. There are a lot of constraints which makes us difficult to accurately detect the fall. In order to overcome the constraints, many features have been extracted from the audio signal and it is given as input to neural networks to increase the accuracy for detecting the fall. In this article, Sect. 2 lays out the existing works in the literature; proposed work is discussed in Sects. 3; Section 4 gives the results and discussions and finally the last section provides the conclusion.

2 Related Work

Most of the fall detection methods employed video surveillance to detect the fall of the person [1, 2], which intrudes on the privacy of their life. Later to mitigate these privacy issues, senior citizens are monitored using a wearable device which has an accelerometer [3, 4]. Fall detection also is done with the help of accelerometers in smartphones [5]. Other than the accelerometer, some of the researchers used Doppler radar sensor to detect the fall of a senior citizen [6]. The cameras and wearable devices are overhead as well as privacy intrusion to the fall detection mechanism. So, researchers used acoustic information with the help of microphones and vibration sensor placed in the environment to detect fall. Features extracted from the microphone are energy, spectral centroid, spectral roll-off, spectral flux and zero crossings. Features extracted from the sensors or microphones are used as input to classifiers to detect the fall. Some of the classifiers employed for fall detection used machine learning techniques [7–9] some are hidden Markov model-based component analysis (HMM-CA) [10] and some are support vector machine (SVM) [11].

3 Proposed System

In this work, fall detection is done using acoustic information with the dataset [11]. In this dataset, the audio data are acquired from using three axial microphones placed on a table and one-floor acoustic sensor (FAS) placed on the floor to sense the vibrations from the floor. By using this acoustic information, the acoustic features like energy, spectral centroid, spectral flux, zero-crossing rate and Mel-frequency cepstral coefficients (MFCC) are extracted. The block diagram of the proposed system design is shown in Fig. 1. The extracted features are used to train SVM and convolutional

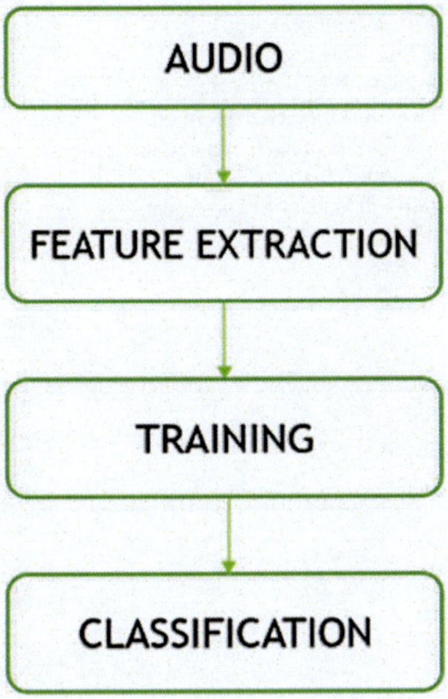

Fig. 1 Generic block diagram of fall detection

neural network (CNN). The trained networks are tested for accuracy of fall detection. In this section, the data set, acoustic features and the classifier are explained in detail.

3.1 Dataset

The dataset comprises recordings of fall events related to everyday objects (book, bag, ball, fork, basket, chair) and to "Rescue Randy", a human mimicking doll. Audio is recorded using the aerial microphone and floor acoustic sensor. The dataset was sampled at 44.1 kHz-24 bit and downsampled to 16 kHz-16 bit. The entire database was recorded using the Presonus AudioBox 44VSL soundcard, FAS placed on the floor and three AKG C400 BL microphones are separated by 4 cm and positioned on a table 80 cm high. The distance of the impact from the sensor (1, 2, 4, 6 m). Height from which the object was dropped (from ground, 50 and 100 cm from the ground). Dropping mode: fall with random angles of impact, free fall, fall caused by a kick, forward fall, backward fall, side fall. In order to add noise, classic and rock music from the speaker is also included during recording the fall. The time-domain waveform is shown in Fig. 2 for dropping off a bag from a distance of 1 m and the doll from 4 m distance.

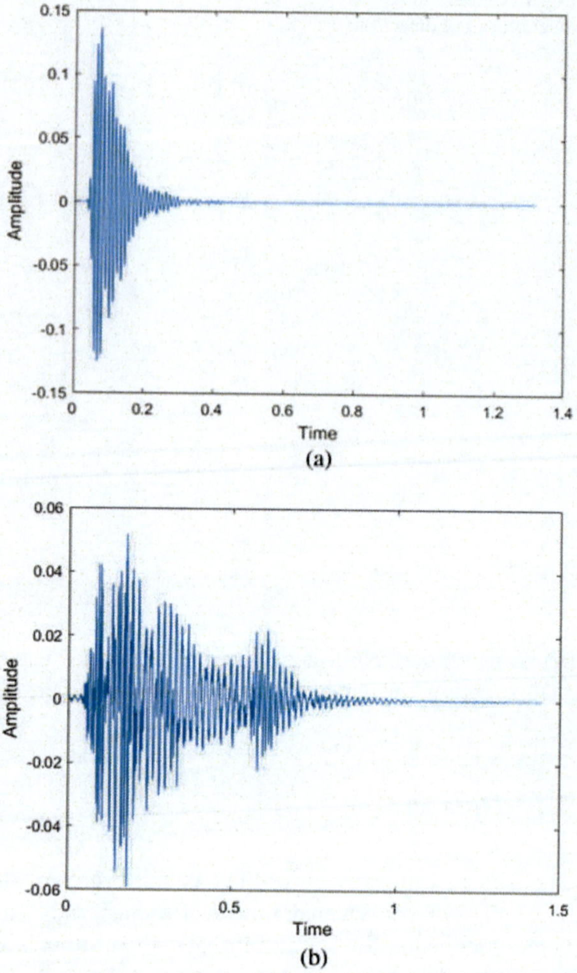

Fig. 2 Sample audio **a** Bag—clean data without music—FAS—random fall—distance (1 m)—height (50 cm), **b** randy—clean data without music—FAS-free fall—distance (4 m)—6 feet tall

3.2 *Feature Extraction*

From the acoustic signal, various features have been extracted such as

1. Energy
2. Spectral centroid
3. Spectral flux
4. Zero-crossing rate
5. Mel-frequency cepstral coefficients (MFCC).

Table 1 Energy of different subjects fall events

Signal	Duration	Energy (dB)
Bag—clean data—FAS—random fall—distance (1 m)—height (50 cm)	1.3130	3.1254e−04
Doll—clean data—FAS—free fall—distance (4 m)—6 feet tall	1.4410	8.3808e−05
Ball—with classic music—mic1—free fall—distance (1 m)—height (50 cm)	1.8610	3.4279e−06
Basket—with rock music—mic2—kick—distance (1 m)—height form ground	1.5810	4.8827e−05

3.2.1 Energy

Energy (E_k) of the signal informs about the strength of the signal $x(n)$ which is written by equation (1). The energy of the various subjects acoustic fall events is shown in Table 1.

$$E_k = \sum_{n=0}^{N} |x[n]|^2 \tag{1}$$

3.2.2 Spectral Centroid

The centroid (C_k) is the middle point of the spectrum of the kth window expressed by the equation (2), the frequency that divides the spectrum into two equal parts, sound signals formed by mainly high-frequency samples have higher centroid values. The spectral centroid is shown in Fig. 3.

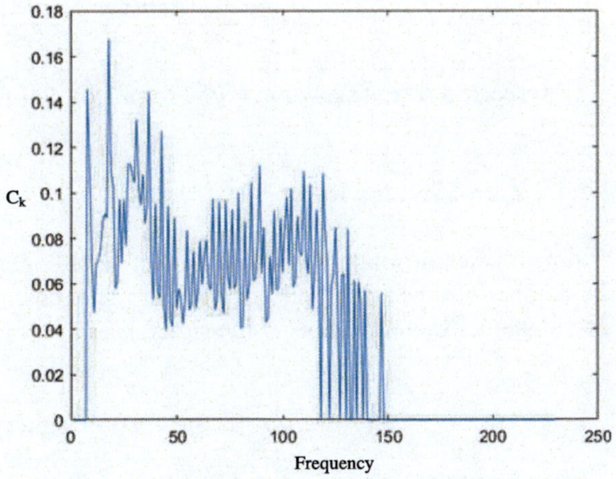

Fig. 3 Spectral centroid of doll

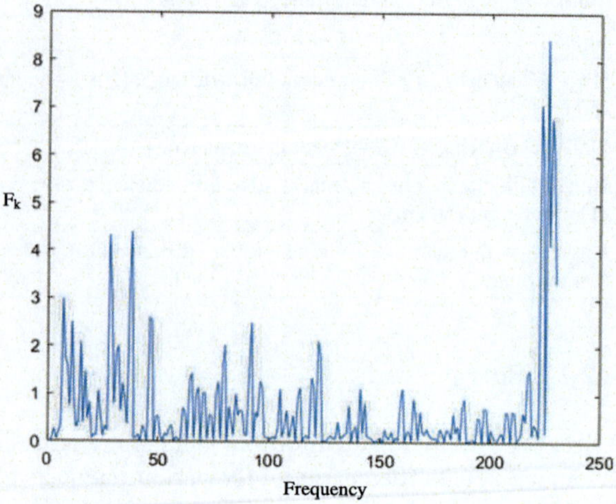

Fig. 4 Spectral flux of doll—clean data—FAS—free fall—distance (4 m)—6 feet tall

$$C_k = \frac{\sum_{n=0}^{N} F_k[n] \cdot n}{\sum_{n=0}^{N} F_k[n]} \tag{2}$$

where $F_k[n]$ is the amplitude of the fast Fourier transform of the n frequency applied to the k window.

3.2.3 Spectral Flux

Spectral flux indicates that the variation in the energy of the spectrum and it is given by equation (3),

$$F_k = \sum_{n=0}^{N} (F_k[n] - F_{k-1}[n])^2 \tag{3}$$

The spectral flux is shown in Fig. 4 for a Doll fall from 6 feet.

3.2.4 Zero-Crossing Rate

This parameter informs about the amount of noise contained in the signal by counting the number of times where the sign of the signal changes from positive to negative. The higher its value, the noisier the signal is and it is given by equation (4),

$$Z_k = \frac{1}{2} \cdot \sum_{n=0}^{N-1} |\text{sign}(x[n]) - \text{sign}(x[n+1])| \tag{4}$$

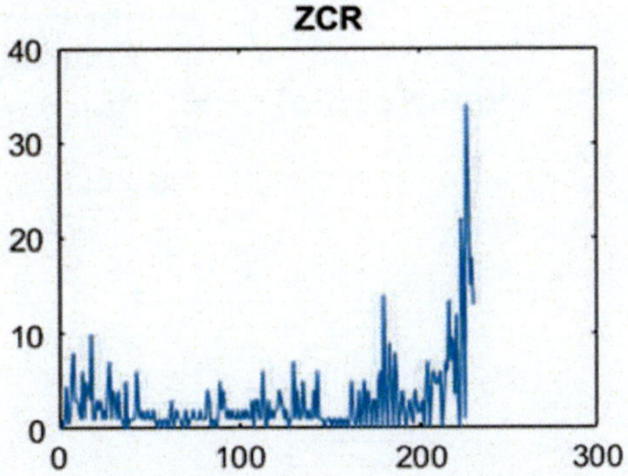

Fig. 5 Zero-crossing rate of doll falling from 6 feet

The zero-crossing rate is shown in Fig. 5.

3.2.5 Mel-Frequency Cepstral Coefficients

MFCC takes human perception sensitivity with respect to frequencies. Frame the signal into short frames (25 ms). Each of the above frames is segmented with a hamming window in order to keep the continuity of the signal. So, to reduce this discontinuity, an overlapping of the segmentation is applied. The time-domain frame is converted to the frequency domain by applying FFT on the frame. A set of Mel-frequency triangular bandpass filters applied to get the spectrum of human perception in various bandwidths. Then, the logarithmic function is applied on the spectrum, motivated by human hearing. Because the framing and windowing are done in overlapping manner, the Mel-filter bank energies are correlated with each other, so discrete cosine transform (DCT) is applied which de-correlates the energies. According to the literature, only 13 DCT coefficients are kept. The MFCC feature extraction method is shown pictorially in Fig. 6.

3.3 Dimensionality Reduction

The features obtained from the signal are 13 MFCC coefficients and 4 other acoustic features. It is learned from the works of literature that dimensionality reduction gets the dominant features by doing feature transformation. Various dimensionality reduction methods exist in literature. We have used principal component analysis (PCA) based on the work [12] in the literatures for speech processing.

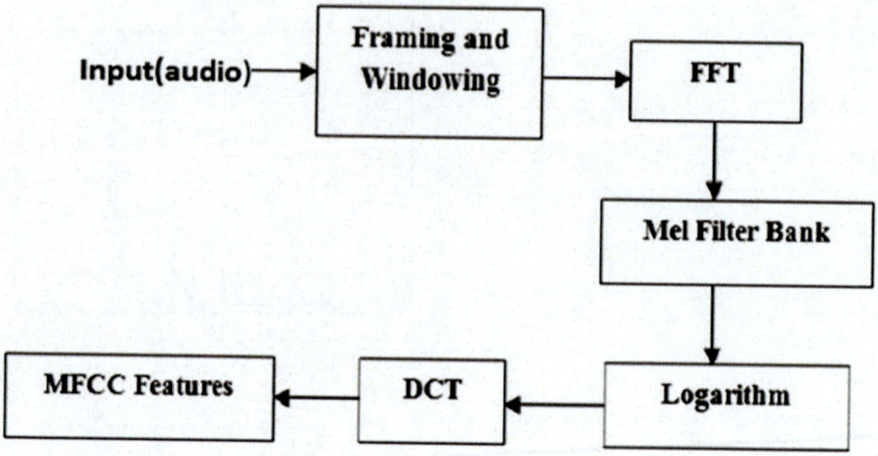

Fig. 6 Block diagram of MFCC feature extraction

3.4 *Deep Neural Network*

In our proposed system, we use the deep neural network (DNN) to train the features and to detect the fall. In our deep neural network architecture, the hidden layers and the number of input nodes are varied to get an optimal accuracy. The activation function used in the network is ReLu.

4 Experimental Setup and Results

Audio features extracted from audio data are one dimensional. The proposed DNN is compared with support vector machine (SVM). It is a binary classifier with a separating hyperplane. SVM is for separation of classes. Here, we take one class as fall (rescue randy) and another class as non-fall (bag, ball, basket, book, chair, fork). In SVM, multilayer perceptron(MLP) kernel function and radial basis function(RBF) are used to compare with DNN.

We tested with 1107 audio files. The proposed system is compared with SVM classifier with multilayer perceptron(MLP) kernel function and radial basis function(RBF). The accuracy obtained by SVM classifier with MLP kernel and RBF kernel is 50% and 83%, respectively, which is lower than DNN. Thus, DNN with 15 hidden layers and 15 input nodes outperformed other DNN configurations as well resulted in 97% accuracy which is shown in Table 2.

The DNN results are compared with SVM with different kernel functions. It is found that DNN always performed better than SVM which is shown in Table 3.

Table 2 Results of DNN with different configurations

Hidden layers	5	6	8	10	15	16
Input nodes	5	6	8	10	15	16
Accuracy in %	92	93	93	95	97	95

Table 3 Results of SVM and DNN

Classifier	Accuracy (%)
SVM (MLP kernel)	50
SVM (RBF kernel)	83
DNN	90

5 Conclusion

Fall detection for elderly people was proposed which uses acoustic information from the microphones and its features are extracted and trained and classified using DNN. Thus, by using a microphone, we can avoid the disadvantages of using wearable devices and cameras. As DNN has been successful in many fields, it is used in our proposed system to increase our accuracy to detect the fall. The accuracy obtained by SVM classifier with MLP kernel and RBF kernel is 50% and 83%, respectively, which is lower than DNN accuracy of 97% with 15 hidden layers. Thus, DNN performs better than SVM in detecting the fall events.

References

1. Foroughi H, Aski BS, Pourreza H (2008) Intelligent video surveillance for monitoring fall detection of elderly in home environments. In: 11th International conference on computer and information technology (ICCIT), pp 219–224
2. Debard G, Karsmakers P, Deschodt M, Vlaeyen E, Vanden Bergh J, Dejaeger E, Milisen K, Goedemé T, Tuytelaars T, Vanrumste B (2011) Camera-based fall detection using multiple features validated with real life video. In: Workshop proceedings of the 7th international conference on intelligent environment, vol 10, pp 441–450
3. Nguyen TT, Cho MC, Lee TS (2009) Automatic fall detection using wearable biomedical signal measurement terminal. In: 2009 annual international conference of the ieee engineering in medicine and biology society, pp 5203–5206. IEEE
4. Dumitrache M, Pasca S (2013) Fall detection algorithm based on triaxial accelerometer data. In: 2013 E-health and bioengineering conference (EHB), Iasi, pp 1–4
5. Aguiar B, Rocha T, Silva J, Sousa I (2014) Accelerometer-based fall detection for smartphones. In: 2014 IEEE international symposium on medical measurements and applications (MeMeA), Lisboa, 2014, pp 1–6
6. Liu L et al (2016) An automatic in-home fall detection system using Doppler radar signatures. JAISE 8:453–466
7. Khan MS, Yu M, Feng P, Wang L, Chambers J (2014) An unsupervised acoustic fall detection system using source separation for sound interference suppression. Signal Processing

8. Almazaydeh L, Al-Otoon K, Al-Dmour A, Elleithy K, Alatoun K (2016) A panoramic study of fall detection technologies. Int J Comput Sci Issues 13:62–65
9. Collado-Villaverde A, R-Moreno MD, Barrero DF, Rodriguez D (2017) Machine learning approach to detect falls on elderly people using sound. In: Benferhat S, Tabia K, Ali M (eds) Advances in artificial intelligence: from theory to practice. IEA/AIE 2017. Lecture Notes in Computer Science, vol 10350. Springer, Cham
10. Irtaza A, Adnan SM, Aziz S, Javed A, Ullah MO, Mahmood MT (2017) A framework for fall detection of elderly people by analyzing environmental sounds through acoustic local ternary patterns. 2017 IEEE international conference on systems, man, and cybernetics (SMC), Banff, AB, 2017, pp 1558–1563
11. Principi E, Droghini D, Squartini S, Olivetti P, Piazza F (2016) Acoustic cues from the floor: a new approach for fall classification. Expert Syst Appl 60:51–61
12. Anand A, Haque MA, Alex JS, Venkatesan N (2018) Evaluation of machine learning and deep learning algorithms combined with dimentionality reduction techniques for classification of Parkinson's disease. In: 2018 IEEE international symposium on signal processing and information technology (ISSPIT), pp 342–347. IEEE

Graphical Model and Model Search for Medical Data Analysis

Naimish P. Mehta, R. Menaka, Arathy S. Prasad, and Thanga Aarthy

Abstract Learning the electrophysiological activities inside the human mind is a significant step toward studying the human brain. Systems, such as electroencephalography, are significant instruments for considering the neurophysiologic activities, in view of their high value of temporal and spatial resolution. In the biomedical research, identifying brain abnormalities such as autism spectrum disorder through electroencephalography (EEG) signals is an extremely exhausting issue for specialists and human services experts. The high volume of data available with EEG will be a useful biomarker for the classification of autism and typical children. Traditional techniques face challenges to deal with such big data. So we present a strategy for autism identification by analyzing the EEG signal through mathematical model. One such modeling using graph theory is applied in this work. The EEG signals are acquired from 3 autism and 3 typical children. The functional connectivity among the neuron regions are plotted through small world networks. From this graphical models using a software tool Gephi, the graphical parameters as betweenness centrality, degree, weighted degree, closeness centrality, modularity, and clustering coefficient are calculated. There is significant difference among these parameters between autistic and typical children.

Keywords Autism · EEG · Graph theory · Gephi

1 Introduction

Autism spectrum disorder (ASD) is a quickly growing brain issue. In current scenario, combined research between psychiatry and neuroscience proposes that diseases due to brain disorders were begun to be researched due to structural and malfunctioned activity of brain [1, 2]. This malfunctioning of brain circuits results in different

N. P. Mehta (✉) · R. Menaka (✉) · A. S. Prasad · T. Aarthy
Vellore Institute of Technology, Chennai, India
e-mail: naimish.mehta1998@gmail.com

R. Menaka
e-mail: menaka.r@vit.ac.in

N. Zhou and S. Hemamalini (eds.), *Advances in Smart Grid Technology*, Lecture Notes in Electrical Engineering 688, https://doi.org/10.1007/978-981-15-7241-8_36

types of abnormality diseases. So autism or autism spectrum disorder is a serious complex condition of brain, which affects a person's social activity, unknowingly restricted and repetitive behavior, and difficulties in communication with others [1–5]. Specialist barely indicates this autism brain malfunctioning in child before 5 year age. Unfortunately, not all people have the equal ability to freely and easily express emotions, where there are many different disabilities, especially those diagnosed with autism spectrum disorder (ASD) [1]. ASD can be recognized through the behavioral of the kid which happen during the first and second year age. Autism affects an expected one out of 125 child in the area of India today, as indicated by the Centre for Disease Control and furthermore an ongoing deliberate survey of the number of inhabitants in South Asia (Bangladesh, India, and Sri Lanka) detailed a rate pervasiveness rate of 0.091–1.072% among kids in the 0–15 year age diagnosed with autism spectrum disorder. Though it is a long-lasting disorder issue, early diagnosis, arranged preparing, and treatment can improve the ability to functions [1]. Different researches are associated with search of an early finding of autism, for example, look measurements, engine development, EEG and different tests, for example, gaze metrics, motor movement, EEG, and other tests such as questionnaires [CARS] and M-CHAT [6]. There has been an expansion in interest in zone of biomarker distinguishing proof and helps people in danger and interventional studies. The probable advantages of earlier identification and interventions encourage us to concentrate on basic neuron behavior instead of visualized social changes. It was understood that the useful functional connectivity systems assessed from brain imaging technologies (MEG, fMRI, and EEG) can be analyzed by methods of graph theory [7]. This autism spectrum disorders have three types: first is autistic disorder; second is pervasive developmental disorder; and at last third one is Asperger's syndrome. Autistic disorder is having more intense level of brain rather than other two. So learning, observing, thinking, reacting, adapting, and problem-solving abilities are challenging and different for autism diagnosed person compared to normal person.

For diagnoses purpose, more than twenty years electroencephalographic (EEG) signals are used in neurophysiology purpose and psychology clinic [4]. Electroencephalography, because of its high transient and spatial resolution, is a magnificent strategy for contemplating neuro-physiological procedures [8, 9]. This EEG signals say the mind activity during various tasks and mentality. EEG signal is one of the bio-electro-signals that is non-station, non-straight and consistently fluctuates with time [4]. Nowadays early recognition and diagnosing of this abnormality autism are possible by extracting noteworthy and increasingly minute data of EEG signals [9].

It is possible that the relation between various kinds of brain region is showed as a graph [10, 2]. In this models, EEG channels are considered as nodes of the graph, and the evaluated connections between the EEG channels represent the edges between these nodes [1]. From EEG data, we are going to build a brain connectivity network and its graph parameters are extracted to detect autism children [1, 2]. An unpredictable network model is utilized for representing to the repeat pattern of EEG signals, based on which the temporal synchronization designs are measured using spectral graph theoretic highlights [11]. This modeling method is a more effective method of studying the complex brain connectivity network [15].

Major contributions of our work are the following.

- We introduce a novel technique to detect autism affected children by developing graphical models using a software called Gephi.
- We examined the parameters got from that graphical models which are related to the nodes and edges of human brain lobes.
- Various examinations performed for various kinds of data sets and the outcome as plots are for all the experiments and additionally recommend that our proposed method is best fitting to separate between autism influenced and ordinary ones.

2 Proposed Methodology

In this work, a proper calculation is proposed to identify autism issue from EEG brain sign dependent on another graphical model made by Gephi. Figure 1 shows the block diagram of our methodology. This methodology is effective and persistence to recognize between autism influenced and typical ones. The whole system of this technique is made out of different areas, utilizing transformation of time series EEG

Fig. 1 Block diagram

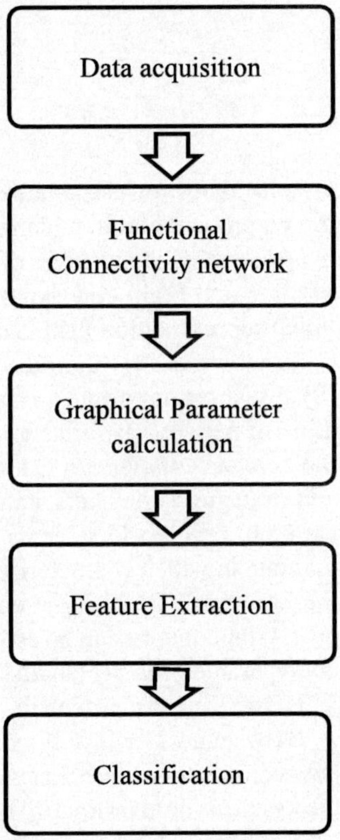

signals into approximately six graph theory parameters of autistic one and normal child.

3 Materials and Methods

3.1 Gephi

Gephi is a software used to create graphical models and all EEG graphical data will be analyzed by this. It is an open-source network for analyzing and visualizing large network graphs and this is written in Java on the NetBeans platform. This product utilizes a three-dimensional render motor to show diagrams continuously and accelerate the investigation. It is principally used to investigate, examine, practice, channel, bunch, control, and fare a wide range of graphs.

The graph network of Gephi contains circles and lines; these are called nodes and edges. Nodes are specks on the graph. It holds metadata (name and some other applicable data). What is more, edges are lines between nodes and additionally called as associations and connection.

3.2 Graph Theory

Graphical network is a mathematical model that is represented by nodes (vertices of the graph) and communication links between each pair of nodes (edges) [1, 2]. The characteristics of the edges determine the details of the network. A single description of structural brain connectivity was introduced through the method of geometrical graph representation [12]. The simple type of graphs is unweighted and undirected brain networks, in which all edges have equivalent power and are undirected networks [2]. The most important advantage of graph theory analysis is the grouping and valuation of network as far as graph parameters, by which it is conceivable to examine the network efficiency [12]. Graphs can be used to show various sorts of relations and strategies in physical, natural, social, and data systems. Underscoring their application to real-world systems, the term system is here and there defined to mean a diagram in which characteristics are related with the node and edges, and the subject that communicates this present reality networks as a system is called network science [14]. Graph theory can be estimated and computed. In the following, main six graph node parameter is explained.

Betweenness Centrality

Betweenness centrality is a measure dependent on the quantity of briefest ways between any two nodes that go through a specific node [13]. Nodes around the edge of the system would ordinarily have a low betweenness centrality. A high betweenness centrality may recommend that the individual is associating different various pieces of the system together [1]. Here, given condition is utilized in Gephi programming to figure betweenness centrality.

$$C_B(ni) = \Sigma_{UW}(\sigma UW(ni)/\sigma UW) \quad (1)$$

$\sigma UW(ni)$—is number of those paths that pass through ni

σUW—is number of shortest paths from node U to node W

Degree

The node degree is the quantity of relations (edges) of the nodes. Estimation Of degree must be less at that point or equivalent to existed nodes [2].

Weighted degree

The weighted level of a lobes resembles the degree. It fully depends on quantity of edge for lobe, however isolated by heaviness of each edge. It is doing the aggregate of heaviness of edges. For instance, a lobe with four edges that weight is 1 (1 + 1+1 + 1=four) is proportionate so a number of nodes directly multiply with weight of that edge.

Closeness Centrality

Closeness centrality of a lobe is a proportion of centrality in that network systems, determined as the equal to entirety of length by the most brief ways between the lobe and every single other lobe in the diagram [2]. In this manner, the more focal a node is, the closer it is to every single other node. Scope of closeness centrality is in the middle of 0–1.

$$C(x) = \frac{1}{\sum_y d(y,x)} \quad (2)$$

Modularity

Modularity is one proportion of the structure of systems or diagrams. It was intended to gauge the quality of division of a system into modules (additionally called gatherings, groups or networks). Systems with high modularity have thick associations between the nodes inside modules however inadequate associations between nodes in various modules. Modularity is regularly utilized in improvement strategies for identifying network structure in systems.

$$Q = \frac{1}{2m}\sum_{vw}\left[A_{vw} - \frac{k_v k_w}{2m}\right]\frac{s_v s_w + 1}{2} \quad (3)$$

Clustering Coefficient

The clustering coefficient (Watts--Strogatz), when applied to a solitary node, is a proportion of how complete the area of a node is. At the point when applied to a whole system, it is the normal bunching coefficient over the majority of the nodes in the system [1]. The clustering coefficient, alongside the mean most brief way, can show a "small world" impact. For the clustering coefficient to be important, it ought to be altogether higher than in form of the system where the majority of the edges have been rearranged. For example, the area of a node, u, is the arrangement of nodes that are associated with *u*. On the off chance that each node in the area of *u* is associated with each other node in the area of *u*, at that point the area of *u* is finished and will have a grouping coefficient of 1. On the off chance that no nodes in the area of *u* are associated, at that point the grouping coefficient will be 0.

$$CCv = 2 * Nv/Kv(Kv - 1) \quad (4)$$

V—Node
Nv—No. of links between neighbors of *v*
Kv—*i*th degree.

3.3 Our EEG Dataset

In this study, EEG system [Nihon kohden MEB9000 version 05-81, sensitivity 7microvolt] was used to acquire EEG data for five [3M, 2F] typically developed children and five [3M, 2F] children with ASD (age 3–7 years) using international 10–20 electrode System. The EEG signals were recorded from 16 channels at the sampling frequency of 500 Hz and filtered with low-pass filter and high-pass filter at a frequency range of [0.53–70 Hz]. These electrodes are FP2-F4(ch1), F4-C4(ch2), C4-P4(ch3), P4-O2(ch4), Fp2-F8(ch5), F8-T4(ch6), T4-T6(ch7), T6-O2(ch8), Fp1-F3(ch9), F3-C3(ch10), C3-P3(ch11), P3-O1(ch12), FP1-F7(ch13), F7-T3(ch14), T3-T5(ch15), and T5- O1(ch16) as shown in Fig. 2.

So, We have four types of datasets, and based on that dataset, we made a graphical model using Gephi software.

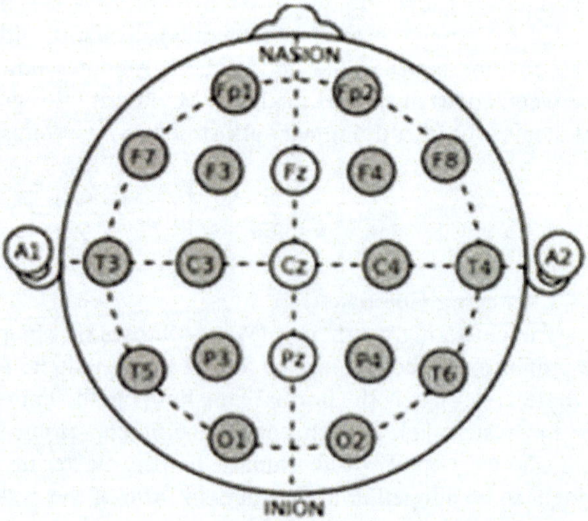

Fig. 2 Electrode positions selected for this research [10-20 system]

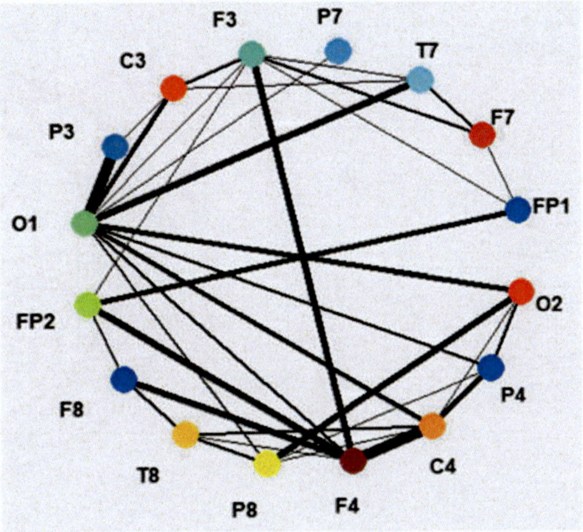

Fig. 3 Normal children's graphical model

1. English audio for total autism, average autism person, and normal person
2. Song with visual for total autism, average autism, and normal person
3. Only video for total autism, average autism, and normal person
4. Tom and Jerry video for total autism, average autism, and normal person

Here, average autism data is for 50% probability that child having autism or not and another two for normal person and autistic person. So that English audio and song with visual were listened to both normal and autistic children as well as only video and video with visual (Tom and Jerry Video) were shown to both normal and abnormal children, and based on that 12 (4×3) graph datasets, we got here 24 plots for each 6 parameters (Figs. 3 and 4).

3.4 Results

Here, we have our observational plots of parameters extracted by Gephi (Figs. 5, 6, 7, 8, 9, 10, 11 and 12).

4 Conclusion and Future Scope

By observing this all parameters, we conclude that for 3,4,5 nodes and 13,14,15 nodes have major difference between autism one and normal one, then middle area nodes like 6,7,8,9,10,11 have almost equal between autism and normal.

Fig. 4 Autistic children's graphical model

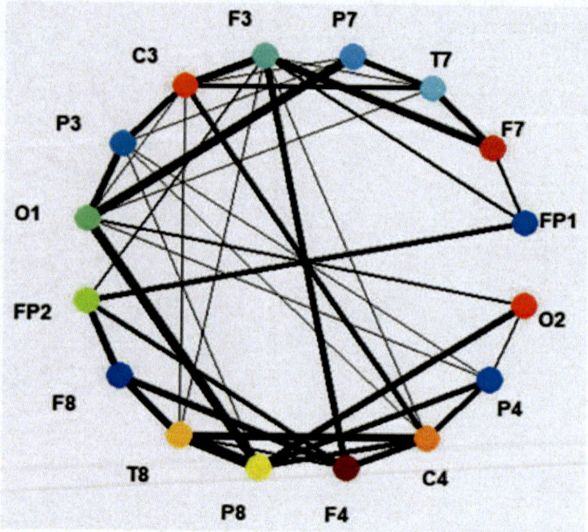

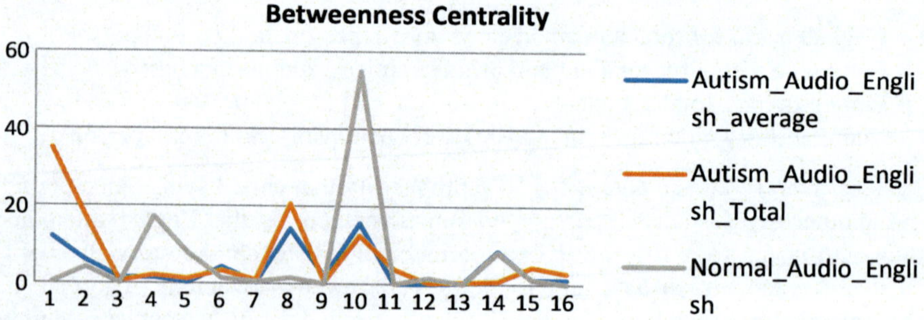

Fig. 5 Plot of C_B for English audio data

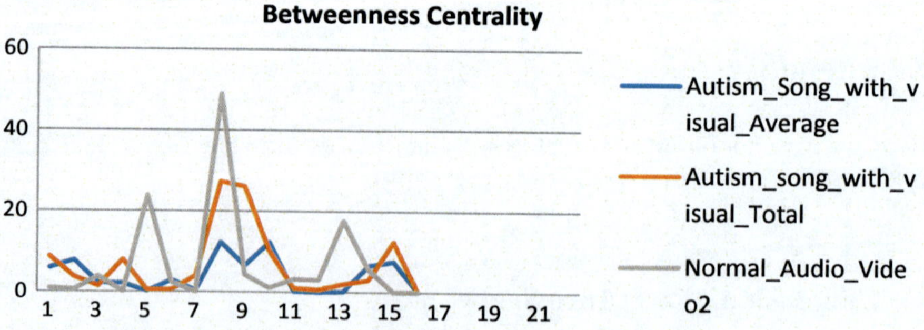

Fig. 6 Plot of C_B for song with visual data

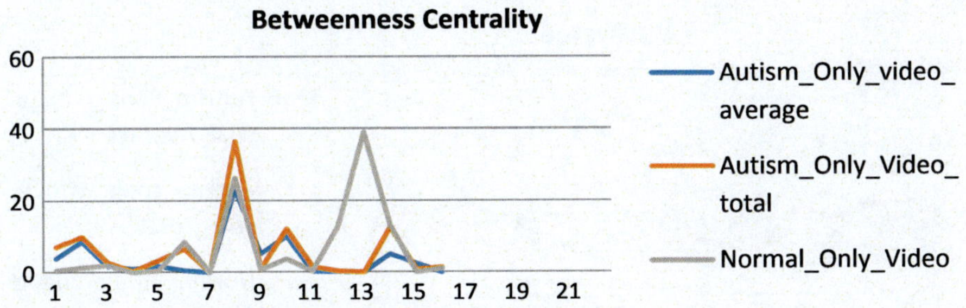

Fig. 7 Plot of C_B for only video data

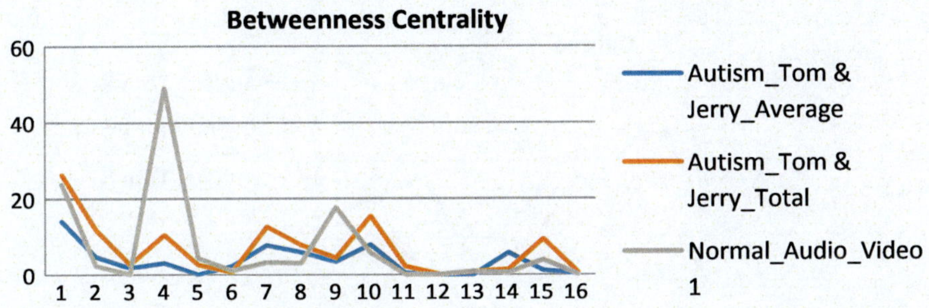

Fig. 8 Plot of C_B for Tom & Jerry video data

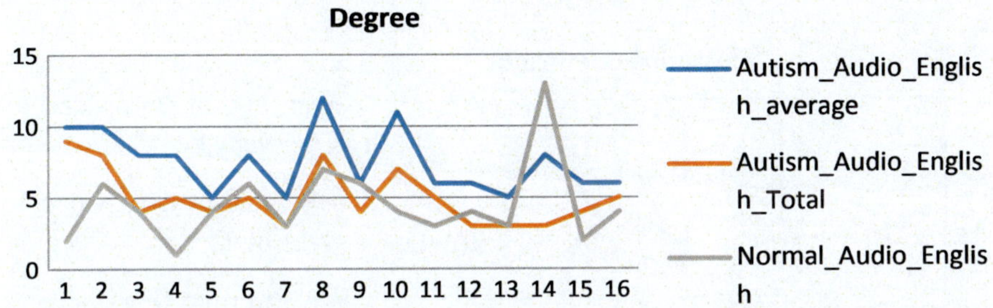

Fig. 9 Plot of degree for English audio data

3-P3(perital node), 4-O1(oxipited node), 5-FP2(frontal node) and 13-FP1(frontal node), 14-F7(frontal node), 15-P7(perital node) are our observation from above plots, and later on we will observe these nodes as physical electrodes. Also, in future, we are working with weighted network of these data types and from that we extract edge weighted parameter.

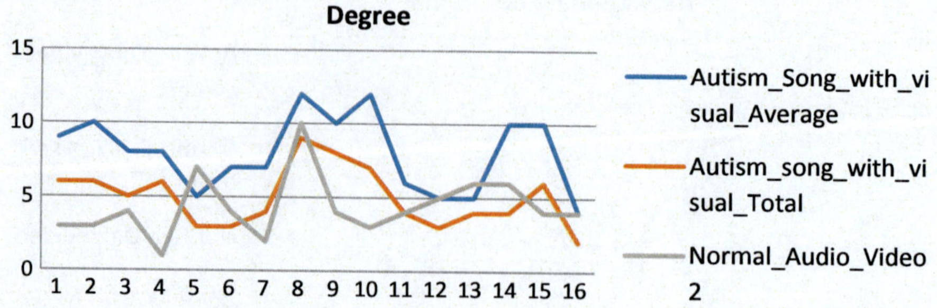

Fig. 10 Plot of degree for song with visual data

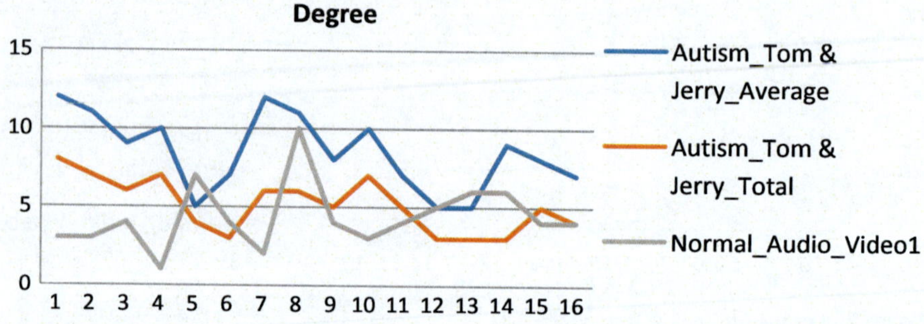

Fig. 11 Plot of degree for Tom and Jerry data

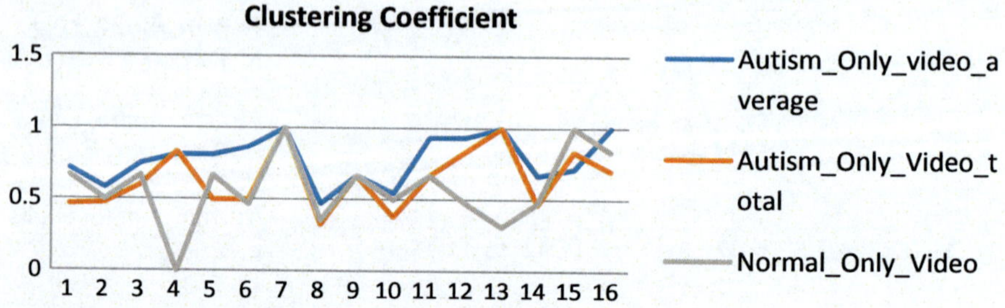

Fig. 12 Plot of clustering coefficient for only video data

References

1. Sato JR, Calebe Vidal M, de Siqueira Santos S, Brauer Massirer K, Fujita A (2018) Complex network measures in autism spectrum disorders. IEEE/ACM Trans Comput Biol Bioinform (TCBB) 15(2):581–587

2. Dejman A, Khadem A, Khorrami A, Comparing the effective connectivity graphs estimated by granger causality index with transfer entropy: a case study on Autism spectrum disorders. In: 2017 24th national and 2nd international Iranian conference on biomedical engineering (ICBME). IEEE, New York, pp 14–15
3. Riedel BC, Jahanshad N, Thompson PM (2017, April) Graph theoretical approaches towards understanding differences in frontoparietal and default mode networks in Autism. In: 2017 IEEE 14th international symposium on biomedical imaging (ISBI 2017). IEEE, New York, pp 460–463
4. Fan J, Wade JW, Key AP, Warren ZE, Sarkar N (2017) EEG-based affect and workload recognition in a virtual driving environment for ASD intervention. IEEE Trans Biomed Eng 65(1):43–51
5. Ali S, Mehmood F, Dancey D, Ayaz Y, Khan MJ, Naseer N, Nawaz R (2019) An adaptive multi-robot therapy for improving joint attention and imitation of ASD children. IEEE Access 7:81808–81825
6. Wang Q, Sourina O (2013) Real-time mental arithmetic task recognition from EEG signals. IEEE Trans Neural Syst Rehabil Eng 21(2):225–232
7. Fallani FDV, Astolfi L, Cincotti F, Mattia D, Tocci A, Salinari S, Babiloni F (2008) Brain network analysis from high-resolution EEG recordings by the application of theoretical graph indexes. IEEE Trans Neural Syst Rehabil Eng 16(5):442–452
8. Varatharajah Y, Chong, MJ, Saboo K, Berry B, Brinkmann B, Worrell G, Iyer R (2017) EEG-GRAPH: a factor-graph-based model for capturing spatial, temporal, and observational relationships in electroencephalograms. In: Advances in neural information processing systems, pp 5371–5380
9. Supriya S, Siuly S, Wang H, Cao J, Zhang Y (2016) Weighted visibility graph with complex network features in the detection of epilepsy. IEEE Access 4:6554–6566
10. Andersen LR, Krebs JH, Andersen JD (1991) STENO: An expert system for medical diagnosis based on graphical models and model search. J Appl Stat 18(1):139–153
11. Fan M, Chou CA (2018) Detecting abnormal pattern of epileptic seizures via temporal synchronization of EEG signals. IEEE Trans Biomed Eng 66(3):601–608
12. Cociu BA, Das S, Billeci L, Jamal W, Maharatna K, Calderoni S, Muratori F (2017) Multimodal functional and structural brain connectivity analysis in autism: a preliminary integrated approach with EEG, fMRI, and DTI. IEEE Trans Cogn Dev Syst 10(2):213–226
13. Storti SF, Galazzo IB, Khan S, Manganotti P, Menegaz G (2016) Exploring the epileptic brain network using time-variant effective connectivity and graph theory. IEEE J Biomed Health Inform 21(5):1411–1421
14. Smith K, Spyrou L, Escudero J (2018) Graph-variate signal analysis. IEEE Trans Signal Process 67(2):293–305
15. Shams WK, Wahab A (2013, March) Source-temporal-features for detection EEG behavior of autism spectrum disorder. In: 2013 5th international conference on information and communication technology for the Muslim world (ICT4M). IEEE, New York, pp 1–5

Forecasting Election Data Using Regression Models and Sentimental Analysis

Saif Gazali and V. Pattabiraman

Abstract Predictive Analytics is emerging as one of the popular branch of Machine Learning with huge amount of money spent on to research, build and test models to improve the accuracy of the outcome. In our paper we have implemented different regression models such as Logistic Regression, Support Vector Machines, Naive Bayes Classifier, and Neural Networks to forecast the result of the US Presidential election 2016. Social media websites have become a very popular communication platform among users. Millions of users share their opinions, extend their support, and vent their anger on various government policies and on different aspects of their life. Hence, social media websites contain huge amount of data for sentimental analysis. We have narrowed our attention toward Twitter, the most popular microblogging website for performing sentimental analysis using the opinions shared by the users on twitter.

Keywords Predictive analysis · Logistic regression · Support vector machine · Naïve Bayes classifier · Neural networks · Sentimental analysis

1 Introduction

Logistic regression analysis is used on a large scale for binary data. The model helps us to effectively analyze the effects of multiple independent variables on a single dependent variable. It is used when there are several explanatory variables and a nominal response variable. SVM (Support Vector Machine) is another method which utilizes the extreme cases that is the extremes of the datasets and draws a decision boundary also known as hyperplane near the extreme points in the dataset. Naive Bayes Classifier is a classification algorithm based on Bayes theorem in which every pair of explanatory variables is independent of each other. Neural Networks is used to process data and can be viewed as similar to the human nervous system. It is made up of interconnected neurons which are used to process the information.

S. Gazali · V. Pattabiraman (✉)
Vellore Institute of Technology, Chennai 600127, India
e-mail: pattabiraman.v@vit.ac.in

N. Zhou and S. Hemamalini (eds.), *Advances in Smart Grid Technology*, Lecture Notes in Electrical Engineering 688, https://doi.org/10.1007/978-981-15-7241-8_37

Table 1 Features within the dataset

ID	Variable	Description	Type
1	State	Name of the 50 states of the United States of America	Qualitative
2	Year	Consists of year of the election—2004, 2008, 2012, 2016	Discrete
3	Rasmussen	An agency in USA which conducts opinion polls	Continuous
4	SurveyUSA	An agency in USA which conducts opinion polls	Continuous
5	DiffCount	Number of polls predicting republican as winner minus the number of polls predicting democrat as Winner	Discrete
6	PropR	Number of polls predicting republican as winner	Discrete
7	Winner	1—Republican win, 0—Republican Lose	Discrete

2 Dataset

To calculate the accuracy of our Election Prediction model, the 'US Presidential election 2016' dataset has been used. The dataset has variable such as 'Rasmussen' which is a renowned American polling firm founded in 2003 and also 'SurveyUSA' which is also a polling firm. The dataset consists of around 7 connection vectors and each of our vectors has six independent variables and also a dependent variable 'Republican' in these cases which is marked as win or loses. Table 1 consist the explanation of each variable present in the dataset.

3 Related Work

Twitter has been used to estimate the popularity of politicians, public sentiment toward the various recently introduced policies like Union budget, tax reforms demonetization etc. Social media is also used to compare people's political preferences expressed online before an election. Social networking sites can be analyzed on a daily or hourly basis to get a detailed insight into emotions of voters during electoral campaigning of various candidates and get public opinion well before declaration of results of the polls. Hung and Mustafaraj [1] claimed that analyzing social networking sites for people discussion on a particular candidate helps to predict the final results in a reliable manner. There have been claims that more the number of Facebook and twitter followers of a particular candidate, more is his/her chance of winning the election. Here the researchers analyzed many social media like Facebook, Twitter, Google and YouTube. Brian and Garret [2] used the US presidential 2008 data to find the consequence of political rumor [3]. They stated that people responded to political rumors and false negative rumor about a candidate which a particular individual oppose was readily accepted by the individual. These rumors led to the significant decrease in the vote share of that candidate Sinha [4] considered various economical such as inflation, exchange rate, gold prices and non-economic factors such as power of period factor (number of years the incumbent president in power), military intervention, the percentage of white and youth voters for predicting the vote share of the candidates. They predicted 48.11% vote share for the Democratic Party and 40.26% vote share for the Republican party with 95% confidence interval.

4 Methodology

4.1 Dataset Selection and Preprocessing

The first step of our implementation is the preprocessing of the data. The preprocessing of the dataset plays a vital role in improving the accuracy of the prediction for training, testing and predicting the outcome as 'Republican' or 'Democrat'. The dataset consists of qualitative variable which are converted to quantitative. Also normalization is performed so that the values of the continuous variables lie between 0 and 1.

4.2 Model Selection

Selecting the appropriate features from the dataset to feed it to the regression models is quintessential. Accuracy of a model can vary drastically depending upon the variables we use. Correlation matrix is a table which shows how the different features are correlated to each other. From our dataset, we have the following correlation matrix which is represented in Table 2.

Table 2 Correlation matrix

	Rasmusssen	SurveyUSA	PropR	DiffCount	Republican
Rasmussen	1.000000	0.9157207	0.8281642	0.5173841	0.7878770
SurveyUSA	0.9157207	1.000000	0.8478491	0.5511599	0.8020143
PropR	0.8281642	0.8478491	1.000000	0.8285177	0.9512832
DiffCount	0.5173841	0.5511599	0.8285177	1.000000	0.8016614
Republican	0.7878770	0.8020143	0.9512832	0.8016614	1.000000

4.3 Regression Models

Regression analysis is used to show how value of a response variable differs when values some of the explanatory variables are changed, while the remaining explanatory variables remains constant.

Logistic Regression

Logistic regression is the suitable regression analysis method to conduct when the dependent variable (target) is categorical. Logistic regression is mainly used for predictive analysis. It can be used to process and garner information about the data and describe the relationship between one dependent dichotomous variable and one or more independent variables. The logistic function is a sigmoid function which takes an input t and outputs between zero and one [5].

The logistic $\sigma(t)$ function can be given as follows:

$$\sigma(t) = \frac{et}{et + 1} = \frac{1}{1 + e - t} \quad (1)$$

If we assume t as a linear function of single independent variable then t can be expressed as follows:

$$t = \beta 0 + \beta 1 x \quad (2)$$

The logistic function can be written as

$$p(x) = \frac{1}{1 + e - (\beta 0 + \beta 1 x)} \quad (3)$$

$p(x)$ is inferred as the probability of the dependent variable equaling a success or case.

$\beta 0$ Represents the intercept from the linear regression $\beta 1$ equation and x represents the regression coefficient multiplied by any value of the predictor.

Support Vector Machine (SVM)

A support vector machine (SVM) processes the given training data and outputs an optimal hyperlane which is a line in more than 3 dimensions that is used to segregate new data particularly test dataset. In 2-D space it is a line dividing a plane in two parts with each class lying on either side. It is used to analyze data used for classification and regression analysis [6].

Figure 1 shows what SVM does, the blue dotted line separates the data into two halves and is known as a Decision Boundary hyperlane in SVM. The other two lines which are also hyperplanes are used to make the right decision boundary. It also represents an SVM model with 2 independent variables PropR and SurveyUSA which are used to predict the binary dependent variable Republican (0 or 1).

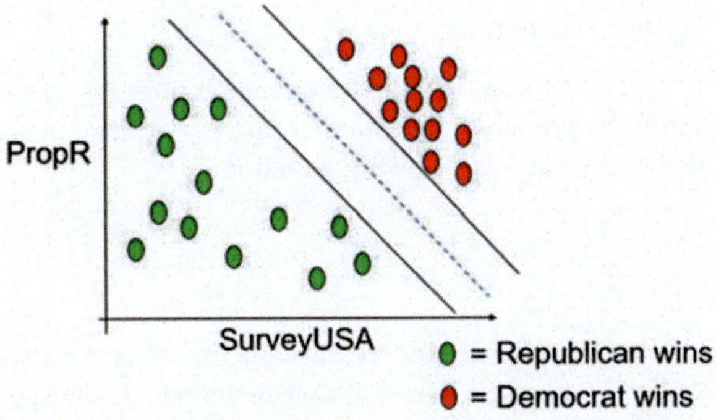

Fig. 1 Support vector machine diagram

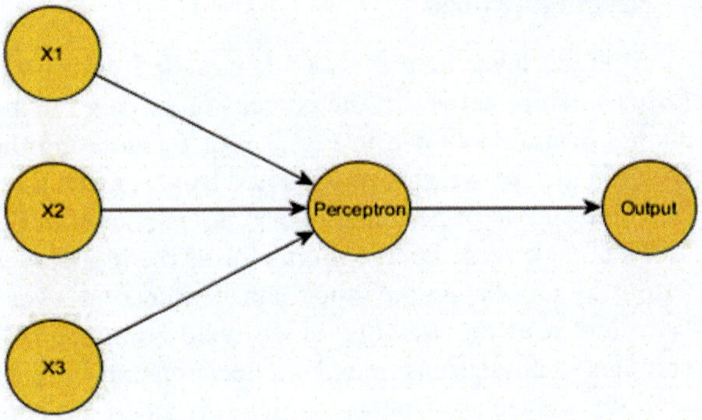

Fig. 2 A perceptron with three inputs $x1$, $x2$ and $x3$

The type of SVM discussed above is known as Linear Support Vector machines or LSVM.

We use Linear Kernel in our SVM model. We tune the kernel to get good performance from the classifier using techniques such as k-fold cross validation. Error estimation using tenfold cross validation came to 0.02645626.

Naïve Bayes Classifier

It is a classification algorithm based on Bayes theorem in which every pair of explanatory variables is independent of each other. The dataset is divided into two parts namely Feature matrix and Response vector. Feature matrix which consists all the rows of the datasets where every vector has the value of the given explanatory variables. In our dataset 'Rasmussen,' 'SurveyUSA,' 'DiffCount' and 'PropR' are features. Response vector includes the value of the Dependent variable for each row of the feature matrix. In our Dataset the class variable is 'Republican.'

It is named as 'naive' because we assume that the explanatory variables are independent of each other but they may have dependence in some way. The reason we are assuming them as independents because they allow us to calculate the probability of the feature by multiplying the individual probability of each piece of feature occurring together using the multiplication rule for independent AND events [7].

Bayes Theorem

It is a theorem which computes the probability of any event which is occurring using the probability of another already occurred event. It is stated mathematically in the form of the following equation:

$$P(y|X) = \frac{P(X|y)P(y)}{P(X)} \tag{4}$$

Here y is the dependent variable and X is the dependent feature with vector of size n. In our case n is 4 which includes 'Rasmussen,' 'SurveyUSA,' 'DiffCount' and 'PropR.'

Neural Networks

A single layer neural network is called perceptron and multi-layer Perceptron forms a neural network. The perceptron having the inputs $x1$, $x2$, $x3$ and outputting a single variable shown in Fig. 2. For formulating the output, Rosenblatt came up with the idea of weights represented by $w1$, $w2$ and so on.

The example of neural network model shown in Fig. 3.

The first layers of perceptron's forms the input layer and carry out simple computations by processing the input data. The second layer of perceptron's form a hidden layer and uses the weights which were computed in the first layer to make more complex computations to make a decision and provide it as output.

If we change the way we defined the algebraic notation for perceptron earlier by replacing threshold with bias where bias(b) = − threshold, we get the following algebraic equation:

$$\text{Output} = \begin{cases} 0 & \text{if} \quad w.x + b \le 0 \\ 1 & \text{if} \quad w.x + b > 0 \end{cases} \tag{5}$$

Fig. 3 Neural network model

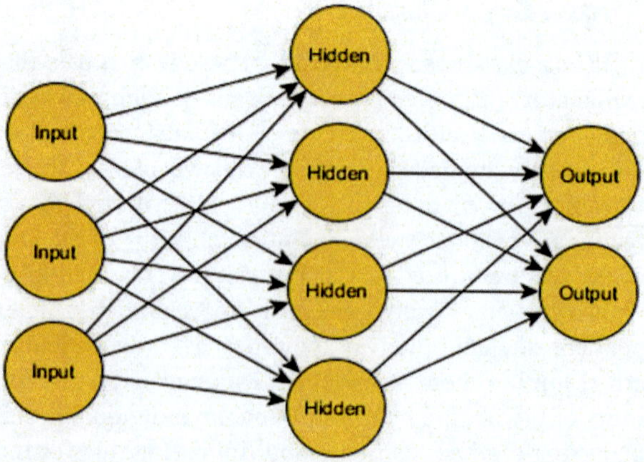

In our case 0 means that democrat has won whereas 1 means that republican has won.

Sentimental Analysis

Twitter has been used to estimate the popularity of politicians [8], public sentiment toward the various recently introduced policies like Union budget, tax reforms demonetization etc. Social media is also used to compare people's political preferences expressed online before an election [9].

Preprocessing of Tweets

Firstly, Tweets are extracted using Twitter API and TwitterR package in R and saved as an excel file for further processing. Then from the excel sheet only the text is extracted using functions of the package dplyr and plyr of R. This text is used to perform sentimental analysis after further cleansing. '@' symbol is removed along with the name using the gsub function in R [10].

Processing of the Text

Here, our text data is processed and converted into a structure called Vcorpus where each line in the text is loaded as the document in the corpus. Then the corpus is converted to Lower case using tm_map function of the text mining package in R. A term-document matrix is created from the corpus. An element in the matrix represents the occurrence of a word in a document of the corpus. The problem with the matrix is that they are very large and extremely sparse [11].

4.4 Model Training

The training part of the Election prediction model is using the appropriate algorithms and implementing on the dataset to construct different regression models such as Logistic regression, SVM etc. The dataset is split into two parts with 60% of the training set and remaining 40% for the testing set.

4.5 Model Testing and Evaluation

The testing part is using the constructed model and implementing it on the test dataset to compute the accuracy of the prediction. Using the confusion matrix we calculate the accuracy and the error. We have true positive and true negative which means perfectly accessed and perfectly denied respectively. Also there are False Positive and False Negative which means incorrectly accessed and incorrectly denied respectively.

The above values can be accessed using 'Confusion Matrix' which is table layout and outputs the performance of a particular model as shown in Table 3.

Table 3 Confusion matrix of a particular model

Model	Predicted: False	Predicted: True
Actual: False	True Negative (TN)	False Positive (FP)
Actual: True	False Negative (FN)	True Positive (TP)

Table 4 Confusion matrix

Model 1, 2	False	True
0	20	0
1	7	23

Table 5 Confusion matrix

Model	False	True
0	20	6
1	0	24

Table 6 Comparison table

Name	Accuracy (%)
Logistic regression	86
Support vector machines	86
Naive Bayes classifier	88
Neural networks	86
Sentimental analysis	86

The confusion matrix for different models such as Logistic Regression, Support Vector Machine, Neural Network and Sentimental Analysis is represented in Table 4 whereas for Naïve Bayes it is represented in Table 5.

We summarize the accuracies of each model implemented in the Table 6.

5 Conclusion

In this paper an Election Prediction System based on different regression models such as Logistic Regression, Support Vector Machines, Naïve Bayes method, Neural networks was implemented and compared with an Election Prediction System based on Sentimental analysis model. We infer from the Table 6 which gives us the comparison between the different models implemented that accuracy of Naive Bayes Classifier is better than other models.

References

1. Chung J, Mustafaraj E (2011) Cam collective sentiment expressed on twitter predict political elections? In: Proceedings of the Twenty-Fifth AAAI conference on artificial intelligence, San Francisco, CA, USA
2. Weeks B, Garrett RK (2014) Electoral consequences of political rumors: motivated reasoning, candidate rumors, and vote choice during the 2008 U.S. presidential election. Int J Public Opinion Res 26:401–422. https://doi.org/10.1093/ijpor/edu005
3. Abramowitz AI (1988) An improved model for predicting the outcomes of presidential elections. PS: Political Sci Politics 21 4:843–847
4. Sinha P, Nagarnaik A, Raj K, Suman V (2016) Forecasting United States Presidential election 2016 using multiple regression models
5. Ng AY, Michael IJ (2002) On discriminative vs. generative classifiers: A comparison of logistic regression and naive bayes. In: Advances in neural information processing systems
6. Hong W-C (2006) A hybrid support vector machine regression for exchange rate prediction. Int J Inf Manage Sci 17(2):19–32
7. Rish I (2001) An empirical study of the naive Bayes classifier. In: IJCAI 2001 workshop on empirical methods in artificial intelligence, vol 3, No 22. IBM, New York
8. Singh P, Sawhney RS, Kahlon KS (2017) Forecasting the 2016 US presidential elections using sentiment analysis. In: Kar A et al (eds) Digital nations—smart cities, innovation, and sustainability. I3E 2017. Lecture notes in computer science, vol 10595. Springer, Cham
9. Sharma P, Moh TS (2016, December) Prediction of Indian election using sentiment analysis on Hindi Twitter. In: 2016 IEEE International Conference on Big Data (Big Data) (pp 1966–1971). IEEE
10. Lindsay R (2008) Predicting polls with Lexicon; Ludvigson SC (2004) Consumer confidence and consumer spending. J Econ Perspect 18(2):29–50
11. Hasan A et al (2018) Machine learning-based sentiment analysis for twitter accounts. Math Comput Appl 23(1)

ICT Technologies

Development and Implementation of the Smart Energy Monitoring System Based on IoT

J. Barsana Banu, J. Jeyashanthi, A. Thameem Ansari, and A. Sathish

Abstract This paper is designed to measure energy consumption in home and buildings and to generate its bill automatically. It is accomplished by utilizing a smart energy meter with the Internet of things (IoT) technology, which will permit the user to successfully observe the energy meter calibrations and verify the electricity bill via the online. Our projected scheme utilizes Arduino to track utilized energy and to send out the units along the cost charged over the Internet. The Liquid Crystal Display (LCD) module is interfaced with Arduino, and the measured voltage, current, power, and the corresponding bill are displayed to the consumers. Arduino also sends data to the Adafruit cloud using the Wi-Fi module NodeMCU ESP-12. This smart meter will permit both the consumer and electricity supplier to ensure the energy usage quickly among the cost charged online. Power cost analysis of a month for a smart home is done in a separate section. Finally, the hardware is implemented with various loads, and results are displayed via LCD.

Keywords Arduino · Ada fruit cloud · Energy meter · Internet of things · LCD · TNEB

1 Introduction

Electricity is one of the essential favors that science has given to humankind. Today, it has become the fundamental necessities of human beings, and one cannot think of a world without electricity. Monitoring and keeping track of electricity consumption are vital for energy conservation. This electricity is generated and distributed by the Tamil Nadu Electricity Board (TNEB) to every place, and hence, it must be measured

J. Barsana Banu (✉)
Mahath Amma Institute of Engineering and Technology, Pudukkottai, India
e-mail: barsanajamal@gmail.com

J. Jeyashanthi · A. Sathish
Sethu Institute of Technology, Kariapatti 626115, India

A. Thameem Ansari
Petroleum Research Center, KISR, Safat 13109, Kuwait

N. Zhou and S. Hemamalini (eds.), *Advances in Smart Grid Technology*, Lecture Notes in Electrical Engineering 688, https://doi.org/10.1007/978-981-15-7241-8_38

whenever consumed to pay the electricity bill. Thus, we require a device to measure the consumed electricity. It is accomplished by an energy meter installed at each place like industries, homes, and organizations where power is consumed. The traditional mechanical power meter offered in [1] depends on the "Magnetic Induction" principle. It comprises of revolving aluminum wheel called Ferri-wheel and several toothed wheels. Depending upon the current flow, the Ferri-wheel revolves which creates a rotation of further wheels. This will be transformed into resultant measurements in the display part. As various mechanical parts are concerned, mechanical defects and breakdown are general. Additional chances of manipulation and current theft will be maximum. The traditional analog energy meter fails to measure the whole energy consumption very precisely, as described in [2].

To conquer this issue, these days, traditional analog meters are changed with digital meters, which are very precise, maximum procession, and consistent as projected in [3]. To precisely compute the measurement deviations of a digital energy meter, the control of factors such as analog to digital converter and communication protocol were analyzed using Monte Carlo method presented in [4]. But, while using these types of energy meters, we require a TNEB employee to visit each place for the measurement of total power consumption of various loads operated in the particular building and for calculation of electricity bill amount. This requires much of manual work with time consumption, and it does not provide user understandable energy data. The conventional metering system is represented in Fig. 1. Smart energy meter results in an advanced metering technology that involves smart meters to read, process, and feedback the information to the consumers. It measures the total energy consumption of the load and calculates the corresponding electricity bill. At the same time, this information is shifted to the clients and electricity supplier simultaneously via a communication network. These smart meters eliminate the need of TNEB employee to visit buildings for calculating electricity statement. Modems are utilized in these smart meters to provide communication systems such as telephone, wireless, fiber cable, and power line communications.

This is represented in Fig. 2. Later, enhanced digital meter is introduced in [5] to reduce manual and calibration errors with the facilitate of IoT and GSM. The enhanced digital energy meters gather detailed consumption datum of a consumer

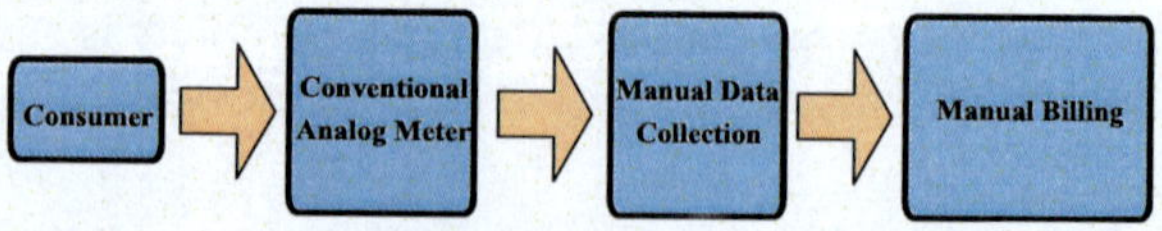

Fig. 1 Conventional metering systems

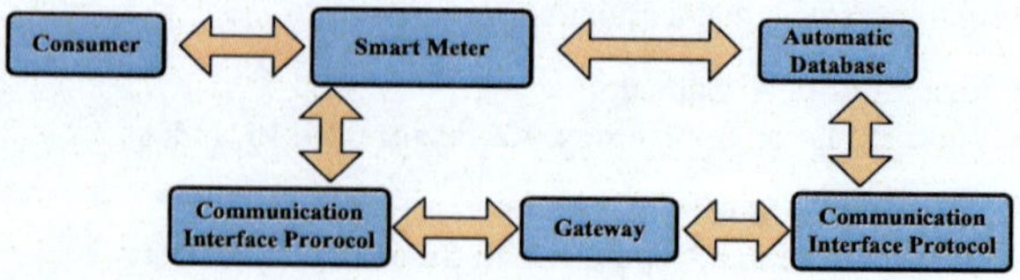

Fig. 2 Smart metering systems

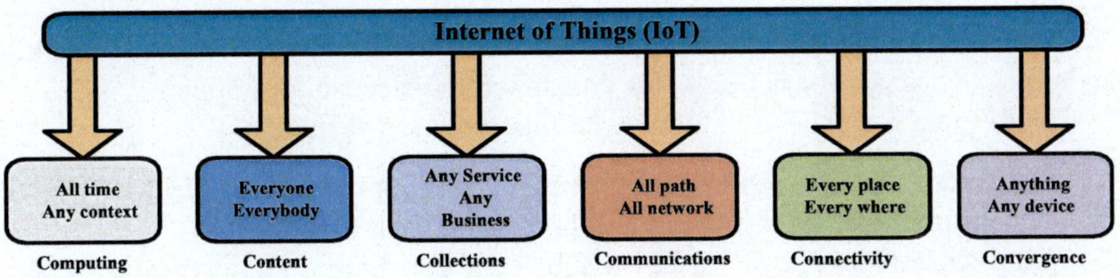

Fig. 3 Internet of things

simply than the traditional metering technique for the reason of the maximum cost of the data logging systems in the latter.

A method for library energy consumption management using LoRa node hardware based on IoT is described in [6]. A low-cost smart power meter is described to determine the energy in an adequate precise approach using two different algorithms, namely the Goertzel algorithm and the zero-crossing algorithm with minimum computational time in [7]. A smart energy meter for power grid-based mobile application is proposed in [8–10]. The framework can gain and process information to get data on energy trade and power quality disorders such as voltage sag, voltage swell, voltage, and current harmonic distortion. A smart meter for home energy management system is described in [11] with experimental results.

Hence, this paper uses IoT for wireless communication and stores the information in a cloud memory. In IoT, everything is configured with Internet protocol addresses, and it can monitor, control, and access remotely following Web technology. The general IOT is illustrated in Fig. 3. Consequently, IoT-based smart energy meter could overcome the issues faced by both the consumers and electricity suppliers.

A smart meter has attractive benefits and features matching the current era of technology modernization. Smart meters give a course to reduce spending money and save it accurately along the continuous estimating, computerized information gathering and avoiding human errors because of manual readings which would, at last, lessen work costs. Smart meters are necessary for efficient energy usage and reduce electricity bill [12, 13]. The smart meter as a critical element for the smart grid is expected to provide economic, social, and environmental benefits for multiple stakeholders. Thus, the research and development in smart meters and their applications will be progressed rapidly in years, and many methods and techniques will be developed. A real-time differential privacy load monitoring (DPLM) algorithm for smart meter is presented in [14]. But security purpose is very limited. IoT-based flying ad hoc network also used for energy monitoring is nowadays described in [15]. IoT-based smart energy calibration configuration for stand-alone photovoltaic configuration is presented in [16].

The main goal of this work is to intend a smart energy meter based on IoT technology that can be installed in buildings to measure total electricity utilization in those buildings. In this projected configuration microcontroller unit, liquid crystal display (LCD), sensors, current sensor, and voltage sensors are present. The traditional analog meter is replaced by a metering module in the projected configuration which consists of a microcontroller that calculates the energy utilized, electricity bill and transmit these information to the consumer and electricity supplier.

The following are the vital features of this proposed system:

(1) No need of TNEB employee to go area by area for calculating electricity bill.
(2) A consumer can set the energy bill budget for each month.
(3) Data can be showed by both the consumer and electricity supplier from anywhere in the world.
(4) Consumers want not remain for the monthly bill.
(5) Needs less preservation, and this energy meter is accurate with compact size.

The IoT-based smart energy meter should be installed at buildings using the technology of ad hoc network as described in [17] where energy is consumed, and the electricity can be monitored. The energy consumption can also be checked by consumers using their smartphone or laptop from any place, and this kind of meters will be a part of a smart city. As today, the majority of people are 24 × 7 online, it will be an extraordinary shelter in the energy that they can monitor their vitality utilization online from anyplace on the globe.

In this paper, Sect. 2 describes the proposed work in detail. Analysis of the conventional type of meters with smart meters and advancements in metering and billing is discussed, in addition, hardware and software modules are described with a flowchart. The block diagram also gives an idea about the overall view of the research. Power cost analysis for simple home and total hardware of the system and its implementation are also presented in Sect. 2. In Sect. 3, the projected configuration is validated among the experimental results. At last, conclusions appear in Sect. 4.

2 Proposed Work

2.1 Working Procedure of Hardware and Software Module

The electricity is generated in power plants, and this generated electricity is circulated to the consumers by transmission and distribution department. TNEB supplies the transmitted electricity to consumers for operating various loads. Here, the conventional meter is replaced with the proposed smart energy meter. The block diagram consists of a controlling unit and a Wi-Fi unit. Figure 4 depicts a figure of the overall system. The major components used are Arduino UNO, current sensor ACS 712, and NodeMCU ESP-12. Current sensor ACS 712 calibrates the current flowing through the loads. Most of the existing methods use PIC microcontroller, whereas here, this project is handled with AtMega328P microcontroller. Arduino receives the sensed value of current and voltage. Then, it calculates power, energy, and the corresponding bill amount. The data is displayed on LCD. At the same time, the information is also transferred to cloud service via the Wi-Fi module NodeMCU ESP-12. NodeMCU is a Firmware on ESP8266, and NodeMCU ESP-12 is a System on Chip (SoC).

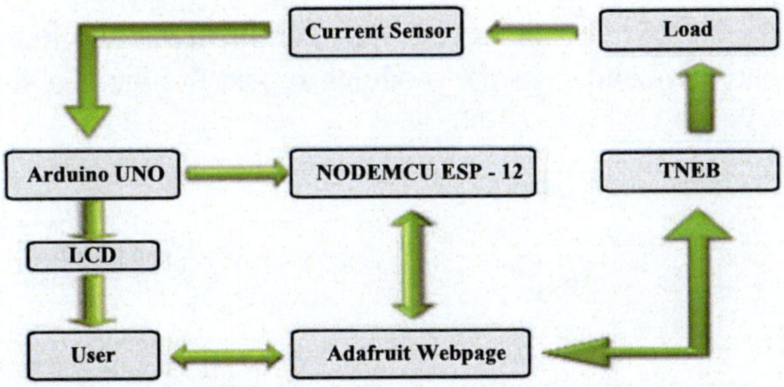

Fig. 4 Block diagram of the system

It includes a micro-USB connector with onboard programmer IC, voltage regulator, reset and program buttons and LED and is utilized to develop IoT-based applications. It has analog and digital pins, which is utilized to interface sensors and to acquire the data over the Internet. Figure 5 shows the detailed pin description of ESP-12.

This allows the user to monitor the total energy consumed with the corresponding bill amount from anywhere from this world. It also sends the electricity bill via email to the consumer using If This Then That (IFTTT). It is free online support to make chains of straightforward conditional statements, known as applets. An applet is triggered by adjust that occur within other Web services such as Gmail, Facebook, Telegram, Instagram, or Pinterest.

The projected scheme calculates the energy consumed at every home automatically, and the values are being displayed to them at their meter box. In general, it helps the consumer to be conscious of their usage level, and they try to minimize if they are using the maximum resource accessible to them. The unit readings and bill are quickly transferred to the EB database through ESP Wi-Fi module. Consequently, minimize the manual labors. The functioning of the proposed system includes two

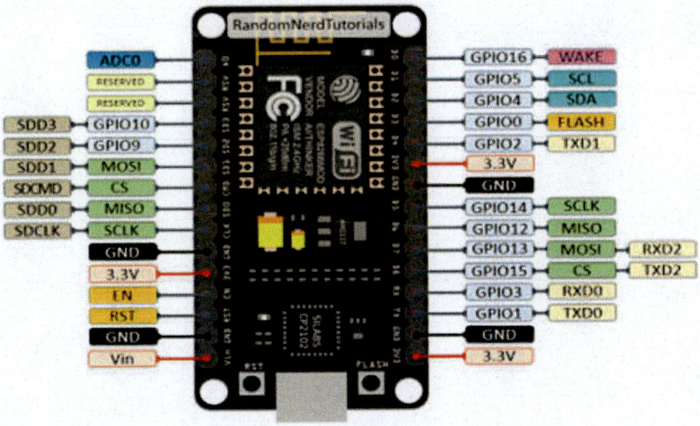

Fig. 5 Pin description of ESP-12

major modules, that is hardware and software modules. The flowcharts of both the hardware and software modules appear in Figs. 6 and 7.

Fig. 6 Working of hardware module

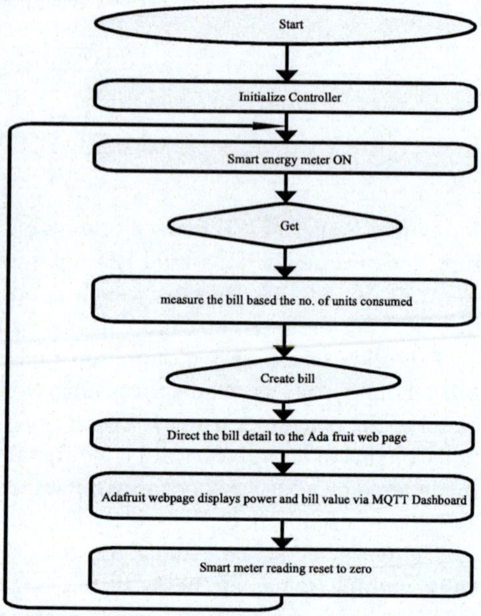

Fig. 7 Working of software module

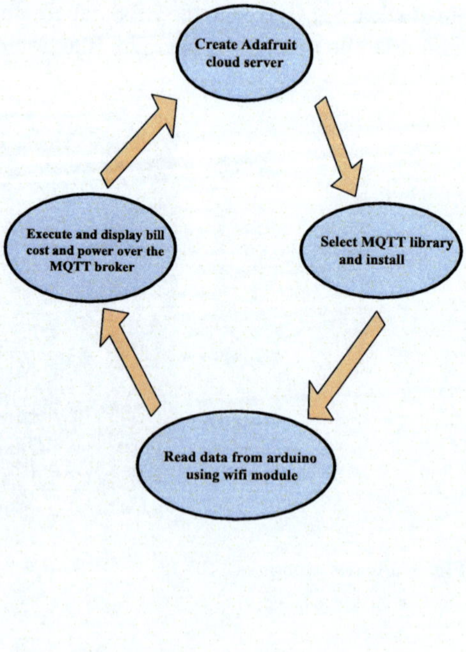

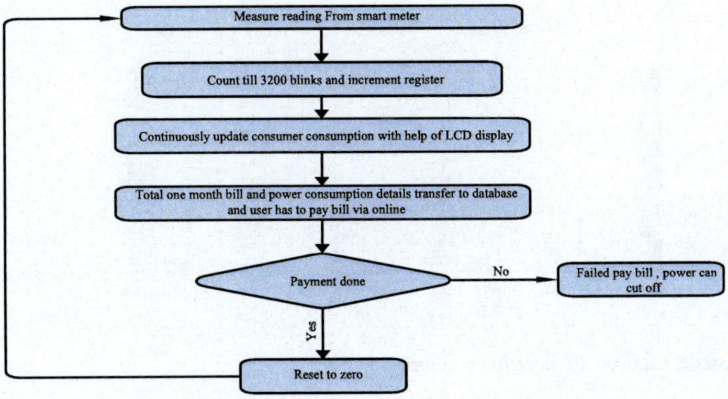

Fig. 8 Data extraction and processing

2.2 *Extraction and Processing Data*

The smart meter associated with the home starts to compute the power consumed by the various load once the switch makes ON. The power meter comprises of LED bulb that blinks 3200 times for one unit consumed. Then, this unit is counted by the Arduino and Arduino contain current and voltage sensors to measure the bill, and the bill is generated. The generated bill and the unit consumed for a month are displayed in the LCD panel and then transferred to the cloud server. The user can now log into their MQTT account to verify their unit consumed or just check the LCD panel connected to the meter. The flow diagram for the extraction and processing of data has appeared in Fig. 8.

The consumer unit readings are transmitted to the cloud with the help of the Wi-Fi module. The meter calibrations, along with the cost, will be viewed in the cloud platform. Every month, the consumer should pay their bill at the local EB station, and the data can be reset. If the consumer failed to pay within a stipulated time period, then the power will be cut off.

2.3 *Power Cost Analysis for a Smart Home*

Our proposed system does not contain enormous and hard calculations. Generally, different meters have different readings. Mostly, 3200 blinks-1 unit depends on the manufacturer. In our case, 3200 blinks of LED is 1 unit. But for practical purpose,

the assumption is different. Let, $V =$ number of flashes of LED, U-number of units of electricity, and $C =$ cost of consumption.

Basically,

Number of units $(U) = (\mathrm{V}/3200)$

Assume that 1 unit cost = 7 Indian Rupee (INR).

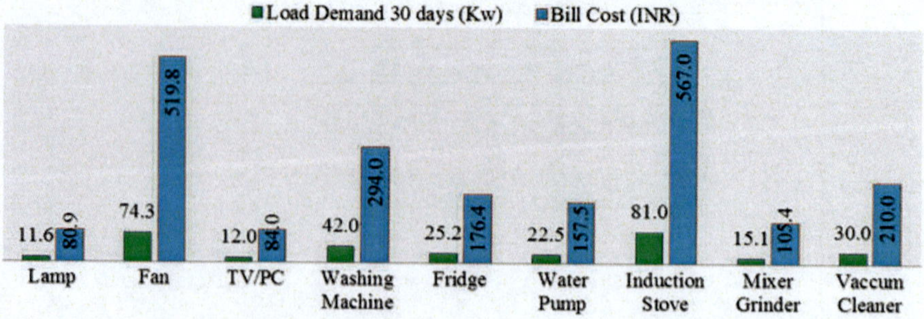

Fig. 9 Load demand vs bill cost for a month

$C = V * 7$ INR

Consider electrical load consumed by a simple home for a month, and finally, both consumer and supplier receive a notification, if the consumer does not react this notification and increase the threshold value, then the meter will obviously turned OFF. Once more to make it ON, consumer wants to revisit the webpage to amplify threshold value.

For realistic reason, decrement and increment of the threshold can be finished by +5 units or −5 units.

Usually, the fundamental unit of electricity is Kilowatt hour (KWh), 1 kWh–1000 W for 1 h.

Example, Ten 100 W bulbs used for 1 h gives 1 kWh.

Load demand versus bill cost for a month for home appliances as described in Fig. 9. Illustrates the daily usage of load, power, and bill amount.

2.4 Hardware Implementation

This proposed work is mainly concentrated on the IoT network. Internet-connected devices are used to enable people to access online data and processes. The primary purpose of IoT devices is to generate real-time data, and then it can be used to analyze and create desired outcomes. IoT system increases and encourages machine-to-machine communication (also known as M2M). Because of this ingenious innovation, physical devices stay in touch with one another, leading to greater efficiency and higher quality. It also allows for full transparency.

Arduino is an open-source electronics platform, and its hardware and software are simple to utilize. Arduino Uno is a microcontroller board that contains the ATmega328P and has 14 digital input/output pins, six analog inputs, a 16 MHz quartz crystal, a USB connection, a power jack, and a reset button. It is relatively inexpensive compared to other microcontroller platforms. Most microcontroller systems are limited to Windows operating systems, but Arduino Software (IDE) runs on Windows, Macintosh OSX, and Linux operating systems. Beginners can easily use

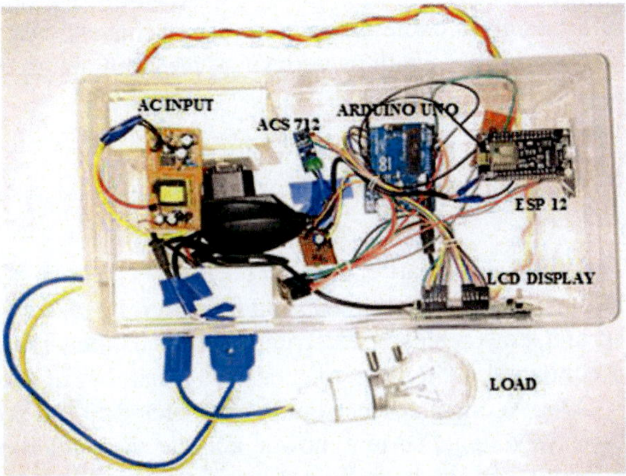

Fig. 10 Overall view of the hardware system

it, yet flexible enough for advanced users as it is accessible for expansion by experienced developers. The language can be extended through C++ libraries, and those individuals who need to comprehend the specific feature can make the jump from Arduino to the AVR C programming language on which it is based. The plans of the Arduino sheets are distributed under a Creative Commons permit. Hence, experienced circuit designers can make their version of the module by extending and improving the module. Therefore, Arduino boards are chosen.

The output of the current sensor AS712 is connected to the analog pin A0 of Arduino UNO. ESP-12 is interfaced with Arduino by connecting RX of ESP-12 with TX of Arduino TX of ESP-12 with RX of Arduino. Arduino is powered with 5 V supply, and ESP-12 is equipped with 3.3 V supply. Arduino and Wi-Fi module NodeMCU ESP-12 are programmed using Arduino IDE. Figure 10 shows the overall view of the hardware system of this work.

Monitor the energy consumption over the Internet, and there is a need for Message Queuing Telemetry Transport (MQTT) broker. MQTT empowers asset obliged IOT gadgets to send, or distribute, data about an offered theme to a server that capacities as an MQTT message merchant. The merchant at that point drives the data out to those customers that have recently bought into the customer's theme. MQTT is a lightweight protocol because all of its messages have a small code footprint. The MQTT protocol is an excellent choice for wireless networks. In this research, the Adafruit IO platform is used as an MQTT broker. Adafruit.io is a cloud service that primarily is used for storing and then retrieving data. Adafruit.io can handle and visualize multiple feeds of data. The dashboard is one of the features integrated into Adafruit IO, which displays a chart, graph, gauge, and log of data. These dashboards can be viewed from anywhere in the world. The trigger in Adafruit IO is also utilized to control and react to all data. In this research, both power and bill data measured using the projected smart energy meter are available in the dashboard. The data is also displayed in the MQTT dashboard android app. Adafruit IO is a system that makes information valuable. Our attention is on convenience and permitting straightforward information associations with small programming required.

3 Result and Discussion

The projected smart energy meter has been tested with a variety of loads such as incandescent bulb, table fan, and induction stove, for an hour to measure the voltage, current, power, and the electricity bill for an hour. The result obtained from each load is compared with others to check the feasibility of the proposed system.

A 40 W incandescent bulb is connected to the proposed system. LCD module displays voltage, current, power, and the corresponding bill. Voltage is 238.2 V and current taken by the bulb is 0.181 A and the power of the lamp is 43.2 W, and the identical bill is Rs. 15. These readings are shown in Fig. 11. Current and voltage measurements of the smart energy meter are verified with multimeter reading. The adafruit webpage and MQTT Dashboard output is displayed in Figs. 12 and 13 respectively.

Next, induction stove is connected as a load to the proposed system. LCD module displays voltage, current, power, and the corresponding bill. Voltage is 232.1 V and current taken by the induction stove is 4.412 A and power of the induction stove is 1024 W, and the identical bill is Rs. 41. These readings are displayed in Fig. 14 and Fig. 15.

Results of various household equipment's are tabulated in Table 1. From these results, it can be concluded that the design and implementation of IoT-based smart energy meter are successfully achieved. The outputs of all the loads are verified with the help of a multimeter.

Keeping in mind of public health and safety, the proposed system used in this research work is designed a way that all the components are kept inside a single box. The elements involving high voltage are displayed with a warning for caution. Thus, the system will be safe and will not affect public health as well as the ingredients used in this system will not disturb the environment.

Fig. 11 LCD output

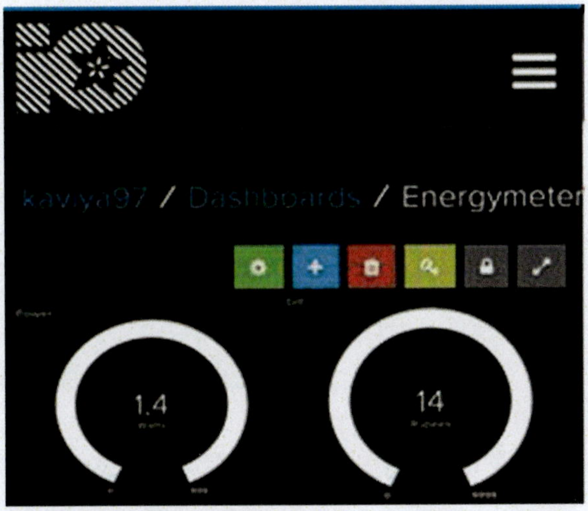

Fig. 12 Adafruit webpage output

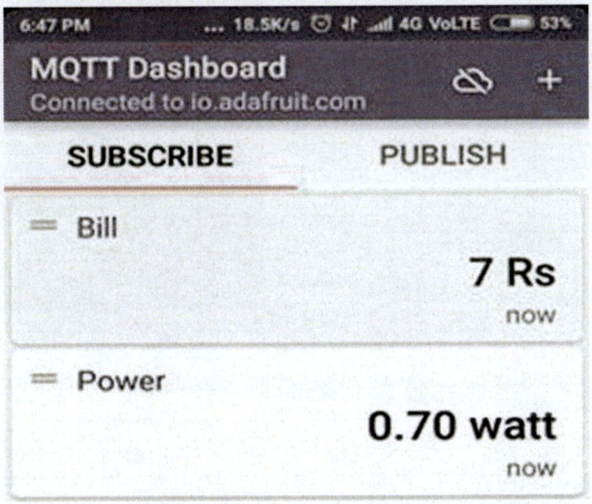

Fig. 13 MQTT Dashboard output

4 Conclusion

The main objective of this work is to develope a smart energy meter based on IoT technology that should be installed in housing and marketable buildings to measure total electricity utilization in that buildings has been fully achieved. As various loads consume energy, this projected smart energy meter reads the readings continuously, and the load utilized has been showed on meter through LCD module as well as the Adafruit webpage with the help of IoT technology. This meter is practically affordable and will act as a primary platform for designing further devices.

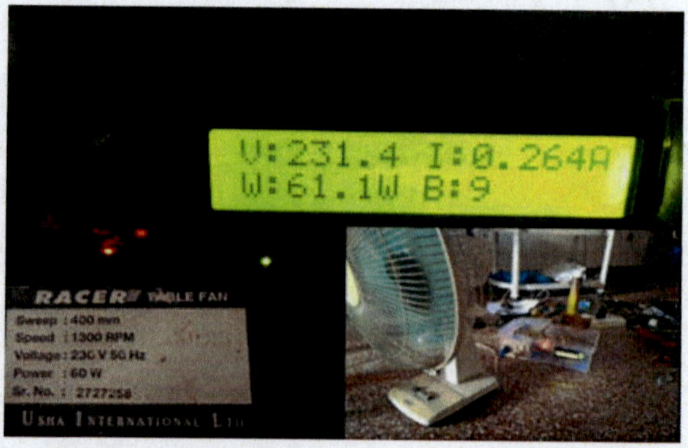

Fig. 14 Fan load output

Fig. 15 Induction stove load output

Table 1 Comparative analysis of different household appliances

Load	Voltage (V)	Current (A)	Power (W)	Bill
Bulb	*234.2*	*0.181*	*43.2*	*Rs. 15*
Table fan	231.4	0.264	61.1	Rs. 9
Induction stove	*232.1*	*4.412*	*1024*	*Rs. 41*

References

1. Sultanem F (1991) Using appliance signatures for monitoring residential loads at meter panel level. IEEE Trans Power Delivery 6(4):1380–1385
2. Tang Y, Chee-Wooi T, Wang C, Parker G (2015) Extraction of energy information from analog meters using image processing. IEEE Trans Smart Grid 6(4):2032–2040
3. Shahidi S, Abdul Gaffar M, Khosru MS (2013) Design and implementation of digital energy meter with data sending capability using GSM network. In: 2nd international conference on advances in electrical engineering (ICAEE), pp 203–206

4. Xiaojuan P, Lei W, Zhengsen J, Xuewei W, Hongtao H, Lijuan L (2016) Evaluation of digital energy meter error by Monte Carlo method. In: Conference on precision electromagnetic measurements, pp 1–2
5. Reni Clenitiaa F, Ilakya E, Preetha GS, Meenakshi B (2017) Enhanced digital energy meter. In: International conference on computation of power, energy information and communication (ICCPEIC), pp 588–591
6. Yang C, Kang H-B, Zhang L, Zhang R (2018) A design of smart library energy consumption monitoring and management system based on IoT. In: Krömer P, Zhang H (eds) Proceedings of the fifth Euro-China conference on intelligent data analysis and applications 2018, AISC, vol 891. Springer, Heidelberg, pp 217–224
7. Abate F, Carratù M, Liguori C, Paciello V (2018) A low cost smart power meter for IoT. Measurement 136:59–66
8. De Capua C, Lipari G, Lugarà M, Morello R (2014) A smart energy meter for power grids. In: IEEE international instrumentation and measurement technology conference (I2MTC), pp 1–6
9. Avancini DB, Joel JPC, Simion GB, Ricardo AL et al (2019) Energy meters evolution in smart grids: a review. J Clean Prod 217:702–715
10. Chhaya L, Sharma P, Kumar A, Govind B (2018) IoT-based implementation of field area network using smart grid communication infrastructure. Smart Cities 1(1):176–189
11. Dalvi SD, Chandrababu D, Kumar V (2016) Design and development of smart electricity meter—the path towards sustainable globe. Int J Ambient Energy 37(4):348–353
12. Shah J, Mishra B (2016) Customized IoT enabled wireless sensing and monitoring platform for smart buildings. Proc Technol 23:256–263
13. Kaur M, Mathew L, Alokdeep, Kumar A (2018) Implementation of smart metering based on internet of things. IOP Conf Ser: Mater Sci Eng 331:1–12
14. Sayed S, Hussain T, Gastli A, Mohieddine B (2019) Design and realization of an open-source and modular smart meter. Energy Sci Eng (Early View), pp 1–18
15. Hassan M, Rehmani M, Kotagiri M et al (2019) Differential privacy for renewable energy resources based smart metering. J Parallel Distrib Comput 131:69–80
16. Banu JB, Moses MB (2018) IoT based augmented perturb-and-observe soft switching boost converters for photovoltaic power systems in smart cities. Wireless Pers Commun 102(4):2619–2641
17. Fadi T, Altrjman C, Sadia D, Anand P (2019) Energy monitoring in IoT-based ad hoc networks: an overview. Comput Electr Eng, pp 133–142

A Collaborative Approach for Course Recommendation System

S. A. Sowmiya and P. Hamsagayathri

Abstract Graduate attributes are the competencies and capabilities set by the college network as benchmark for its students during their time with the organization. The principle point of the paper is to assist students with their scholastic choices, by suggesting elective courses dependent on the graduating attributes scores. Students rate the significance of each course in improving their graduating qualities toward the end of every semester. The elective courses taken by peer students and rated similarly are identified by a collaborative filtering algorithm and prescribed to students. This paper predicts the rating for each course and graduating attribute. The elective courses with the greatest expected ratings are shortlisted for proposal. Also, the aggregated GA score is determined for each student at end of every semester. The shortlisted courses that can improve the feeble GA scores are at prescribed to students.

Keywords Graduating attributes · Collaborative filtering algorithm · K-nearest neighbors

1 Introduction

A recommendation system is a type of information filtering system that helps to predict the "rating" or "preference" provided by a user for a given item. The predicted rating will be helpful to recommend a new item or product to the user. The real-time examples for recommendation systems are e-commerce like Amazon or Flipkart, which recommend products to the user and entertainment systems like Netflix, which recommend movies to the users [1].

S. A. Sowmiya (✉)
Department of Computer Science and Engineering, Bannari Amman Institute of Technology, Sathyamangalam, Erode Dt, Tamil Nadu 638401, India
e-mail: sowmiya26@gmail.com

P. Hamsagayathri (✉)
Department of Electronics and Communication Engineering, Bannari Amman Institute of Technology, Sathyamangalam, Erode Dt, Tamil Nadu 638401, India
e-mail: palanisamy.hamsagayathri@gmail.com

N. Zhou and S. Hemamalini (eds.), *Advances in Smart Grid Technology*, Lecture Notes in Electrical Engineering 688, https://doi.org/10.1007/978-981-15-7241-8_39

This paper aims to develop the course recommendation system that recommends appropriate elective courses to graduate students. The university community decides the graduating attributes, which are the traits and competencies, students should have excelled while completing their degree course. They are traits that prepare graduates as better personalities in an unknown future. Some of the attributes are communication, problem analysis, design and development of solutions to complex problems, ethics, individual and team work, etc. The educational institutions select courses to accomplish these attributes in consistent manner. Each course is assigned a general rating based on its impact on every graduating attribute (GA). The courses having greater impact in improving GAs are recognized by the students. These impacts can be converted to ratings and used in the recommendation systems.

There are numerous data mining systems developed in the field of education [2, 3] and especially on how these recommender systems can be used for proposing elective courses [4, 5]. Some of the recommendation engines use the course content based on content-based filtering techniques [6, 7]. Here, the recommendation is mainly based on course comparison between the contents and students profile [8]. Few systems use the grades achieved by similar students for recommendations. These systems use collaborative filtering techniques that identify the similarity among students and recommend courses already chosen by similar students [9, 10]. Some systems predict the scores based on the similarity among students and provide recommendations [11, 12].

The existing system identifies similarity based on students' ratings that vary widely depending on student's interest in particular subject [13, 14]. Recommendations based on similarity among students performance are not related to individual's competency profile or skill set [12, 15]. Students are unaware of the importance of every course and its impact in shaping their careers.

The rating of the GAs can be utilized to create course recommendation. Each course specifically affects each GA and graduates should enter their rating based on their impact through survey. A recommender system will utilize the information provided by the senior students and prescribe courses to junior students based on the similar ratings. Toward the end of every semester, score will be produced for each GA and understudies will be told to concentrate on low GA scores. Students will be notified to focus on low GA scores. So the system will further refine the recommendation to suggest courses that can help in improving low GA scores.

The recommendations from the proposed system are more refined and personalized for every student. Students are well aware of the competency profiles and are able to compare their skill set with benchmarks set by the institutions. They can easily identify their area of improvements from GA scores and focus on the courses that can improve the scores. The application with its graphical user interface provides a user-friendly environment to rate existing courses and looking for new recommendations. Students can effectively shape their career path by choosing right courses at right time.

2 Related Work

The authors of [16] "A collaborative approach for research paper recommender system" presents a collaborative approach for research paper recommender system. A source research paper is recommended only if it referred any of the target paper's references and there should exist a third paper that refers both the source and target paper. The algorithm then measures the percentage of similarity between the target paper and the qualified source papers. The top-N most similar papers are then recommended based on the assumption that if there exists significant recurrence between the target paper and the qualified source papers, then there exists some percentage of similarities between them. But this algorithm did not consider the relevance based on the titles and abstracts of the recommending papers for better performances.

The authors of "Recommender Systems for university elective course recommendation" [17] proposed recommender system that targets on student evaluation done in the past course. Student assessments are conventional strategies to predict student performance such as failing or passing or forecasting successful completion of the course, in this continuation anticipating the classification of degree or achievement. The main idea in this system is based on student's mastery and the faculty's experience in elective subject. The assessment strategy developed with the presumption that learner will never learn anything new after the tests. But learning can also be developed from some factors like curiosity, self-efficiency, belonging, interest, faculty expertise in subject, etc.

The authors of "A recommender for improving the student academic performance" [18] focus on developing a framework of clever recommender system that can predict tertiary students' first year academic performance. This system also recommends most important actions to be taken to help the students to improve their academics. These measures can help the institution management in their decision making about early intervention strategies. A multi-dimensional methodological approach was introduced. A thorough interview would be conducted in identifying each students' background factors like parent's educational qualification, sponsor, family size, both father's and mother's occupation, etc. The recommendations are based on the student's background factors.

3 Algorithm Description

3.1 Data Preparation

There are several ways for collecting information on graduating attributes [19]. Students can do self-assessment at the end of each semester. In this self-assessment, students can provide best rating to the courses that had a great impact in developing their GAs. These ratings are then used in the recommender systems. For a new GA or a course, the instructor's assessment and the student's performance can be used to provide ratings.

The rating should be given as either 0 (no impact), 1 (minor impact), or 2 (major impact). Few students may rate even no impacts as 1 but some others may start giving the rating of 1 to even major impacts. For each semester, students may take from 4 to 6 elective courses that are not taken earlier. Students should be aware of the ratings and their impacts. Any bias in the ratings would definitely impact the recommendation for the next year students. So it is the responsibility of the institution to educate the students about the entire system and the importance of their ratings. A reference guide can be included in the rating page for student's reference.

3.2 *Algorithm Design*

Collaborative filtering depends on the user's historical preference on set of items [15, 20]. The core assumption here is that the users who have agreed in the past tend to agree in the future also. In this paper, each student is represented by a pair of the courses taken and the assessments (ratings) they provided for that course for each graduating attributes (GAs). In this paper, we are considering each GA of the course as an item. Here, each item is composed of both GA and a course. In this paper, we recommend course that helps to increase the score in all GAs. The similarity between students is calculated based on Pearson's correlation. It is known as the best method of measuring the association between variables of interest because it is based on the method of covariance. It gives information about the magnitude of the association, or correlation, as well as the direction of the relationship. The similarity metric is mentioned using the following equation.

$$\mathrm{sim}(a, b) = \frac{\Sigma_{g \in G_a \cap G_b}(R_{a,g} - \tilde{R}_a) \cdot (R_{b,g} - \tilde{R}_b)}{\Sigma_{g \in G_a \cap G_b}(R_{a,g} - \tilde{R}_a)^2 \cdot \sqrt{\Sigma_{g \in G_a \cap G_b}(R_{b,g} - \tilde{R}_b)^2}} \tag{1}$$

where

- g indicates a individual element---combination of a GA and course,
- G_a indicates the courses taken by student a,
- G_b indicates the courses taken by student b,
- $R_{a,g}$ indicates the rating provided by student a for element g,
- $R_{b,g}$ indicates the rating provided by student b for element g,
- $G_a \cap G_b$ indicates the similar courses taken by students a and b,
- $\tilde{R}_a$ indicates the mean rating provided by student a.

The next step would be to find the students similar to the current student in order to generate predictions. The easiest way is to use the k-nearest neighbors' algorithm. Here, we should not simply pick the best n neighbors, but we should make sure that selected similar and the active student have rated a common number of course competences. Finally, ratings can be predicted by doing a weighted sum of ratings given to each item by similar students to our target student.

4 System Architecture

This project involves developing a graphical user interface to interact with the course recommendation system. The GUI will have screens to login with user credentials, register new user details, and provide rating to finished courses specific to graduating attributes. Then a collaborative filtering algorithm will be used to find similarity among students based on their rating. Additionally, a GA score will be calculated at the end of each semester to check the student's improvement in each attribute and suggest best course to students.

Here, the project assumes students provide genuine rating to each course independent of their independent of their likes, dislikes, or personal hatred. The rating may be influenced by student interest toward the subjects and the instructor. The academic proficiency of the advisors also has a major impact. So, it is better idea to educate the students well before providing rating.

The following algorithm is used to predict the ratings for students. This calculation is done only after the students provide necessary ratings. Else the students should be advised to do it and then look for recommendations.

1. The input dataset should contain course, graduating attribute, and rating provided by the students.
2. The input dataset is then divided into training dataset and test dataset.
3. The value of k for KNN is then set to square root of number of records in training dataset.
4. The Euclidean distance, which is defined as the straight line distance between two points in Euclidean space, is calculated between test data and each row of training data.
5. The calculated distance is then arranged in ascending order.
6. The first k rows are then short listed from the sorted distance array.
7. The mean ratings provided by these short listed students are identified.
8. The average of these ratings for each course is then identified as the predicted rating for the target student.

The accumulated GA score is calculated based on the students GA score history. The courses by similar students are again filtered. Those courses that can improve the proficiency are recommended to the students (refer Fig. 1).

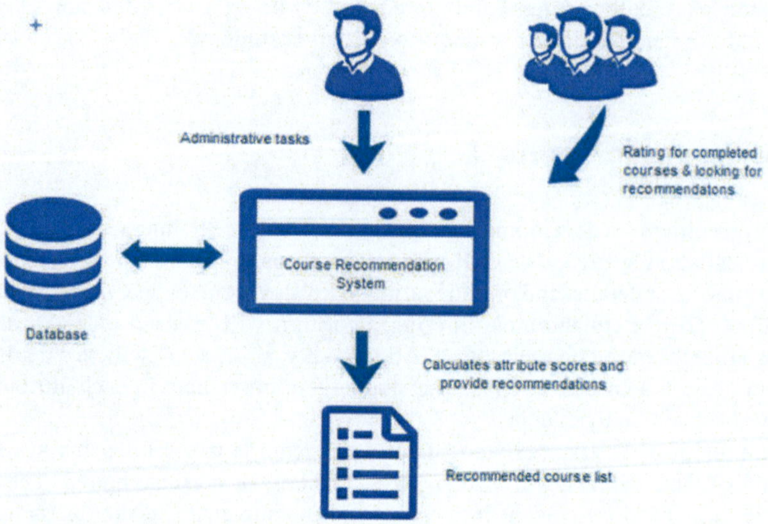

Fig. 1 System architecture of course recommendation system

5 Experimentation and Results

The project is implemented using R language and MySQL database. The mock data of more than 100 students for 20 courses have been updated in the database. There are two major types of users and each has specific set of modules: Student and Admin.

Students may wish to view their previous semester results. These results can be viewed in Academic Profile tab. The application fetches the semester results from semester_result table. This tab displays the semester result details in a tabular format for each semester (Fig. 2).

This is the core component of this application. This page will display two data tables. One will contain the list of elective courses for the next semester. The second table will display the recommended courses specific to that student. For example, if student has to select 3 elective courses from a list of 15, then the application will display the 15 courses in one table and 3 recommended courses in other table (Fig. 3).

Once student completes his survey during every semester, the GA scores are calculated and saved in graduating attribute score table. The graduating attribute score helps in identifying the gray areas and student can start working on those attributes (Fig. 4).

If the student has already submitted the survey, then only a message displaying "You have already completed t he survey. Thank You" will be displayed in the survey page. If the student is yet to complete the survey, the previous semester courses will be displayed to students.

1. For each course, 5 graduating attribute and respective radio buttons to choose No Impact, Minor Impact, and Major Impact will be displayed.

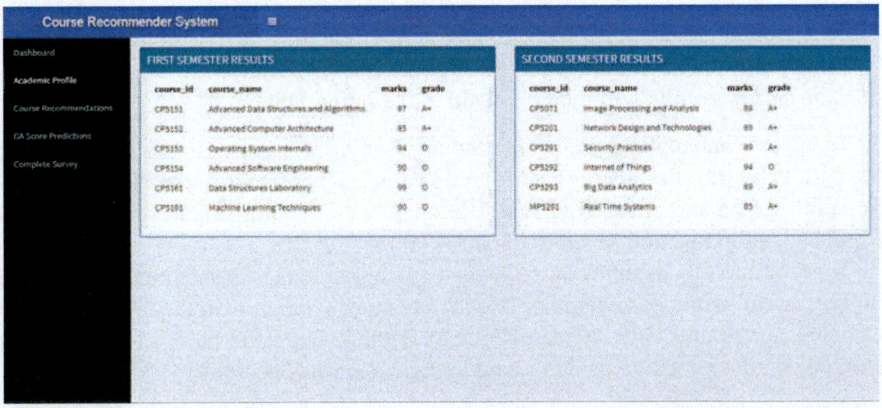

Fig. 2 Academic profile page

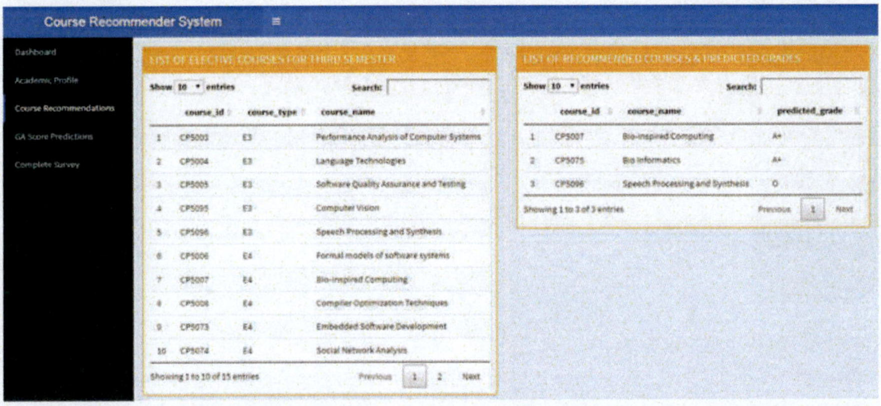

Fig. 3 Course recommendation page

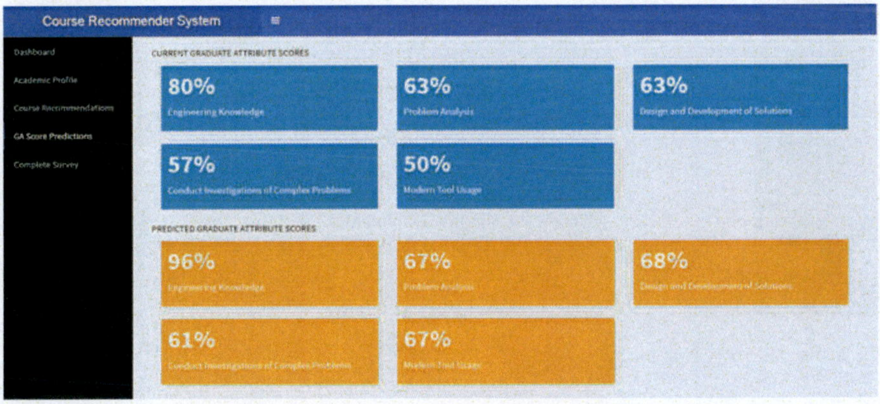

Fig. 4 GA score prediction page

2. The student should click the ratings for all graduating attributes in each course. Then, the student should click the "Submit Survey" Button.
3. The survey results will then be saved in the rating table

To access this application, every student should have been already registered in the system by the administrator. He should enter all the details in the "Register New Student" screen and click "Register" Button. Once the button is clicked, all entered details are validated and saved in database (Figs. 5 and 6).

To recommend a specific course, every course should have been already registered in the system by the administrator. He should enter all details in the "Register New Course" screen and click "Create Course" Button. Once the button is clicked, all entered details are validated and saved in the database (Fig. 7).

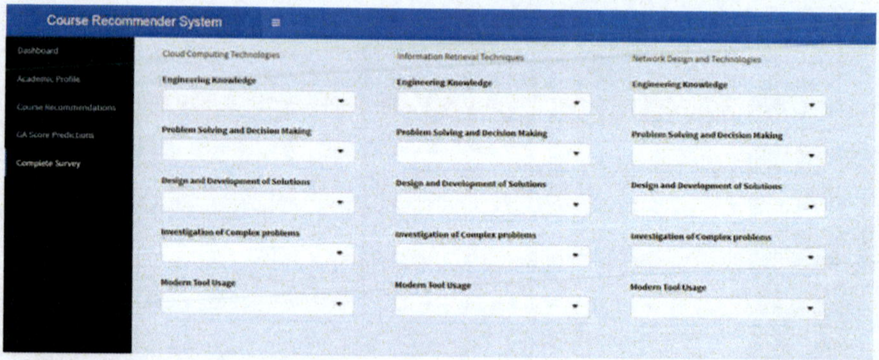

Fig. 5 Complete survey page

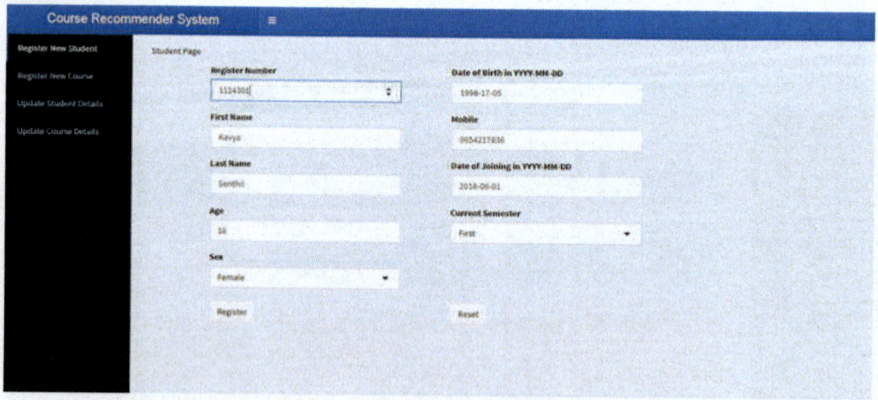

Fig. 6 Register new student page

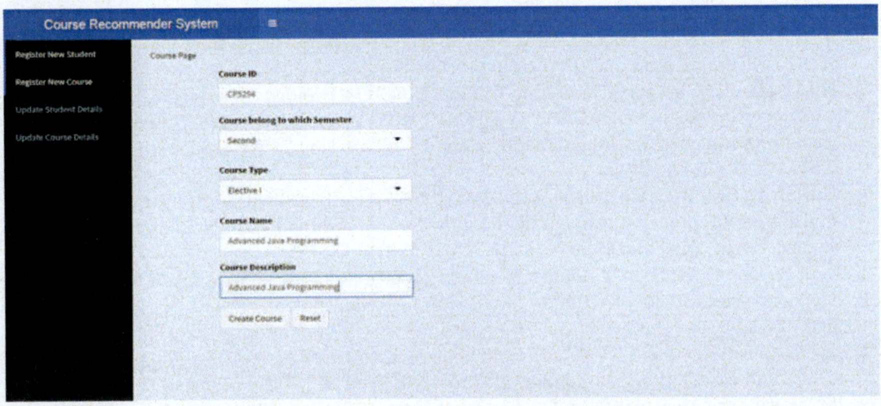

Fig. 7 Register new course page

6 Conclusion and Future Work

This paper proposed an elective course recommendation system based on graduating attributes, which universities set as the benchmarks for their students. This will benefit the students to choose the right elective course. The experiment is conducted on a user generated dataset that could have biased the results on certain factors. So testing on the real dataset may detect new challenges. This will help to improve the proposed approach. Also, the other factors like importance of each course, student's interest in subjects can also be considered while recommending courses in future approaches. Additional care should be taken while recommending a new course whose ratings is not yet available or to the new students who have not provided any ratings so far. This cold-start problem is generally prevalent in many recommendation systems and hence the future work can concentrate on this problem.

References

1. Abbas M, Riaz MU, Rauf A, Khan MT, Khalid S (2017) Context-aware Youtube recommender system. In: 2017 international conference on information and communication technologies (ICICT), Karachi, Pakistan, 2017, pp 161–164
2. Bydzovska H, Course enrollment recommender system. In: Proceeding of the 9th international conference on educational data mining, pp 312–317
3. Romero C, Ventura S, Pechenizkiy M, Baker RS (2010) Handbook of educational data mining. Chapman and Hall/CRC
4. Feghali T, Zbib I, Hallal S (2011) A web-based decision support tool for academic advising. Educ Technol Soc 14(1):82–94

5. Al-Badarenah A, Alsakran J (2016) An automated recommender system for course selection. Int J Adv Comput Sci Appl (IJACSA), pp 166–175
6. Thai-Nghe N, Horváth T, Schmidt-Thieme L (2011) Factorization models for forecasting student performance. In: Pechenizkiy M, Calders T, Conati C, Ventura S, Romero C, Stamper J (eds) Proceedings of the 4th international conference on educational data mining (EDM 2011). ISBN 978-90-386-2537-9
7. Mostafa L, Oately G, Khalifa N, Rabie W (2014) A case based reasoning system for academic advising in Egyptian educational institutions. In: 2nd international conference on research in science, engineering and technology (ICRSET '2014)
8. Binh N, Duong H, Hieu T, Nhuan N, Son N (2008) An integrated approach for an academic advising system in adaptive credit-based learning environment. VNU J Sci Nat Sci Technol 24:110–121
9. Ali Z, Khusro S, Ullah I (2016) A hybrid book recommender system based on table of contents (TOC) and association rule mining. In: Proceedings of the 10th International Conference on informatics and systems (INFOS '16), Cairo, Egypt, 2016, pp 68–74
10. Toscher A, Jahrer M (2010) Collaborative filtering applied to educational data mining. In: KDD Cup 2010: improving cognitive models with educational data mining
11. Takacs G, Pil´aszy I, N´emeth B, Tikk D (2007) On the gravity recommendation system. In: Proccedings of KDD cup workshop at SIGKDD'07, 13th 175 references ACM international conference on knowledge discovery and data mining, 22–30, San Jose, CA, USA, p 42
12. Thai-Nghe N, Horváth T, Schmidt-Thieme L (2011) Personalized forecasting student performance. In: Proceedings of the 11th IEEE international conference on advanced learning technologies (ICALT 2011), pp 412–414. IEEE Xplore
13. Thai-Nghe N, Drumond L, Krohn-Grimberghe A, Schmidt-Thieme L (2010) Recommender system for predicting student performance. In: Proceedings of the 1st workshop on recommender systems for technology enhanced learning (RecSysTEL 2010), vol 1, Elsevier's Procedia CS, pp 2811–2819
14. Thai-Nghe N, Janecek P, Haddawy P (2007) A comparative analysis of techniques for predicting academic performance. In: Proceedings of the 37th ASEE/IEEE frontiers in education (FIE 2007), pp T2G-7–T2G-12. ISSN: 0190-5848. E-ISBN: 978-1-4244-1084-2. Print ISBN: 978-1-4244-1083-5. IEEE Xplore
15. Thai-Nghe N, Drumond L, Horváth T, Schmidt-Thieme L (2012) Using factorization machines for student modeling. In: Proceedings of FactMod 2012 at the UMAP 2012, vol 872, CEUR-WS, ISSN: 1613–0073
16. Haruna K, Ismail MA, Damiasih D, Sutopo J, Herawan T (2017) A collaborative approach for research paper recommender system. PLoS ONE 12:e0184516
17. Bhumichitr K, Channarukul S, Saejiem N, Jiamthapthaksin R, Nongpong K (2017) Recommender systems for university elective course recommendation. In: 2017 14th international joint conference on computer science and software engineering (JCSSE), Nakhon Si Thammarat, 2017, pp 1–5
18. Goga Maria, Kuyoro Shade, Goga Nicolae (2015) A recommender for improving the student academic performance. Soc Behav Sci 180:1481–1488
19. Feng M, Heffernan N, Koedinger K (2009) Addressing the assessment challenge with an online system that tutors as it assesses. User Model User-Adap Inter 19(3):243–266
20. Al-Sarem M (2015) Building a decision tree model for academic advising affairs based on the Algorithm C4. 5. Int J Comput Sci Issues (IJCSI) 12(5) ISSN (Print): 1694-0814 ISSN (Online): 1694-0784

Vision-Based Lane Detection for Advanced Driver Assistance Systems

Rakesh Acharya Dharoori and Sathiya Narayanan

Abstract Lane lines play a key role in indicating traffic flow and directing vehicles; lane detection serves as a core component in most of the modern-day advanced driver assistance systems (ADASs). Computer vision-based lane detection is an essential technology for self-driving cars. This paper proposes a lane detection system to detect lane lines in urban streets and highway roads under complex background. In order to nullify the distortions caused by the camera lenses, we generate a distortion model by calibrating images against a known object, and apply a generalized filtering approach using Sobel operator (Canny edge detection) in HLS color space. A bird eye view of image is generated using perspective transformation. A special search strategy using sliding window algorithm is used to detect lane lines, and later, curve fitting is done using polynomial regression. Thus, the obtained lane detector is overlaid upon a video to fill the detected portion of the lane. Then, it is applied to the video to detect lane lines. The image processing pipeline is written in Python using OpenCV libraries, and video processing is done using MoviePy. In this paper, the system developed is tested by applying it on a video taken from a camera mounted over the car. The environment used to implement the system is Anaconda. The results obtained show that the proposed system for lane detection, self-calibration and vehicle offset estimation is effective, accurate for both straight and curved lanes and robust to challenging environments.

Keywords Advanced driver assistance systems (ADASs) · Distortion model · Sobel operator color space

R. A. Dharoori · S. Narayanan (✉)
School of Electronics Engineering, Vellore Institute of Technology, Chennai 600127, India
e-mail: sathiyanarayanan.s@vit.ac.in

N. Zhou and S. Hemamalini (eds.), *Advances in Smart Grid Technology*, Lecture Notes in Electrical Engineering 688, https://doi.org/10.1007/978-981-15-7241-8_40

1 Introduction

1.1 Background

The World Health Organization (WHO) as part of a study estimated that 1.35 million people die each year as a result of road traffic. Most of the time, traffic accidents are a result of human blunders such as inattention, diversion and preoccupation. Techniques like road cognizance and lane marking play vital role in helping humans avoid mistakes. Lane detection is the fundamental aspect of the most current advanced driver assistance systems (ADASs) like lane departure warning system (LDWS), lane change assistance system (LCAS) and lane centering. Vision-based techniques are being used extensively these days due to exhaustive knowledge base and low-cost camera devices. Most of the current driver assistance products ranging from the Mobileye Series ADAS to Tesla autopilot are using the computer vision-based lane detection systems. This is because of lane marking that is painted for human visual perception. These markings are utilized by vision-based algorithms to detect lanes from camera systems and prevent driver from inadvertent lane distractions. It is clear from the above statement that accuracy and robustness are two vital characteristics for a lane detection system. Therefore, the system must be capable to detect unreasonable changes and adjust the tracking accordingly.

1.2 Literature Review

There are numerous techniques developed over the past few years on computer vision-based lane detection systems. Xing et al. [1] discuss advances in lane detection algorithms, integration and assessment. Different lane detection methods based on image processing and machine learning are proposed in it. Most of them use similar pipelines with difference in filtering and vision-based algorithms [1–7]. Lim et al. [8] proposes a lane detection system with an additional module of lane tracking. They use Kalman filter-based method to perform tracking. He et al. [9] describes a lane departure system in addition to lane detection system using Hough transform. Gupta and Choudhary [10] presents a camera-based framework for the LDS. References [11, 12] discuss the methods to increase the driver's ability with the help of ADAS and role of lane detection in it. However, these did not take into consideration the distortion effect caused by cameras which in turn results in an inaccurate measurement of the vision algorithms, and they are mostly limited to straight lane detection. They tend to fail for curved lane detection. We propose a lane detection system with a pipeline that has distortion correction module included in it. This makes the system accurate and robust even if there are distortions in the images.

1.3 Paper Organization

In this paper, the proposed lane detection system can be designed using two pipelines, image processing and video processing. The image processing pipeline discussed in Sect. 2 includes different vision techniques applied to the images of the roads with different types of lane markings (white, yellow, left, right, etc.) to detect the lane lines. But, firstly, the images are processed for lens distortion correction. This method is clearly explained using distortion equations and camera matrix. Once the detector is ready, we simply overlay the detected lane on the video. Once any video from real-world scenario (with lane markings) is put through the video pipeline, the lane detector outputs the video with lane lines marked suggesting the detected lane lines. The experimental results obtained are discussed in Sect. 3. This paper concludes with Sect. 4 that discusses the future work to extend the system in building a better version.

2 Lane Detection System (Image Processing and Video Processing Pipelines)

The image processing pipeline comprises six steps that makes use of different computer vision techniques to detect lane lines in the given input image. Each step reduces the image details by removing unwanted details. The six techniques are distortion correction, perspective warping, Canny edge detection, dominant peak detection in histogram, sliding window search algorithm and curve fitting using regression (Fig. 1).

2.1 Image Processing Pipeline

Distortion Correction: In ray optics, distortion is a deviation from rectilinear projection in which straight lines in a scene do not remain straight in the image. Some less than ideal optical systems (cheaper cameras) distort incoming light to focus it on sensor. Even though this is useful in capturing images of environment, they might result in inaccurate measurements in computer vision applications. There are two major kinds of distortions, radial distortion and tangential distortion. Radial distortion causes straight lines in the images appear curved and bulged out. Tangential distortion occurs when image taking lens is not aligned perfectly parallel to imaging plane. Below equations represent both radial and tangential distortion.

$$x_{\text{distorted}} = x * \left(1 + k_1 * r^2 + k_2 * r^4 + k_3 * r^6\right) \tag{1}$$

$$y_{\text{distorted}} = y * \left(1 + k *_1 r^2 + k_2 * r^4 + k_3 * r^6\right) \tag{2}$$

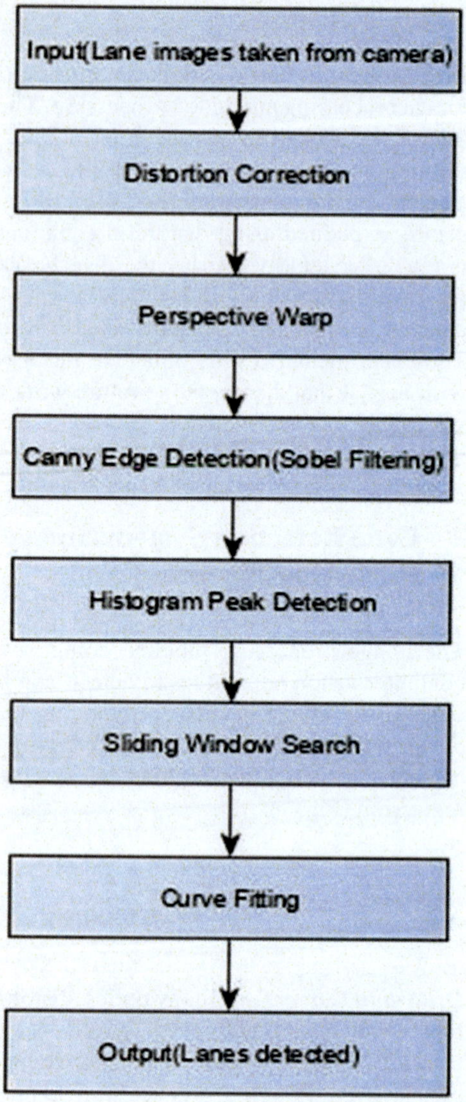

Fig. 1 Image processing pipeline

$$x_{\text{distorted}} = x + \left[(2 * p_1 * x * y) + \left(p_2 * \left(r^2 + 2 * x^2\right)\right]\right. \tag{3}$$

$$y_{\text{distorted}} = y + \left[p_1\left(r^2 + 2 * y^2\right) + (2 * p_2 * x * y)\right] \tag{4}$$

$(x_{\text{distorted}}, y_{\text{distorted}})$ = image coordinates with distortion
(x, y) = distortion free image coordinates
k_n = nth radial distortion coefficient

$p_n = n$th tangential distortion coefficient
$r =$ distance from$(x, y)$to center of radial distortion

From Eqs. (1)–(4), it is clear that for any stereo application, these distortions need to be corrected before using them for any computer vision application. This can be corrected only if values of corresponding distortion coefficients are available. Intrinsic parameters of camera are used to calculate distortion coefficients. These parameters include focal length (f_x, f_y) and optical centers (c_x, c_y). The matrix formed with the corresponding values is called camera matrix, and it depends only on the camera. To calculate camera matrix and distortion coefficients, we collect images of a well-defined pattern; in this paper, we considered asymmetric chessboard taken by the camera that has to be calibrated and find the object points (3D) and image points (2D). These points are used to calculate the camera matrix and distortion coefficients of the camera using *cv.calibrateCamera()* function present in OpenCV library. Thus, the obtained distortion coefficients and camera matrix are used to get the new camera matrix with distortion correction using *cv.getOptimalNewCameraMatrix()* function. The obtained camera matrix is distortion free and is used to remove distortion in the images using *cv.undistort()*.

$$\text{Distortion coefficients} = (k_1 k_2 p_1 p_2 k_3)$$

$$\text{Camera matrix} = \begin{bmatrix} f_x & 0 & c_x \\ 0 & f_y & c_y \\ 0 & 0 & 1 \end{bmatrix}$$

Though the difference might not be noticeable, it has a huge impact on the accuracy of computer vision algorithms (Fig. 2).

Perspective Warp: In camera space, it is difficult to detect curved lanes. Therefore, we apply geometric transformations to get a bird eye view of the road. It is a simple conversion of 3D world into a 2D image. This is done to assume the lane on a 2D surface and then to fit a polynomial that accurately represents the lane (Fig. 3).

Sobel Filtering: Filtering out lanes is an important task in lane detection. It helps in detecting the lanes by differentiating them from the road. A color filter is not a good idea because there are cases where lanes use light-colored concrete, which makes the road to pass the filter and pipeline intuits a white lane. In this paper, a method similar to edge detector is used in HLS color space to avoid the above-mentioned perception, and operator that is used in Canny edge detector is used to get the gradient of image function.

Histogram Peak detection: Before applying an algorithm to detect lane lines, it is needed to find an ideal beginning point for the algorithm. The algorithm works effectively if it begins at a region where lane pixels are present. This can be achieved by creating the histogram plot of binary image with respect to *X*-axis. With the help of this plot, lane line pixels in each column of the image are detected. Dominant peaks are observed at each side of the image. They represent the lane line present in the image.

Fig. 2 **(i)** Distorted image (barrel distortion), **(ii)** original image, distortion correction applied to driving scene, **(iii)** distorted image, **(iv)** after distortion correction

Sliding window search algorithm: A sliding window is a rectangular area of fixed width and height that slides over an image. These window regions are generally used to apply image classifier to determine if window has the object of interest, in this case lane lines. It is used to differentiate between left and right lanes to fit curves corresponding to the lane boundaries. The search begins from the initial point from histogram peak detection, measuring pixels present in the window. At the point when the measure of pixels arrives at a specific limit, window is moved to average horizontal situation of the distinguished pixels. In the event that enough pixels are not identified, the next window begins in a similar parallel position. This calculation proceeds until the windows arrive at the other edge of the picture. To separate among right and left path, the pixels that fall inside the windows are checked. Right and left lanes are marked using blue and red colors, respectively.

Curve fitting: The blue and red pixels marked using sliding window algorithm are just data points and do not provide a proper representation of the lane lines. A proper curve is needed to best fit the data points (blue and red marked pixels) to form a lane that they actually represent. This is done using polynomial regression method.

Fig. 3 **(i)** Original image and **(ii)** after perspective warp

A best polynomial that is least squares fit to the data points is found, and this gives the lane line.

Overlay: In this part, an overlay that fills in the detected portion of the lane is created. Thus, created lane lines are overlaid upon the original image of the road with lane lines.

2.2 Video Processing Pipeline

The image processing pipeline detects lane lines accurately in the images, but in the real-time implementation, the data obtained from the camera mounted upon an automobile is a video. The lane lines detected using the image processing pipeline should be applied to a video stream. The lane lines oscillate and remain unstable in a video stream. This can be resolved by averaging the slope and intercepts of the lane lines from frames. This technique of interframe averaging makes the lane lines stable and slows down their movement. In this paper, MoviePy library tools are used for video processing. The video pipeline works as follows: video obtained from the camera mounted on the automotive is clipped to several images. Thus, the obtained images are subjected to the image processing pipeline, and the resultant lanes detected are written back to the video file. Thus, we obtain a video with lane lines detected and marked.

3 Results and Discussions

The system initially develops a detector, using images of different lane line scenarios using the image processing pipeline. Before the images are subjected to the pipeline, they undergo lens distortion correction. This requires a distortion model obtained by calibrating the image with known object. In this paper, an asymmetric chessboard is used. Images of chessboard with different orientations are considered to generate a distortion model. Once the images are subjected to image processing pipeline, they undergo different transforms. Sliding window search and curve fitting are used to detect the lane lines and mark them (Fig. 4). The lane detector is ready. The obtained detector is applied on to a video to detect the lane lines using interframe averaging. This lane detection procedure is computationally more efficient compared to several existing lane detection techniques.

4 Conclusion

The lane detector presented in this paper gives good results for different lane line scenarios including curved lanes. It is robust to changes in the environment. The lane detection system discussed in this paper has an inbuilt lens distortion correction module, which is capable of correcting the distorted images. This reduces the error that can occur due to inaccurate measurements in computer vision algorithms. However, the system is affected by shadows that intervene the environment to some extent and also with the drastic changes in the texture of the road. The pipeline can be updated using deep learning techniques to create an extremely robust lane detector. In addition, it can include a module for vehicle detection and traffic sign recognition. By adding these modules, the lane detection system can be upgraded to a small-scale driver assistance system. The future work will focus on (i) applying deep learning techniques to make the detector more efficient and robust; and (ii) implementing the system on an embedded platform like Raspberry Pi by building a mini vehicle model to test the accuracy of the system.

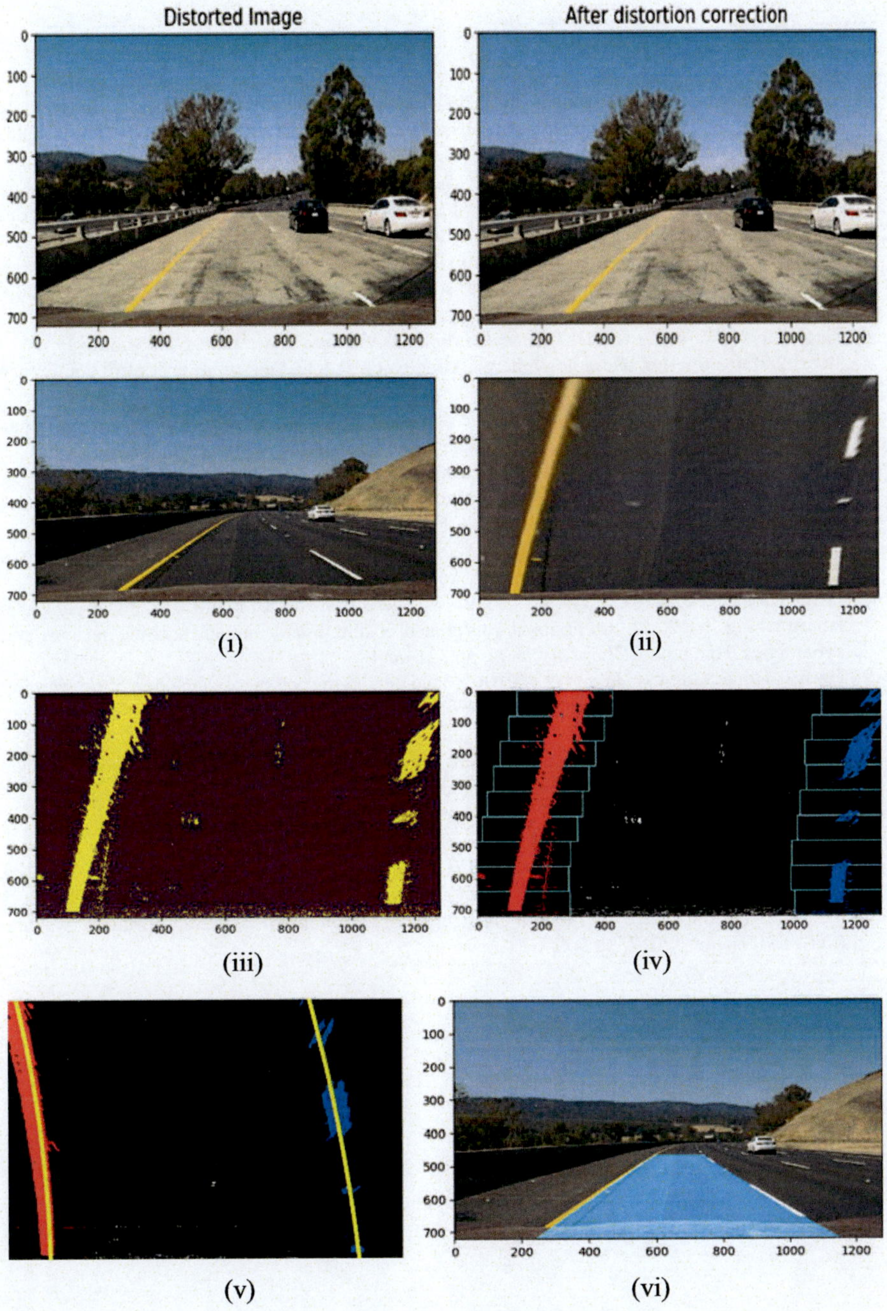

Fig. 4 **(i)** Original image, **(ii)** image after applying perspective transform, **(iii)** gradient of the image using Sobel operator, **(iv)** sliding window search algorithm, **(v)** curve fitting using polynomial regression, **(vi)** overlaid lanes on the original image

References

1. Xing Y et al (2018) Advances in vision-based lane detection: algorithms, integration, assessment, and perspectives on ACP-based parallel vision. IEEE/CAA J Autom Sinica 5(3):645–661
2. Shirke S, Rajabhushanam C (2017) A study of lane detection techniques and lane departure system. In: 2017 international conference on algorithms, methodology, models and applications in emerging technologies *(ICAMMAET)*, Chennai, 2017, pp 1–4
3. Feniche M, Mazri T (2019) Lane detection and tracking for intelligent vehicles: a survey. In: 2019 international conference of computer science and renewable energies (ICCSRE*)*, Agadir, Morocco, 2019, pp 1–*4*
4. Gao Q, Feng Y, Wang L (2017) A real-time lane detection and tracking algorithm. In: 2017 IEEE 2nd information technology, networking, electronic and automation control conference (ITNEC), Chengdu, 2017, pp 1230–1234
5. de Paula MB, Jung CR (2015) Automatic detection and classification of road lane markings using onboard vehicular cameras. IEEE Trans Intell Transp Syst 16(6):3160–3169
6. Wang H, Wang Y, Zhao X, Wang G, Huang H, Zhang J (2019) Lane detection of curving road for structural highway with straight-curve model on vision. IEEE Trans Veh Technol 68(6):5321–5330
7. Yuan C, Chen H, Liu J, Zhu D, Xu Y (2018) Robust lane detection for complicated road environment based on normal map. IEEE Access 6:49679–49689
8. Lim KH, Seng KP, Ang L, Chin SW (2009) Lane detection and Kalman-based linear-parabolic lane tracking. In: 2009 international conference on intelligent human-machine systems and cybernetics, Hangzhou, Zhejiang, 2009, pp 351–354
9. He J, R Hong, Gong J, Huang W (2010) A lane detection method for lane departure warning system. In: 2010 international conference on optoelectronics and image processing, Haikou, 2010, pp 28–31
10. Gupta A, Choudhary A (2018) A framework for camera-based real-time lane and road surface marking detection and recognition. IEEE Trans Intell Veh 3(4):476–485
11. Bosetti P, Da Lio M, Saroldi A (2015) On curve negotiation: from driver support to automation. IEEE Trans Intell Transp Syst 16(4):2082–2093
12. Rossi R, Gecchele G, Gastaldi M, Biondi F, Mulatti C (2017) An advanced driver assistance system for improving driver ability. Design and test in virtual environment. In: 2017 5th IEEE international conference on models and technologies for intelligent transportation systems (MT-ITS), Naples, 2017, pp 509–513

IoT-Based Energy Management System with Data Logging Capability

O. V. Gnana Swathika, G. Kanimozhi, E. Umamaheswari, Soj Rujay, and Soudeep Saha

Abstract With increasing energy demand and the necessity to fulfill the energy requirement, it is mandatory to increase the energy generation. However, shortage of supply resources stands as a blockade in this present scenario. Hence, an efficient energy management system is required. This paper demonstrates a working prototype of intelligent energy management system. It is an IoT-based energy management system with Web server-type manual control. It involves serial communication between microcontrollers such as Arduino Uno and NodeMCUs, sensors and computational intelligence. The sensors used in this case are LDR, thermistor, and PIR motion sensor. The system includes LED lights and fans as load. The raw data from the sensors is read by the Arduino and it serially communicates the raw data to both the NodeMCUs. The data is then converted to standard formats and uploaded to the ThingSpeak cloud server for data logging and analysis by one of the NodeMCU. Using the second NodeMCU, a Web server is created and used for manual control of the loads. This Wi-Fi server sends data to the NodeMCU as per the user input from the browser and this NodeMCU controls the load accordingly. A current sensor is also connected along the supply line for power measurement. The current sensor is an ACS712 20 A sensor which is connected to the NodeMCU and is responsible for uploading data to the ThingSpeak cloud server. The working model has three rooms for demonstration purpose but can be increased accordingly as per need with certain changes in system hardware model.

Keywords Energy management system (EMS) · Internet of things (IoT) · Data logging

1 Introduction

Developing countries like India desperately need energy for pushing forward its economic growth [1]. However, such countries do not have sufficient energy resources

O. V. Gnana Swathika (✉) · G. Kanimozhi · E. Umamaheswari · S. Rujay · S. Saha
VIT University, Chennai, India
e-mail: gnanaswathika.ov@vit.ac.in

N. Zhou and S. Hemamalini (eds.), *Advances in Smart Grid Technology*, Lecture Notes in Electrical Engineering 688, https://doi.org/10.1007/978-981-15-7241-8_41

to cater to the increasing energy demand [2]. The option to increase the commercial energy generation is held back by the depletion of non-sustainable fossil fuels and environment degradation. It is mandatory to separate the economic growth from fossil fuel dependency [3]. It is need of the hour to manage the limited energy sources in a better way. Lighting accounts for 19% of the world's electricity consumption [4]. With recent advances in LED technology, a better LED lighting solution is developed which can substitute CFLs and incandescent lamps. The key purpose of the energy management and automation system is focused upon better comfort of the user at reduced operation cost and increased energy efficiency. Energy efficiency improvement also contributes to environmental protection [5]. By implementing proper intelligent residential energy management system, it is possible to reduce the electricity bill while maintaining the power demand under the maximum demand limit as well as integrate battery energy storage [6]. It is possible to integrate this data to IoT cloud server. With IoT implementation in energy management system, the promotion of energy efficiency and power management of the residential areas is achieved. It is also possible to log the data into IoT servers and use it for further analysis at any given point of time [7].

The system proposed in this paper aims to successfully achieve the points mentioned above. The design process of the system is further discussed and implemented in this paper.

1.1 Objectives

The objective of this paper is summarized into two parts as shown in Fig. 1. First part deals with the load and its control, whereas the second part deals with data logging into the IoT cloud server. The aim of this paper is to achieve a system which is able to control the loads to provide maximum energy efficiency and also able to integrate the obtained sensor data into the IoT cloud server along with providing wireless manual control of the loads to the user for personal convenience.

For load control as per the required situation with energy efficiency, light and temperature sensors are used to provide optimal comfort along with required energy efficiency. An occupancy sensor is also used to provide the occupancy status of

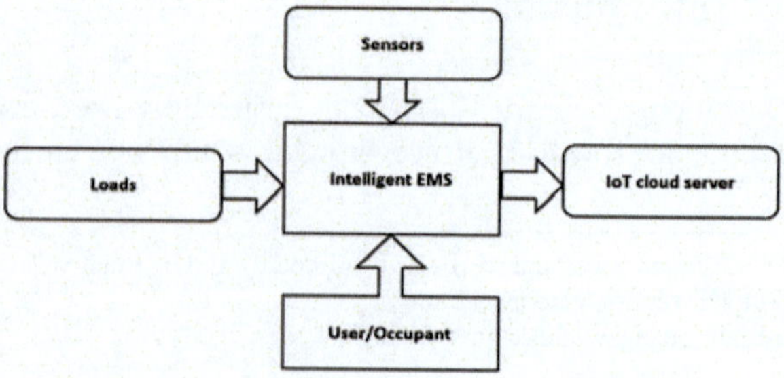

Fig. 1 General layout of intelligent energy management system

the room for turning the loads ON and OFF as the situation permits. The system mentioned is autonomous in nature but for the convenience of the user or occupant a Web browser-based wireless manual control of the loads is provided which is accessed using any electronic device with browser support. Lastly, the data from the sensors namely light, temperature, occupancy, and current are utilized to calculate the luminance, temperature, occupancy status, and power consumption, respectively. This data is hence used for monitoring and further analysis.

2 Proposed System

2.1 Overview

The energy management system proposed in this paper aims to provide optimal energy efficiency along with cost effectiveness. The microcontrollers used in the hardware fabrication are open source and readily available in the market. Both the Arduino Uno and NodeMCU are easily programmable using ArduinoIDE which is an open-source platform and may be operated in numerous devices. The NodeMCU has only one analog input pin which is connected to the ACS712 20A current sensor. The block diagram of the proposed system is as shown in Fig. 2. It shows the connection between Arduino, NodeMCU, sensors, load, cloud server, and Wi-Fi Server.

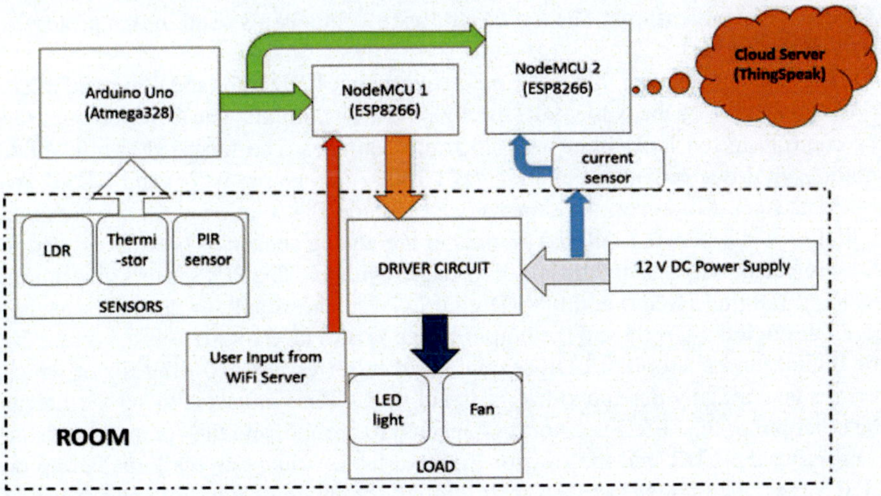

Fig. 2 Block diagram of the proposed system

Table 1 List of hardware components

Component	Specification
Arduino Uno	
NodeMCU ESP8266	
Dot matrix board	10 × 6 inch
DC power adapter	12 V–1 A
DC motor	9 V
LED module	12 V
4N25 opto-coupler	–
BC547 transistors	–
Resistor bank	¼ W
IC L7805	5 V
IC L7809	9 V
PIR motion sensor	–
LDR 5 MM	5 V
Thermistor	5 V, 10 kΩ
ACS712	20 A

2.2 *Hardware and Software Specifications of the Proposed System*

Hardware specifications. The list of hardware components used in this prototype is shown in Table 1.

The loads included in this real-time network are: LED light and a fan which are PWM controlled by the NodeMCU through the driver circuit. The driver circuit used for controlling the LED light is 4N25 (opto-coupler). The fan used is a 9 V DC motor with driver circuit which includes L7809 (voltage regulator) and BC547 (n-p-n transistor). Moreover, the sensors used include PIR motion sensor, which is a digital sensor, whereas LDR and thermistor are analog sensors. The sensors used in this case are powered using the L7805 voltage regulator. The PIR sensors give output as +5 V (Digital HIGH) and 0 V (Digital LOW). The supply to the PIR sensor is given through the L7805 and the output which is provided by PIR sensor is read by the digital pin of Arduino. LDR is connected in series with 10 kΩ resistor which forms a potential divider network. One pin of the LDR is connected to +5 V whereas the other pin of the 10 kΩ resistor is connected to ground, and the common junction connecting the LDR and the resistor gives variable voltage as the light falling on LDR varies. The voltage is then read through the analog-to-digital convert pin of the Arduino as a 10-bit input. The thermistor follows the same analogy as that of the LDR. Arduino receives data from the sensors and serially communicates it to NodeMCU. The data points which are received by the NodeMCU from the Arduino are then mapped to the PWM pins with required logic which is then used to control loads accordingly. The power supply for the load used here is 12 V DC adapter.

The interfacing between the NodeMCU and loads is done via the driver circuits. As mentioned earlier, the LED light used here is 12 V DC with driver circuit 4N25 and 9 V DC motor with L7809 (voltage regulator) and BC547 (n-p-n transistor).

Software specifications. The data which is received by the NodeMCU through the Arduino is converted into proper readable units using proper factors. For example, the raw data from the temperature sensor is converted to degree Celsius reading and the raw data from the LDR in converted to cd/m^2 (candela per square meter). This converted data is then sent to the cloud server. The cloud server is provided by ThingSpeak. The NodeMCU is responsible for collecting and uploading the data into the cloud server for monitoring and data logging purpose. A current sensor is also connected along the supply line for power measurement. The current sensor is an ACS712 20A sensor which is connected to the NodeMCU and is responsible for uploading data to the ThingSpeak cloud server. Using the second NodeMCU, a Web server is created and used for manual control of the loads. This Wi-Fi server sends data to the NodeMCU as per the user input from the browser and this NodeMCU controls the load accordingly. In the NodeMCU which is used for load control through the Wi-Fi server, HTML coding is used which is supported by almost all the Web browsers. This enables the wireless manual control to be platform independent, i.e., it is accessible by any device with Web browser support irrespective of its type.

3 Design of Intelligent Energy Management System

Figure 3 depicts the design approach taken in the implementation of an intelligent energy management system.

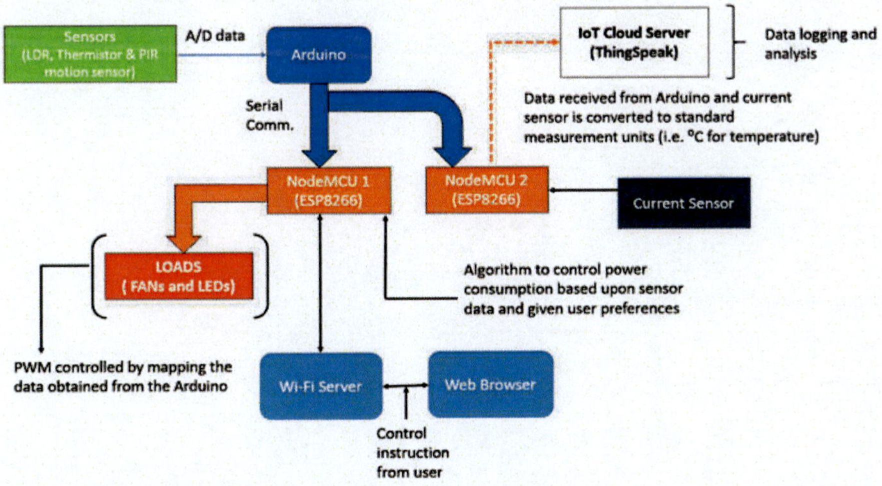

Fig. 3 Design of intelligent EMS

This paper demonstrates a working prototype of intelligent energy management system. It is an IoT-based energy management system with Web server-type manual control. It involves serial communication between microcontrollers such as Arduino Uno and NodeMCUs in this case, sensors and computational intelligence. The sensors used in this case are LDR, thermistor, and PIR motion sensor. The system includes LED lights and fans as load. The raw data from the sensors is read by the Arduino and it serially communicates the raw data to both the NodeMCUs. The data is then converted to standard formats and uploaded to the ThingSpeak cloud server for data logging and analysis by one of the NodeMCU. Using the second NodeMCU, a Web server is created and used for manual control of the loads. This Wi-Fi server sends data to the NodeMCU as per the user input from the browser and this NodeMCU controls the load accordingly. A current sensor is also connected along the supply line for power measurement. The current sensor is an ACS712 20A sensor which is connected to the NodeMCU is responsible for uploading data to the ThingSpeak cloud server. The working model has three rooms for demonstration purposes.

4 Results and Discussion

The raw data which was received by the NodeMCU via serial communication from the Arduino is then converted in proper standard unit as shown in Fig. 4.

The room occupancy condition which is sensed by the PIR sensor is also shown at the ThingSpeak cloud server using indicator lights as the PIR sensor sends digital data of either 1 (Occupied) or 0 (Unoccupied) as shown in Figs. 5 and 6. Figure 7

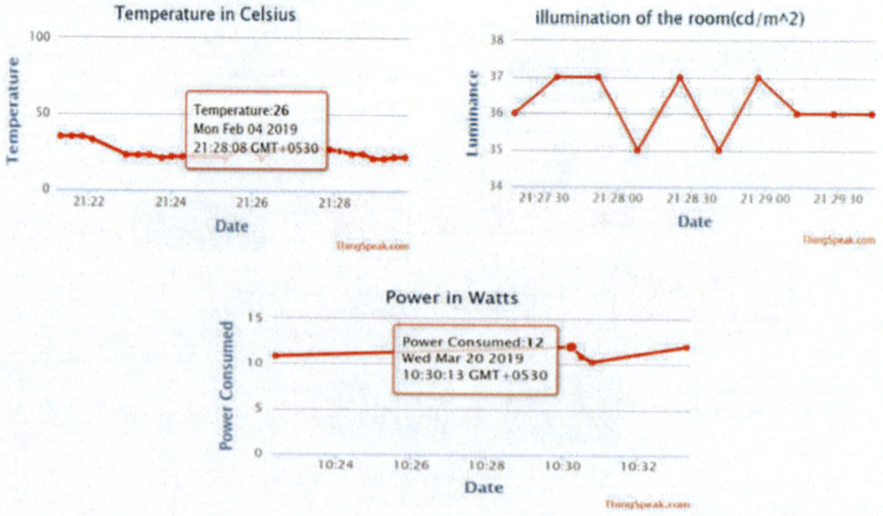

Fig. 4 Display of temperature, illuminance, and power consumption data received by ThingSpeak

Fig. 5 Occupancy indicator (occupied)

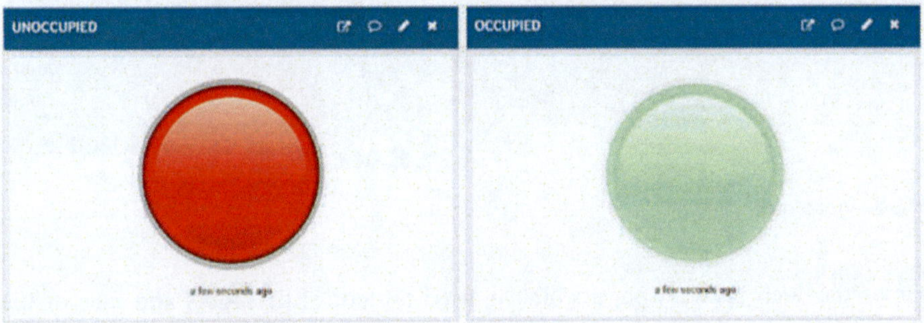

Fig. 6 Occupancy indicator (unoccupied)

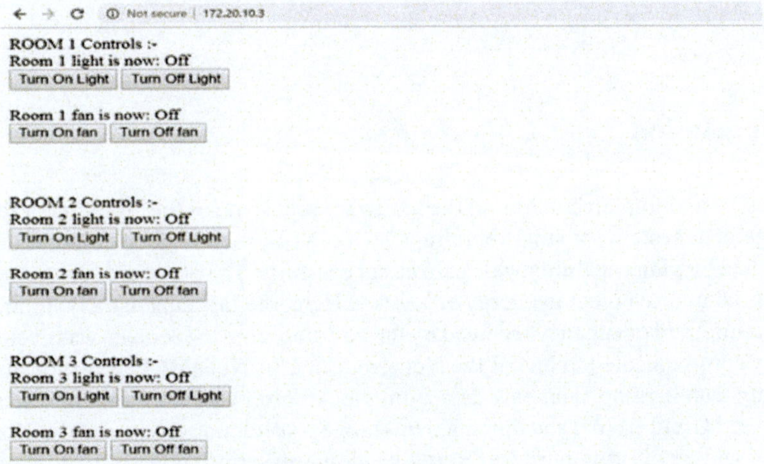

Fig. 7 HTML page with control options for the system

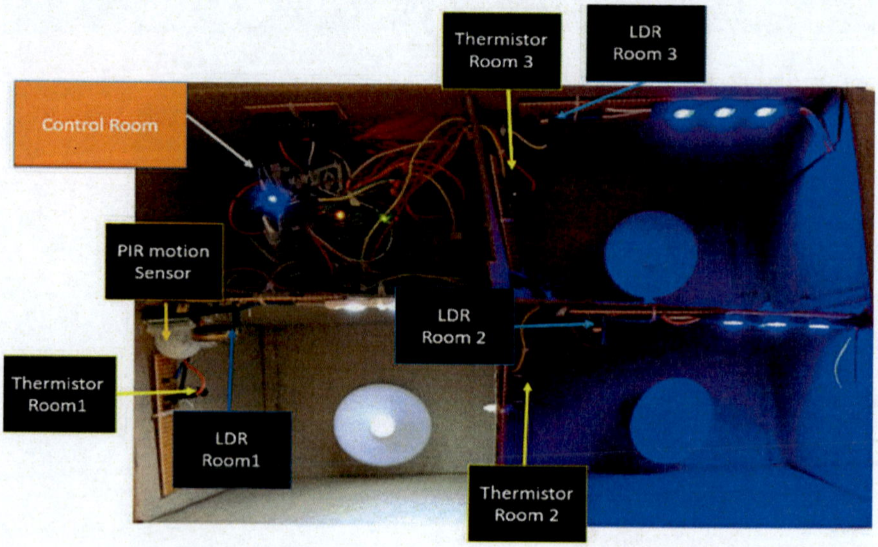

Fig. 8 Prototype image for Case 1

shows the Web browser page which is used to control the lights and fans of the respective rooms.

Hardware Results.

Case 1: Loads in all the rooms are ON with occupancy data of room 1. The HTML page and the prototype are shown in Figs. 8 and 9.

5 Conclusion

The design and implementation of the hardware prototype of the "Intelligent Energy Management system for smart buildings" with Wi-Fi server-based control and IoT-based data logging capability was realized successfully. The proposed system helps in substantial improvement in energy efficiency. Here, the raw data namely luminance, temperature, and occupancy received by the Arduino from the sensors was eventually serially communicated to one of the NodeMCUs. This NodeMCU is responsible for handling data logging. This raw data from the Arduino was converted into standard units (i.e., °C and cd/m^2) and uploaded to ThingSpeak cloud server using IoT for data logging and monitoring purpose. Here, the cloud server provided by ThingSpeak is for visual representation of the data in graphical formats which can also be used for further analysis. The same raw data is also serially communicated to the second NodeMCU which is responsible for controlling the loads. The data received by this second NodeMCU was mapped accordingly to its PWM pins using proper algorithms in order to control the brightness of the light and speed of fan. Using the Wi-Fi

Fig. 9 HTML page for Case 1

capabilities of this NodeMCU, we also create a local area Wi-Fi server which enables wireless control of the loads by the user through Web browser.

References

1. Andrei Nardelli M, Deuschle E, Dalpaz L, de Azevedo J, Pessoa LN, Ghisi E (2017) Assessment of light emitting diodes technology for general lighting: a critical review. Renew Sustain Energy Rev 75:368–379
2. Chasta R, Singh R, Gehlot A, Mishra RG, Choudhury S (2016) A SMART building automation system. Int J Smart Home 10(9)
3. Arun SL, Selvan MP (2018) Intelligent residential energy management system for dynamic demand response in smart buildings. IEEE Syst J 12(2):1329–1340
4. Marinakis V, Doukas H (2018) An Advanced IoT-based system for intelligent energy management in buildings. Sensors 18:610
5. Prachchhak G, Bhatt C, Thik J (2019) Data logging and visualization using bolt IoT. In: Kamal R, Henshaw M, Nair P (eds) International conference on advanced computing networking and informatics. Advances in intelligent systems and computing, vol 870. Springer, Singapore
6. Borkowski P, Pawlowski M (2011) Home/building management systems (HMS/BMS) to protect environment by control modern lighting installations. In: International conference on chemical engineering and applications, IPCBEE, vol 23 (2011)
7. Doukas H, Patlitzianas KD, Iatropoulos K, Psarras J (2007) Intelligent building energy management system using rule sets. Sci Direct 42:3562–3569